Pearls and Pitfalls in
ABDOMINAL IMAGING

Pearls and Pitfalls in ABDOMINAL IMAGING

Pseudotumors, Variants and Other Difficult Diagnoses

Fergus V. Coakley M.D.

Professor of Radiology and Urology
Section Chief of Abdominal Imaging
Vice Chair for Clinical Services
Department of Radiology and Biomedical Imaging
University of California, San Francisco

CAMBRIDGE
UNIVERSITY PRESS

CAMBRIDGE UNIVERSITY PRESS
Cambridge, New York, Melbourne, Madrid, Cape Town, Singapore,
São Paulo, Delhi, Dubai, Tokyo

Cambridge University Press
The Edinburgh Building, Cambridge CB2 8RU, UK

Published in the United States of America by Cambridge University Press, New York

www.cambridge.org
Information on this title: www.cambridge.org/9780521513777

First published 2010

Printed in the United Kingdom at the University Press, Cambridge

A catalogue record for this publication is available from the British Library

Library of Congress Cataloging-in-Publication Data

Coakley, Fergus.
Pearls and pitfalls in abdominal imaging : pseudotumors, variants and other difficult diagnoses / Fergus V. Coakley.
p. ; cm.
Includes bibliographical references and index.
ISBN 978-0-521-51377-7 (hardback)
1. Abdomen–Imaging. I. Title.
[DNLM: 1. Radiography, Abdominal–methods. 2. Digestive System Diseases–radiography. 3. Pelvis–radiography. WI 900 C652p 2010]
RC944.C53 2010
617.5'507572–dc22

2010016392

ISBN 978-0-521-51377-7 Hardback

This book is dedicated to my parents, Dermot and Maeve, for their constant support and guidance in my early years, and to my wonderful wife, Sara, and our delightful children, Declan and Fiona, who keep me grounded, happy, and in love now that I have reached my later years!

Contents

Section 9 Gastrointestinal tract

Section 10 Peritoneal cavity

Section 11 Ovaries

Section 12 Uterus and vagina

Section 13 Bladder

Section 14 Pelvic soft tissues

Section 15 Groin

Section 16 Bone

Preface

This book represents the convergence of three related themes which have occupied a large part of my professional life. First, ever since I started training as a radiologist almost 20 years ago, I have been intrigued by the "pattern recognition" that lies at the heart of our specialty. This approach to diagnosis can be very powerful, but also prone to error if different entities look the same. As a first year resident reading out the overnight Emergency Department plain films at Leicester Royal Infirmary, hardly a fracture went reported without checking our heavily thumbed and coffee-stained edition of Keats [1] for possible mimics or confounders. Second, one of my most popular postgraduate lectures is entitled "Pearls and pitfalls in abdominal CT," and this talk grew out of my early interest in normal variants simulating disease. It is clear that all radiologists struggle with the basic questions as to whether a study is normal or abnormal, or whether findings of a given diagnosis can be due to anything else. Third, most physicians are perfectionists and dislike making mistakes, especially when those mistakes can be harmful to patients. We are entrusted with caring for patients who are often at their sickest and most miserable. Anything we can do to improve their care fulfills our duty to them, and also helps address ongoing and legitimate public concern regarding medical errors and patient safety [2, 3]. The literature consistently suggests that 1.0 to 2.6% of radiology reports contain serious errors [4–6]. My experience in clinical practice, in running a quality assurance program, and in medical malpractice work has convinced me that many of these interpretative mistakes in abdominal imaging are avoidable. These convergent processes motivated me to write this book.

In a nutshell, the core concept of this work is to bring together those abdominal imaging entities that can cause confusion and mismanagement in daily radiological practice, and provide a tightly focused textbook that can be readily used as bench-side reference to avoid these problems. The "pearls and pitfalls" include technical artifacts, anatomic variants, mimics, and a miscellany of diagnoses that are under-recognized (e.g., adenomyomatosis of the gallbladder) or only recently described (e.g., pseudocirrhosis of fulminant hepatic failure). The common denominator is that these entities present real problems for the practicing radiologist. I have attempted to cover all major modalities within the contemporary practice of abdominal imaging, including ultrasound, CT, PET/CT, and MRI. Pitfalls at radiography and fluoroscopy are largely excluded, in order to reflect the reality of current practice. This is not a value judgment, but simply reflects the evolving nature of radiology–this book would have been very different if written 50 or even 25 years ago. My aim is to provide an easily used resource when a practicing radiologist sees something odd or confusing, and also to provide examples of common medicolegal pitfalls (e.g., mistaking perforated colon cancer for diverticulitis, or missing strangulated obstructed bowel). The conditions were selected based on my experience working in a busy academic tertiary referral center. As far as possible, I have tried to include diagnoses that are clinically important (e.g., benign conditions that can look malignant, malignant conditions that can look benign, and normal variants that may prompt unnecessary additional tests) rather than including mimics that may be interesting but clinically unimportant (e.g., confusing one benign condition for another is usually of no great clinical consequence). Similarly, I have tried to include pitfalls that occur with some reasonable frequency and are not extreme exotica – as a rough rule of thumb, I have only included a given entity if I have seen it more than once. Inevitably, as a single author trying to pull together a group of thematically linked but diverse diagnoses, the result is eclectic and reflects my personal experience. Hopefully, any resulting omissions or bias will be offset by some uniformity of thought and approach. But if I have omitted any item that merits inclusion or committed any other errors, please let me know, in anticipation of a second edition!

In order to provide structure to the book content, the imaging entities are presented in approximate anatomic order from the diaphragm to the symphysis pubis, with grouping by location and organ system. Within each group, I have also tried to arrange items anatomically – for example, in the gastrointestinal tract, the items begin with the stomach and proceed to the large bowel. Other things being equal, I have tried to order by frequency, so that rarer entities or conditions that are only seen on one modality are described after more common items. The book is heavily illustrated, with a relatively small amount of text, since I am a strong believer in the teaching power of images over words. I have tried to make the text user-friendly, with an informal tone. The text for each entity follows the same format (imaging description, importance, typical clinical scenario, differential diagnosis, and teaching point). As such, each entity stands alone and can be read in isolation. A busy reader could probably make do by reading the teaching point and looking at the figures.

In summary, the overarching goal of this work is to provide a resource for the practicing radiologist when they see something that makes them think "that's weird" or "what else could that be?" Ultimately, the intent is to provide a bench book that assists any radiologist reading out abdominal imaging studies and improves the interpretation of such studies so that patient care is improved. The book is intended for any radiologist that reports abdominal imaging studies as part of their daily practice. I will feel satisfied if anything in this book facilitates a diagnosis that might otherwise not have been made, or prevents a misdiagnosis.

REFERENCES

1 Keats TE. *Atlas of normal roentgen variants that may simulate disease*, 4th edition. St. Louis, New York, London: Mosby, 1988.

2 Kohn LT, Corrigan JM, Donaldson MS, eds. *To err is human: building a safer health system*. Washington, D.C.: National Academy Press, 2000.

3 Milstein A. Ending extra payment for "never events" – stronger incentives for patients' safety. *N Engl J Med* 2009; **360**: 2388–2390.

4 Bechtold RE, Chen MY, Ott DJ, et al. Interpretation of abdominal CT: analysis of errors and their causes. *J Comput Assist Tomogr* 1997; **21**: 681–685.

5 Carney E, Kempf J, DeCarvalho V, Yudd A, Nosher J. Preliminary interpretations of after-hours CT and sonography by radiology residents versus final interpretations by body imaging radiologists at a level 1 trauma center. *Am J Roentgenol* 2003; **181**: 367–373.

6 Ruchman RB, Jaeger J, Wiggins EF 3rd, et al. Preliminary radiology resident interpretations versus final attending radiologist interpretations and the impact on patient care in a community hospital. *Am J Roentgenol* 2007; **189**: 523–526.

Acknowledgements

I have been extraordinarily fortunate to have worked with some of the most talented and outstanding academic radiologists in the country, who have served as exceptional mentors and role models. Dr. David Panicek first inspired me to think about variants and pitfalls in abdominal imaging, and he did his best to make me write plain and understandable English. Dr. Hedvig Hricak and Dr. Alexander Margulis selflessly provided advice and guidance when I moved to San Francisco in 1997, and I consider myself lucky to have worked with them. Dr. Ronald Arenson continues to be the fairest and most supportive Chair that one could hope to have. Without their influence and assistance, this book would not exist – and I am deeply indebted to them all. In addition, I would like to specifically thank Drs. Peter Callen, Peter Cooperberg, Diego Ruiz, and Judy Yee for their graciousness in sharing images from their case material for reproduction in this book.

CASE 1

Pseudolipoma of the inferior vena cava

Imaging description

Pseudolipoma of the inferior vena cava refers to the apparent presence of a fatty mass in the lumen of the inferior vena cava as it passes through the diaphragm from the liver into the right atrium. The appearance is a partial volume artifact due to a layer of fat that sits above the caudate lobe next to the inferior vena cava. The cava deviates to the midline as it passes from the liver into the right atrium, and depending on local anatomy and the phase of respiration, the fat above the caudate lobe can be partial volumed in such a way that it appears to be within the vessel (Figure 1.1) [1].

Importance

Pseudolipoma of the inferior vena cava may be mistaken for a true fat-containing tumor of the inferior vena cava, such as a lipoma or liposarcoma [2], resulting in unnecessary follow-up investigations and patient anxiety.

Typical clinical scenario

Pseudolipoma of the inferior vena cava has a reported frequency of 0.5% at abdominal CT [3], but this seems far higher than I would have expected based on my clinical experience. While pseudolipoma of the inferior vena cava can be seen in anyone, it is commoner in cirrhosis, presumably because there is a greater degree of anatomic distortion and potential for partial volume artifact due to shrinkage of the liver and greater deviation of the inferior vena cava as it passes through the diaphragm in these patients (Figure 1.2).

Differential diagnosis

Theoretically, a true lipoma or liposarcoma could arise as a primary intraluminal caval mass, but this has not been reported. Venous invasion by locally aggressive angiomyolipoma may cause a fatty tumor thrombus in the cava [4], but the presence of a renal mass with contiguous spread into the cava is distinctive and should not result in confusion with a pseudolipoma.

Teaching point

The appearance of fat in the lumen of the inferior vena cava as it passes through the diaphragm is a normal variant due to partial volume artifact and does not require additional testing.

REFERENCES

1 Han BK, Im JG, Jung JW, Chung MJ, Yeon KM. Pericaval fat collection that mimics thrombosis of the inferior vena cava: demonstration with use of multi-directional reformation CT. *Radiology* 1997; **203**: 105–108.

2 Perry JN, Williams MP, Dubbins PA, Farrow R. Lipomata of the inferior vena cava: a normal variant? *Clin Radiol* 1994; **49**: 341–342.

3 Miyake H, Suzuki K, Ueda S, et al. Localized fat collection adjacent to the intrahepatic portion of the inferior vena cava: a normal variant on CT. *Am J Roentgenol* 1992; **158**: 423–425.

4 Moulin G, Berger JF, Chagnaud C, Piquet P, Bartoli JM. Imaging of fat thrombus in the inferior vena cava originating from an angiomyolipoma. *Cardiovasc Intervent Radiol* 1994; **17**: 152–154.

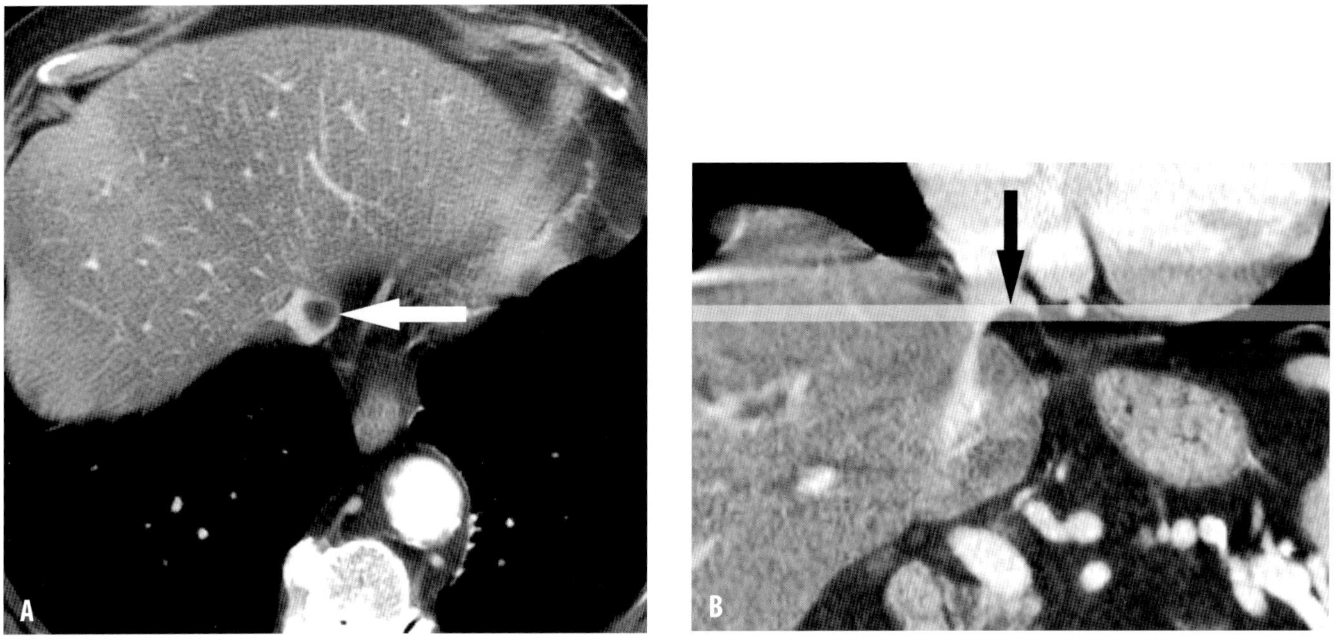

Figure 1.1 **A.** Axial contrast-enhanced CT image in a 70 year old man with prostate cancer shows an apparent fatty mass (arrow) in the lumen of the inferior vena cava as it passes through the diaphragm. **B.** Coronal reformatted CT image demonstrates the mechanism of this partial volume artifact; fat (arrow) above the caudate lobe is partial volumed into the lumen of the cava on the corresponding axial section (at the level indicated by the shaded rectangle).

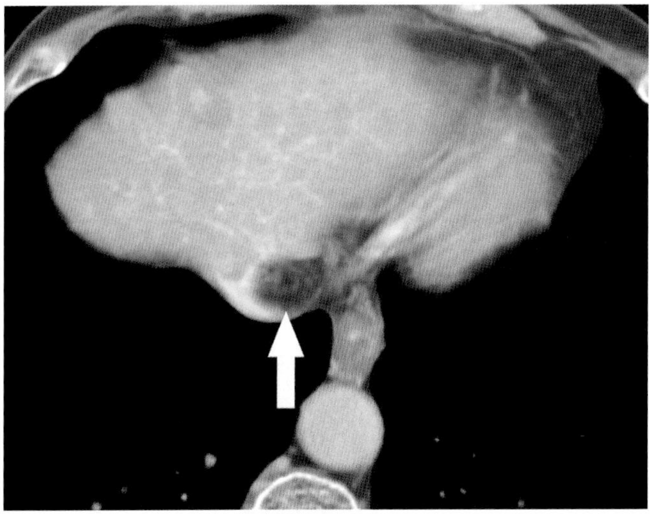

Figure 1.2 Axial contrast-enhanced CT image in a 67 year old woman with alcoholic cirrhosis (note the irregular liver contour) shows a pseudolipoma (arrow).

CASE 2

Superior diaphragmatic adenopathy

Imaging description

The superior diaphragmatic (or cardiophrenic or epicardiac) lymph nodes are in the mediastinum, but are routinely included on the upper slices of abdominal CT or MRI studies because they lie on the superior surface of the diaphragm in the fat adjacent to the heart. They are divided into anterior (paracardiac) and lateral (juxtaphrenic) groups [1, 2]. The anterior group lies posterior to the lower sternum. The lateral group abuts the entrance of the phrenic nerve into the diaphragm, adjacent to the inferior vena cava on the right and the cardiac apex on the left. The normal superior diaphragmatic lymph nodes are usually small and often not visible by CT imaging. Pathological enlargement is generally defined as a short axis diameter greater than 5 mm [2, 3], although some use a short axis threshold of 8 or 10 mm [4, 5]. Enlarged superior diaphragmatic nodes are seen as nodular soft tissue structures lying just superior to the diaphragm and posterior to the sternum, adjacent to the cardiac apex, or abutting the supradiaphragmatic inferior vena cava (Figure 2.1).

Importance

The superior diaphragmatic lymph nodes receive lymph from the peritoneal cavity and the anterosuperior part of the liver. Enlargement of these nodes may be seen in:

Liver disease. In practice, cirrhosis and chronic hepatitis [6] are probably the commonest causes of superior diaphragmatic adenopathy. In chronic hepatitis, the degree of nodal enlargement (but not the level of serum liver enzymes) correlates with disease severity on biopsy [7].

Peritoneal disease. The principal peritoneal cause of superior diaphragmatic adenopathy is ovarian cancer. In general, studies of these nodes do not have a histopathological standard of reference because these nodes are not easily accessible for tissue sampling and outcome is used as an alternative endpoint. In the case of ovarian cancer, superior diaphragmatic nodes greater than 5 mm in short axis diameter confer a worse prognosis [3] and are presumably metastatic in nature (Figure 2.2).

Other malignancy. Superior diaphragmatic adenopathy may also be seen in other cancers, with widespread, hepatic (Figure 2.3), or peritoneal spread. In at least some oncologic settings, it is possible that superior diaphragmatic adenopathy is reactive rather than metastatic. For example, in patients with resectable hepatic metastases from colorectal cancer, superior diaphragmatic nodes greater than 5 mm in short axis diameter do not confer a worse prognosis, which may indicate they are reactive and not metastatic [8]. With the greater utilization of PET, more data on the likely pathological basis of superior diaphragmatic adenopathy may emerge.

Typical clinical scenario

The identification of superior diaphragmatic adenopathy should prompt a careful search for hepatic or peritoneal disease (Figures 2.4 and 2.5). Reactive superior diaphragmatic adenopathy in cirrhosis or chronic hepatitis is frequently accompanied by portal, portacaval, or retroperitoneal adenopathy (which I call "liver pattern adenopathy") [7].

Differential diagnosis

The appearance of superior diaphragmatic adenopathy is usually distinctive, although occasionally large nodal deposits may be difficult to distinguish from pleural or pulmonary masses (Figure 2.6).

Teaching point

Superior diaphragmatic adenopathy can be a useful diagnostic clue to hepatic and peritoneal disease.

REFERENCES

1 Williams PL, Warwick R. *Gray's anatomy*, 38th edition. Edinburgh, London, Melbourne, New York: Churchill Livingstone, 1995; 1624.

2 Aronberg DJ, Peterson RR, Glazer HS, et al. Superior diaphragmatic lymph nodes: CT assessment. *J Comput Assist Tomogr* 1986; **10**: 937–941.

3 Rouviere H. *Anatomy of the human lymphatic system: a compendium.* Ann Arbor, MI: Edwards Brothers, 1938; 86–88.

4 Holloway BJ, Gore ME, A'Hern RP, et al. The significance of paracardiac lymph node enlargement in ovarian cancer. *Clin Radiol* 1997; **52**: 692–697.

5 Graham NJ, Libshitz HI. Cascade of metastatic colorectal carcinoma from the liver to the anterior diaphragmatic lymph nodes. *Acad Radiol* 1995; **2**: 282–285.

6 Wechsler RJ, Nazarian LN, Grady CK, et al. The association of paracardial adenopathy with hepatic metastasis found on CT arterial portography. *Abdom Imaging* 1995; **20**: 201–205.

7 Dodd GD 3rd, Baron RL, Oliver JH 3rd, et al. Enlarged abdominal lymph nodes in end-stage cirrhosis: CT-histopathologic correlation in 507 patients. *Radiology* 1997; **203**: 127–130.

8 Zhang XM, Mitchell DG, Shi H, et al. Chronic hepatitis C activity: correlation with lymphadenopathy on MR imaging. *Am J Roentgenol* 2002; **179**: 417–422.

9 Aslam R, Coakley FV, Williams G, et al. Prognostic importance of superior diaphragmatic adenopathy at computed tomography in patients with resectable hepatic metastases from colorectal carcinoma. *J Comput Assist Tomogr* 2008; **32**: 173–177.

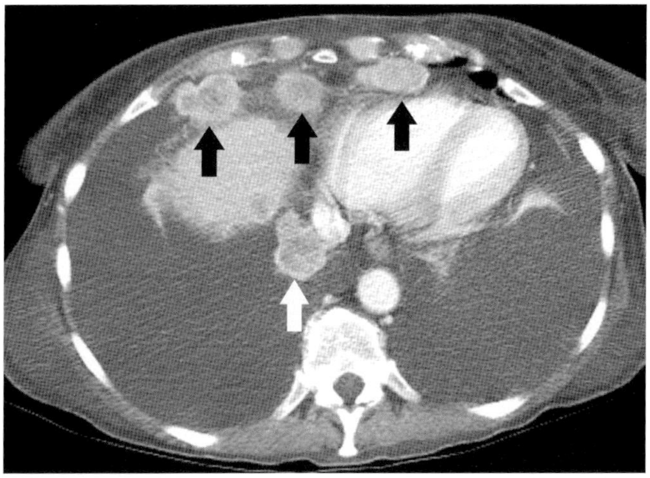

Figure 2.1 Axial contrast-enhanced CT image in a 55 year old woman with advanced ovarian cancer shows marked superior diaphragmatic adenopathy involving both the anterior or paracardiac (black arrows) and lateral or juxtaphrenic (white arrow) groups of nodes.

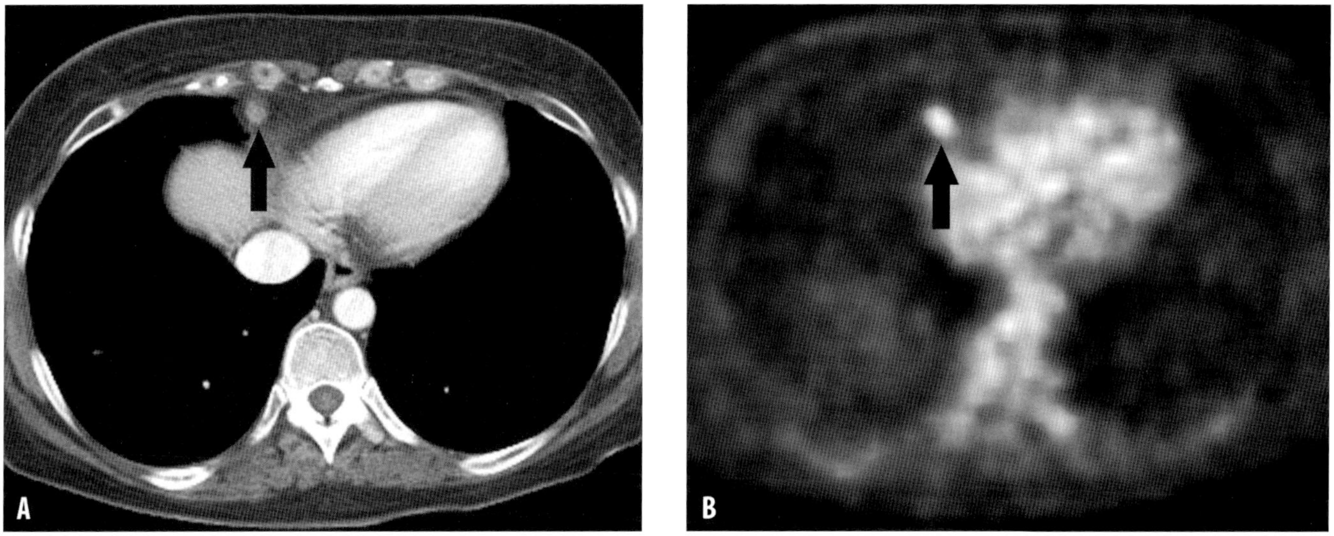

Figure 2.2 A. Axial contrast-enhanced CT image in a 48 year old woman with ovarian cancer shows superior diaphragmatic adenopathy (arrow). **B.** Axial FDG PET image shows increased uptake in the node (arrow), confirming the metastatic nature of the enlargement.

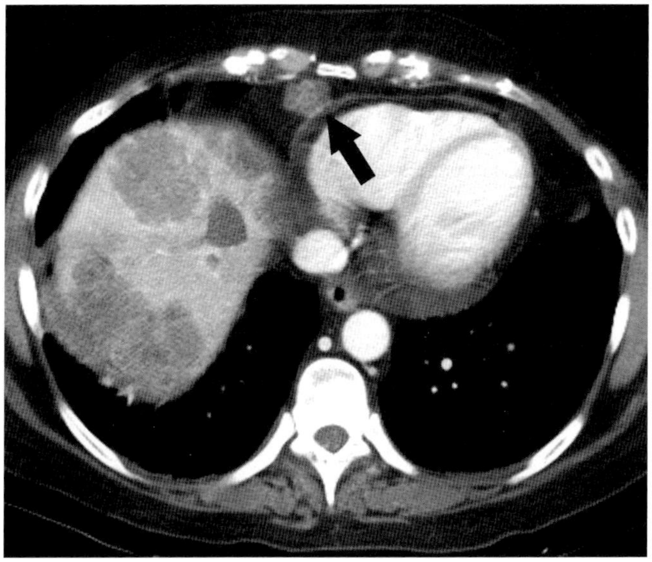

Figure 2.3 Axial contrast-enhanced CT image in a 72 year old woman with breast cancer metastatic to the liver shows malignant-appearing superior diaphragmatic adenopathy (arrow).

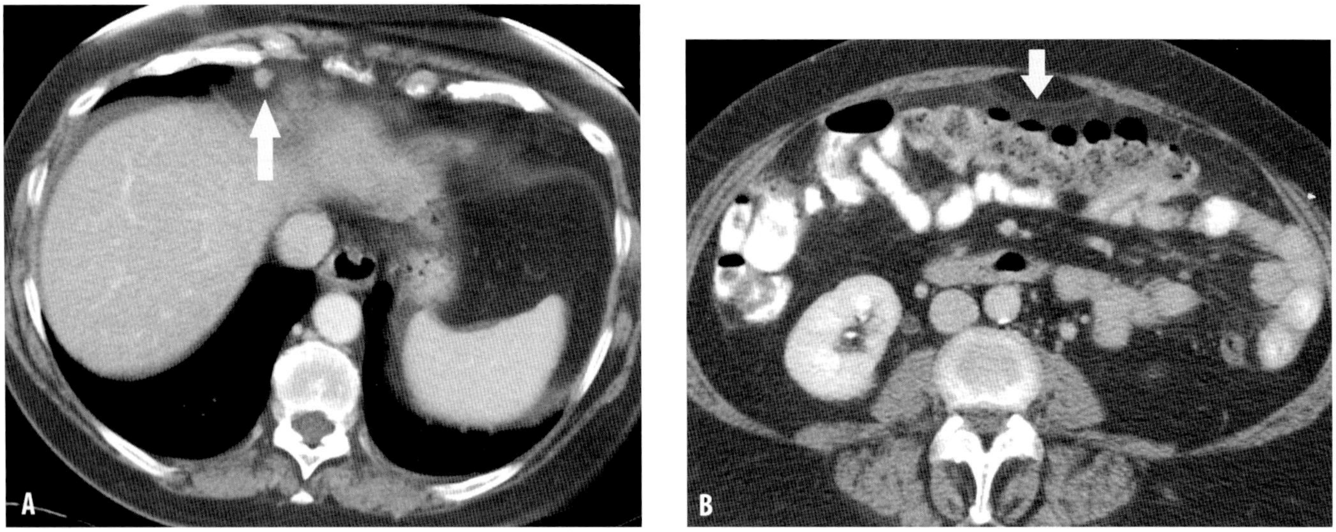

Figure 2.4 A. Axial contrast-enhanced CT image in a 62 year old woman with newly diagnosed ovarian cancer shows superior diaphragmatic adenopathy (arrow). B. Axial contrast-enhanced CT image at a more inferior level shows subtle infiltration (arrow) of the greater omentum. This is particularly concerning for peritoneal spread, given the co-existence of superior diaphragmatic adenopathy. Malignant infiltration of the omentum was confirmed at surgery.

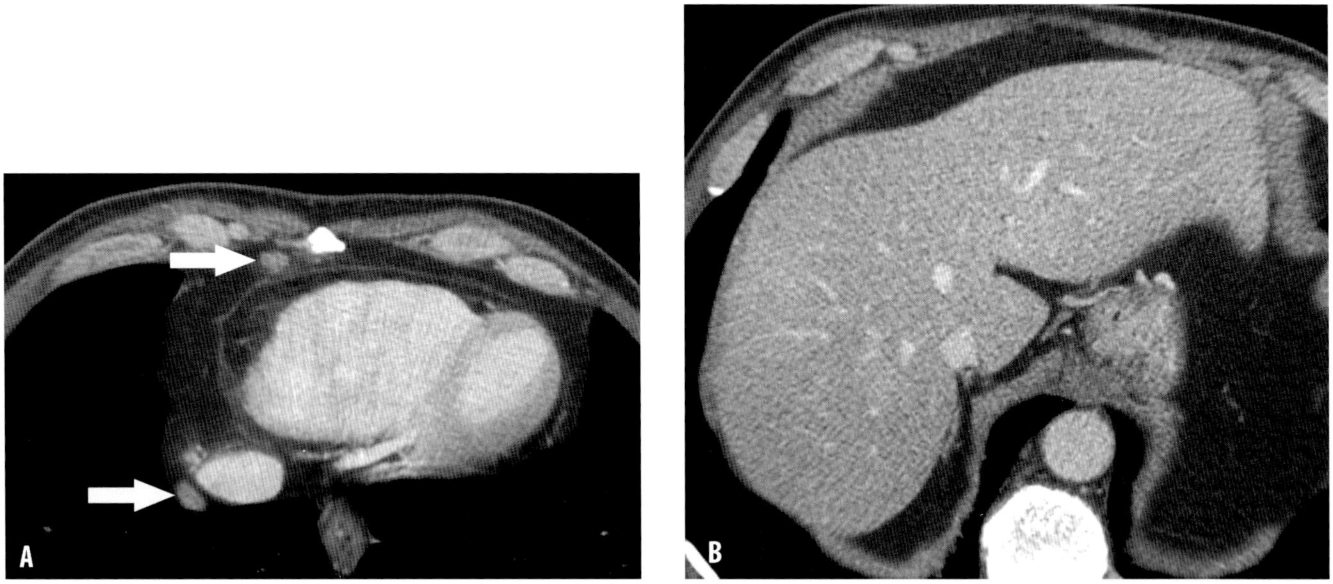

Figure 2.5 A. Axial contrast-enhanced CT image in a 58 year old man with chronic hepatitis C shows superior diaphragmatic adenopathy (arrows). B. Axial contrast-enhanced CT image at a more inferior level shows a relatively large left hepatic lobe, but a smooth liver surface. Biopsy showed grade 3 inflammatory change and stage 3 fibrosis but no definite cirrhosis. Superior diaphragmatic adenopathy can be an indicator of clinically important liver disease even when the liver appears relatively normal at imaging.

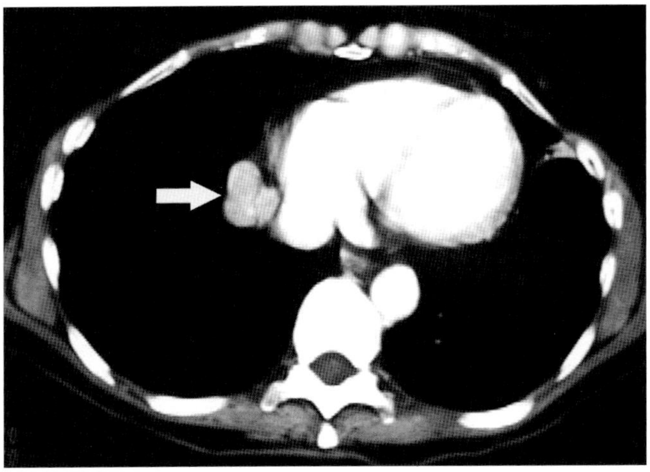

Figure 2.6 Axial contrast-enhanced CT image in a 56 year old woman with ovarian cancer shows an enlarged paracaval superior diaphragmatic node (arrow). This could potentially be confused for a pleural or pulmonary mass.

CASE **3**

Lateral arcuate ligament pseudotumor

Imaging description

The diaphragmatic crura fuse with each other medially to form the single midline median arcuate ligament, behind which the aorta passes from the thorax into the abdomen. Laterally, the crura extend in front of the psoas muscles as the paired medial arcuate ligaments, which provide a ligamentous attachment for the diaphragm. The medial arcuate ligament is classically described as attaching to the transverse process of L1, although a dissection study suggests it actually attaches to the transverse process of L2 [1]. More laterally still, the crura continue in front of the quadratus lumborum muscles as the paired lateral arcuate ligaments, which pass from the spinal attachment to the 12th rib. Prominent lateral arcuate ligaments may be seen as distinct soft tissue nodules of 1 cm or more in diameter in continuity with the diaphragm and projecting into the posterior pararenal space of the retroperitoneum on cross-sectional imaging (Figure 3.1) [2].

Importance

A prominent lateral arcuate ligament may simulate a retroperitoneal mass, or suggest peritoneal metastases in the hepatorenal pouch (if right-sided).

Typical clinical scenario

Nodular projections into the retroperitoneum due to prominent lateral arcuate ligaments were seen in 5 of 100 unselected CT scans, and were bilateral in 3 patients [2]. No particular association with age, sex, or respiratory position has been described.

Differential diagnosis

The usual appearance of the lateral arcuate ligaments is that of soft tissue nodules anterior to the lower posterior ribs. When paired, the bilateral symmetric arrangement allows for easy differentiation from disease [3]. When unilateral, band-like curvilinear continuity with the diaphragm is a useful clue. Occasionally, a pleural metastasis deep in the costophrenic recess may be difficult to distinguish from the lateral arcuate ligament, although correlation with prior imaging or PET scan may clarify (Figure 3.2).

Teaching point

An apparent tumor implant abutting the diaphragm anterior to the lower posterior ribs is likely to represent a prominent lateral arcuate ligament.

REFERENCES

1 Deviri E, Nathan H, Luchansky E. Medial and lateral arcuate ligaments of the diaphragm: attachment to the transverse process. *Anat Anz* 1988; **166**: 63–67.

2 Silverman PM, Cooper C, Zeman RK. Lateral arcuate ligaments of the diaphragm: anatomic variations at abdominal CT. *Radiology* 1992; **185**: 105–108.

3 Panicek DM, Benson CB, Gottlieb RH, Heitzman ER. The diaphragm: anatomic, pathologic, and radiologic considerations. *Radiographics* 1988; **8**: 385–425.

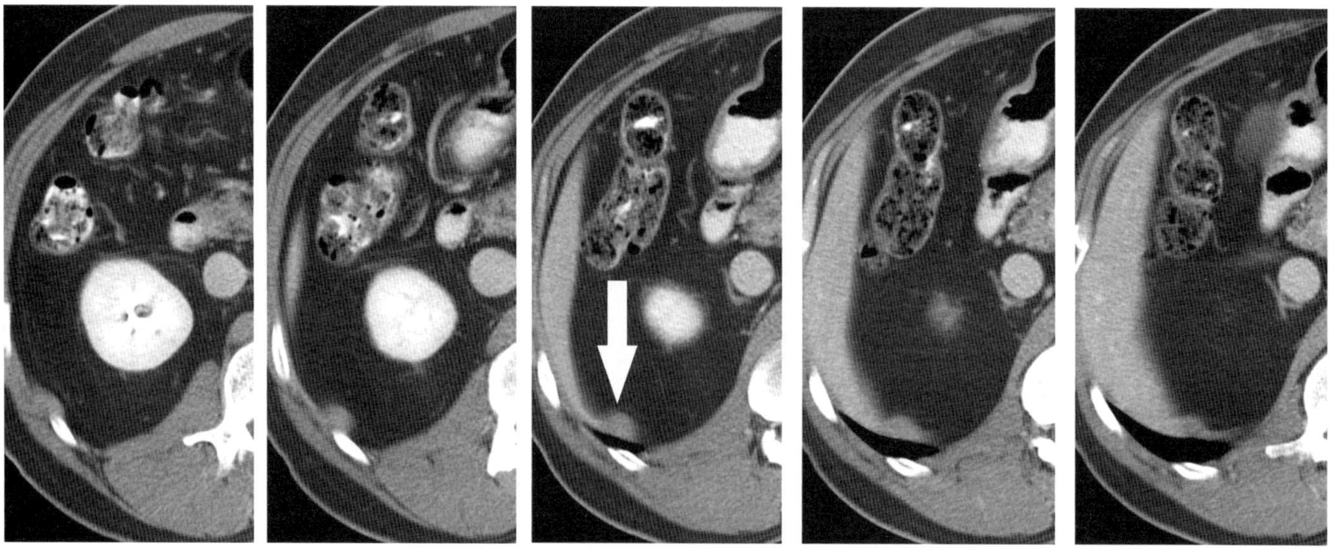

Figure 3.1 Montage of five axial contrast-enhanced CT images arranged from superior to inferior and obtained in a 50 year old man with abdominal pain. The central image shows a soft-tissue nodule (arrow) adjacent to the liver that mimics a peritoneal implant, but curvilinear continuity with the diaphragm on serial images confirms the structure is the lateral arcuate ligament.

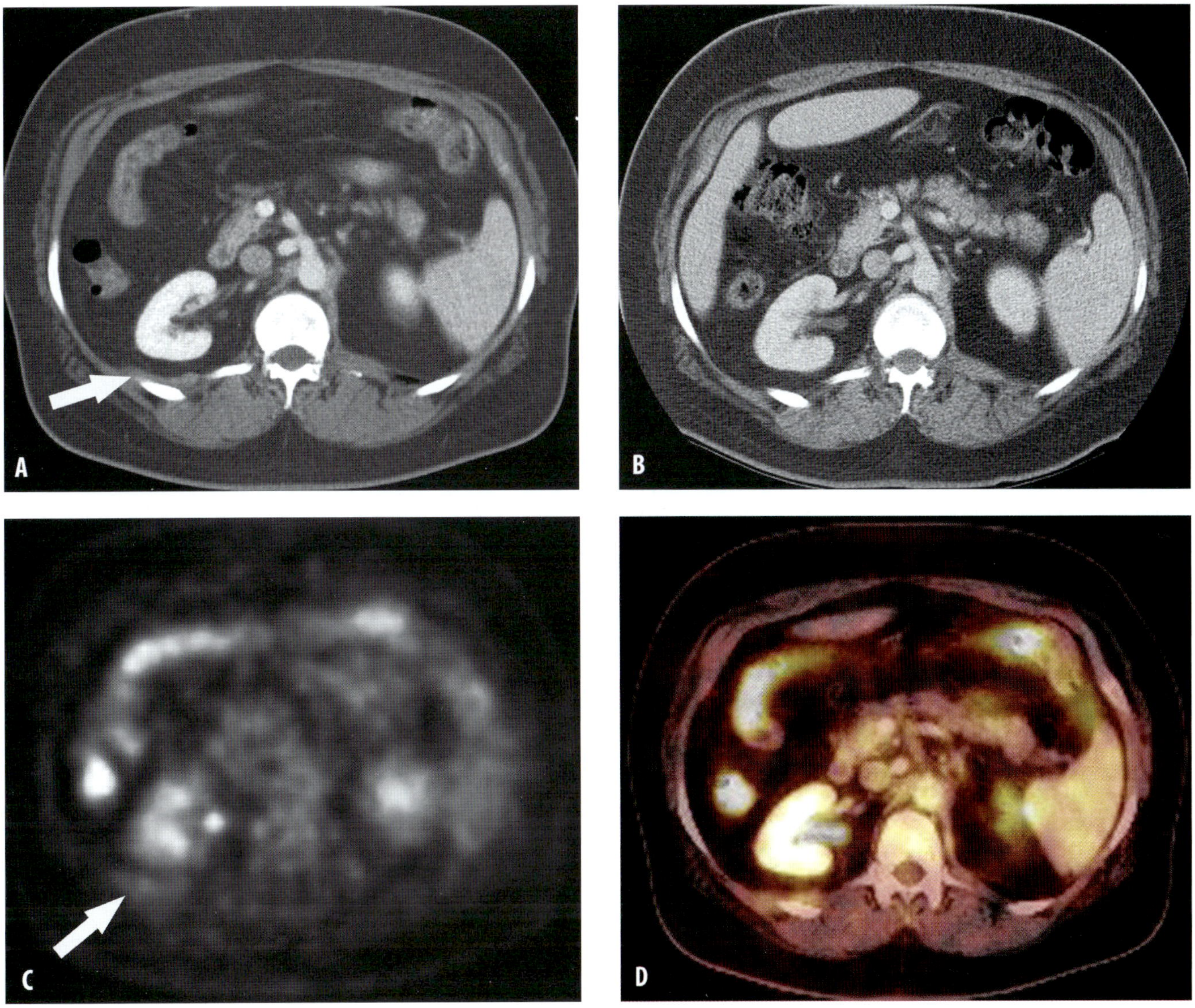

Figure 3.2 **A.** Axial contrast-enhanced CT image in a 46 year old man with recurrent malignant thymoma. A plaque-like focus of soft-tissue thickening (arrow) abutting the front of the lower right ribs resembles a prominent lateral arcuate ligament. **B.** Axial contrast-enhanced CT image performed five years before does not show the plaque-like focus of soft-tissue thickening. **C.** Axial PET image at the corresponding level shows increased FDG uptake (arrow) in the soft-tissue thickening. **D.** Fused PET/CT image verifies the increased uptake is within the soft-tissue thickening, confirming the diagnosis of a pleural metastasis deep in the costophrenic recess. Occasionally, such a metastasis may be difficult to distinguish from the lateral arcuate ligament.

CASE 4 Diaphragmatic slip pseudotumor

Imaging description

Prominent muscular slips of the diaphragm may be seen as soft-tissue nodules in contiguity with the diaphragm on CT or MRI (Figures 4.1 and 4.2) [1, 2].

Importance

Prominent diaphragmatic slips may mimic perihepatic metastatic implants, resulting in unnecessary follow-up investigations and patient anxiety.

Typical clinical scenario

Prominent diaphragmatic slips are described as being more frequent in deep inspiration [1]. Such diaphragmatic pseudotumors are also commoner in elderly or emphysematous patients [3].

Differential diagnosis

The distinction of prominent diaphragmatic slips from true peritoneal implants is based on their continuity peripherally with the diaphragm, curvilinear course when tracked over serial slices, and separation from adjacent viscera by subdiaphragmatic fat. Decubitus and expiratory CT sections are also said to help [1].

Teaching point

An apparent peritoneal implant abutting the diaphragm should be examined closely in order to make an accurate distinction from a prominent diaphragmatic slip.

REFERENCES

1 Rosen A, Auh YH, Rubenstein WA, et al. CT appearance of diaphragmatic pseudotumors. *J Comput Assist Tomogr* 1983; **7**: 995–999.

2 Schwartz EE, Wechsler RJ. Diaphragmatic and paradiaphragmatic tumors and pseudotumors. *J Thorac Imaging* 1989; **4**: 19–28.

3 Caskey CI, Zerhouni EA, Fishman EK, Rahmouni AD. Aging of the diaphragm: a CT study. *Radiology* 1989; **171**: 385–389.

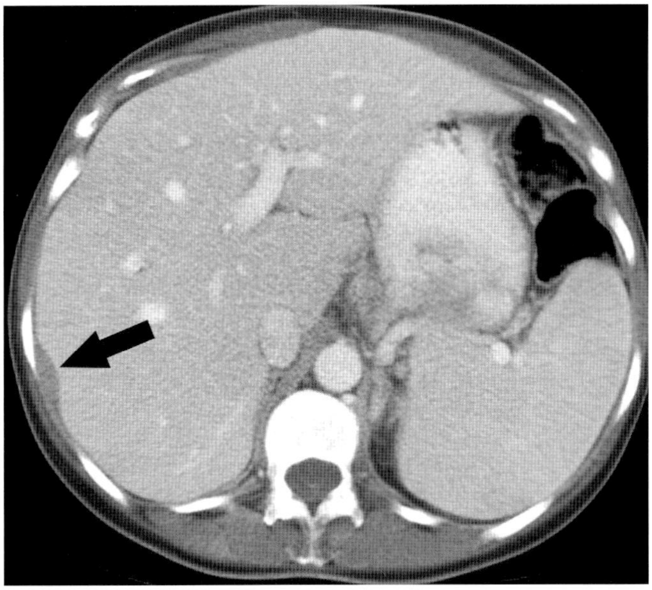

Figure 4.1 Axial contrast-enhanced CT image in a 63 year old woman with ovarian cancer shows a soft-tissue nodule (arrow) adjacent to the liver that mimics a peritoneal implant. Curvilinear continuity with the diaphragm was evident on serial images (not shown), confirming the diagnosis of a diaphragmatic slip.

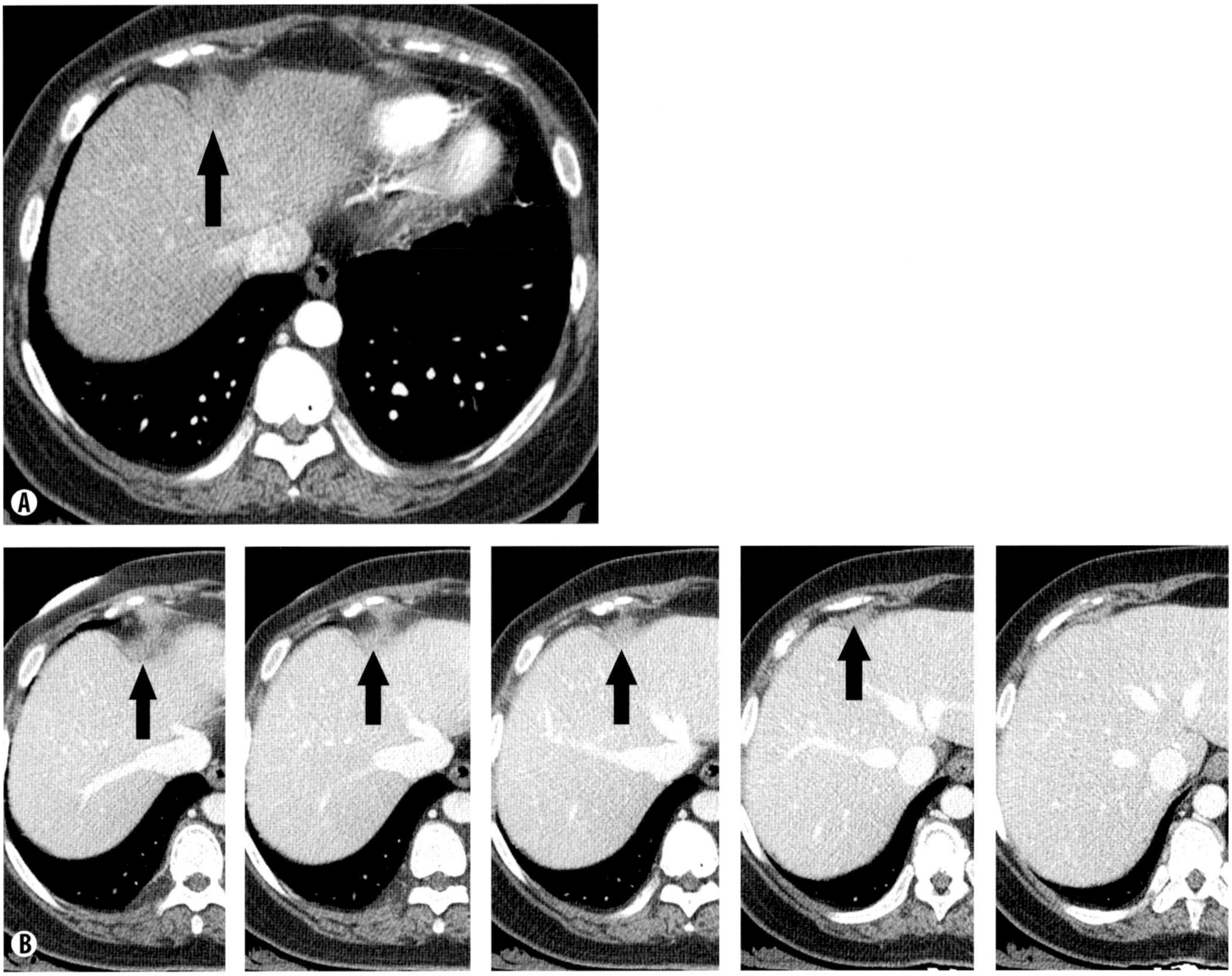

Figure 4.2 **A.** Axial contrast-enhanced arterial phase CT image in a 32 year old woman with gastrointestinal bleeding shows an apparent mass (arrow) anterior to the liver. **B.** Montage of five axial contrast-enhanced portal venous phase CT images arranged from superior to inferior shows the apparent mass (arrows) is in curvilinear continuity with the diaphragm, confirming the structure is a prominent diaphragmatic slip.

CASE 5 Diaphragmatic crus mimicking adenopathy

Imaging description

The diaphragmatic crura are paired tendinous structures that extend downward from the diaphragm to attach to the upper three lumbar vertebrae on the right and the upper two lumbar vertebrae on the left. The crura are variable in size, and may measure up to 2.1 cm in thickness [1]. The right diaphragmatic crus is generally longer and thicker than the left. Prominent crura may appear as soft tissue nodules in contiguity with the upper lumbar vertebrae on cross-sectional imaging (Figure 5.1) [2].

Importance

Misinterpretation of the diaphragmatic crura as retroperitoneal adenopathy is a recognized diagnostic pitfall, and may result in unnecessary follow-up investigations and patient anxiety.

Typical clinical scenario

Prominence of the diaphragmatic crura is more frequent when scans are obtained in deep inspiration but is largely unaffected by age or gender [2], with the exception that the diaphragmatic crura are larger and more nodular relative to body size in children under five years of age [3].

Differential diagnosis

The distinction of the crura from true retroperitoneal disease can usually be made by close examination of serial axial images, which confirms their continuity with the diaphragm and curvilinear course. Obtaining scans at full expiration and full inspiration has also been described as helpful, because the crura increase in thickness on inspiration when compared with the size on expiration (Figure 5.2) [4]. In difficult cases, correlation with prior studies or PET may be helpful (Figure 5.3).

Teaching point

Apparent retroperitoneal adenopathy abutting the upper lumbar spine should be scrutinized to evaluate the possibility of prominent diaphragmatic crura as a confounding mimic.

REFERENCES

1 Dovgan DJ, Lenchik L, Kaye AD. Computed tomographic evaluation of maximal diaphragmatic crural thickness. *Conn Med* 1994; **58**: 203–206.

2 Callen PW, Filly RA, Korobkin M. Computed tomographic evaluation of the diaphragmatic crura. *Radiology* 1978; **126**: 413–416.

3 Brengle M, Cohen MD, Katz B. Normal appearance and size of the diaphragmatic crura in children: CT evaluation. *Pediatr Radiol* 1996; **26**: 811–814.

4 Williamson BR, Gouse JC, Rohrer DG, Teates CD. Variation in the thickness of the diaphragmatic crura with respiration. *Radiology* 1987; **163**: 683–684.

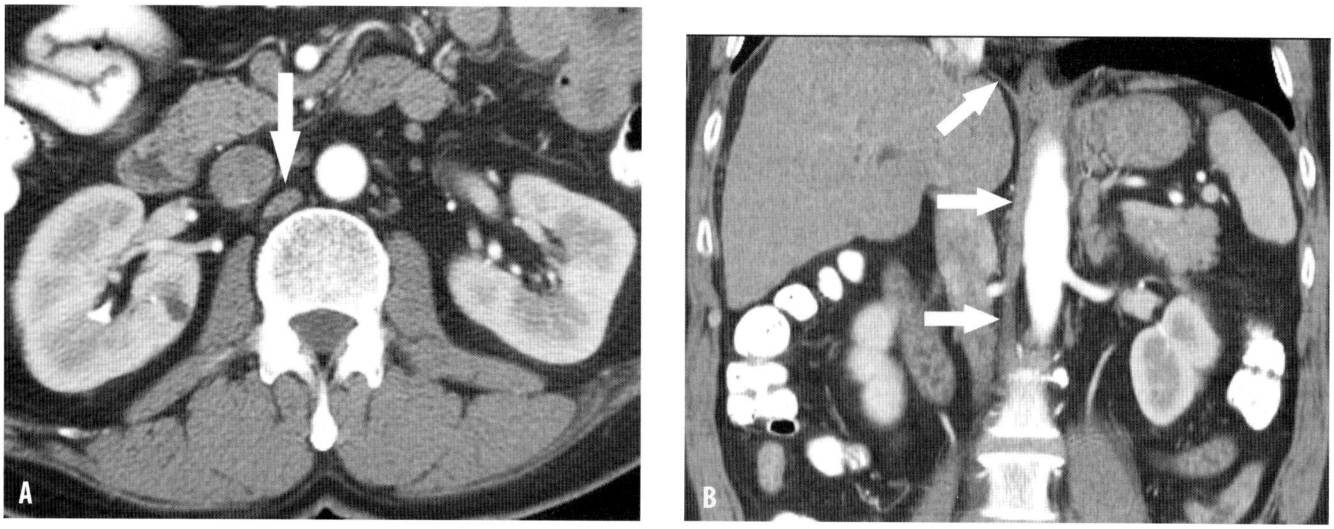

Figure 5.1 **A.** Axial contrast-enhanced CT image in a 58 year old man with cirrhosis and hepatocellular carcinoma shows an ovoid soft-tissue structure (arrow) anterior to the lumbar spine that could be interpreted as an interaortocaval lymph node. **B.** Coronal reformatted contrast-enhanced CT image demonstrates the structure is actually a prominent right diaphragmatic crus (arrows).

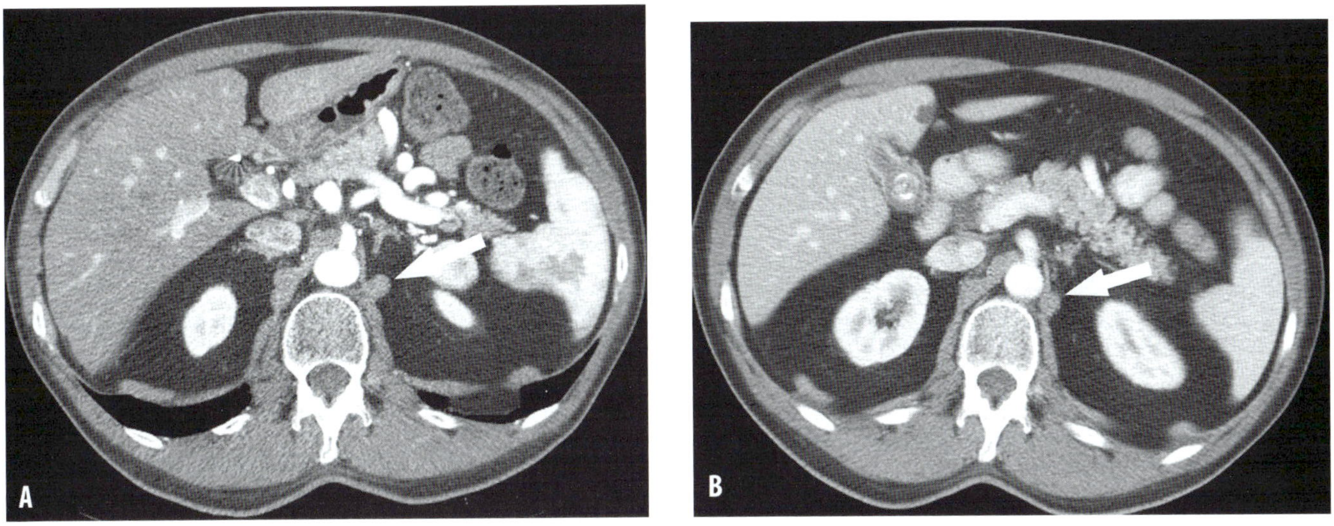

Figure 5.2 **A.** Axial contrast-enhanced CT image obtained during full inspiration shows a nodule (arrow) anterior to the lumbar spine. **B.** Axial contrast-enhanced CT image obtained during full expiration shows the nodule (arrow) has decreased in size. Such a reduction in size is characteristic of the diaphragmatic crura.

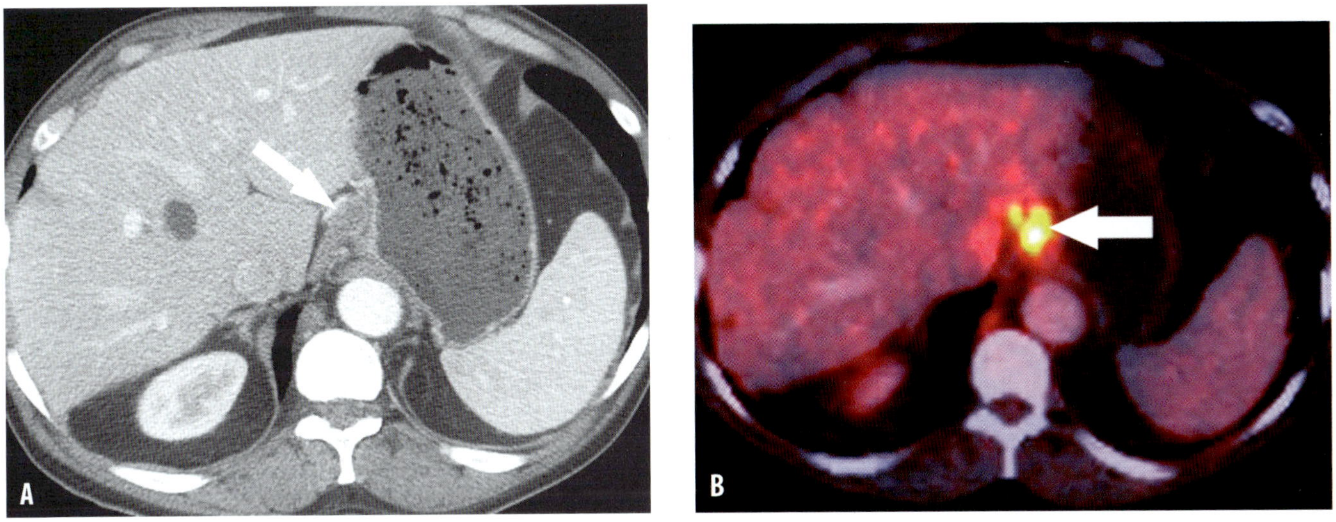

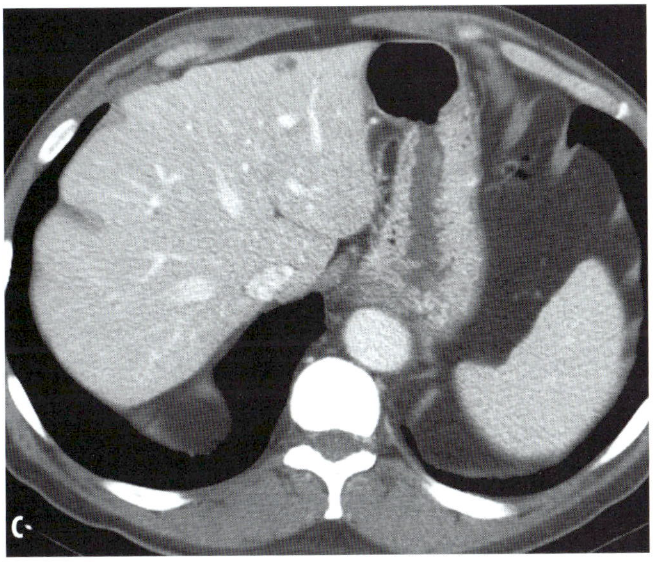

Figure 5.3 **A.** Axial contrast-enhanced CT image in a 73 year old man with metastatic lung cancer shows an ovoid structure (arrow) at the level of the diaphragmatic esophageal hiatus that could reasonably be interpreted as either metastasis or a prominent crus. **B.** Axial fused PET/CT image shows increased uptake in the structure (arrow) consistent with a metastasis. **C.** Axial contrast-enhanced CT image at the corresponding level performed one year earlier shows the soft-tissue structure was not present previously, again confirming the diagnosis of metastasis.

CASE

6 Epiphrenic diverticulum mimicking hiatal hernia

Imaging description

Epiphrenic diverticula are outpouchings of the distal esophagus just above the diaphragm that appear as thin-walled, air or air-fluid filled structures adjacent to the distal esophagus. An epiphrenic diverticulum can mimic a hiatal hernia at CT (Figures 6.1 and 6.2) [1].

Importance

Misdiagnosis of an epiphrenic diverticulum as a hiatal hernia is unlikely to have serious consequences, but might result in a missed opportunity to recognize a treatable esophageal disorder. Occasionally, epiphrenic diverticula can be complicated by malignancy, obstruction, bleeding, or perforation and are then of greater clinical importance [2–6]. An epiphrenic diverticulum has been described as a cause of false positive uptake at iodine-131 scintigraphy, potentially resulting in a misdiagnosis of metastatic thyroid cancer [7].

Typical clinical scenario

Epiphrenic diverticula are believed to be pulsion diverticula generated by underlying esophageal dysmotility [8], although not all patients complain of dysphagia or have dysmotility evident on esophagography [9]. While most patients can be treated conservatively, some may require surgery. Operative correction requires both a diverticulectomy and a myotomy to address the underlying motility disorder [8].

Differential diagnosis

The primary differential is a hiatal hernia, which is a far commoner abnormality of the distal esophagus. A diverticular neck is rarely seen at CT in cases of epiphrenic diverticula. Two other observations are more helpful in the distinction of epiphrenic diverticula from hiatal hernias (Figures 6.3 and 6.4):

- Epiphrenic diverticula are thin-walled while hiatal hernias are thick-walled and contain gastric mucosa and rugae.
- Hiatal hernias are usually associated with widening of the esophageal hiatus, which has been defined as "whenever the diaphragmatic crura were not tightly opposed and in intimate association with the esophageal wall" [10].

Teaching point

The possibility of an epiphrenic diverticulum should be considered before making a diagnosis of a hiatal hernia at CT.

REFERENCES

1 Kim KW, Berkmen YM, Auh YH, Kazam E. Diagnosis of epiphrenic esophageal diverticulum by computed tomography. *J Comput Tomogr* 1988; **12**: 25–28.

2 Lai ST, Hsu CP. Carcinoma arising from an epiphrenic diverticulum: a frequently misdiagnosed disease. *Ann Thorac Cardiovasc Surg* 2007; **13**: 110–113.

3 Hung JJ, Hsieh CC, Lin SC, Wang LS. Squamous cell carcinoma in a large epiphrenic esophageal diverticulum. *Dig Dis Sci* 2009; **54**: 1365–1368.

4 Niv Y, Fraser G, Krugliak P. Gastroesophageal obstruction from food in an epiphrenic esophageal diverticulum. *J Clin Gastroenterol* 1993; **16**: 314–316.

5 Abul-Khair MH, Khalil A, Mohsen A. Bleeding from an epiphrenic oesophageal diverticulum. *Eur J Surg* 1992; **158**: 377–378.

6 Stalheim AJ. Spontaneous perforation of diverticulum of distal esophagus. *Minn Med* 1978; **61**: 424–426.

7 Nguyen BD, Roarke MC. Epiphrenic diverticulum: potential pitfall in thyroid cancer iodine-131 scintigraphy. *Clin Nucl Med* 2005; **30**: 631–632.

8 Rivkin L, Bremner CG, Bremner CH. Pathophysiology of mid-oesophageal and epiphrenic diverticula of the oesophagus. *S Afr Med J* 1984; **66**: 127–129.

9 Fasano NC, Levine MS, Rubesin SE, Redfern RO, Laufer I. Epiphrenic diverticulum: clinical and radiographic findings in 27 patients. *Dysphagia* 2003; **18**: 9–15.

10 Ginalski JM, Schnyder P, Moss AA, Brasch RC. Incidence and significance of a widened esophageal hiatus at CT scan. *J Clin Gastroenterol* 1984; **6**: 467–470.

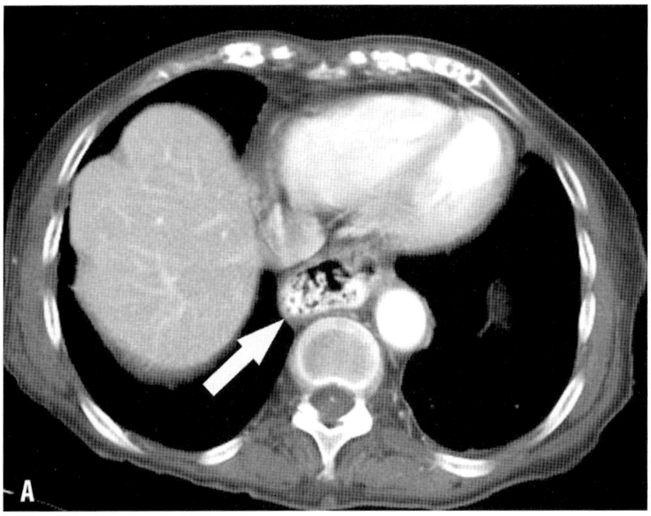

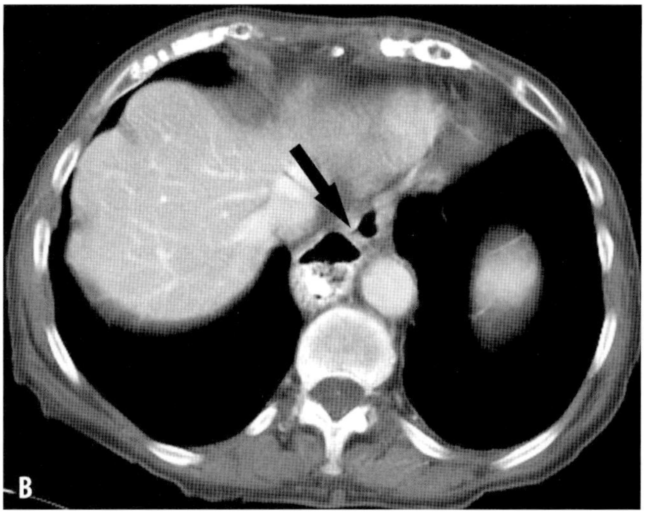

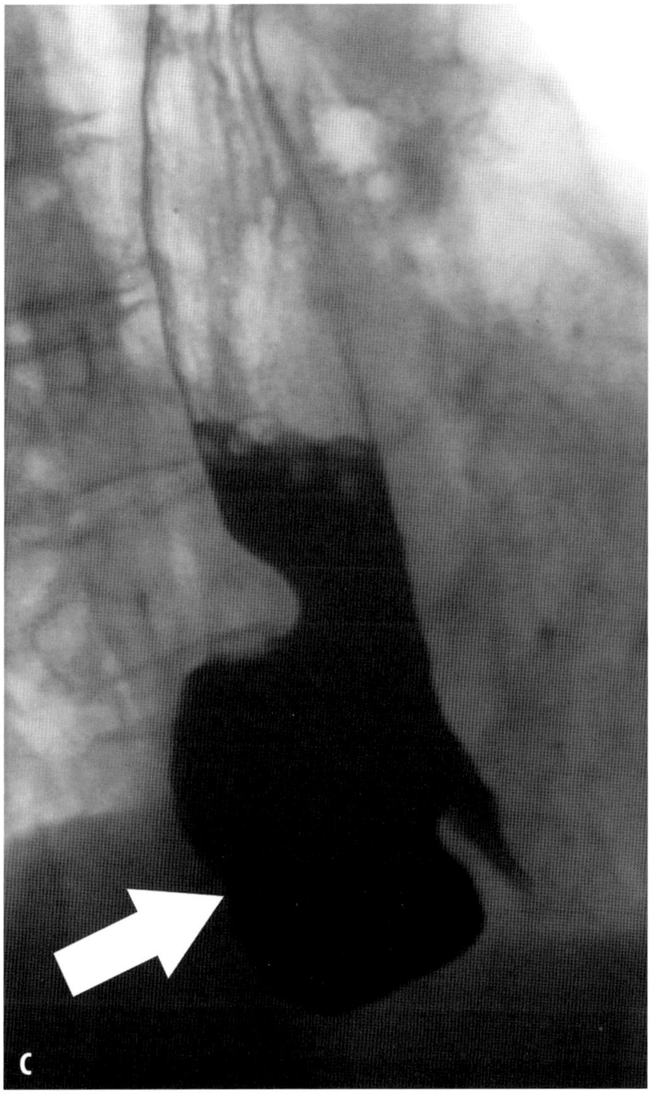

Figure 6.1 **A.** Axial contrast-enhanced CT image in an 83 year old woman with a suspected 8 mm right upper lobe lung nodule seen on chest radiograph performed because of cough (no nodule was seen on CT). An air and fluid-filled structure (arrow) in the posterior mediastinum was reported as a hiatal hernia. **B.** Axial contrast-enhanced CT image at a more superior level. Note that no gastric rugae can be seen in the structure, and there is a subtle beak-like extension (arrow) of the esophageal lumen towards the structure. In retrospect, these features suggest the diagnosis of an epiphrenic diverticulum. **C.** Oblique image from a barium esophagram performed three years later because of dysphagia confirms the presence of an epiphrenic diverticulum (arrow). Esophageal dysmotility and spontaneous gastroesophageal reflux were also seen. The patient required diverticulectomy and myotomy.

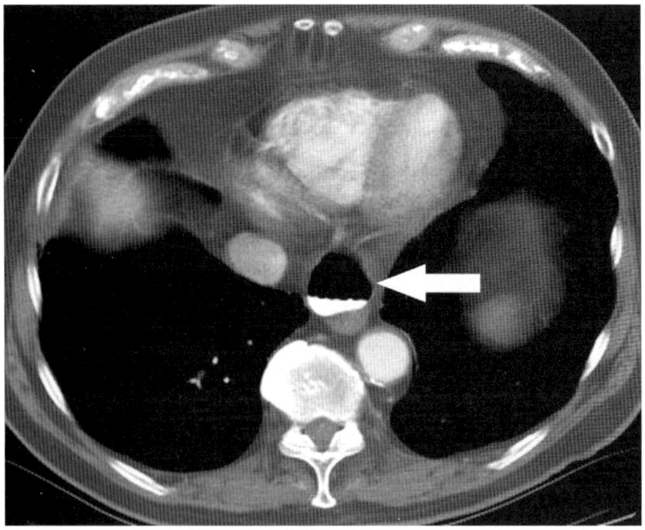

Figure 6.2 Axial contrast-enhanced CT image in an 86 year old man with metastatic prostate cancer shows an incidental and asymptomatic epiphrenic diverticulum (arrow).

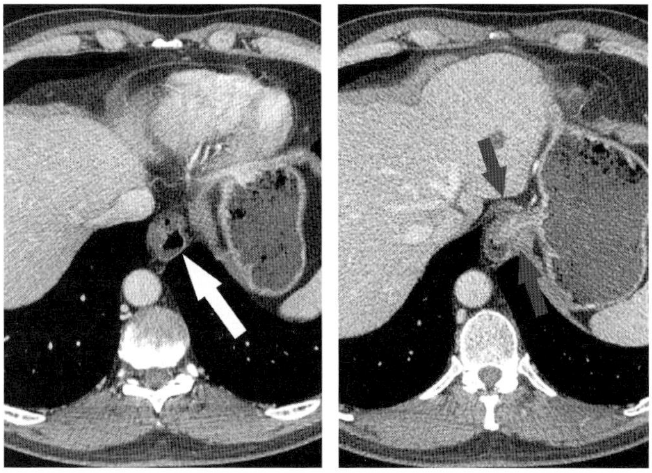

Figure 6.3 Photomontage of axial contrast-enhanced CT images in a 62 year old man being staged for newly diagnosed prostate cancer shows an incidental epiphrenic diverticulum (white arrow). Note the diverticulum is thin-walled and lacks gastric rugae or mucosa, and that the diaphragmatic crura (grey arrows) are closely opposed.

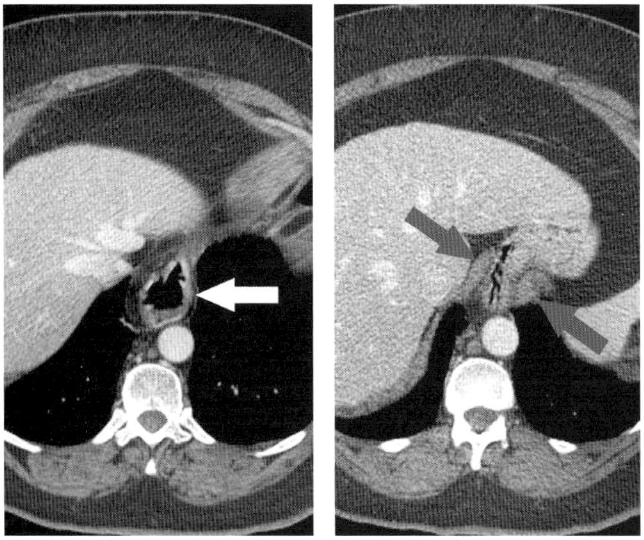

Figure 6.4 Photomontage of axial contrast-enhanced CT images in a 61 year old woman prior to nephrectomy for a left renal cell carcinoma shows an incidental hiatal hernia (white arrow). Note the hernia is thick-walled with gastric rugae, and that the diaphragmatic crura (grey arrows) are splayed apart with widening of the esophageal hiatus.

CASE 7

Mediastinal ascites

Imaging description

In a hiatal hernia, the stomach protrudes into the chest through the esophageal hiatus of the diaphragm. The stomach is an intraperitoneal organ, and so herniation of the stomach through the diaphragm is inevitably accompanied by herniation of the adjacent peritoneal recesses [1]. In a patient with a hiatal hernia and ascites, this can lead to ascitic fluid filling the peritoneal recesses around the herniated stomach in the chest, resulting in a fluid collection in the posterior mediastinum above the esophageal hiatus that has been termed "mediastinal ascites" (Figure 7.1) [2]. The anatomy of peritoneal herniation in hiatal hernia is such that fluid first accumulates to the left of and anterior to the esophagus and later surrounds the esophagus bilaterally.

Importance

On CT or MRI, mediastinal ascites may simulate fluid-filled mediastinal pathology such as a foregut cyst, mediastinal abscess, necrotic tumor, or pancreatic fluid collection [1].

Typical clinical scenario

Mediastinal ascites can occur in any patient with ascites and a hiatal hernia. My experience is that it occurs primarily in older patients with large volume ascites due to ovarian cancer or cirrhosis.

Differential diagnosis

The primary distinction is between true fluid-filled pathology in the posterior mediastinum and mediastinal ascites. Identification of a hiatal hernia and continuity of the thoracic fluid with intra-abdominal ascites are helpful signs in making the correct diagnosis [1, 2]. Rarely, the omentum alone can be herniated through the esophageal hiatus, resulting in a fatty mass above the diaphragm [3, 4]. Such an omental hernia can also be associated with a mediastinal ascites (Figure 7.2). Diagnostic findings in omental herniation include a fatty bilobed mass in the posterior mediastinum that is in continuity with subdiaphragmatic fat and contains omental blood vessels passing through the esophageal hiatus.

Teaching point

Fluid collecting in the posterior mediastinum above the esophageal hiatus can be due to herniation of intraperitoneal ascites surrounding a hiatus hernia and is known as mediastinal ascites; identification of a hiatal hernia and continuity of the fluid with intra-abdominal ascites should help establish this diagnosis and prevent confusion with true fluid-filled mediastinal pathology.

REFERENCES

1 Godwin JD, MacGregor JM. Extension of ascites into the chest with hiatal hernia: visualization on CT. *Am J Roentgenol* 1987; **148**: 31–32.

2 Pandolfo I, Gaeta M, Scribano E, et al. Mediastinal pseudotumor due to passage of ascites through the esophageal hiatus. *Gastrointest Radiol* 1989; **14**: 209–211.

3 Lee MJ, Breathnach E. CT and MRI findings in paraoesophageal omental herniation. *Clin Radiol* 1990; **42**: 207–209.

4 Kubota K, Ohara S, Yoshida S, Nonami Y, Takahashi T. Intrathoracic omental herniation through the esophageal hiatus: a case report. *Radiat Med* 2001; **19**: 307–311.

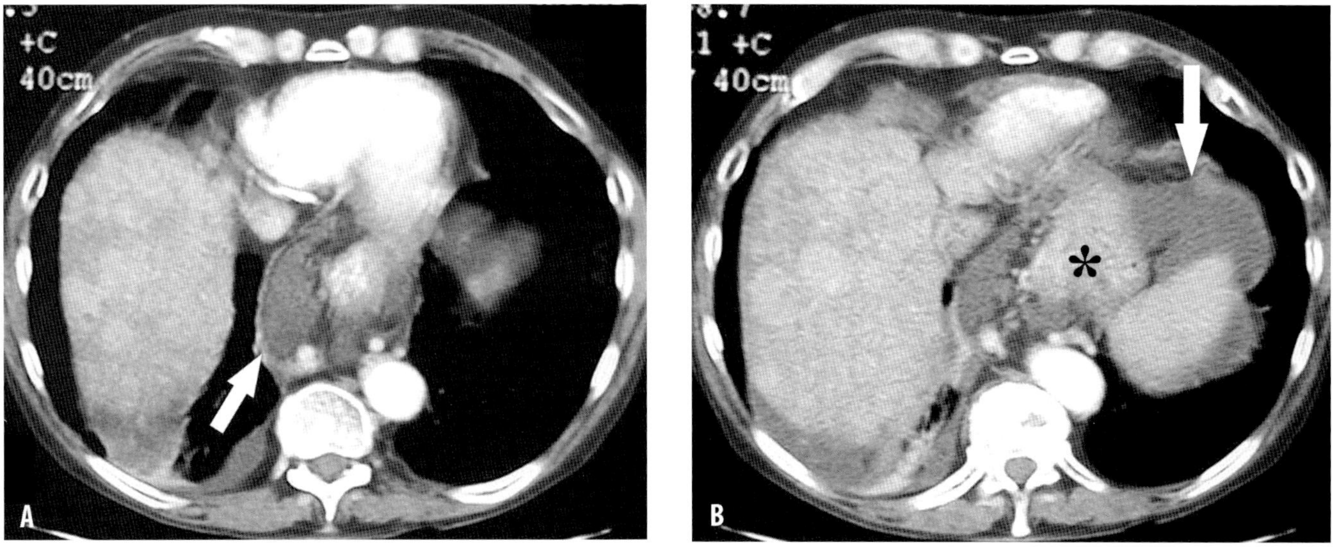

Figure 7.1 A. Axial contrast-enhanced CT image through the lower chest in a 62 year old woman with malignant ascites due to ovarian cancer shows fluid (arrow) in the posterior mediastinum. B. Axial contrast-enhanced CT image at a more inferior level shows the fluid is in continuity with intra-abdominal ascites (arrow) and also shows a large hiatal hernia (asterisk), establishing the diagnosis of so-called mediastinal ascites.

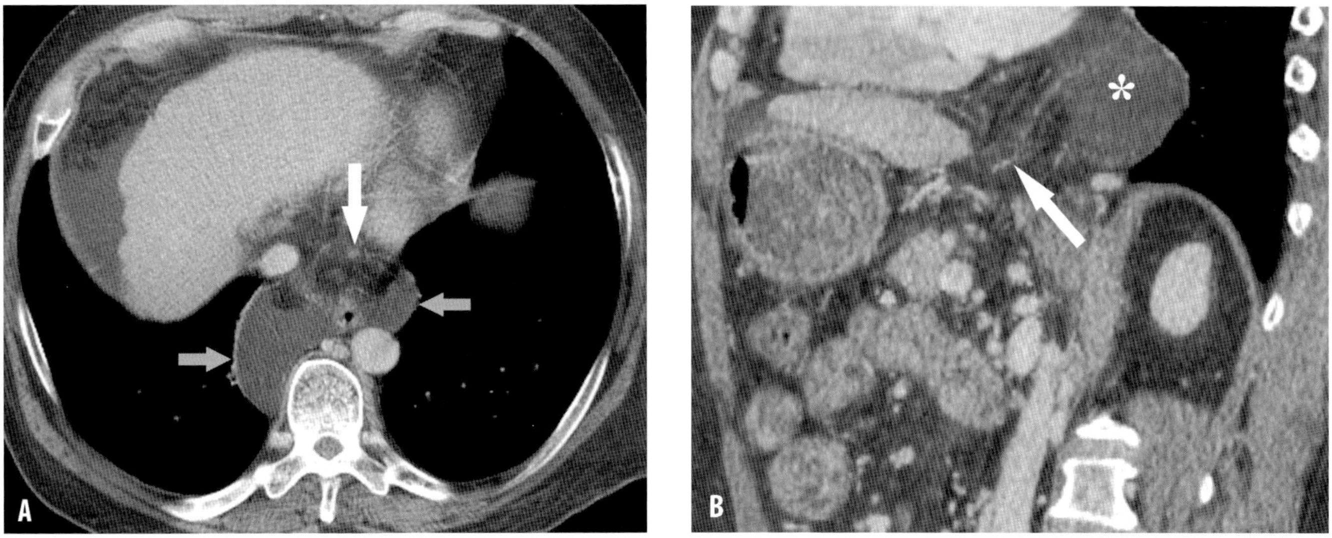

Figure 7.2 A. Axial contrast-enhanced CT image through the lower chest in a 51 year old man with a large volume of ascites secondary to cirrhosis shows fluid (grey arrows) accumulating on both sides of the esophagus. A fatty mass (white arrow) is seen in the posterior mediastinum anterior to the esophagus. Note the presence of perihepatic ascites. B. Curved planar sagittal reformatted contrast-enhanced CT image shows the fluid (asterisk) is associated with a fatty mass that is continuous with fat below the diaphragm and contains vessels (arrow) that pass through the esophageal hiatus. The findings are consistent with omental herniation associated with mediastinal ascites.

CASE 8 Diaphragmatic PET/CT misregistration artifact

Imaging description

The process of attenuation correction at PET/CT is applied to the PET images in order to account for differences in tissue density. For example, lung is less attenuating and so is made relatively "colder" after correction. However, the process depends on accurate co-registration between the CT and PET images. Because CT images are generally acquired in held inspiration and PET images are acquired during quiet respiration, there is frequently a mismatch between the two datasets. This occurs particularly near the diaphragm, such that the liver is more superior on the PET images than on the CT images. As a result, portions of the liver are corrected as if they were lung, and become too "cold" (Figure 8.1). The net result is that "hot spots" in the liver may initially appear to be in the lung when reviewing the attenuation corrected images. Review of the non-corrected images is the key to recognizing this artifact (Figure 8.2) [1, 2].

Importance

Incorrect localization of hepatic FDG "hot spot" foci to the lungs could have several adverse consequences. At a minimum, the error is confusing, since there will be no anatomic correlate for the foci of increased FDG uptake in the lungs on CT. Worse, a patient may be incorrectly assumed to have pulmonary metastases so that, for example, a patient with resectable colorectal hepatic metastases could be denied surgery because of apparent extrahepatic disease.

Typical clinical scenario

This artifactual misregistration typically occurs when a patient has metastases in the superior portion of the liver; attenuation over-correction of these parts of the liver results in an appearance that simulates "hot spots" in the lung.

Differential diagnosis

Review of the non-corrected PET images allows confident identification of this artifact, and there is no real differential.

Teaching point

Apparent pulmonary metastases seen near the diaphragm on only the PET portion of a PET/CT study should prompt careful review of the non-corrected images, since the finding may represent misregistration with attenuation over-correction of metastases that are actually in the liver.

REFERENCES

1 Papathanassiou D, Becker S, Amir R, Menéroux B, Liehn JC. Respiratory motion artefact in the liver dome on FDG PET/CT: comparison of attenuation correction with CT and a caesium external source. *Eur J Nucl Med Mol Imaging* 2005; **32**: 1422–1428.

2 Sarikaya I, Yeung HW, Erdi Y, Larson SM. Respiratory artefact causing malpositioning of liver dome lesion in right lower lung. *Clin Nucl Med* 2003; **28**: 943–944.

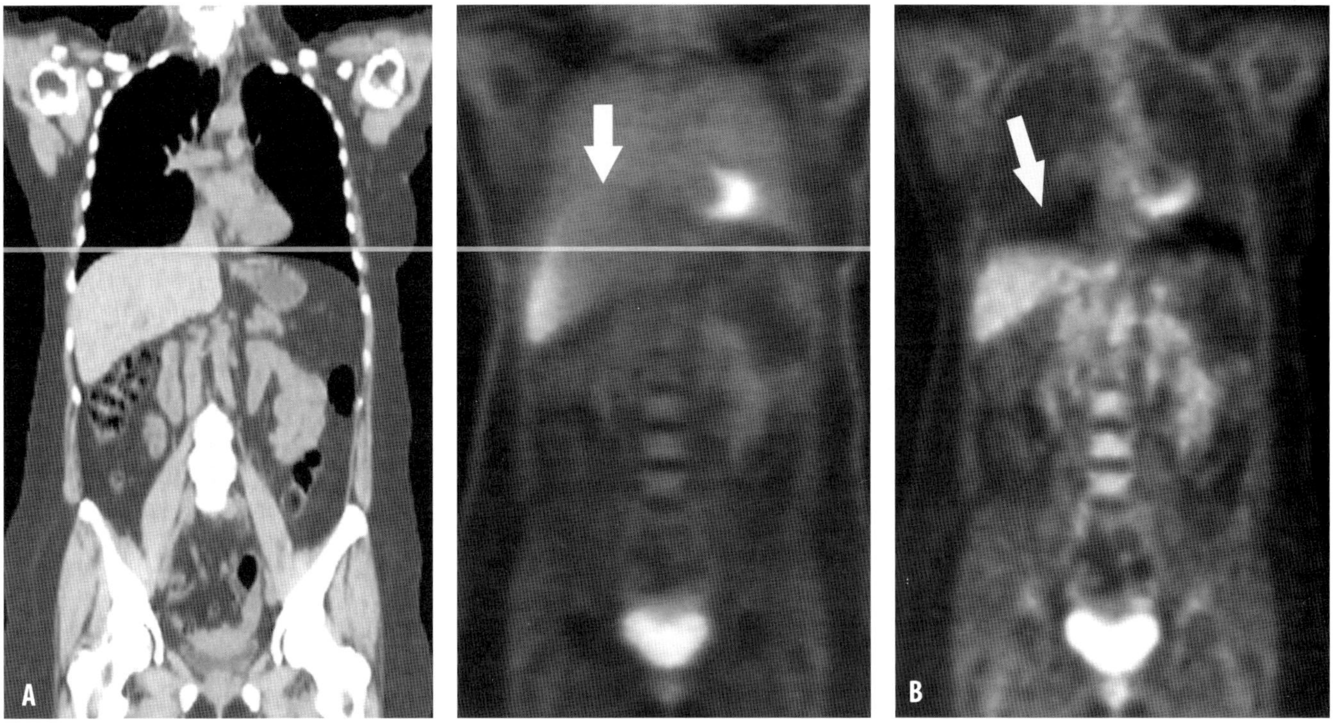

Figure 8.1 **A.** Montage of coronal reformatted non-enhanced CT images obtained during suspended respiration and the corresponding non-attenuation corrected coronal FDG PET image obtained during quiet respiration as part of a PET/CT examination in a 56 year old man with melanoma. Note that the diaphragm (arrow) lies at a higher level on the PET image than on the CT image (line) because the diaphragm is relatively "pushed down" on the CT image by the inspiratory effort. Accordingly the upper liver appears to correspond to lung tissue when the PET data are electronically registered with the CT data. **B.** Corresponding attenuation corrected coronal FDG PET image shows that the upper liver has been over-corrected and appears dark (arrow), because it is treated as if it were lung tissue during the attenuation correction process (note that lung is made "blacker" during attenuation correction to account for its lower tissue density).

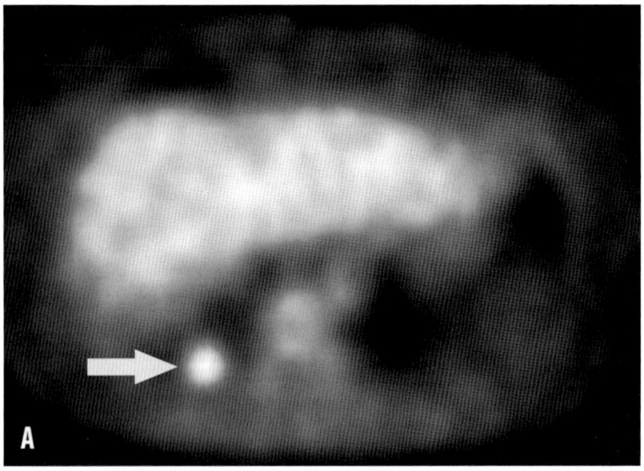

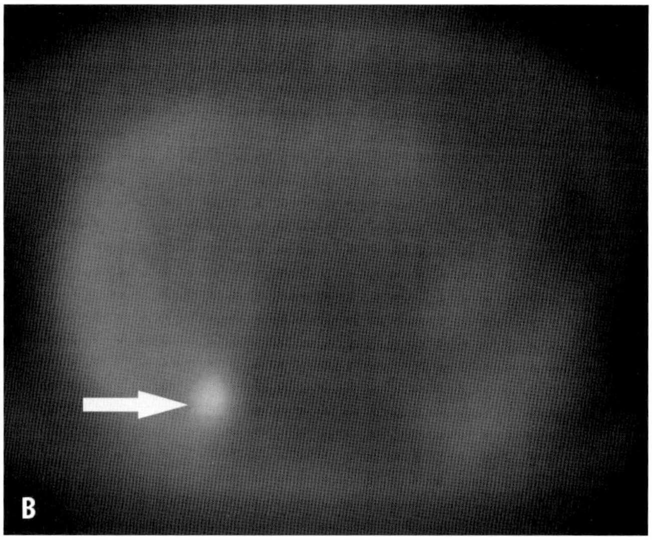

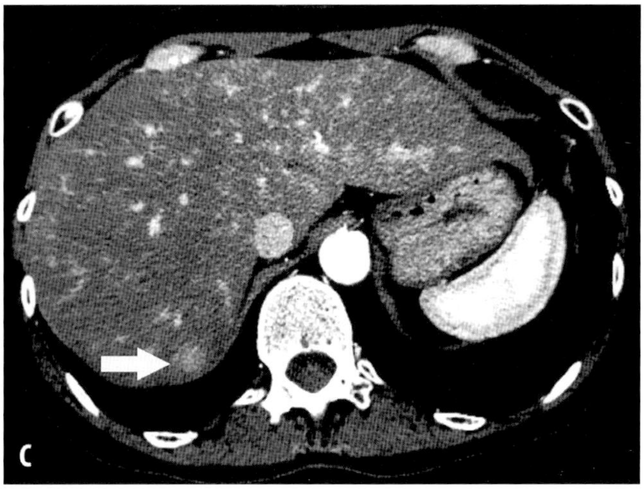

Figure 8.2 **A.** Attenuation corrected axial FDG PET image in a 62 year old woman with melanoma shows a focus of increased uptake (arrow) that appears to be in the base of the right lung, but CT images (not shown) of the lungs were unremarkable. **B.** Corresponding non-attenuation corrected axial FDG PET image shows the increased focus of uptake (arrow) is actually in the liver. **C.** Corresponding axial contrast-enhanced CT image confirms the presence of a hypervascular metastasis (arrow) in the posterior right hepatic lobe.

CASE 9 Lung base mirror image artifact

Imaging description

Ultrasound images are based on the assumption that sound waves pass directly from the transducer to an object, are reflected, and return directly to the transducer. This assumption is not always correct. For example, when the sound wave strikes an obliquely oriented reflecting surface such as the diaphragm, reflected echoes from an object offset from or even outside the beam can create a mirror image of an object in the "straight ahead" view of the transducer (Figure 9.1). Such "displaced" mirror image artifacts are commonly seen above the diaphragm when scanning the upper abdomen (Figures 9.2–9.4) [1].

Importance

"Full blown" mirror images of the liver, kidneys, or spleen are generally easy to recognize and ignore. Occasionally, mirror images of a hepatic cyst or solid tumor may suggest supradiaphragmatic fluid or lung mass. This is particularly problematic when the source object is not in the image (remember, the source object does not have to be in the primary beam to create a mirror image, and so may not be visible in the image).

Typical clinical scenario

Mirror image artifacts are commonly seen during upper abdominal ultrasound.

Differential diagnosis

Awareness is the key to the recognition of mirror image artifacts, and the distinction of pseudolesions created by such artifacts from true supradiaphragmatic pathology.

> **Teaching point**
>
> The possibility of mirror image artifacts should be considered for abnormalities seen just above the diaphragm at ultrasound.

REFERENCE

1 Feldman MK, Katyal S, Blackwood MS. US artifacts. *Radiographics* 2009; **29**: 1179–1189.

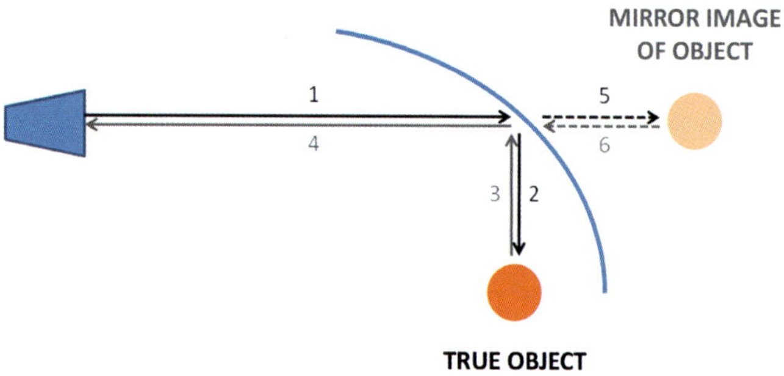

Figure 9.1 Schematic diagram illustrating the formation of mirror image artifacts at ultrasound. A sound wave (1) leaves the transducer and is reflected away from the primary beam by a reflective surface such as the diaphragm. The reflected beam (2) hits a real object and returns to the transducer along the same reflected pathway (3 and 4). The ultrasound scanner "thinks" the sound has travelled directly along the path of the primary beam (i.e., along 5 and 6), and so creates a "mirror image" of the object in that direction.

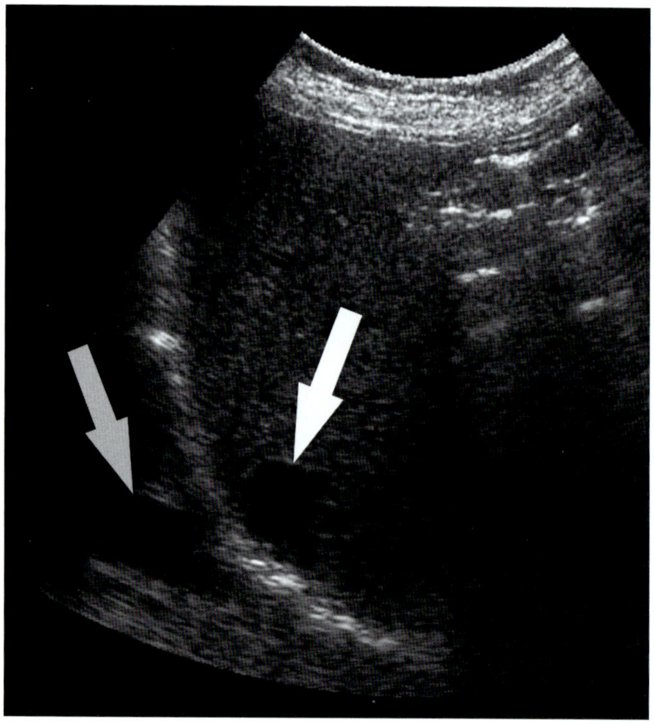

Figure 9.2 Sagittal ultrasound image of the right hepatic lobe in a 45 year old woman with chronic upper abdominal pain shows an incidental liver cyst (white arrow). A mirror image (grey arrow) of the cyst just above the diaphragm could mistakenly be interpreted as a pleural effusion or other focal supradiaphragmatic fluid-containing abnormality.

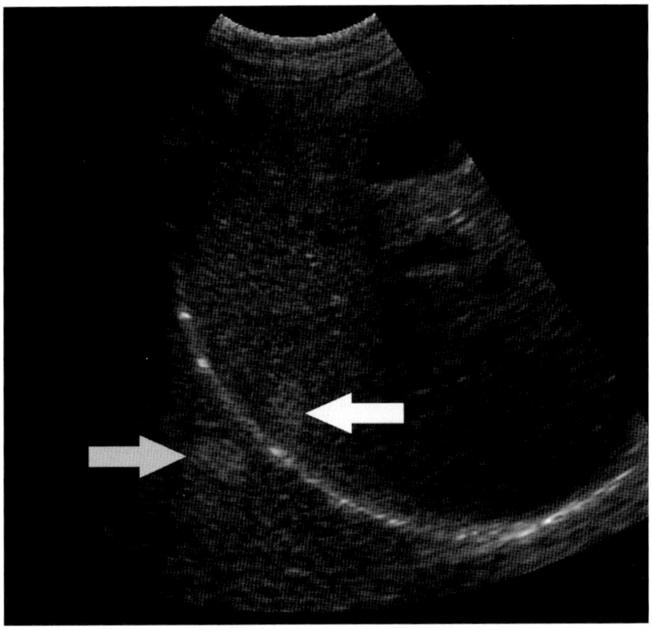

Figure 9.3 Sagittal ultrasound image of the right hepatic lobe in a 52 year old woman with an incidental hemangioma (white arrow) shows a mirror image (grey arrow) of the hemangioma just above the diaphragm that could conceivably be interpreted as a pulmonary mass within a consolidated or collapsed lung.

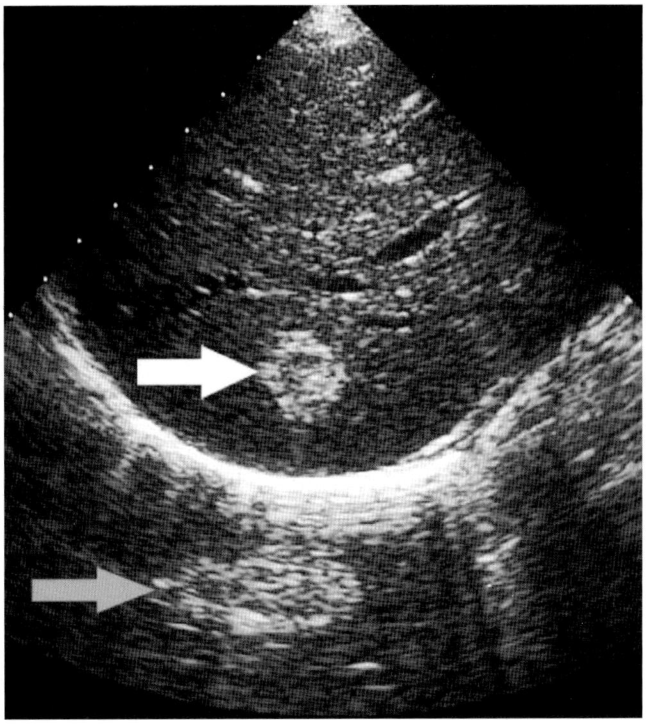

Figure 9.4 Sagittal ultrasound image of the right hepatic lobe in a 36 year old man with an incidental hemangioma (white arrow) showing a mirror image (grey arrow) of the hemangioma just above the diaphragm. Note that the mirror lesion is not identical to the source lesion. The term "mirror" refers to the mechanism by which the artifact is produced, and does not imply that the source and misplaced objects will appear the same on the ultrasound image. In fact, the source object may not be in the image at all.

Images for Figures 9.2–9.4 graciously contributed by Dr Peter Cooperberg, Vancouver.

CASE **10**

Peridiaphragmatic pseudofluid

Imaging description

Crescentic foci of increased T2 signal that mimic fluid may be seen at frequency-selective fat-saturated T2-weighted imaging adjacent to the diaphragm, and can be misinterpreted as small pockets of pleural fluid or ascites (Figure 10.1). The artifact is due to failed fat saturation secondary to local field inhomogeneity at the air-tissue interface between the lung and the diaphragm [1].

Importance

Unsuppressed fat may mimic fluid around the diaphragm, and falsely suggest the presence of pleural fluid or ascites.

Typical clinical scenario

This artifact is common, and in one study was seen in 81% (42/52) of unselected consecutive patients undergoing fat-suppressed T2-weighted fast spin-echo MR imaging [1].

Differential diagnosis

The correct diagnosis of failed fat suppression can be made by cross-registration with other sequences. For example, fat will be of high signal on unsuppressed T1-weighted images, while water will be of low signal (Figure 10.2). Another useful clue that is often seen is co-existent failed fat suppression in the subcutaneous tissue related to field inhomogeneities created by surface coil elements (Figure 10.3).

Teaching point

Apparent fluid around the diaphragm on frequency-selective fat-saturated T2-weighted imaging should be inspected closely, since it frequently represents pseudofluid due to failed fat suppression.

REFERENCE

1 Yoshimitsu K, Varma DG, Jackson EF. Unsuppressed fat in the right anterior diaphragmatic region on fat-suppressed T2-weighted fast spin-echo MR images. *J Magn Reson Imaging* 1995; **5**: 145–149.

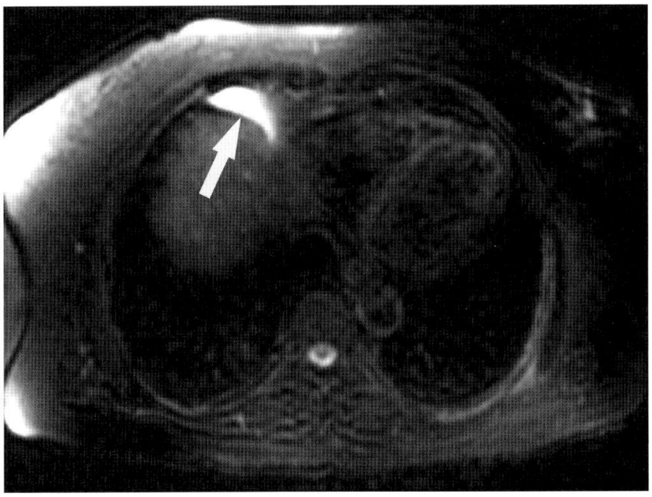

Figure 10.1 Axial T2-weighted fat-saturated MR image shows a crescentic focus (arrow) of increased T2 signal intensity adjacent to the diaphragm, suggesting a small pocket of pleural fluid or ascites.

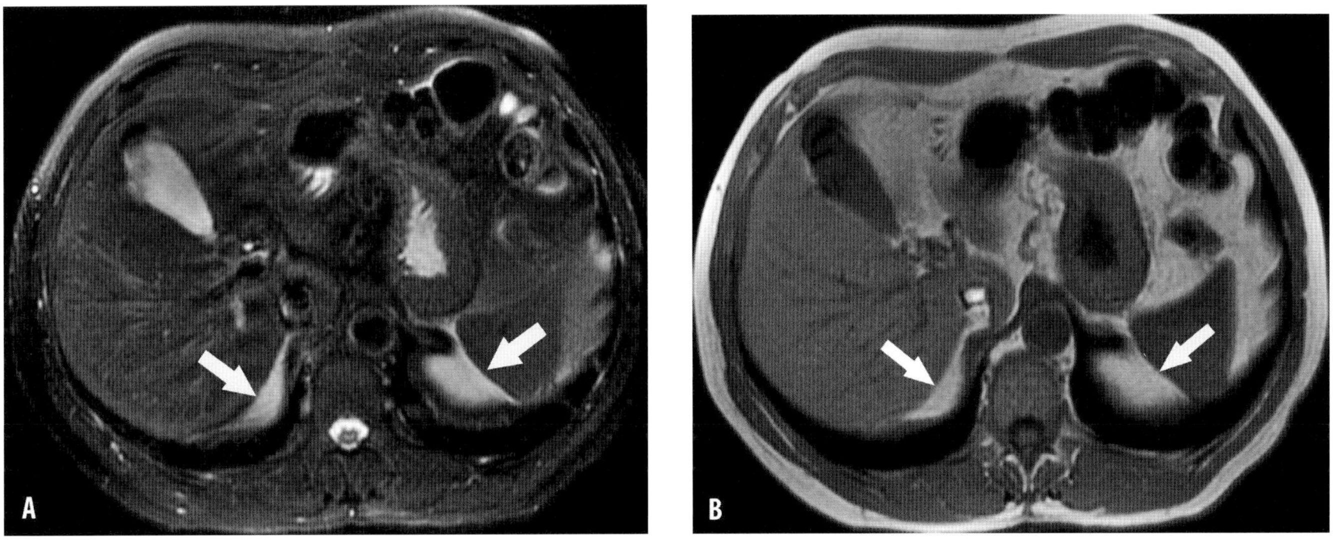

Figure 10.2 **A.** Axial T2-weighted fat-saturated MR image showing crescentic foci (arrows) of increased T2 signal intensity adjacent to the diaphragm. **B.** Examination of an axial T1-weighted MR image at the same level shows high T1 signal intensity in the corresponding areas (arrows), confirming the T2 findings are due to local failure of fat saturation (water would be of low T1 signal intensity).

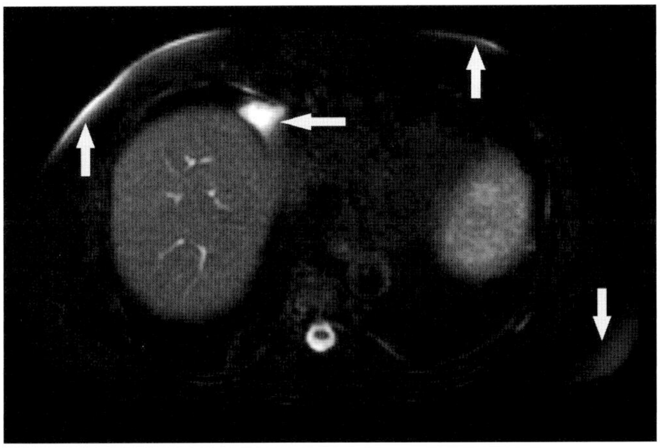

Figure 10.3 Axial T2-weighted fat-saturated MR image showing crescentic focus (horizontal arrow) of increased T2 signal intensity adjacent to the diaphragm. Areas of failed fat saturation are seen on the body surface (vertical arrows), and these co-existent findings provide a useful clue that the peridiaphragmatic finding is also due to failed fat suppression.

CASE 11 Pseudocirrhosis of treated breast cancer metastases

Imaging description

In patients with metastases to the liver from breast cancer, treatment with chemotherapy can result in diffuse hepatic nodularity (Figure 11.1). This entity is referred to as "pseudocirrhosis" because it resembles cirrhosis at cross-sectional imaging [1]. Features of portal hypertension such as portosytemic venous collaterals, splenomegaly, and bland ascites may also develop (Figure 11.2) [2]. This suggests that the prefix "pseudo" may itself be a misnomer, and that this condition may progress to more closely resemble true cirrhosis.

Importance

The erroneous diagnosis of cirrhosis in a patient with metastatic breast cancer could result in unnecessary workup or treatment. In addition, changes of pseudocirrhosis may greatly complicate or even preclude meaningful evaluation of the underlying metastases in the liver, and radiological therapeutic monitoring may depend on evaluating the response of extrahepatic disease sites. It is not known if pseudocirrhosis indicates treatment response and supports continuation of chemotherapy, or if these changes are harbingers of therapeutic toxicity that merit discontinuation or substitution of drug treatment.

Typical clinical scenario

Hepatic contour abnormalities were seen after a median follow-up interval of 15 months in 68 of 91 women (75%) with breast cancer metastatic to the liver who received chemotherapy [2]. Contour abnormalities consisted of limited retraction (n = 42), widespread retraction (n = 10), or diffuse nodularity (n = 16). Even if the term pseudocirrhosis is restricted to those with diffuse nodularity, this would indicate a frequency of at least 18% (16 of 91) for the development of pseudocirrhosis in the population at risk. It is unclear why the phenomenon of pseudocirrhosis seems almost specific for breast cancer metastatic to the liver, with only sporadic reports of such changes in other primary malignancies such as colon or pancreas (Figures 11.3 and 11.4) [3, 4]. No correlation has been found between specific chemotherapy regimens and hepatic contour changes [1, 2]. The histopathological and pathophysiological basis of pseudocirrhosis is not well understood, since obtaining tissue in these patients with incurable malignancy is rarely clinically indicated. The available evidence suggests desmoplastic fibrosis, tumor infiltration, and nodular regenerative hyperplasia secondary to drug hepatotoxicity may all be contributing factors [1, 5].

Differential diagnosis

In isolation, pseudocirrhosis may resemble true cirrhosis, but awareness of the condition combined with review of prior imaging and clinical history should prevent this misdiagnosis.

Teaching point

Diffuse surface nodularity with or without signs of portal hypertension in patients receiving chemotherapy for breast cancer metastases to the liver is likely due to pseudocirrhosis rather than true cirrhosis.

REFERENCES

1 Young ST, Paulson EK, Washington K, et al. CT of the liver in patients with metastatic breast carcinoma treated by chemotherapy: findings simulating cirrhosis. *Am J Roentgenol* 1994; **163**: 1385–1388.

2 Qayyum A, Lee GK, Yeh BM, et al. Frequency of hepatic contour abnormalities and signs of portal hypertension at CT in patients receiving chemotherapy for breast cancer metastatic to the liver. *Clin Imaging* 2007; **31**: 6–10.

3 Kang SP, Taddei T, McLennan B, Lacy J. Pseudocirrhosis in a pancreatic cancer patient with liver metastases: a case report of complete resolution of pseudocirrhosis with an early recognition and management. *World J Gastroenterol* 2008; **14**: 1622–1624.

4 Hubert C, Sempoux C, Horsmans Y, et al. Nodular regenerative hyperplasia: a deleterious consequence of chemotherapy for colorectal liver metastases? *Liver Int* 2007; **27**: 938–943.

5 Nascimento AB, Mitchell DG, Rubin R, Weaver E. Diffuse desmoplastic breast carcinoma metastases to the liver simulating cirrhosis on MR imaging: report of two cases. *Radiology* 2001; **221**: 117–121.

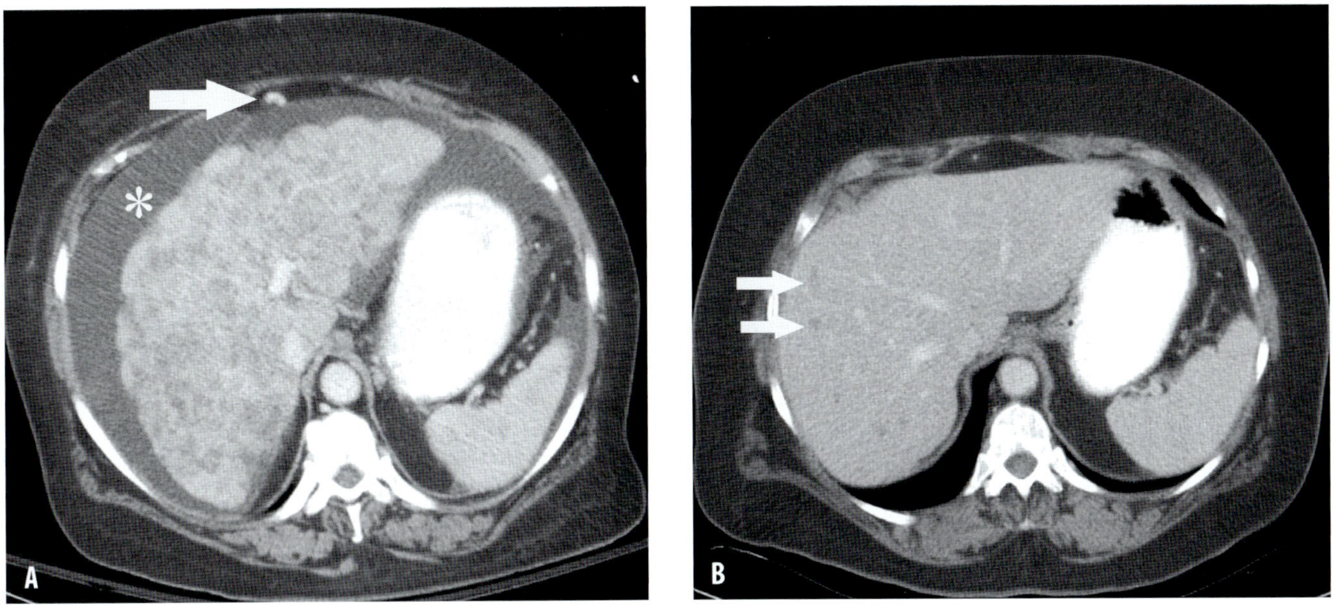

Figure 11.1 A. Axial contrast-enhanced CT section in a 69 year old woman with breast cancer treated with chemotherapy. The liver demonstrates diffuse parenchymal heterogeneity and surface nodularity, with ascites (asterisk) and a recanalized umbilical vein (arrow). The appearances resemble cirrhosis with portal hypertension. B. Axial contrast-enhanced CT section performed 11 months earlier (before starting chemotherapy) shows two small hepatic metastases (arrows). The liver is otherwise normal. The setting of breast cancer metastases treated with chemotherapy indicates that the rapid development of the diffuse changes in Figure 11.1A likely represent pseudocirrhosis of treated breast cancer.

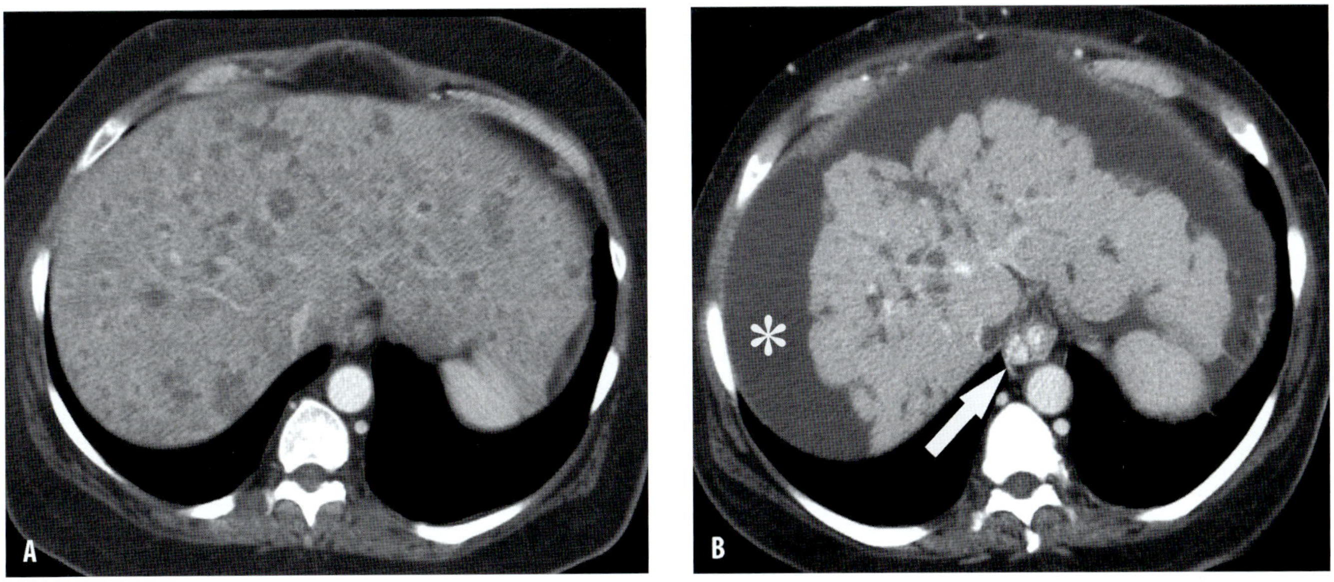

Figure 11.2 A. Axial contrast-enhanced CT section in a 59 year old woman with multiple hypodense hepatic metastases from breast cancer. B. Axial contrast-enhanced CT section performed six months after starting chemotherapy shows diffuse hepatic nodularity, bland ascites (asterisk), esophageal varices (arrow), and partial regression of hepatic metastases. The findings are those of pseudocirrhosis of treated breast cancer metastases, but in the absence of prior studies and clinical history could suggest a diagnosis of cirrhosis.

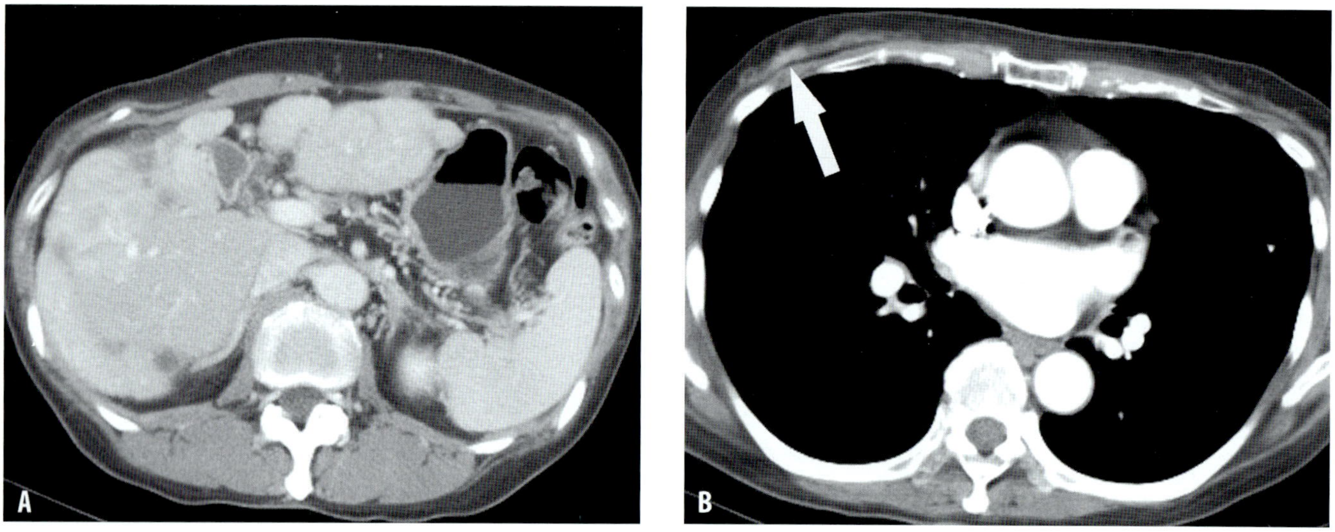

Figure 11.3 **A.** Axial contrast-enhanced CT section in a 68 year old woman after two cycles of chemotherapy for hepatic metastases thought to be from a primary pancreatic carcinoma. The liver surface is coarsely lobulated with several irregular hypodense parenchymal lesions. The appearance was considered suggestive of pseudocirrhosis, casting doubt on the diagnosis of pancreatic cancer. **B.** Axial contrast-enhanced CT section at a more superior level demonstrates a small hypervascular lesion (arrow) in the right breast. Further work-up including resection of the breast mass confirmed a diagnosis of metastatic breast cancer.

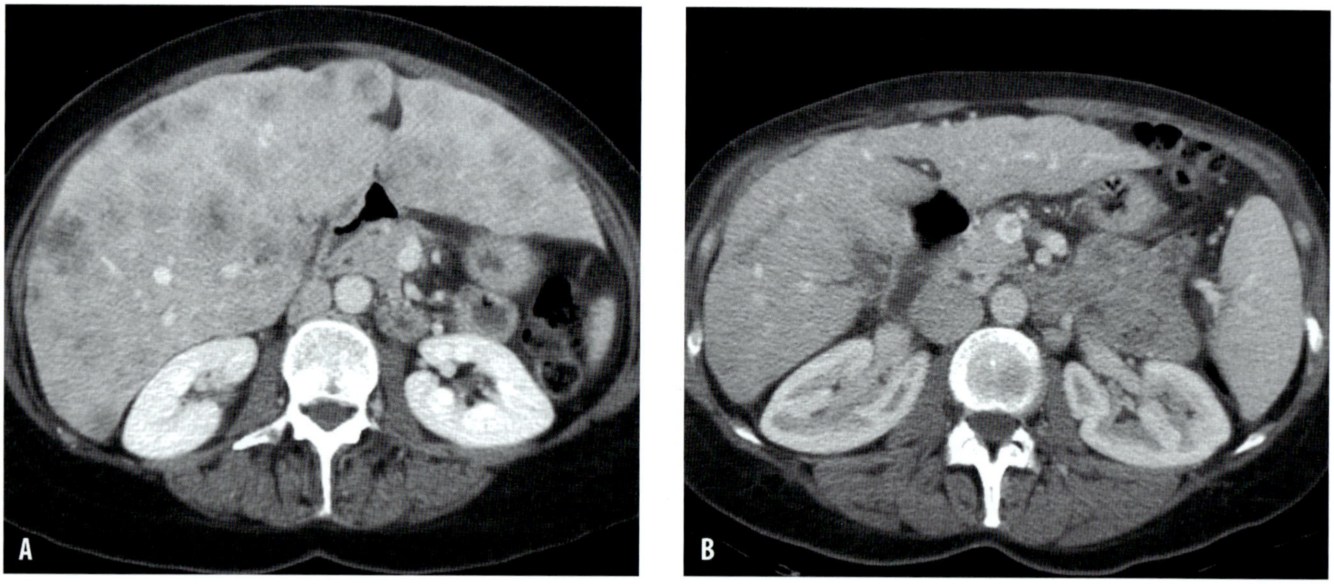

Figure 11.4 **A.** Axial contrast-enhanced CT section in a 45 year old woman with colorectal cancer metastatic to the liver. Multiple metastases are visible. **B.** Axial contrast-enhanced CT section 6 months later, after 12 cycles of FOLFOX chemotherapy (combination of folinic acid, 5-fluorouracil, and oxaliplatin). The metastases have nearly completely resolved while the liver has shrunk and developed a diffusely nodular contour. The latter findings are those of pseudocirrhosis, which is only rarely seen outside the setting of metastatic breast cancer.

CASE 12 Pseudocirrhosis of fulminant hepatic failure

Imaging description

Fulminant hepatic failure may result in fine diffuse nodularity of the hepatic surface (Figure 12.1), and should not be interpreted as indicating underlying cirrhosis.

Importance

The erroneous diagnosis of underlying cirrhosis in a patient with fulminant hepatic failure could adversely impact transplantation status, since true fulminant hepatic failure receives higher priority than acute-on-chronic liver failure.

Typical clinical scenario

In the first study to report this pitfall, 15 of 35 (43%) patients with fulminant hepatic failure demonstrated hepatic surface nodularity at pre-transplantation imaging [1]. A combination of alternating foci of confluent regenerative nodules and necrosis was seen throughout the liver in most of these patients, suggesting this is the histopathological correlate of the imaging finding (Figure 12.2).

Differential diagnosis

The nodularity associated with fulminant hepatic failure appears characteristically fine and diffuse, which is to be expected, given its histopathological basis. This particular type of nodularity has a fairly limited number of causes. The primary differential consideration is cirrhosis, and outside of the special circumstance of fulminant hepatic failure, hepatic surface nodularity is generally the most accurate and specific sign of cirrhosis [2, 3]. In clinical practice, the only condition that mimics cirrhosis with any real frequency is pseudocirrhosis of treated breast cancer metastases to the liver, but again the clinical setting is distinctive [4]. Fine diffuse nodularity has occasionally been reported in patients with miliary metastases and sarcoidosis [5–7].

Teaching point

In the specific setting of fulminant hepatic failure, fine diffuse surface nodularity of the liver is not a reliable sign of cirrhosis and should not be interpreted as indicating underlying cirrhosis.

REFERENCES

1 Poff JA, Coakley FV, Qayyum A, et al. Frequency and histopathologic basis of hepatic surface nodularity in patients with fulminant hepatic failure. *Radiology* 2008; **249**: 518–523.

2 Keedy A, Westphalen AC, Qayyum A, et al. Diagnosis of cirrhosis by spiral computed tomography: a case-control study with feature analysis and assessment of interobserver agreement. *J Comput Assist Tomogr* 2008; **32**: 198–203.

3 Simonovsky V. The diagnosis of cirrhosis by high resolution sonography of the liver surface. *Br J Radiol* 1999; **72**: 29–34.

4 Qayyum A, Lee GK, Yeh BM, et al. Frequency of hepatic contour abnormalities and signs of portal hypertension at CT in patients receiving chemotherapy for breast cancer metastatic to the liver. *Clinical Imaging* 2007; **31**: 6–10.

5 Di Lelio A, Cestari C, Lomazzi A, Beretta L. Cirrhosis: diagnosis with sonographic study of the liver surface. *Radiology* 1989; **172**: 389–392.

6 Warshauer DM, Molina PL, Hamman SM, et al. Nodular sarcoidosis of the liver and spleen: analysis of 32 cases. *Radiology* 1995; **195**: 757–762.

7 Brancatelli G, Federle MP, Ambrosini R, et al. Cirrhosis: CT and MR imaging evaluation. *Eur J Radiol* 2007; **61**: 57–69.

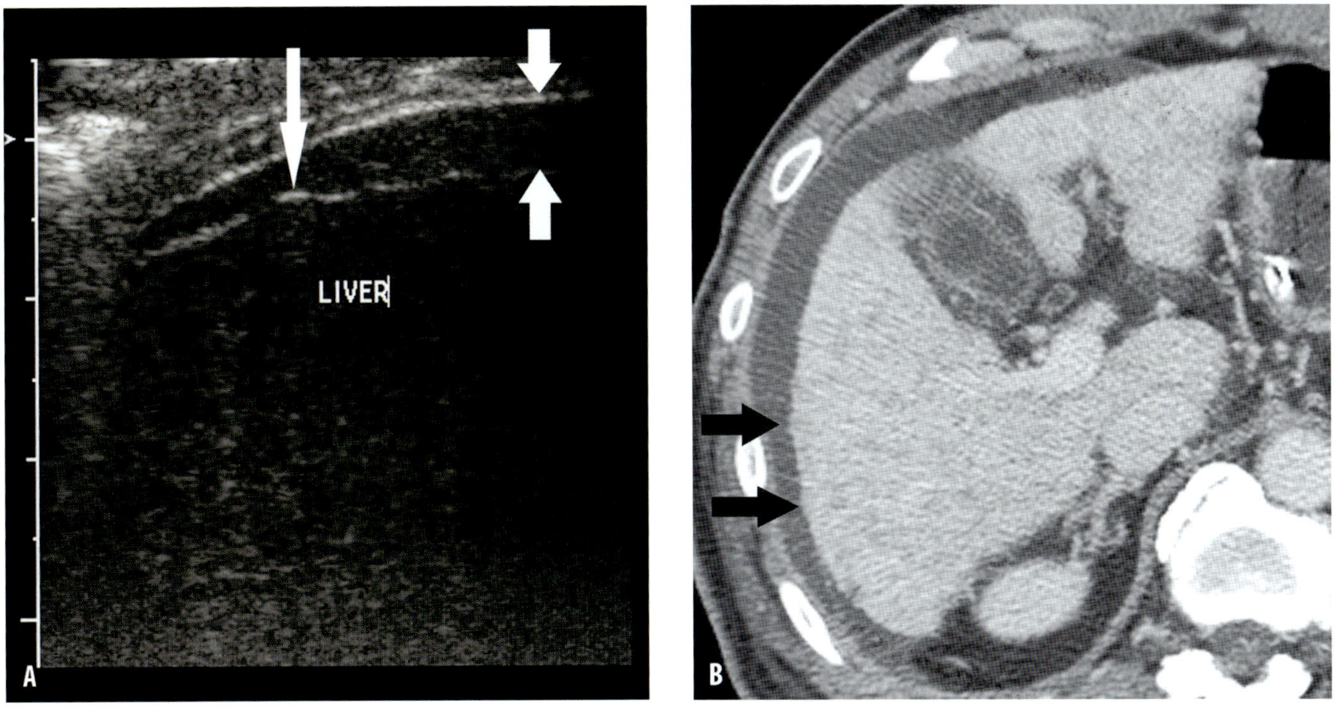

Figure 12.1 **A.** Ultrasound image of the liver surface in a 61 year old man with fulminant hepatic failure due to hepatitis. A layer of ascitic fluid (between short wide arrows) outlines the liver surface, which is nodular (long thin arrow). **B.** Contrast-enhanced CT image performed one day later again shows nodularity (arrows) of the liver surface outlined by ascites, suggestive of cirrhosis. Histopathological examination of the explanted liver five days later showed confluent regenerative nodules surrounded by large areas of necrosis, but no cirrhosis.

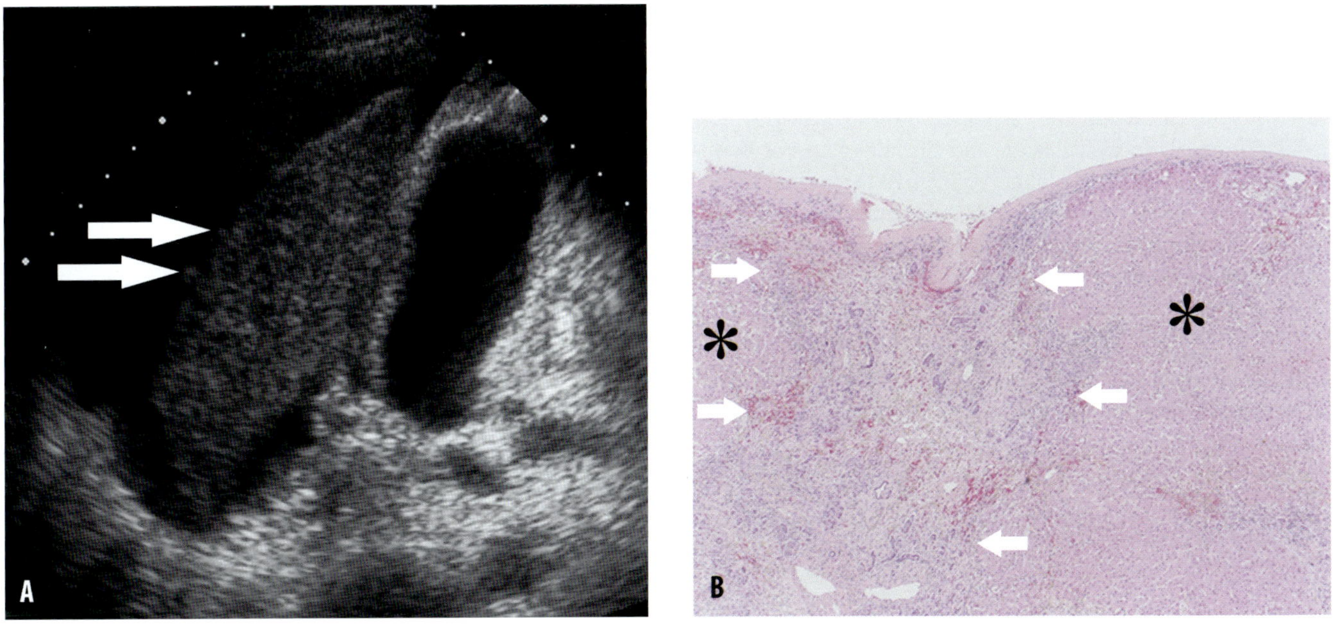

Figure 12.2 **A.** Ultrasound image of the liver in a 67 year old woman with fulminant hepatic failure that developed shortly after commencing methyldopa for longstanding hypertension. The liver surface demonstrates a finely nodular appearance (arrows). **B.** Photomicrograph (20x) of a hematoxylin and eosin-stained slide of the explanted hepatic surface from transplant surgery performed three days later shows the irregularity of the liver surface and reflects a combination of confluent regenerative nodules (asterisks) with alternating bands of necrosis (between arrows).

CASE 13 Nutmeg liver

Imaging description

Pathologically, the term nutmeg liver refers to the speckled appearance of the cut liver in chronic venous congestion, due to dilated and congested red central veins surrounded by paler, unaffected liver tissue (resembling a grated nutmeg kernel) [1]. Radiologically, the term refers to the distinctive pattern of diffuse reticular heterogeneous enhancement seen in the arterial and early portal venous phases of enhancement at CT or MRI in patients with venous congestion (Figures 13.1–13.4), due either to Budd-Chiari syndrome or cardiac or pericardial disease with elevated right heart pressure [2–5]. The terms mosaic perfusion and shattered glass have also been used to describe to this distinctive hepatic enhancement pattern.

Importance

Nutmeg liver detected at CT or MRI suggests either Budd-Chiari syndrome or heart disease. Recognition of passive hepatic congestion on CT or MRI may help explain liver function abnormalities in patients with heart failure [3] and may be an important clue to the diagnosis of constrictive pericarditis, which might otherwise go unrecognized [6]. Rarely, nutmeg liver can progress to cardiac cirrhosis [1].

Typical clinical scenario

Nutmeg liver is most frequently seen in right heart failure, and this diagnosis may be suggested by ancillary imaging findings such as retrograde hepatic venous opacification, hepatomegaly, cardiomegaly, pleural effusions, ascites, and periportal edema [3, 4]. Pericardial disease may be suggested by pericardial effusion, thickening, or calcification [4, 6]. Nutmeg liver is seen primarily in the subacute and chronic phases of Budd-Chiari syndrome [2, 7], when it may be accompanied by non-visualization or narrowing of the hepatic veins and intrahepatic inferior vena cava, caudate lobe hypertrophy, peripheral atrophy, and intrahepatic collateral veins [8].

Differential diagnosis

Diffuse heterogeneous enhancement of the liver can also be seen in hereditary hemorrhagic telangiectasia (Osler-Weber-Rendu disease) [8], cholangitis [9], cirrhosis, miliary metastases, and sarcoidosis (Figures 13.5–13.9). Of these, only cholangitis and hereditary hemorrhagic telangiectasia cause heterogeneity that is transient, while the others typically cause persistent heterogeneity across all phases of enhancement. Neither cholangitis nor hereditary hemorrhagic telangiectasia causes a reticular pattern of enhancement, so confusion with true nutmeg liver is unlikely. In addition, cholangitis is usually distinctive clinically, while hereditary hemorrhagic telangiectasia is typically accompanied by enlargement of the hepatic artery, which can serve as a useful clue to the diagnosis.

Teaching point

Nutmeg liver indicates hepatic venous congestion, either from heart disease or Budd-Chiari syndrome. Ancillary clinical and imaging features should allow for diagnosis of the underlying cause.

REFERENCES

1 Giallourakis CC, Rosenberg PM, Friedman LS. The liver in heart failure. *Clin Liver Dis* 2002; **6**: 947–967, viii–ix.

2 Mathieu D, Vasile N, Menu Y, et al. Budd-Chiari syndrome: dynamic CT. *Radiology* 1987; **165**: 409–413.

3 Moulton JS, Miller BL, Dodd GD 3rd, Vu DN. Passive hepatic congestion in heart failure: CT abnormalities. *Am J Roentgenol* 1988; **151**: 939–942.

4 Holley HC, Koslin DB, Berland LL, Stanley RJ. Inhomogeneous enhancement of liver parenchyma secondary to passive congestion: contrast-enhanced CT. *Radiology* 1989; **170**: 795–800.

5 Mauro MA, Stackhouse DJ, Parker LA, Schiebler ML. Computed tomography of hepatic venous hypertension: the reticulated-mosaic pattern. *Gastrointest Radiol* 1990; **15**: 35–38.

6 Johnson KT, Julsrud PR, Johnson CD. Constrictive pericarditis at abdominal CT: a commonly overlooked diagnosis. *Abdominal Imaging* 2008; **33**: 349–352.

7 Camera L, Mainenti PP, Di Giacomo A, et al. Triphasic helical CT in Budd-Chiari syndrome: patterns of enhancement in acute, subacute and chronic disease. *Clin Radiol* 2006; **61**: 331–337.

8 Torabi M, Hosseinzadeh K, Federle MP. CT of nonneoplastic hepatic vascular and perfusion disorders. *Radiographics* 2008; **28**: 1967–1982.

9 Arai K, Kawai K, Kohda W, et al. Dynamic CT of acute cholangitis: early inhomogeneous enhancement of the liver. *Am J Roentgenol* 2003; **181**: 115–118.

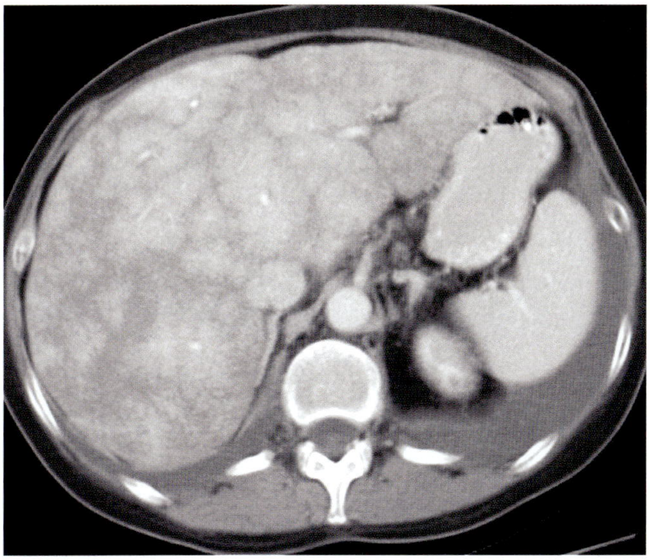

Figure 13.1 Axial contrast-enhanced CT image in an 84 year old man with chronic congestive heart failure shows the typical mottled reticular pattern of diffuse heterogeneous hepatic enhancement known as nutmeg liver.

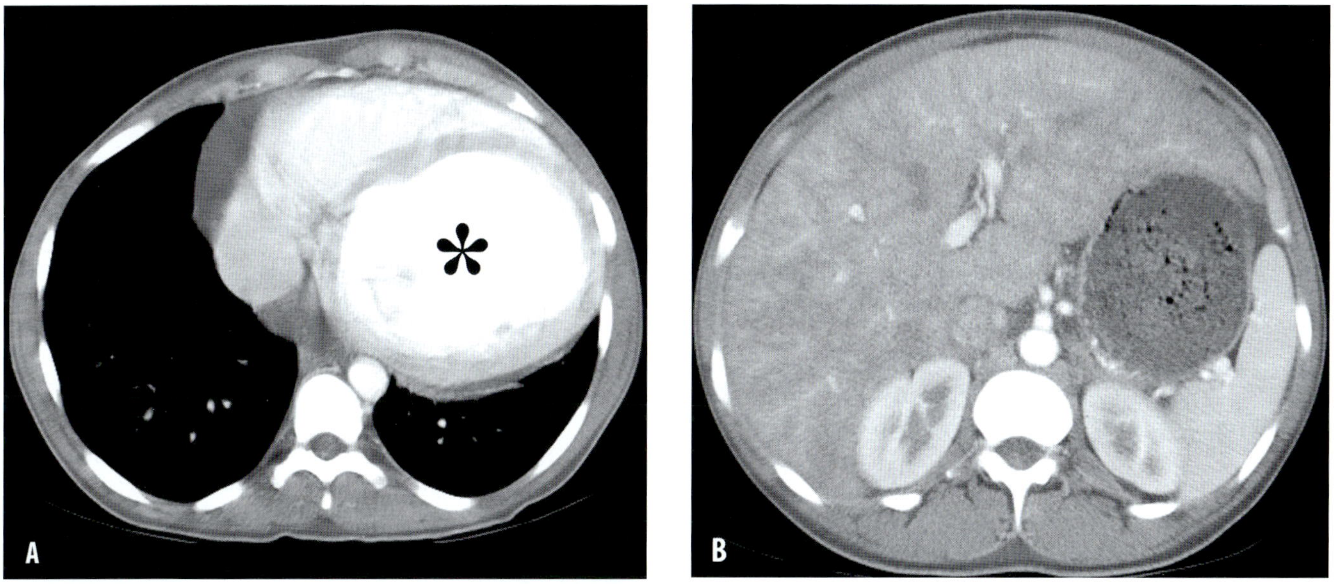

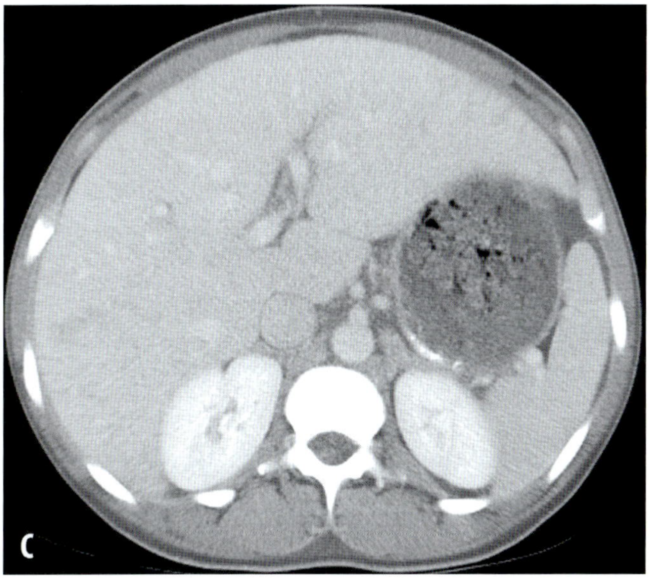

Figure 13.2 **A.** Axial contrast-enhanced CT image in a 21 year old woman with dilated cardiomyopathy shows marked dilatation of the left ventricle (asterisk). **B.** Axial contrast-enhanced CT image at a more inferior level shows the typical diffuse heterogeneous reticular enhancement of nutmeg liver, due to venous congestion secondary to heart failure. The liver is also enlarged. **C.** Axial contrast-enhanced CT image at the same level as Figure 13.2B in a later phase of enhancement shows the mottled enhancement of the liver has largely disappeared. Nutmeg liver is a transient phenomenon that is most marked in the early phases of enhancement after intravenous contrast administration.

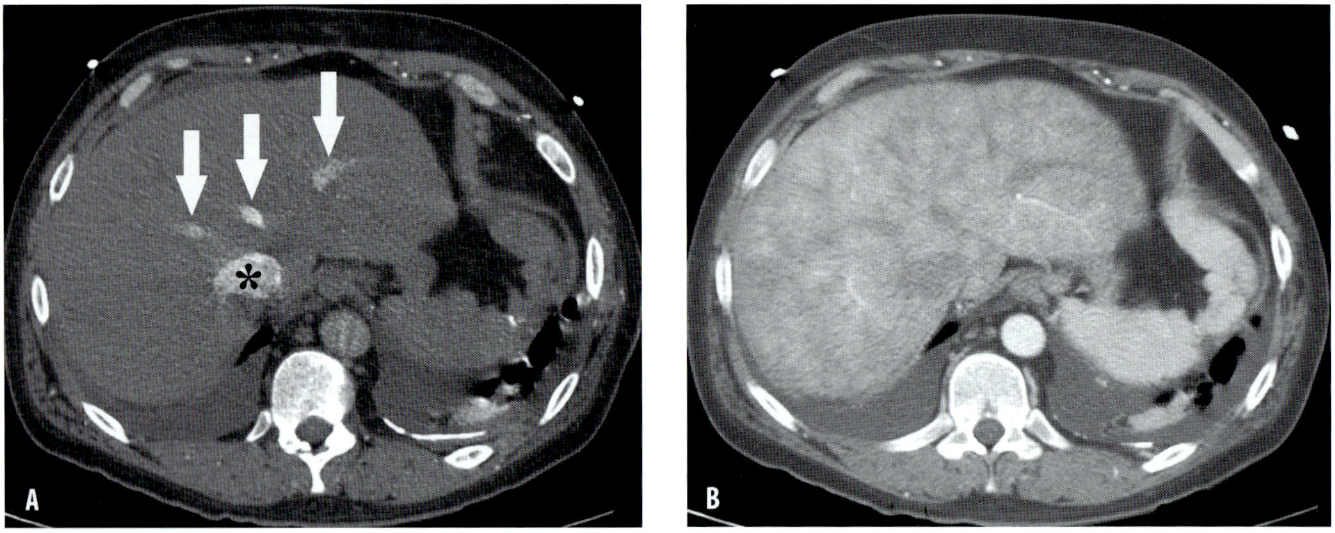

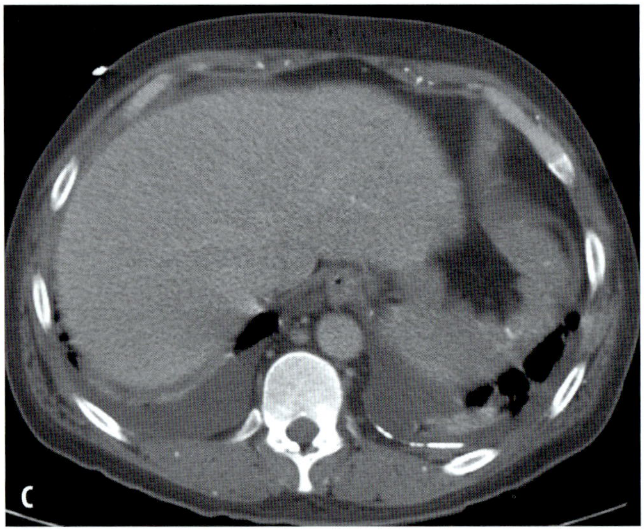

Figure 13.3 **A.** Axial early arterial phase contrast-enhanced CT image in a 32 year old man with ischemic heart failure as a complication of end-stage renal disease related to systemic lupus erythematosus shows reflux of contrast into the inferior vena cava (asterisk) and hepatic veins (arrows), consistent with cardiac failure and elevated right heart pressure. **B.** Axial late arterial phase contrast-enhanced CT image at the same level as Figure 13.3A shows the diffuse heterogeneous enhancement of nutmeg liver due to venous congestion. **C.** Axial delayed phase contrast-enhanced CT image at the same level as Figures 13.3A and 13.3B shows resolution of the heterogeneous liver texture seen during the early phase of enhancement.

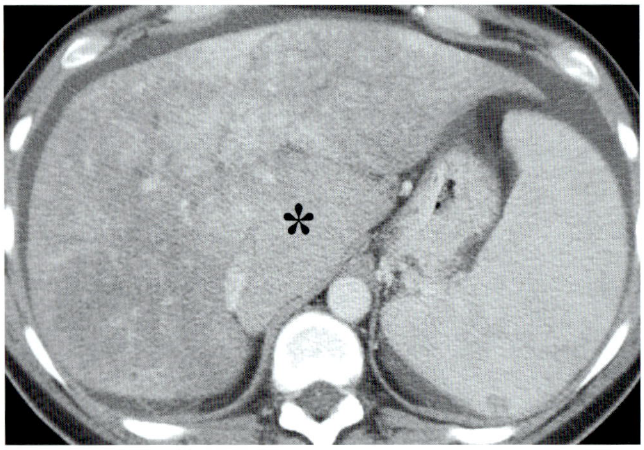

Figure 13.4 Axial contrast-enhanced CT image in a 38 year old man with idiopathic Budd-Chiari syndrome shows the diffuse reticular enhancement of nutmeg liver. Note the caudate lobe (asterisk) is relatively unaffected and hypertrophied.

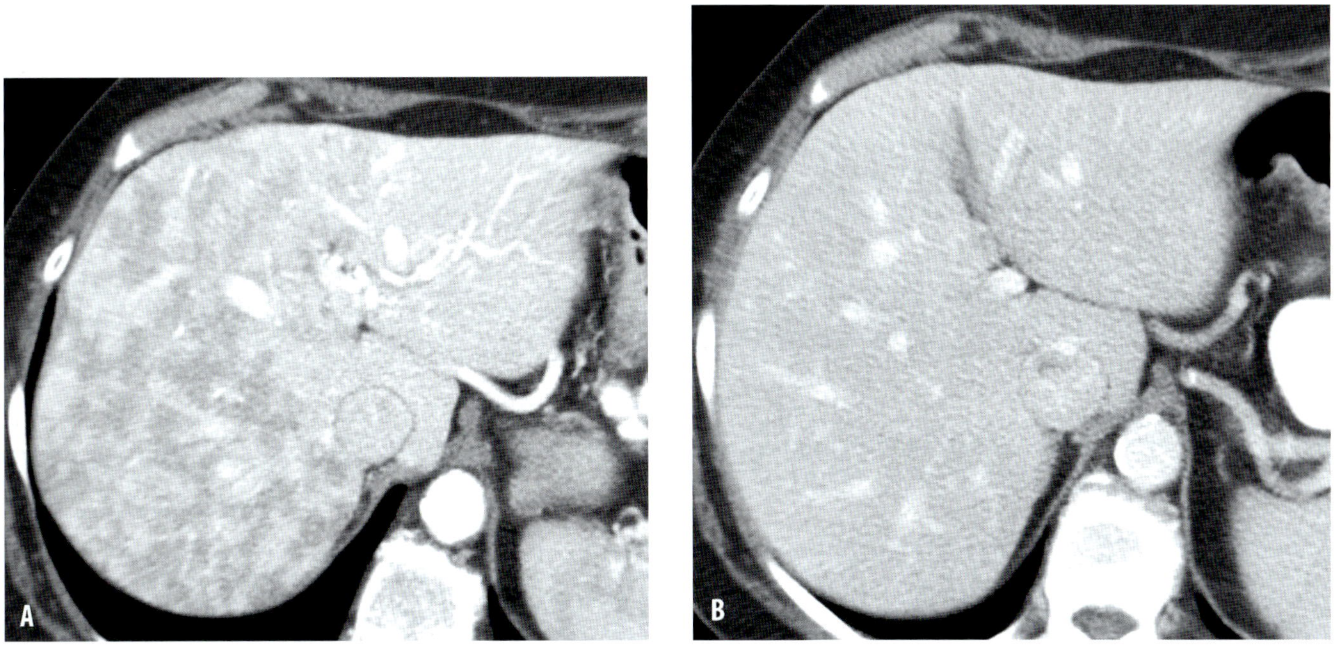

Figure 13.5 A. Axial arterial phase contrast-enhanced CT image in a 73 year old woman with hereditary hemorrhagic telangiectasia shows diffuse heterogeneous enhancement of the liver, particularly in the right lobe. **B.** Axial delayed phase contrast-enhanced CT image shows resolution of the heterogeneity. Hereditary hemorrhagic telangiectasia can result in transient heterogeneous hepatic enhancement that resembles nutmeg liver.

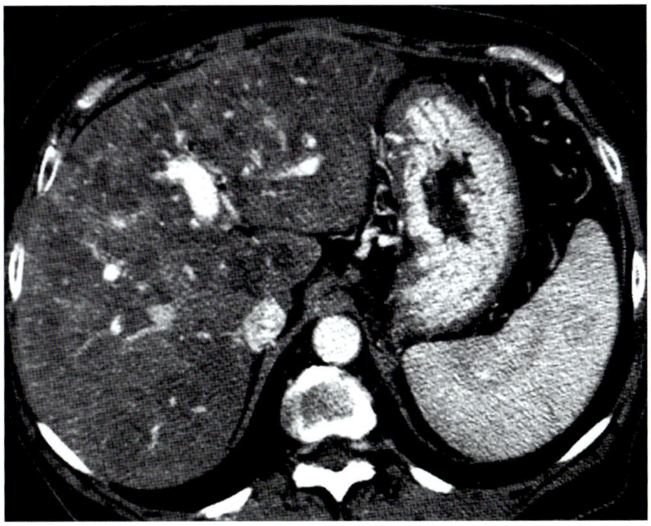

Figure 13.6 Axial contrast-enhanced CT image in a 67 year old woman with cholangitis occurring after a Whipple's procedure for pancreatic carcinoma shows diffuse heterogeneous hepatic enhancement. The heterogeneous enhancement resolved after antibiotic treatment.

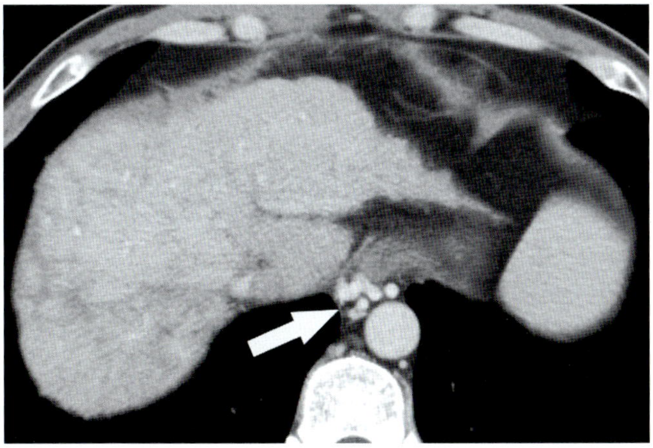

Figure 13.7 Axial contrast-enhanced CT image in a 56 year old man with alcoholic cirrhosis shows a heterogeneous pattern of enhancement resembling nutmeg liver. Note the presence of gastroesophageal varices (arrow) due to portal hypertension.

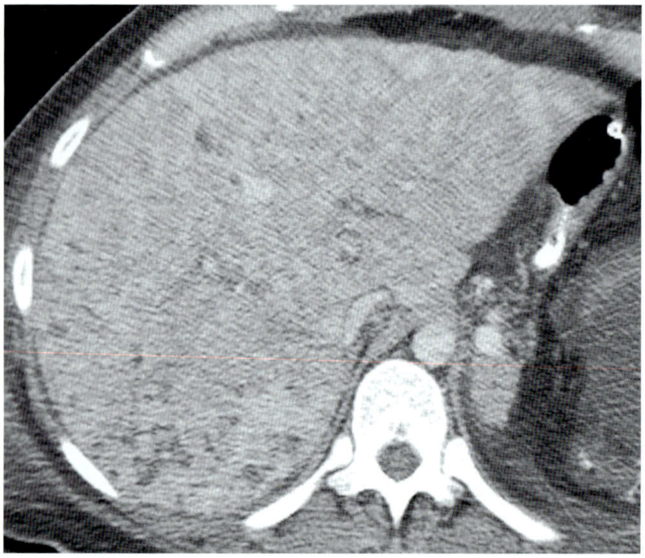

Figure 13.8 Axial contrast-enhanced CT image in a 35 year old woman with breast cancer shows diffusely irregular parenchymal enhancement due to biopsy-proven miliary metastases.

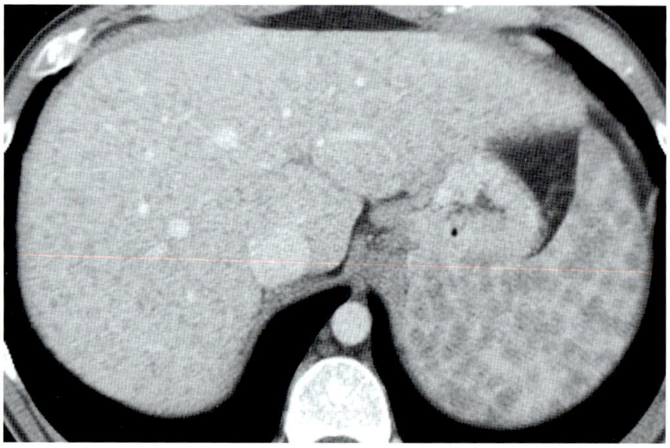

Figure 13.9 Axial contrast-enhanced CT image in a 32 year old man shows diffuse nodular heterogeneous enhancement of the liver, with multiple hypodense nodules in the spleen. Retroperitoneal adenopathy (not shown) was also present. A diagnosis of sarcoidosis was established by nodal biopsy.

CASE 14 Nodular regenerative hyperplasia

Imaging description

Nodular regenerative hyperplasia is an increasingly recognized condition characterized by transformation of normal liver parenchyma into hyperplastic regenerative nodules in the absence of fibrosis (the absence of fibrosis distinguishes nodular regenerative hyperplasia from cirrhosis) [1, 2]. The condition has been reported in up to 3% of unselected patients at autopsy [3]. From a radiological perspective, there appear to be two forms of the condition: a diffuse form in which the nodules are small and widespread, and a focal form in which the nodules are few in number, scattered throughout the liver, and measure up to a few centimeters in size. In the diffuse form, imaging findings can be subtle across all modalities [2, 4, 5]. At US, multiple masses that are hypo-, iso-, or hyperechoic may be seen (Figure 14.1). AT CT the lesions are usually hypodense with little enhancement. At MRI, nodular regenerative hyperplasia is usually of similar signal intensity to the liver on T1, T2, and post-gadolinium sequences. While the literature on this topic is limited, it seems the nodules may be more obvious on ultrasound than CT or MRI [4], and this has been my experience (Figures 14.1 and 14.2). A periportal distribution has been described [3]. In the focal form of nodular regenerative hyperplasia, multiple hypervascular masses are seen at CT or MRI (Figure 14.3).

Importance

Diffuse nodular regenerative hyperplasia may be radiologically occult or suggest diffuse liver disease such as cirrhosis (Figure 14.4) or fungal infection. Focal nodular regenerative hyperplasia may be mistaken radiologically for multiple hypervascular metastases or multifocal hepatocellular adenoma or carcinoma. Biopsy may be confusing, since the simple finding of hepatocytes may suggest normal liver or adenoma, and radiological correlation may be critical in facilitating a correct diagnosis.

Typical clinical scenario

Diffuse nodular regenerative hyperplasia results in non-cirrhotic portal hypertension in up to 50% of affected patients [1], and is one of the principal causes of non-cirrhotic portal hypertension. Clinical consequences include variceal bleeding and hypersplenism. Diffuse nodular regenerative hyperplasia may also be detected incidentally. The pathogenesis of diffuse nodular regenerative hyperplasia is unknown, but there are well-recognized associations with systemic cardiovascular, autoimmune, and myeloproliferative diseases, the administration of certain drugs including chemotherapy, solid organ and bone marrow transplantation, and human immunodeficiency virus infection [1, 2, 6–10]. The frequency of diffuse nodular regenerative hyperplasia has been reported as 22.5% (23 of 103) in bone marrow transplant recipients [7], 13.5% (5 of 37) in renal transplant recipients with liver dysfunction [8], and 1.2% (14 of 1191) in liver transplant recipients [9]. Vascular abnormalities in the liver may also be responsible [11], and it is possible that nodular regenerative hyperplasia is a response to altered hemodynamics. Focal nodular regenerative hyperplasia is seen primarily in longstanding Budd-Chiari syndrome, although similar masses have been reported in autoimmune hepatitis [12, 13].

Differential diagnosis

Findings of diffuse nodular regenerative hyperplasia may resemble cirrhosis with portal hypertension, although the liver surface is usually not nodular, and marked surface nodularity would strongly favor cirrhosis. The appearance of multiple hypervascular masses in a patient with longstanding Budd-Chiari syndrome or autoimmune hepatitis may suggest metastases or hepatocellular carcinoma, although stability on follow-up studies favors nodular regenerative hyperplasia. Biopsy should also help distinguish, provided histopathological findings are correlated with imaging.

Teaching point

Diffuse nodular regenerative hyperplasia should be considered when multiple small masses are seen in the liver of a patient with known risk factors, particularly if the clinical picture does not favor malignancy and if the lesions are more prominent at ultrasound than other modalities. Focal nodular regenerative hyperplasia should be considered when multiple hypervascular hepatic masses are seen in a patient with longstanding Budd-Chiari syndrome or autoimmune hepatitis.

REFERENCES

1 Dachman AH, Ros PR, Goodman ZD, Olmsted WW, Ishak KG. Nodular regenerative hyperplasia of the liver: clinical and radiologic observations. *Am J Roentgenol* 1987; **148**: 717–722.

2 Leung VK, Ng WL, Luk IS, et al. Unique hepatic imaging features in a patient with nodular regenerative hyperplasia of the liver associating with systemic lupus erythematosus. *Lupus* 2007; **16**: 205–208.

3 Wanless IR. Micronodular transformation (nodular regenerative hyperplasia) of the liver: a report of 64 cases among 2,500 autopsies and a new classification of benign hepatocellular nodules. *Hepatology* 1990; **11**: 787–797.

4 Casillas C, Marti-Bonmati L, Galant J. Pseudotumoral presentation of nodular regenerative hyperplasia of the liver: imaging in five patients including MR imaging. *Eur Radiol* 1997; **7**: 654–658.

5 Clouet M, Boulay I, Boudiaf M, et al. Imaging features of nodular regenerative hyperplasia of the liver mimicking hepatic metastases. *Abdom Imaging* 1999; **24**: 258–261.

6 Hubert C, Sempoux C, Horsmans Y, et al. Nodular regenerative hyperplasia: a deleterious consequence of chemotherapy for colorectal liver metastases? *Liver Int* 2007; **27**: 938–943.

7 Snover DC, Weisdorf S, Bloomer J, McGlave P, Weisdorf D. Nodular regenerative hyperplasia of the liver following bone marrow transplantation. *Hepatology* 1989; **9**: 443–448.

8 Allison MC, Mowat A, McCruden EA, et al. The spectrum of chronic liver disease in renal transplant recipients. *Q J Med* 1992; **83**: 355–367.

9 Devarbhavi H, Abraham S, Kamath PS. Significance of nodular regenerative hyperplasia occurring de novo following liver transplantation. *Liver Transpl* 2007; **13**: 1552–1556.

10 Tateo M, Sebagh M, Bralet MP, et al. A new indication for liver transplantation: nodular regenerative hyperplasia in human immunodeficiency virus-infected patients. *Liver Transpl* 2008; **14**: 1194–1198.

11 Grazioli L, Alberti D, Olivetti L, et al. Congenital absence of portal vein with nodular regenerative hyperplasia of the liver. *Eur Radiol* 2000; **10**: 820–825.

12 Rha SE, Lee MG, Lee YS, et al. Nodular regenerative hyperplasia of the liver in Budd-Chiari syndrome: CT and MR features. *Abdom Imaging* 2000; **25**: 255–258.

13 Qayyum A, Graser A, Westphalen A, et al. CT of benign hypervascular liver nodules in autoimmune hepatitis. *American Journal of Roentgenology* 2004; **183**: 1573–1576.

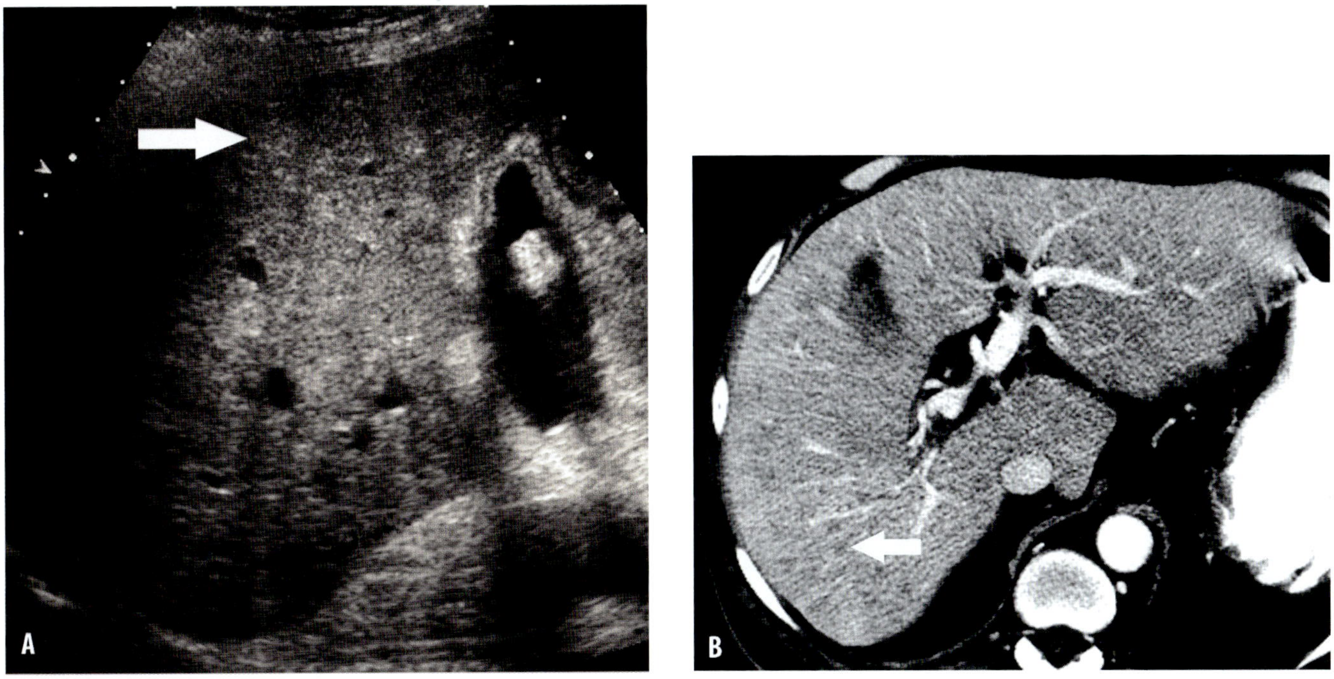

Figure 14.1 **A.** Sagittal ultrasound image of the right hepatic lobe in a 58 year old man with persistent sepsis one day after right hemicolectomy for colonic volvulus, and a history of renal transplantation for human immunodeficiency virus nephropathy. Subtle echogenic nodules are seen throughout the liver. A representative nodule (arrow) is visible anteriorly. **B.** Axial contrast-enhanced CT image shows subtle parenchymal heterogeneity consisting of small hypodense nodules. A representative nodule (arrow) is visible posteriorly. Subsequent liver biopsy confirmed a diagnosis of diffuse nodular regenerative hyperplasia.

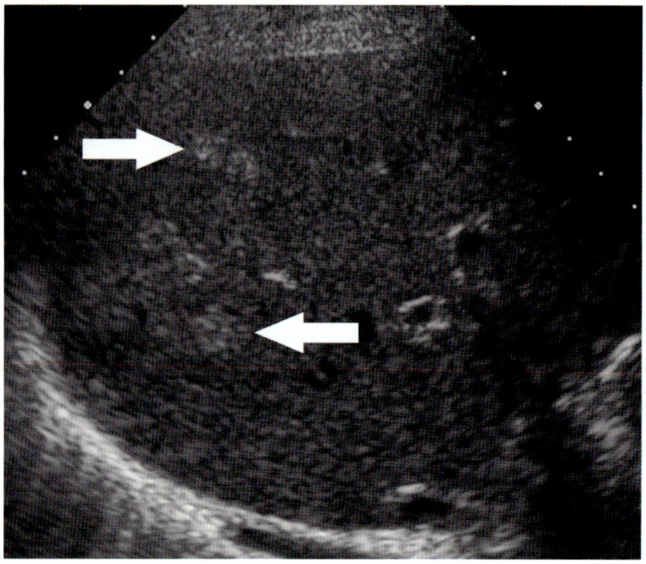

Figure 14.2 Sagittal ultrasound image of the right hepatic lobe in a 64 year old woman with elevated liver enzymes detected while receiving chemotherapy for acute lymphoblastic leukemia. Past medical history included a sibling allogeneic bone marrow transplant performed 21 years previously for chronic myeloid leukemia. Scattered subtle echogenic nodules (arrows) are visible. Contrast-enhanced CT and MRI, including in and out of phase gradient echo images (not shown), were unremarkable. A presumptive diagnosis of diffuse nodular regenerative hyperplasia was made.

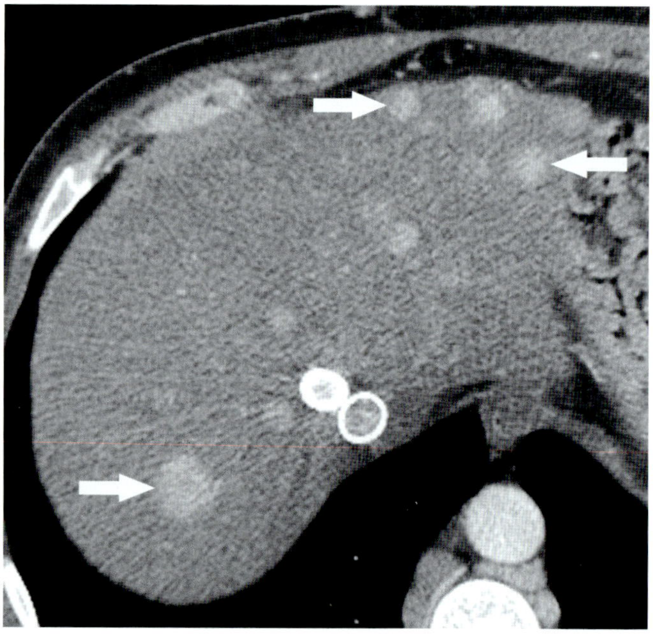

Figure 14.3 Axial contrast-enhanced CT image in a 46 year old man with a 3-year history of Budd-Chiari syndrome managed by transjugular intrahepatic portosystemic shunt creation. Multiple hypervascular masses (arrows) are visible. The masses were stable on serial studies, and biopsy confirmed the diagnosis of focal nodular regenerative hyperplasia.

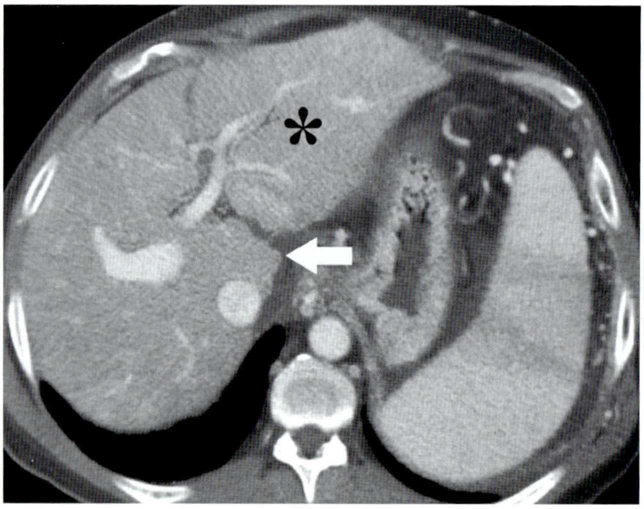

Figure 14.4 Axial contrast-enhanced CT image in a 60 year old man with portal hypertension leading to gastrointestinal bleeding, ascites, and thrombocytopenia due to biopsy-proven diffuse nodular regenerative hyperplasia in a liver transplant performed 18 years earlier for primary sclerosing cholangitis. The liver surface is irregular (arrow) with relative hypertrophy of the left hepatic lobe (asterisk). These findings mimic cirrhosis.

CASE 15 Pseudoprogression of treated hepatic metastases

Imaging description

In patients with metastases to the liver, treatment response to chemotherapy may result in a relative reduction of enhancement in the metastases such that they become more conspicuous and erroneously suggest disease progression (Figures 15.1–15.3). Such "pseudoprogression" has been primarily described in metastases from gastrointestinal stromal tumors treated with imatinib (Figure 15.4) [1], but can occur in other malignancies.

Importance

Misdiagnosis of treatment response as treatment failure may result in an unwarranted cessation or change of successful treatment.

Typical clinical scenario

Pseudoprogression can be seen in patients with hepatic metastases treated by chemotherapy in which treatment response is accompanied by a reduction in enhancement within the lesions. While the phenomenon is not well described and has been primarily reported in neuroradiology [2], my experience suggests it may be more frequent with contemporary chemotherapy regimens that incorporate novel anti-angiogenic and other biological agents that can affect tumor perfusion or cause tumor necrosis.

Differential diagnosis

When hepatic metastases become more conspicuous on treatment, the primary differential consideration is true disease progression. Correlation with disease evolution at other sites, with tumor markers or with PET imaging, may help in this distinction. Pseudoprogression due to a change in CT density secondary to necrosis or reduced perfusion is distinct from and should not be confused with the phenomena of tumor flare (primarily used to describe the transient testosterone surge and symptomatic exacerbation that can accompany the initiation of androgen deprivation with luteinizing hormone-releasing hormone analogs in prostate cancer) or tumor lysis syndrome (the clinical or metabolic derangement caused by rapid tumor lysis with the abrupt hematogenous release of cellular components; observed most frequently after the start of chemotherapy in patients with acute lymphoblastic leukemia or Burkitt's lymphoma) [3, 4]. Finally, hepatic metastases may change in conspicuity while on treatment due to the development or regression of chemotherapy-associated diffuse fatty liver [5, 6], in which the density of the background liver changes rather than the density of the metastases themselves (Figure 15.5). Diffuse fatty liver is particularly associated with the use of 5-fluorouracil [6–8].

Teaching point

The development of new or more conspicuous hepatic metastases on treatment can sometimes reflect treatment response with reduced tumor enhancement or necrosis resulting in so-called pseudoprogression. Correlation with other disease sites, tumor markers, or PET may be helpful in making the distinction from true tumor progression.

REFERENCES

1 Linton KM, Taylor MB, Radford JA. Response evaluation in gastrointestinal stromal tumours treated with imatinib: misdiagnosis of disease progression on CT due to cystic change in liver metastases. *Br J Radiol* 2006; **79**: e40–e44.

2 Clarke JL, Chang S. Pseudoprogression and pseudoresponse: challenges in brain tumor imaging. *Curr Neurol Neurosci Rep* 2009; **9**: 241–246.

3 Bubley GJ. Is the flare phenomenon clinically significant? *Urology* 2001; **58**(2 Suppl 1): 5–9.

4 Coiffier B, Altman A, Pui CH, Younes A, Cairo MS. Guidelines for the management of pediatric and adult tumor lysis syndrome: an evidence-based review. *J Clin Oncol* 2008; **26**: 2767–2778.

5 Robinson PJ. The effects of cancer chemotherapy on liver imaging. *Eur Radiol* 2009; **19**: 1752–1762.

6 Peppercorn PD, Reznek RH, Wilson P, Slevin ML, Gupta RK. Demonstration of hepatic steatosis by computerized tomography in patients receiving 5-fluorouracil-based therapy for advanced colorectal cancer. *Br J Cancer* 1998; **77**: 2008–2011.

7 Moertel CG, Fleming TR, Macdonald JS, Haller DG, Laurie JA. Hepatic toxicity associated with fluorouracil plus levamisole adjuvant therapy. *J Clin Oncol* 1993; **11**: 2386–2390.

8 Sørenson P, Edal AL, Madsen EL, et al. Reversible hepatic steatosis in patients treated with interferon alfa-2a and 5-fluorouracil. *Cancer* 1995; **75**: 2592–2596.

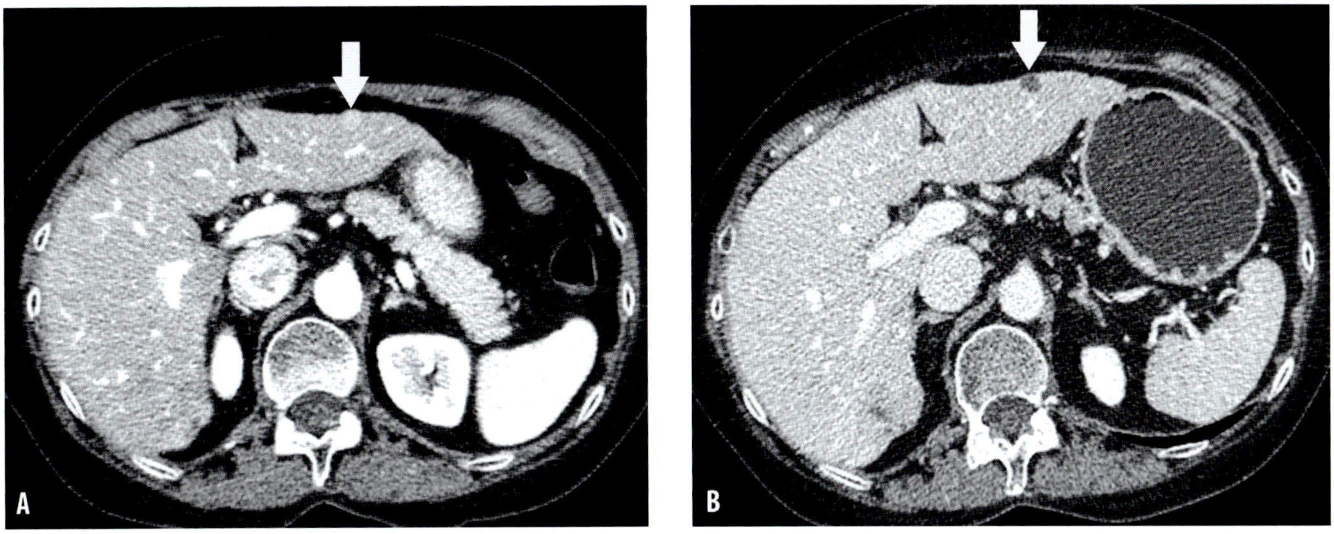

Figure 15.1 A. Axial contrast-enhanced CT image in a 62 year old woman with metastatic melanoma shows a subtle hypervascular metastasis in the left hepatic lobe. **B.** Axial contrast-enhanced CT image obtained three months later after interval treatment with an anti-CTLA-4 antibody (a novel promoter of anti-tumor T cell activity) shows marked reduction of enhancement within the tumor, presumably due to necrosis. This change could be misinterpreted as progression if the subtle hypervascular metastasis in Figure 15.1A was not appreciated.

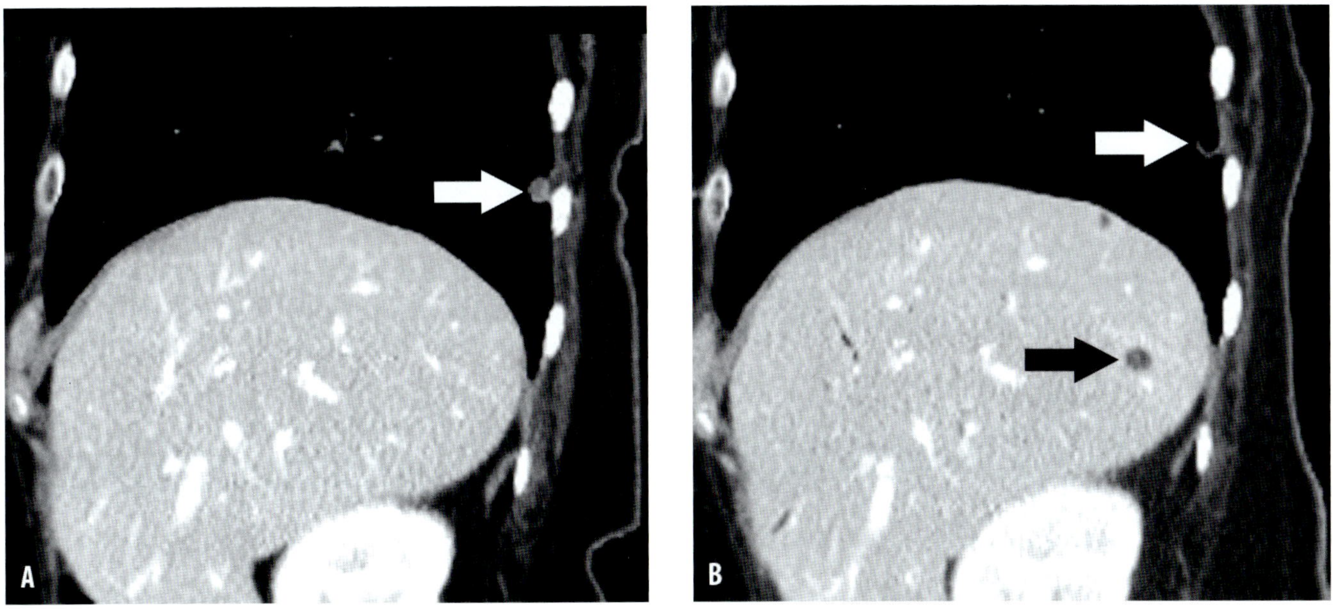

Figure 15.2 A. Sagittal reformatted contrast-enhanced CT image in a 63 year old woman with metastatic pancreatic cancer shows a pulmonary metastasis (arrow) and an apparently normal liver. **B.** Sagittal reformatted contrast-enhanced CT image after two full cycles of chemotherapy, including erlotinib and an investigational monoclonal antibody directed against the insulin-like growth factor receptor. A hypoattenuating lesion (black arrow) in the liver appears new, and could be considered indicative of disease progression. However, over the same interval the pulmonary metastasis had cavitated (white arrow), other pulmonary metastases (not shown) had shrunk, and the serum CA 19–9 level decreased from 231 to 198 units/mL. Based on these other indices of disease response, the appearance in the liver was considered most consistent with pseudoprogression due to necrosis within a previously isoattenuating metastasis.

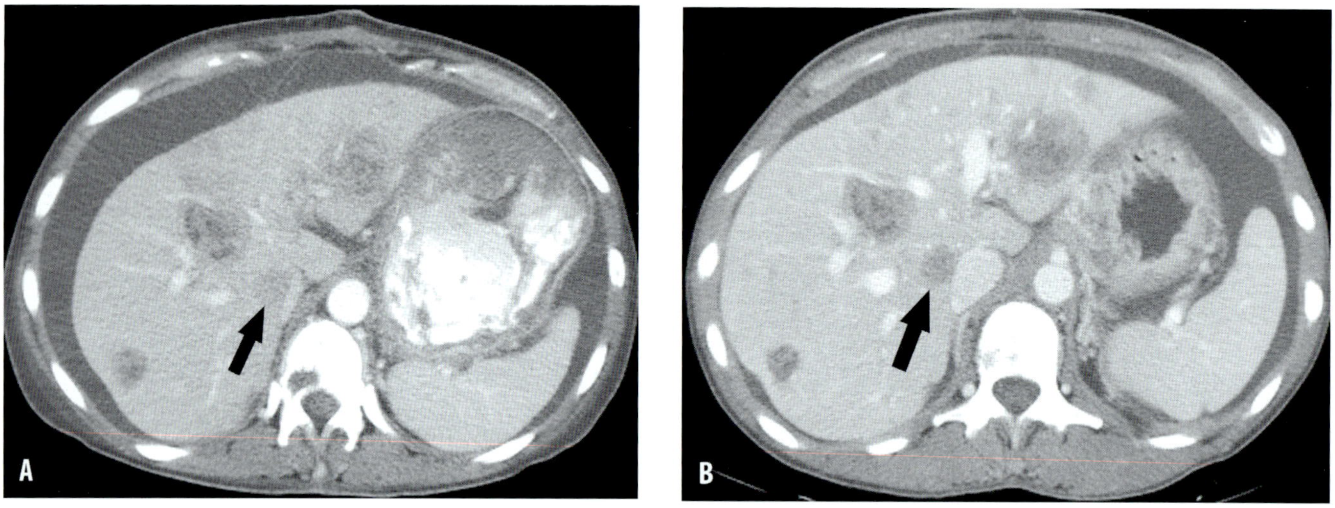

Figure 15.3 A. Axial contrast-enhanced CT image in a 49 year old woman with breast cancer shows multiple hepatic metastases. Note a subtle metastasis (arrow) adjacent to the inferior vena cava. **B.** Axial contrast-enhanced CT image obtained after five months of chemotherapy appears initially to show progression of disease, with new and more conspicuous metastases. On closer examination, all the metastases were present on the baseline study, including the previously subtle lesion (arrow) adjacent to the inferior vena cava, and the appearances are an example of pseudoprogression.

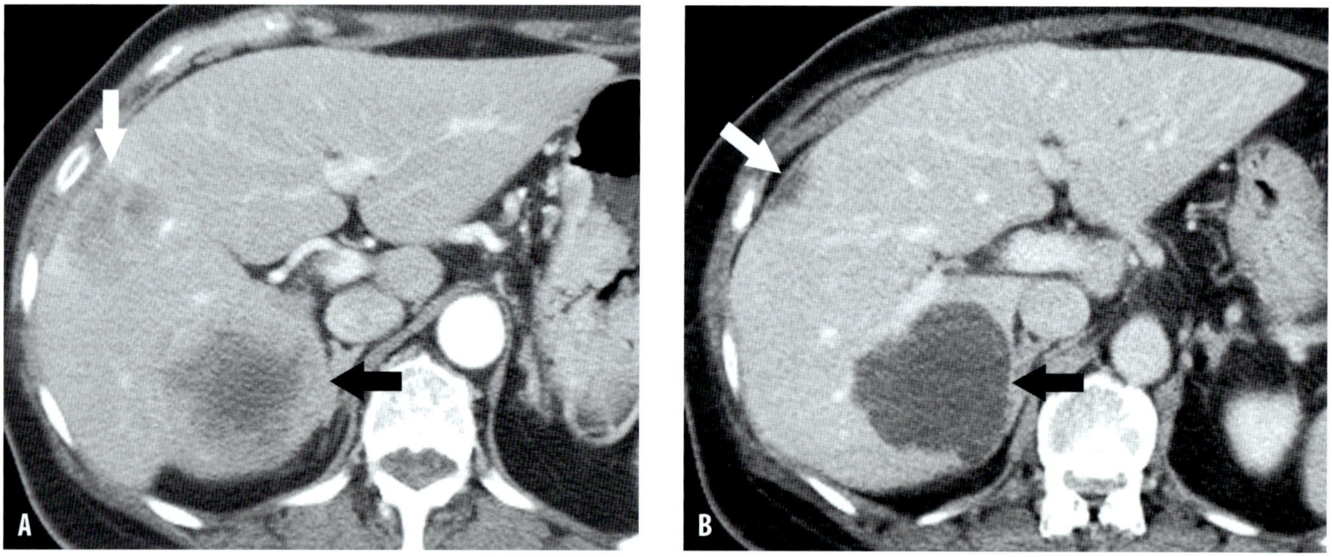

Figure 15.4 A. Axial contrast-enhanced CT image in a 61 year old woman with a gastrointestinal stromal tumor of the stomach shows two heterogeneously enhancing liver metastases (arrows). **B.** Axial contrast-enhanced CT image after five months of chemotherapy with imatinib shows marked reduction of enhancement in the metastases (arrows), presumably due to necrosis. Note that on initial inspection, the posterior right lobe metastasis (black arrow) appears larger than in Figure 15.4A, because of necrosis of the peripheral enhancing solid tumor.

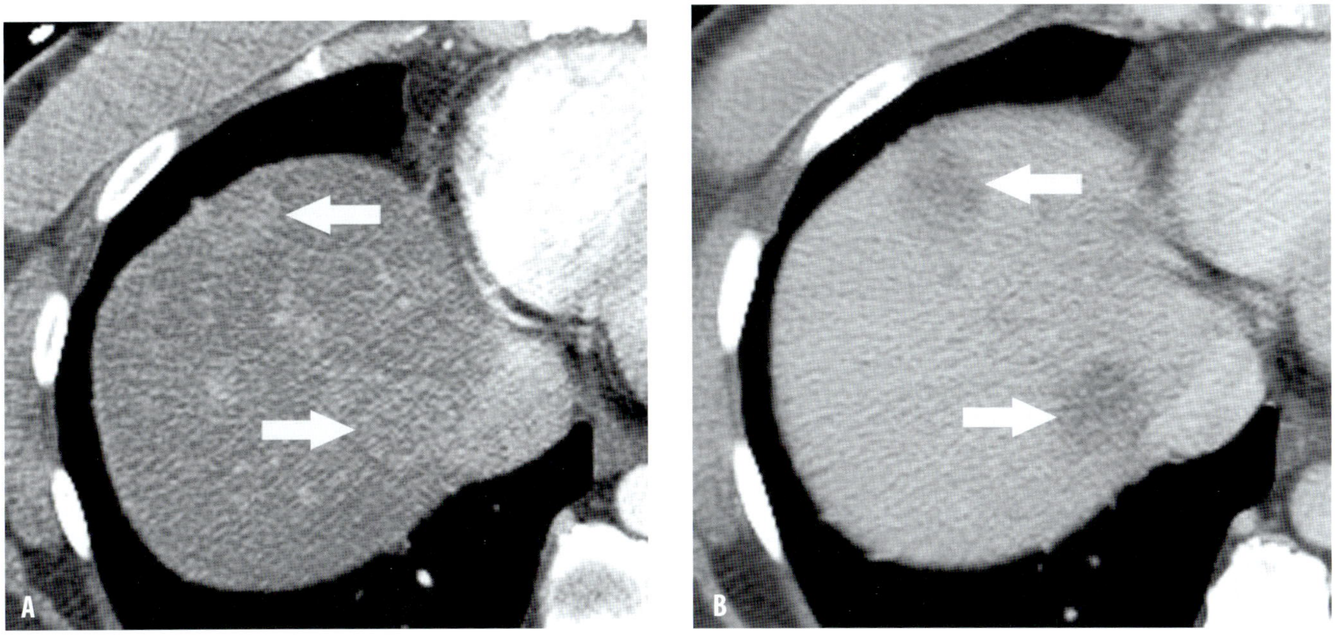

Figure 15.5 **A.** Axial contrast-enhanced CT image in a 50 year old woman receiving chemotherapy for pancreatic cancer shows two subtle liver metastases (arrows). The metastases are difficult to appreciate because the background liver is diffusely fatty and low-density. **B.** Axial contrast-enhanced CT image one month later again shows two hepatic metastases (arrows). The metastases are more conspicuous because diffuse fatty infiltration of the liver has resolved. At initial inspection, these metastases could be mistaken for new lesions and result in an erroneous diagnosis of disease progression.

CASE 16 Pseudothrombosis of the portal vein

Imaging description

On early post-contrast CT studies of the abdomen, the portal vein sometimes appears to contain a central ill-defined filling defect that disappears on more superior images. This "pseudothrombosis" is due to the laminar mixing of enhanced blood from the splenic vein with less enhanced blood from the superior mesenteric vein (Figure 16.1) [1]. This pseudolesion resolves on later phases of enhancement.

Importance

This may be mistaken for a true tumor or bland thrombus of the portal vein, resulting in unnecessary follow-up investigations and patient anxiety.

Typical clinical scenario

Pseudothrombosis of the portal vein is generally only seen on early post-contrast CT images, such as CT arteriography.

Differential diagnosis

Occasionally arterioportal shunting in the liver can cause premature opacification of a portal vein branch, and the resulting mixing of opacified and unopacified portal venous blood can result in a "pseudothrombus" (Figure 16.2). Other than these considerations, cross-sectional imaging has high accuracy in the identification of portal vein thrombus, and the only main limitation is that a diminutive portal vein can sometimes be mistaken for chronic thrombosis [1].

Teaching point

The appearance of an apparent filling defect in the portal vein on early post-contrast CT images should be correlated with later phase images, because occasionally mixing artifacts may result in pseudothrombosis in the arterial phase.

REFERENCE

1 Shah TU, Semelka RC, Voultsinos V, et al. Accuracy of magnetic resonance imaging for preoperative detection of portal vein thrombosis in liver transplant candidates. *Liver Transpl* 2006; **12**: 1682–1688.

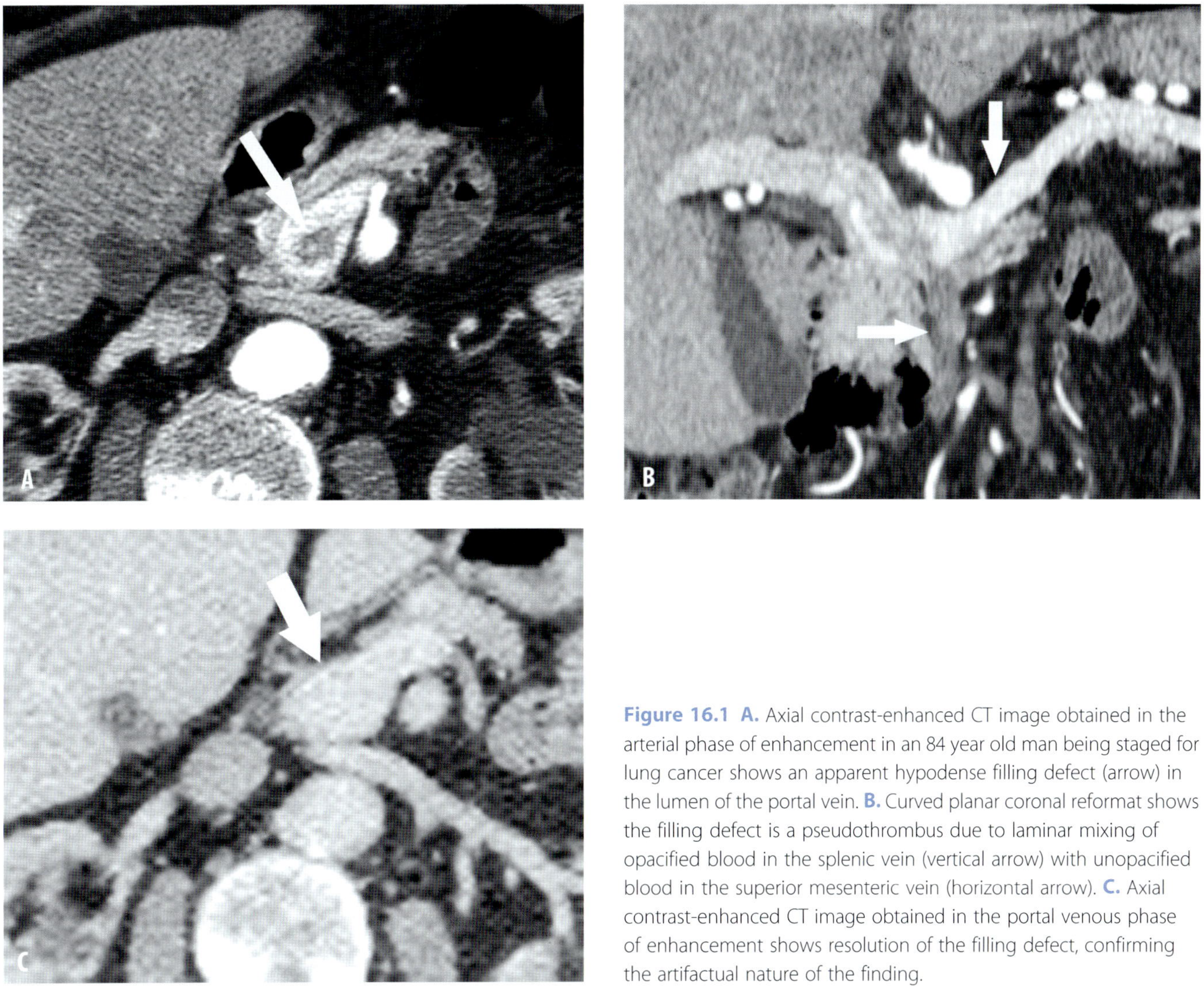

Figure 16.1 **A.** Axial contrast-enhanced CT image obtained in the arterial phase of enhancement in an 84 year old man being staged for lung cancer shows an apparent hypodense filling defect (arrow) in the lumen of the portal vein. **B.** Curved planar coronal reformat shows the filling defect is a pseudothrombus due to laminar mixing of opacified blood in the splenic vein (vertical arrow) with unopacified blood in the superior mesenteric vein (horizontal arrow). **C.** Axial contrast-enhanced CT image obtained in the portal venous phase of enhancement shows resolution of the filling defect, confirming the artifactual nature of the finding.

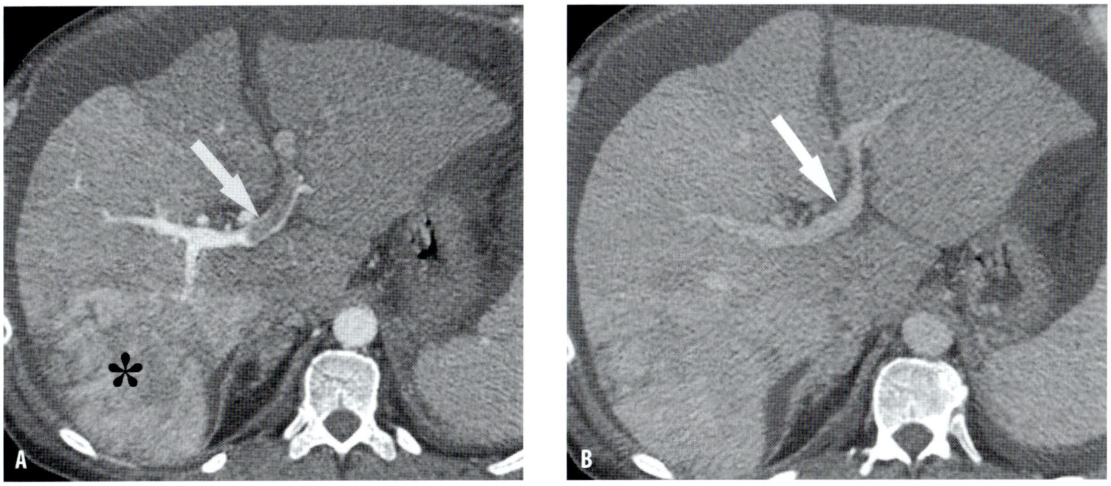

Figure 16.2 **A.** Axial contrast-enhanced CT image obtained in the arterial phase of enhancement in a 64 year old woman with chronic hepatitis B and a large hypervascular hepatocellular carcinoma (asterisk) in the posterior right lobe shows an apparent hypodense filling defect (arrow) in the lumen of the left portal vein. **B.** Axial contrast-enhanced CT image obtained in the portal venous phase of enhancement shows the left portal vein is patent and without any true filling defect, indicating the arterial phase finding was due to arterioportal shunting through the tumor with mixing of prematurely opacified right portal vein blood with non-opacified left portal vein blood.

CASE 17 Biliary hamartomas

Imaging description

Biliary hamartomas (also known as von Meyenburg complexes) are small benign nodules composed of disordered bile ducts in a fibrous stroma that are variably solid to cystic at pathological examination [1]. At imaging, they appear as multiple, small, randomly distributed nodules that are hypoechoic to hyperechoic with or without a characteristic "ring down" artifact at ultrasound, non-enhancing and hypodense at CT, and non-enhancing and T2 hyperintense at MRI (Figures 17.1–17.3).

Importance

Biliary hamartomas are usually an incidental finding. The lesions are most problematic when they are detected in an oncologic patient at CT, when they may be mistaken for possible metastases. While there have been eight reported cases of biliary hamartomas associated with cholangiocarcinoma in the pathological literature, it is unclear whether this is more than mere coincidence and certainly no special follow-up is required for patients with unequivocal biliary hamartomas [2].

Typical clinical scenario

In population based studies, biliary hamartomas have a reported autopsy incidence of 0.7 to 2.8% [3, 4], although the diagnosis is not made with this frequency at imaging. It is possible that some small low-density or T2 hyperintense lesions at CT and MRI, respectively, that are dismissed as "possible cysts" are in fact biliary hamartomas, although mistaking one benign incidental diagnosis for another would seem to be of little clinical consequence.

Differential diagnosis

The random distribution of multiple small nodules in the liver with no larger masses and no extrahepatic disease is suggestive of biliary hamartomas independent of the imaging modality. More specifically, "ring down" artifact at ultrasound and marked T2 hyperintensity at MRI help suggest the diagnosis. Metastases, with the exception of neuroendocrine metastases [5], are usually only moderately and not markedly hyperintense on T2-weighted images.

Teaching point

The diagnosis of biliary hamartomas should be considered for multiple, small, randomly distributed nodules in the liver, especially if there is "ring down" artifact at ultrasound or marked T2 hyperintensity at MRI.

REFERENCES

1 Lev-Toaff AS, Bach AM, Wechsler RJ, et al. The radiologic and pathologic spectrum of biliary hamartomas. *Am J Roentgenol* 1995; **165**: 309–313.

2 Röcken C, Pross M, Brucks U, Ridwelski K, Roessner A. Cholangiocarcinoma occurring in a liver with multiple bile duct hamartomas (von Meyenburg complexes). *Arch Pathol Lab Med* 2000; **124**: 1704–1706.

3 Thommesen N. Biliary hamartomas (von Meyenburg complexes) in liver needle biopsies. *Acta Pathol Microbiol Scand A* 1978; **86**: 93–99.

4 Chung EB. Multiple bile-duct hamartomas. *Cancer* 1970; **26**: 287–296.

5 Soyer P, Gueye C, Somveille E, Laissy JP, Scherrer A. MR diagnosis of hepatic metastases from neuroendocrine tumors versus hemangiomas: relative merits of dynamic gadolinium chelate-enhanced gradient-recalled echo and unenhanced spin-echo images. *Am J Roentgenol* 1995; **165**: 1407–1413.

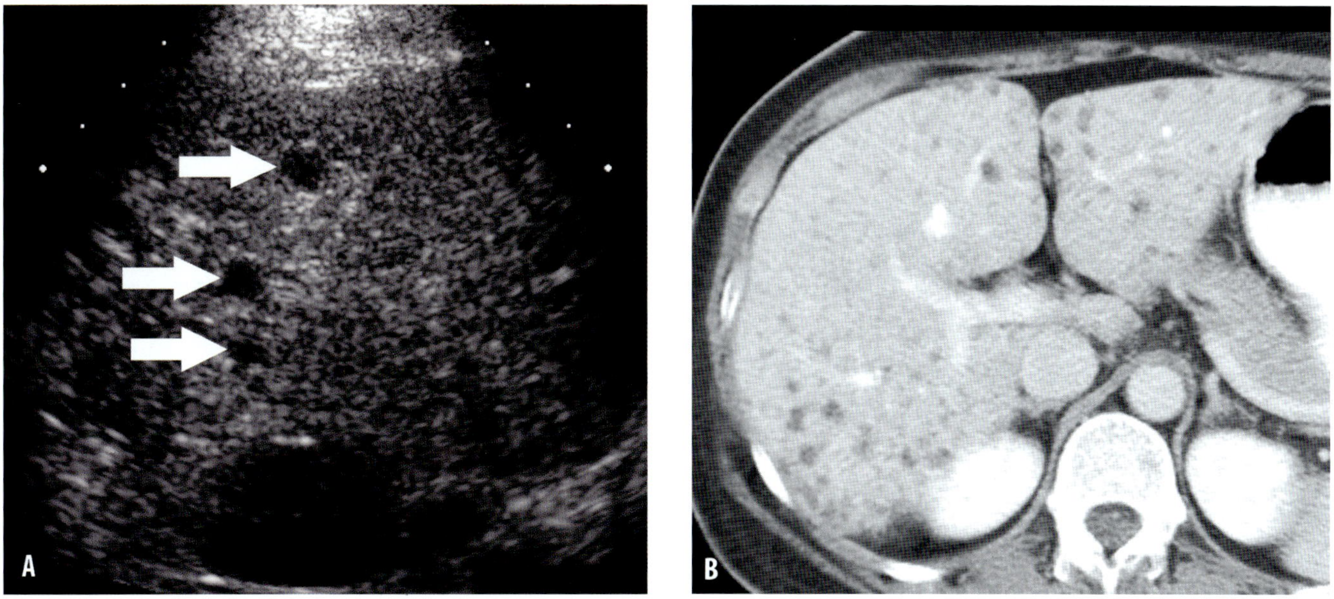

Figure 17.1 **A.** Ultrasound image in a 52 year old woman with chronic abdominal pain showing multiple small hypoechoic nodules (arrows) in the liver. **B.** Axial contrast-enhanced CT image in the same patient shows the characteristic appearance of biliary hamartomas, with multiple small randomly distributed hypodense nodules. The appearances were unchanged on a CT performed five years later for unrelated reasons.

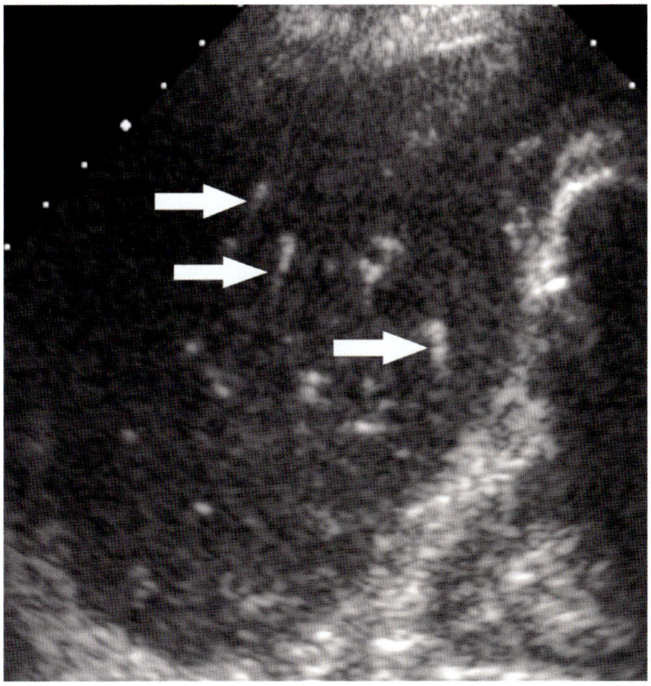

Figure 17.2 Ultrasound image in a 79 year old man with chronic abdominal discomfort showing multiple scattered foci of "ring down" artifact (arrows) in the liver, consistent with biliary hamartomas.

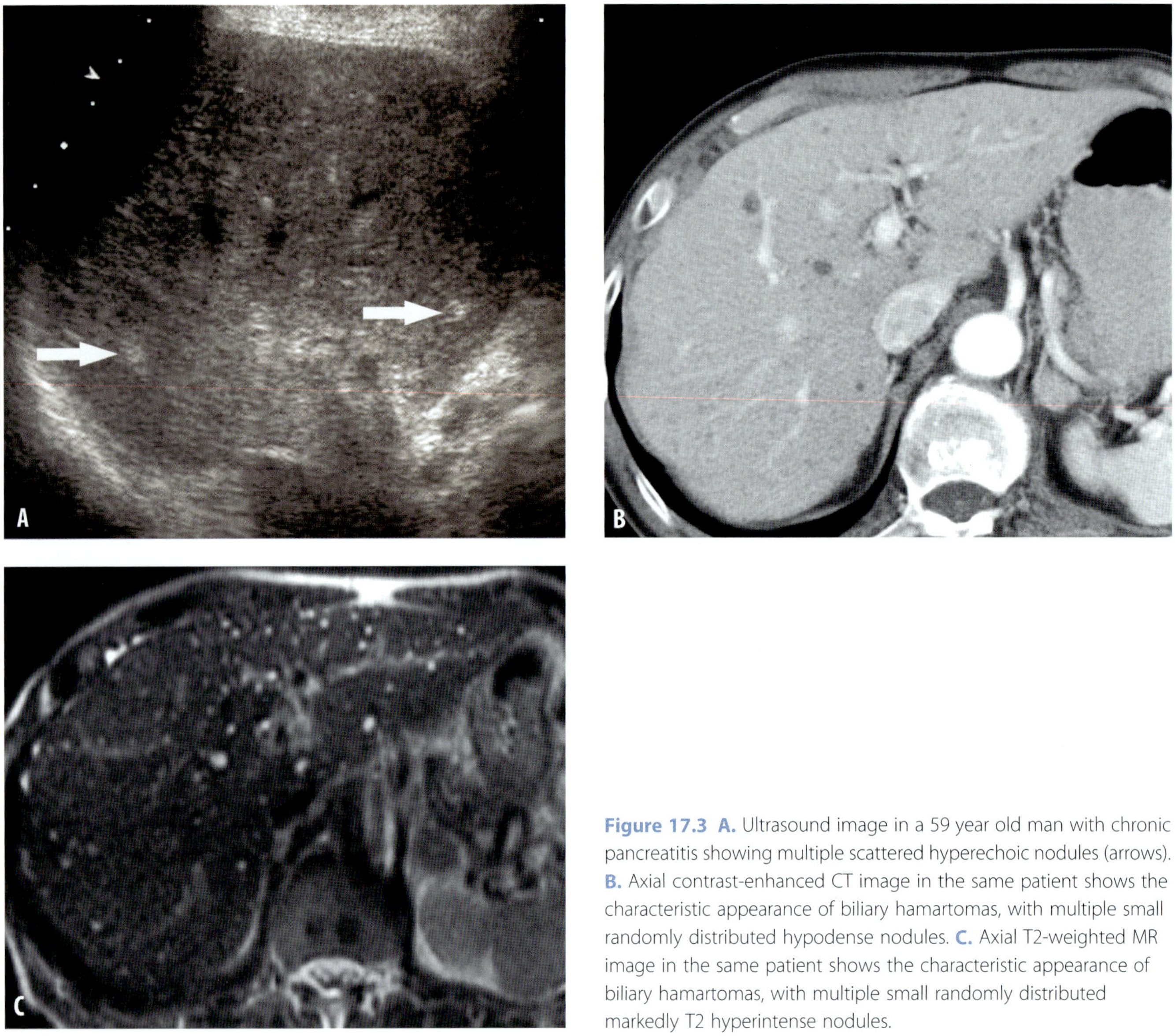

Figure 17.3 A. Ultrasound image in a 59 year old man with chronic pancreatitis showing multiple scattered hyperechoic nodules (arrows). **B.** Axial contrast-enhanced CT image in the same patient shows the characteristic appearance of biliary hamartomas, with multiple small randomly distributed hypodense nodules. **C.** Axial T2-weighted MR image in the same patient shows the characteristic appearance of biliary hamartomas, with multiple small randomly distributed markedly T2 hyperintense nodules.

CASE 18 Nodular focal fatty infiltration of the liver

Imaging description

Focal fatty infiltration of the liver is usually easily recognized based on the characteristic findings of a focal lesion with a geographic shape adjacent to the porta hepatis or fissure for the ligamentum teres that may contain non-distorted traversing blood vessels [1]. Occasionally, focal fat is nodular and located in other locations in the liver. Such atypical nodular focal fatty infiltration may result in an appearance of echogenic lesions at ultrasound or hypodense lesions at CT that mimic metastases (Figures 18.1–18.2) [1–6].

Importance

Nodular focal fatty infiltration may mimic metastases, leading to patient anxiety and unnecessary investigations in pursuit of a non-existent primary site.

Typical clinical scenario

Nodular focal fatty infiltration is typically seen as an incidental finding in patients being imaged for unrelated reasons, but may occasionally be seen with patients with known causes of diffuse fatty infiltration such as diabetes or alcohol abuse (Figure 18.3).

Differential diagnosis

A target-like appearance with central echogenicity on ultrasound and central hyperdensity on contrast-enhanced CT has been described [2–4] but would not appear sufficiently distinctive to allow a confident diagnosis. MRI is often critical to the diagnosis by demonstrating signal loss on opposed phase imaging, and the absence of abnormal enhancement on post-contrast imaging that might suggest other focal hepatic lesions that may contain microscopic fat, such as focal nodular hyperplasia, adenoma, or hepatocellular carcinoma [1, 4]. With respect to these other focal lesions, it should be noted that fat occurs in 35 to 77% of adenomas, up to 35% of small hepatocellular carcinomas, and is extremely rare in focal nodular hyperplasia and usually patchy rather than uniform [7–10]. All of these lesions are typically hypervascular (Figures 18.4 and 18.5), and most hepatocellular carcinomas occur in patients with cirrhosis.

Teaching point

Nodular focal fatty infiltration should be considered when incidental echogenic or hypodense hepatic nodules suggestive of metastases are seen in a patient without a primary malignancy. Opposed phase gradient-echo MRI can be critical in making the correct diagnosis.

REFERENCES

1 Prasad SR, Wang H, Rosas H, et al. Fat-containing lesions of the liver: radiologic-pathologic correlation. *Radiographics* 2005; **25**: 321–331.

2 Kemper J, Jung G, Poll LW, et al. CT and MRI findings in multifocal hepatic steatosis mimicking malignancy. *Abdom Imaging* 2002; **27**: 708–710.

3 Rubaltelli L, Savastano S, Khadivi Y, et al. Targetlike appearance of pseudotumors in segment IV of the liver on sonography. *Am J Roentgenol* 2002; **178**: 75–77.

4 Yates CK, Streight RA. Focal fatty infiltration of the liver simulating metastatic disease. *Radiology* 1986; **159**: 83–84.

5 Kröncke TJ, Taupitz M, Kivelitz D, et al. Multifocal nodular fatty infiltration of the liver mimicking metastatic disease on CT: imaging findings and diagnosis using MR imaging. *Eur Radiol* 2000; **10**: 1095–1100.

6 Flournoy JG, Potter JL, Sullivan BM, Gerza CB, Ramzy I. CT appearance of multifocal hepatic steatosis. *J Comput Assist Tomogr* 1984; **8**: 1192–1194.

7 Chung KY, Mayo-Smith WW, Saini S, et al. Hepatocellular adenoma: MR imaging features with pathologic correlation. *Am J Roentgenol* 1995; **165**: 303–308.

8 Arrive L, Flejou JF, Vilgrain V, et al. Hepatic adenoma: MR findings in 51 pathologically proved lesions. *Radiology* 1994; **193**: 507–512.

9 Kutami R, Nakashima Y, Nakashima O, Shiota K, Kojiro M. Pathomorphologic study on the mechanism of fatty change in small hepatocellular carcinoma of humans. *J Hepatol* 2000; **33**: 282–289.

10 Stanley G, Jeffrey RB Jr, Feliz B. CT findings and histopathology of intratumoral steatosis in focal nodular hyperplasia: case report and review of the literature. *J Comput Assist Tomogr* 2002; **26**: 815–817.

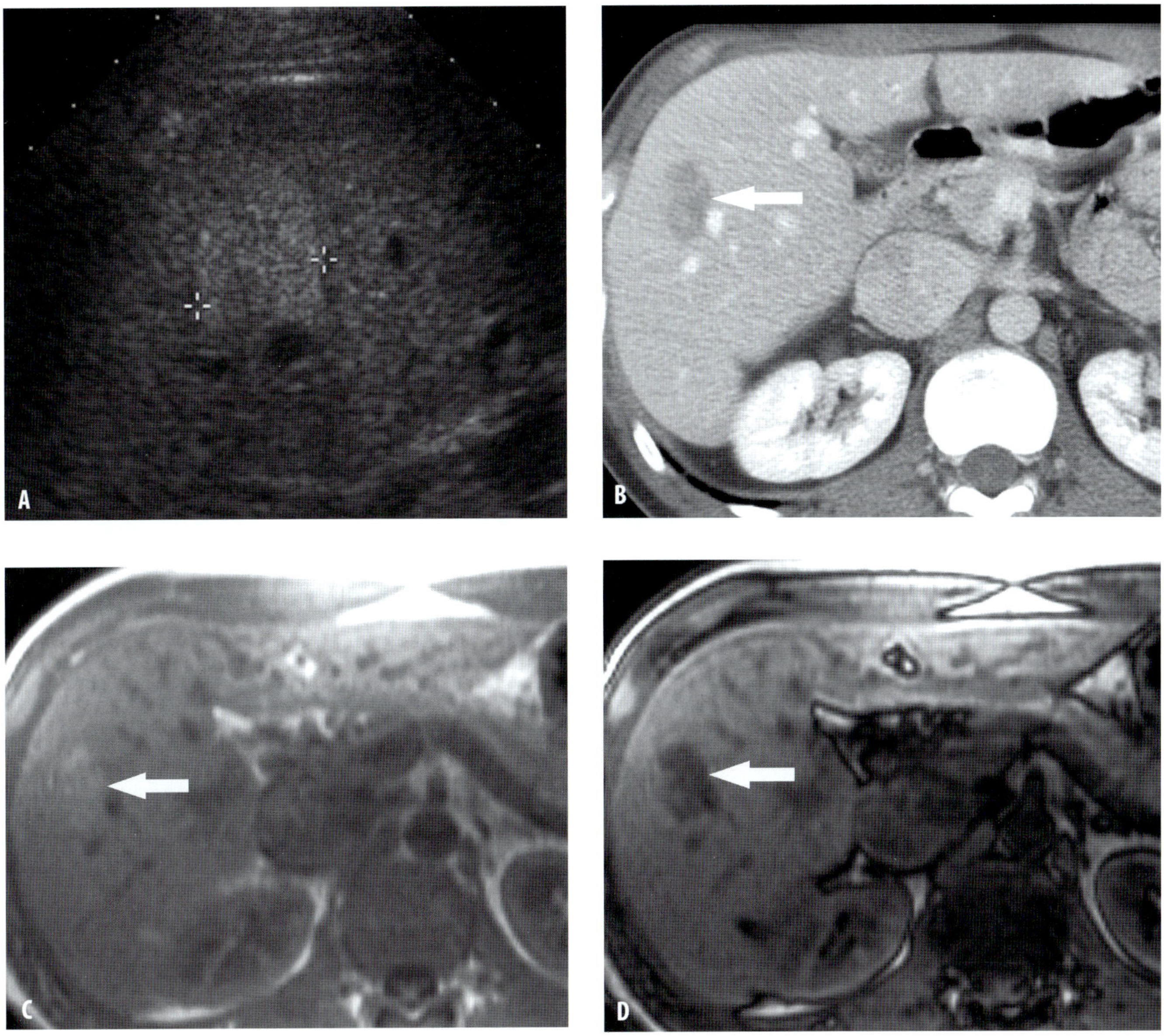

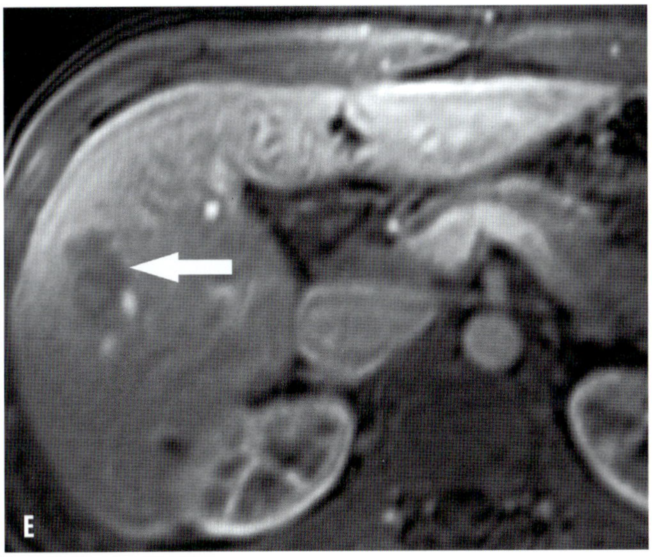

Figure 18.1 A. Ultrasound image in a 31 year old Asian man with chronic hepatitis B shows a non-specific 2.2 cm echogenic nodule (between calipers) in the liver. **B.** Axial contrast-enhanced portal venous phase CT image shows the nodule (arrow) is hypodense. No peripheral nodular globular enhancement is seen to suggest a hemangioma. **C.** Axial in phase gradient-echo T1-weighted MR image shows the nodule (arrow) is of slightly increased T1 signal intensity. **D.** Axial opposed phase gradient-echo T1-weighted MR image shows marked signal intensity loss throughout the nodule (arrow), indicating the presence of microscopic fat. **E.** Axial gradient-echo fat-saturated T1-weighted MR image in the arterial phase of enhancement after intravenous gadolinium administration shows no hypervascularity to suggest the diagnosis of hepatocellular carcinoma, focal nodular hyperplasia, or adenoma. The MRI findings are those of nodular focal fatty infiltration.

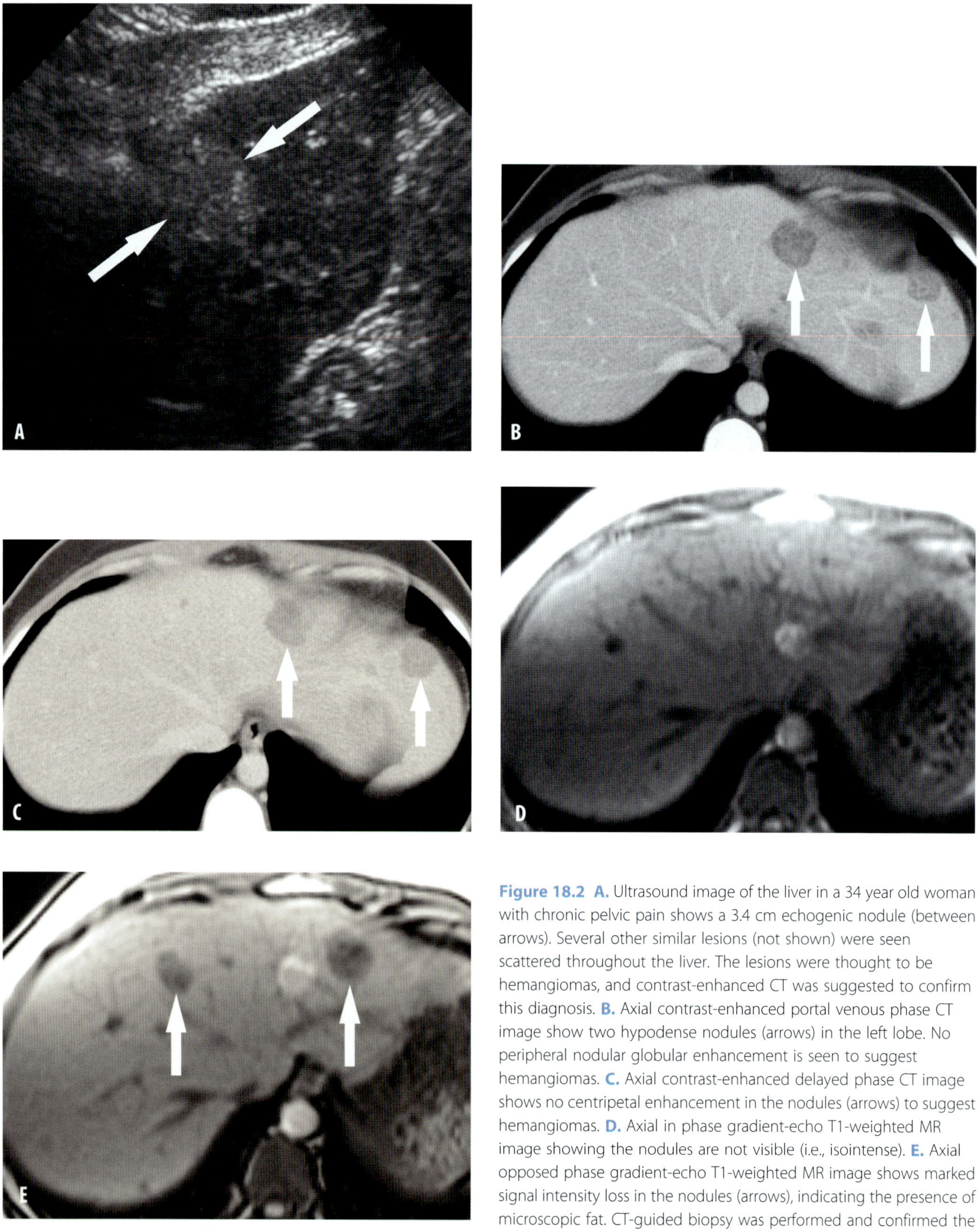

Figure 18.2 **A.** Ultrasound image of the liver in a 34 year old woman with chronic pelvic pain shows a 3.4 cm echogenic nodule (between arrows). Several other similar lesions (not shown) were seen scattered throughout the liver. The lesions were thought to be hemangiomas, and contrast-enhanced CT was suggested to confirm this diagnosis. **B.** Axial contrast-enhanced portal venous phase CT image show two hypodense nodules (arrows) in the left lobe. No peripheral nodular globular enhancement is seen to suggest hemangiomas. **C.** Axial contrast-enhanced delayed phase CT image shows no centripetal enhancement in the nodules (arrows) to suggest hemangiomas. **D.** Axial in phase gradient-echo T1-weighted MR image showing the nodules are not visible (i.e., isointense). **E.** Axial opposed phase gradient-echo T1-weighted MR image shows marked signal intensity loss in the nodules (arrows), indicating the presence of microscopic fat. CT-guided biopsy was performed and confirmed the diagnosis of nodular focal fatty infiltration.

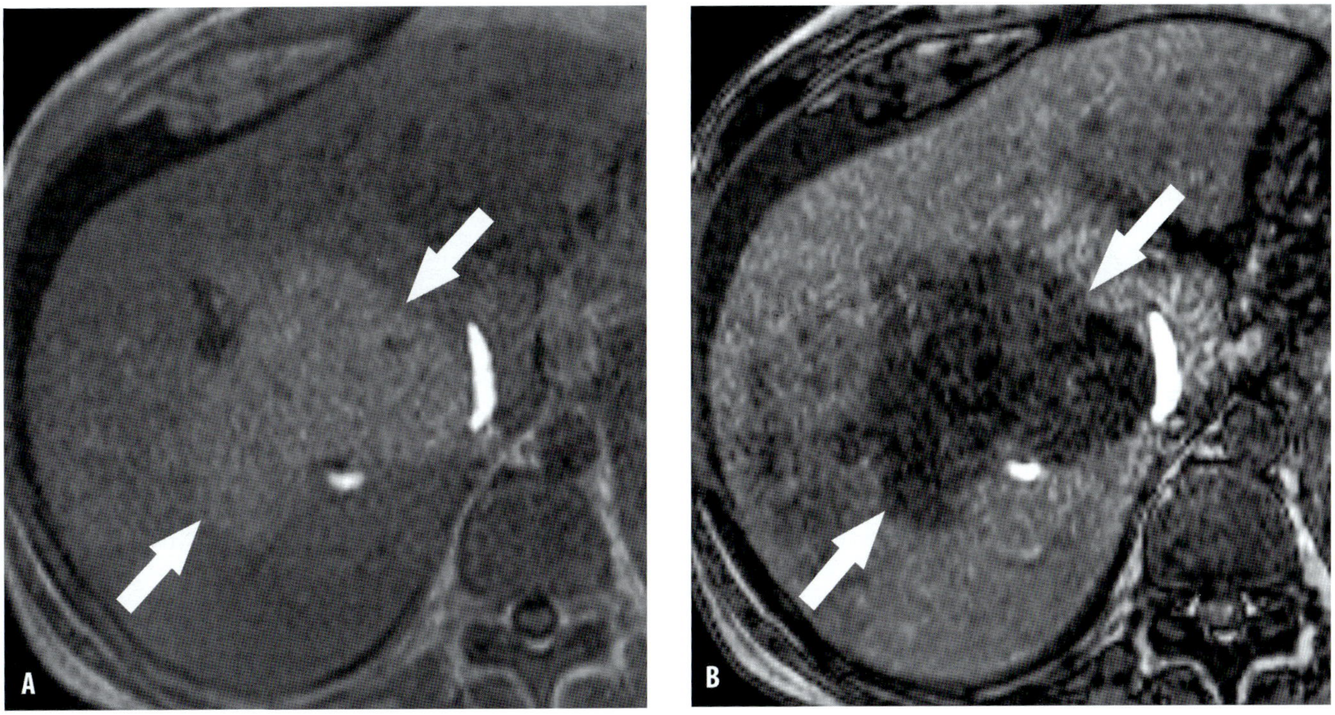

Figure 18.3 A. Axial in phase gradient-echo T1-weighted MR image in a 40 year old alcoholic showing a masslike lesion (between arrows) of increased T1 signal intensity in the right hepatic lobe. **B.** Axial opposed phase gradient-echo T1-weighted MR image shows marked signal intensity loss throughout the lesion (between arrows), confirming the diagnosis of focal fatty infiltration.

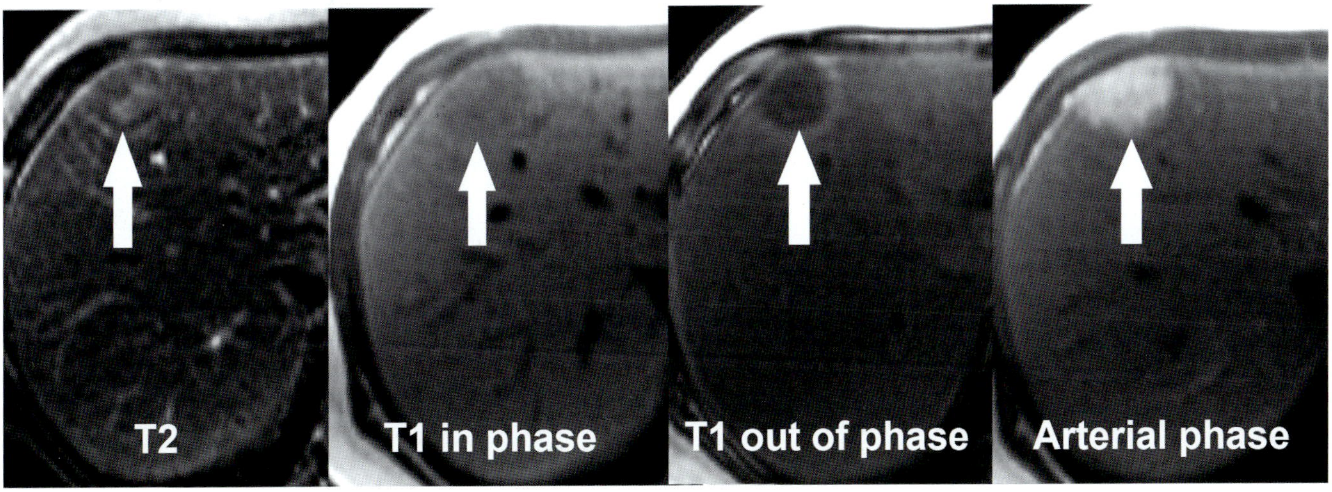

Figure 18.4 Photomontage of fat-saturated T2, gradient-echo T1 in phase, gradient-echo T1 out of phase, and arterial phase gradient-echo T1 post-gadolinium MR images in a 32 year old woman with a hepatic adenoma (arrows). The lesion is of subtly hyperintense T2 signal and demonstrates marked signal loss on out of phase versus in phase images. These findings might suggest a diagnosis of nodular focal fatty infiltration. However, note the tumor shows marked hypervascular enhancement, which is not compatible with a diagnosis of nodular focal fatty infiltration.

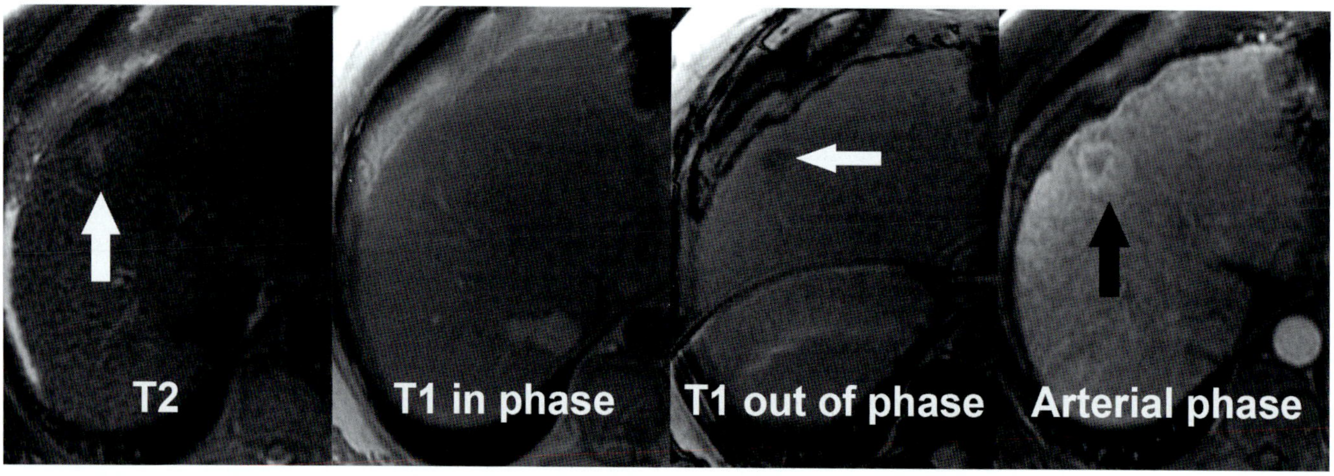

Figure 18.5 Photomontage of fat-saturated T2, gradient-echo T1 in phase, gradient-echo T1 out of phase, and arterial phase gradient-echo T1 post-gadolinium MR images in a 65 year old man with hepatocellular carcinoma arising in a cirrhotic liver. The tumor is of subtly hyperintense T2 signal (vertical white arrow) and demonstrates focal signal loss on out of phase versus in phase images (horizontal white arrow). These findings might suggest a diagnosis of nodular focal fatty infiltration. However, note the tumor shows hypervascular enhancement (black arrow), which is not compatible with a diagnosis of nodular focal fatty infiltration.

CASE 19 Nodular focal fatty sparing of the liver

Imaging description

Focal fatty sparing is usually geographic in shape and located near the gallbladder fossa and porta hepatis [1]. Occasionally, focal sparing is nodular and located elsewhere in the liver. Such atypical nodular focal fatty sparing may result in lesions that are hypoechoic at ultrasound and hyperdense at CT and may be mistaken for metastases or other tumors (Figures 19.1 and 19.2) [2–4]. MRI can be very helpful in establishing the diagnosis of nodular focal fatty sparing [2].

Importance

Failure to recognize the characteristic MR findings of nodular fatty sparing could lead to a misdiagnosis of hepatic tumor and inappropriate biopsy.

Typical clinical scenario

By definition, fatty sparing occurs in patients with diffuse fatty infiltration of the liver, which may be idiopathic or secondary to obesity, starvation, parenteral nutrition, steroid therapy, diabetes mellitus, alcohol, and hepatitis.

Differential diagnosis

Nodular or mass-like lesions in a diffusely fatty liver that are only well seen on out of phase gradient-echo MR imaging, when visualized during an MR study that includes T2-weighted and dynamic gadolinium-enhanced sequences, should suggest focal fatty sparing. It is possible that true primary or secondary tumors in a diffusely fatty liver could also be better seen on opposed phase imaging, but such lesions would almost certainly demonstrate increased T2 signal intensity or altered vascularity relative to the liver parenchyma on dynamic gadolinium-enhanced images. As such, alternative diagnoses should be considered for focal lesions that are seen in a diffusely fatty liver on sequences other than out of phase gradient-echo MR imaging (Figure 19.3).

Teaching point

Nodular focal fatty sparing should be considered when incidental hypoechoic or hyperdense hepatic nodules suggestive of metastases or other tumors are seen in a patient without a primary malignancy and with signs of diffuse fatty infiltration of the liver. Opposed phase gradient-echo MRI can be critical in making the correct diagnosis.

REFERENCES

1 Chong VF, Fan YF. Ultrasonographic hepatic pseudolesions: normal parenchyma mimicking mass lesions in fatty liver. *Clin Radiol* 1994; **49**: 326–329.

2 Tom WW, Yeh BM, Cheng JC, et al. Hepatic pseudotumor due to nodular fatty sparing: the diagnostic role of opposed-phase MRI. *Am J Roentgenol* 2004; **183**: 721–724.

3 Zezos P, Tatsi P, Nakos A, et al. Focal fatty liver sparing lesion presenting as a "pseudotumour": case report. *Acta Gastroenterol Belg* 2006; **69**: 323–326.

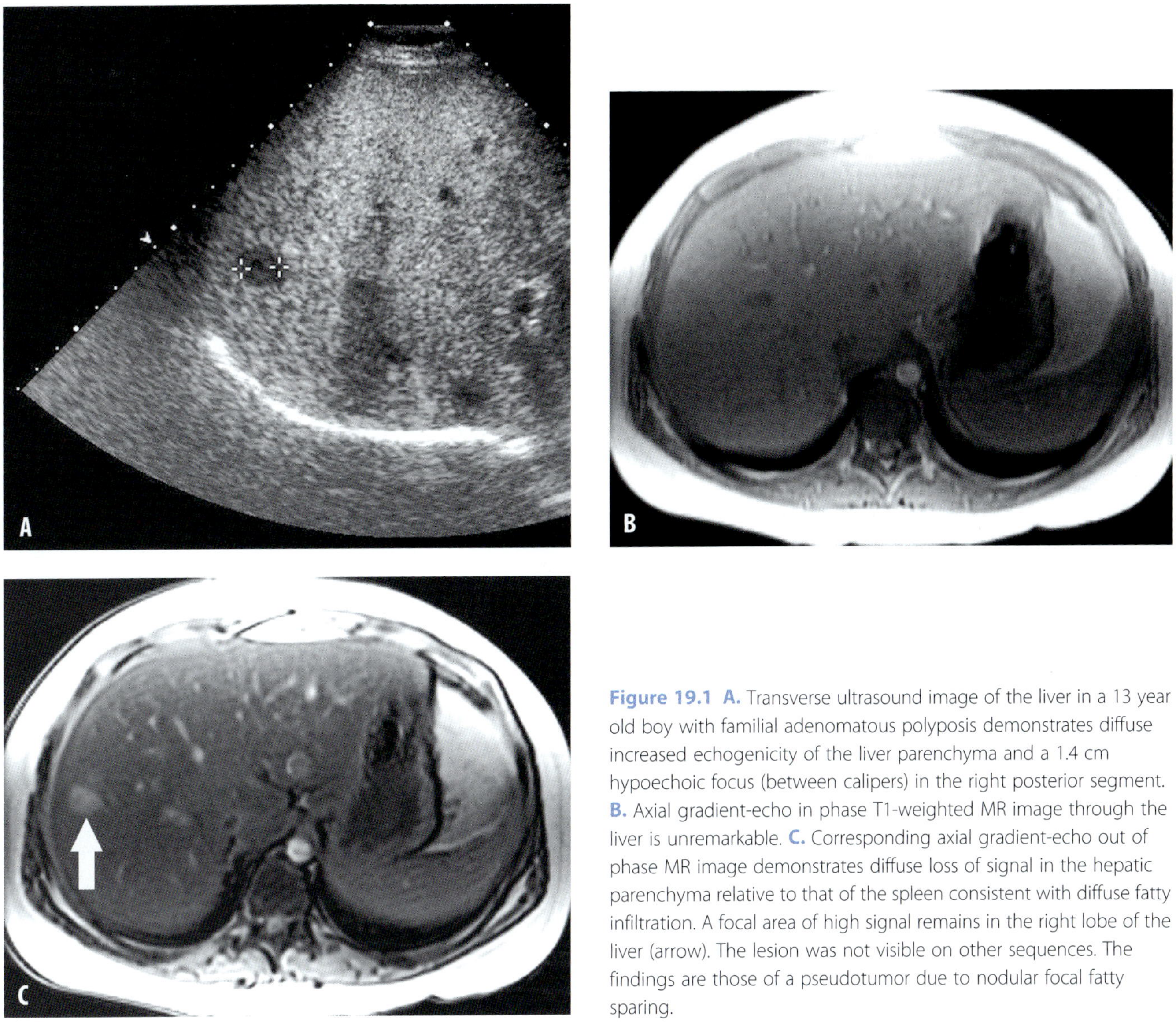

Figure 19.1 A. Transverse ultrasound image of the liver in a 13 year old boy with familial adenomatous polyposis demonstrates diffuse increased echogenicity of the liver parenchyma and a 1.4 cm hypoechoic focus (between calipers) in the right posterior segment. **B.** Axial gradient-echo in phase T1-weighted MR image through the liver is unremarkable. **C.** Corresponding axial gradient-echo out of phase MR image demonstrates diffuse loss of signal in the hepatic parenchyma relative to that of the spleen consistent with diffuse fatty infiltration. A focal area of high signal remains in the right lobe of the liver (arrow). The lesion was not visible on other sequences. The findings are those of a pseudotumor due to nodular focal fatty sparing.

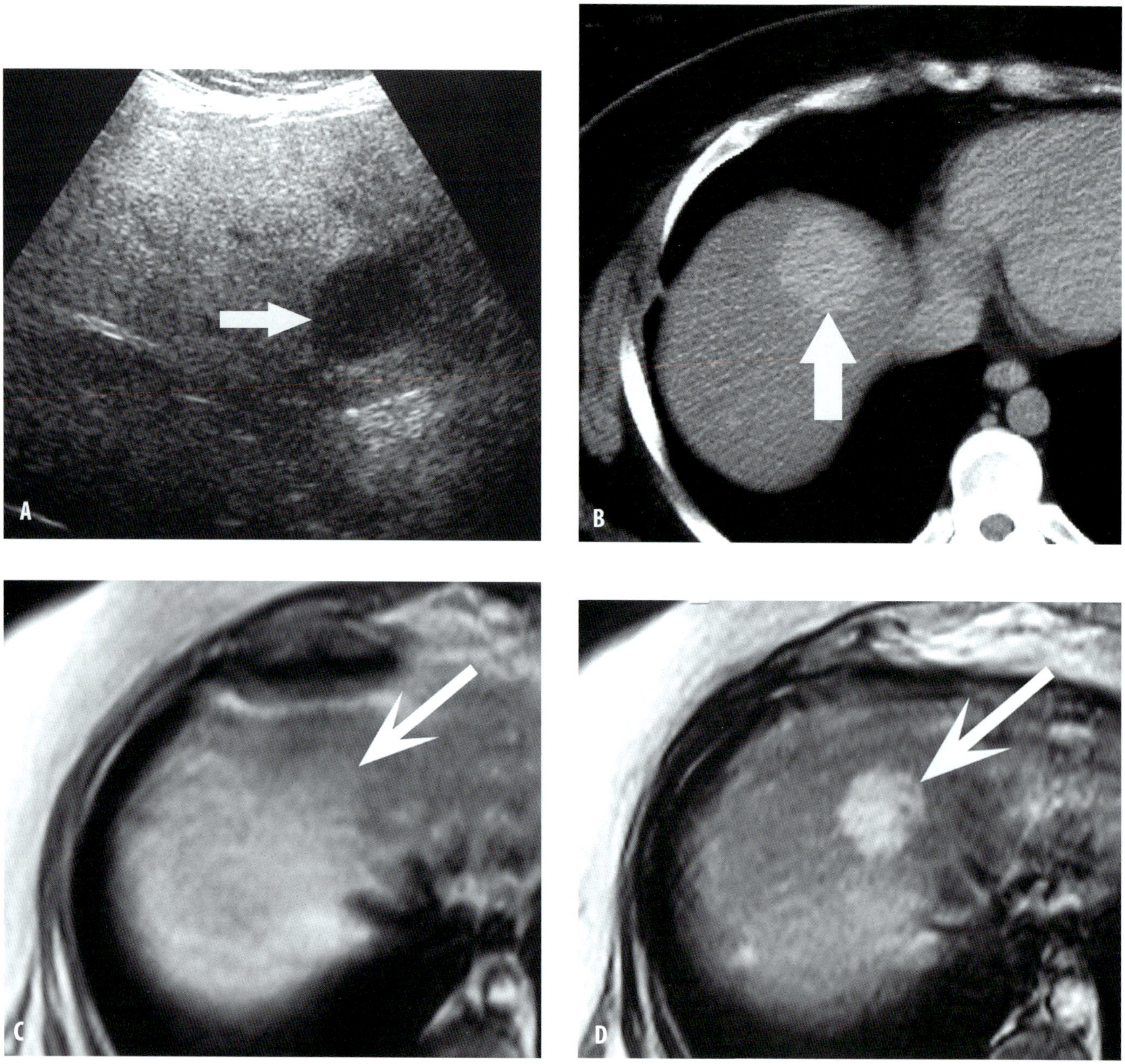

Figure 19.2 **A.** Ultrasound image of the right hepatic lobe in an asymptomatic 20 year old man being scanned because of a strong family history of cancer. The liver parenchyma is diffusely hyperechoic, suggesting diffuse fatty infiltration, and contains a focal hypoechoic lesion (arrow). **B.** Axial non-enhanced CT image shows a diffusely low-attenuation liver with a relatively dense focus (arrow) near the liver dome. **C.** Axial gradient-echo in phase T1-weighted MR image through the liver shows a subtle isointense nodule in the right hepatic lobe. **D.** Corresponding axial gradient-echo out of phase MR image demonstrates diffuse loss of signal in the hepatic parenchyma consistent with diffuse fatty infiltration. A focal area of high signal remains in the right lobe of the liver (arrow). The lesion was not visible on other sequences. The findings are those of a pseudotumor due to nodular focal fatty sparing.

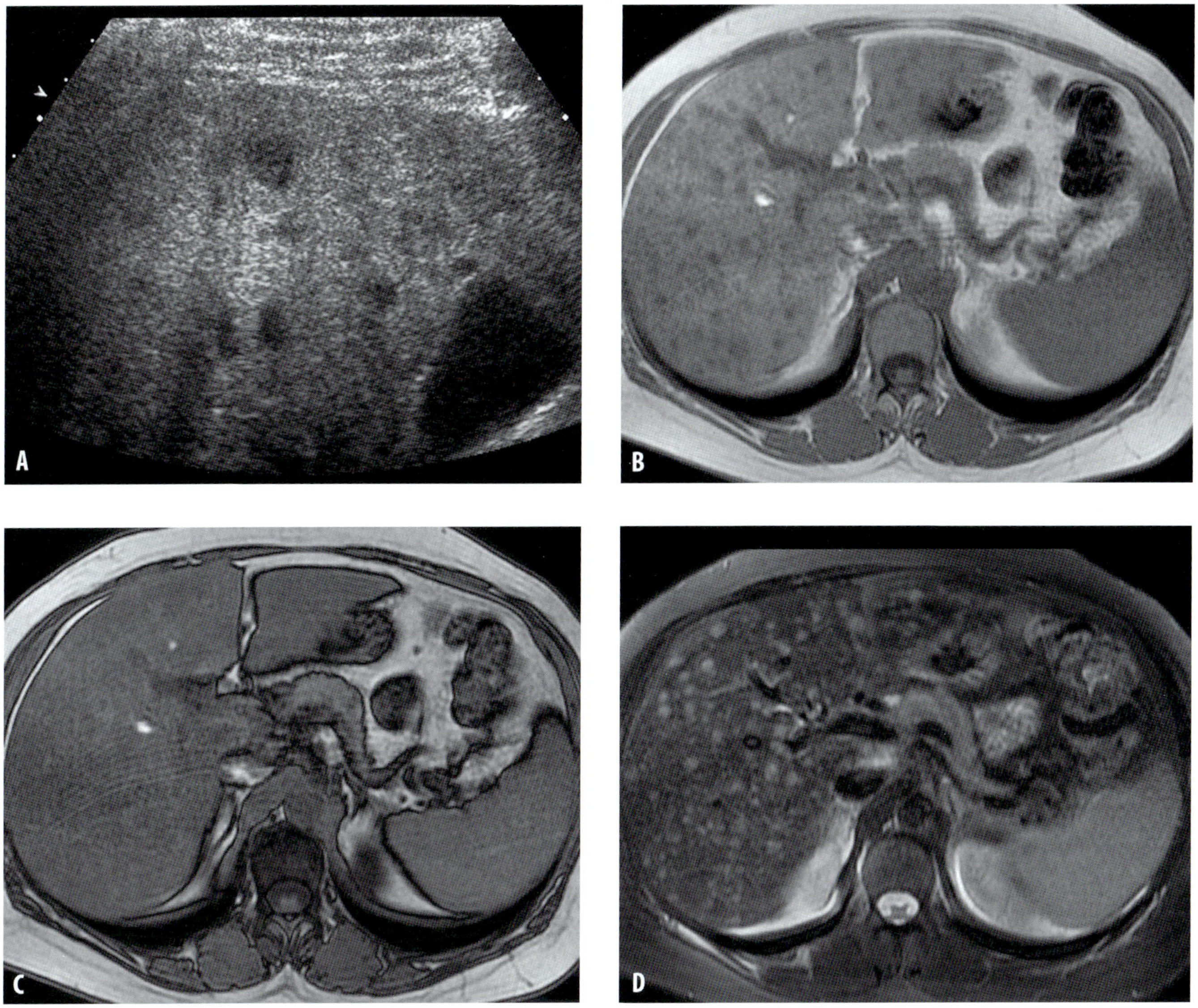

Figure 19.3 A. Ultrasound image of the liver in a 45 year old woman with newly diagnosed breast cancer and biliary hamartomas. The liver parenchyma is diffusely hyperechoic, suggesting diffuse fatty infiltration, and contains multiple hypoechoic nodules. **B.** Axial gradient-echo in phase T1-weighted MR image through the liver shows multiple hypointense nodules in the liver. **C.** Corresponding axial out of phase gradient-echo MR image demonstrates diffuse loss of signal in the hepatic parenchyma consistent with diffuse fatty infiltration. The multiple nodules are not visible. However, this should not be misinterpreted as indicating multifocal fatty sparing – true nodules of fatty sparing should be hyperintense relative to the liver on out of phase gradient-echo MRI. **D.** Axial fat-saturated T2-weighted MR image shows multiple hyperintense nodules scattered randomly throughout the liver, consistent with biliary hamartomas. Again, this signal intensity would not be consistent with multifocal fatty sparing, which would be expected to display isointensity with the background liver parenchyma on this sequence.

CASE 20

Hepatocellular carcinoma mimicking focal nodular hyperplasia

Imaging description

At MRI, focal nodular hyperplasia appears as a mass that is mildly hypointense on T1-weighted images, mildly hyperintense on T2-weighted images, hypervascular after gadolinium administration, and has a central scar that is of high T2 signal with delayed enhancement after gadolinium [1–4]. When all these findings are present, the diagnosis of focal nodular hyperplasia can usually be made with high accuracy [2, 5]. However, hepatocellular carcinoma arising in a cirrhotic liver can demonstrate all of these features and so a diagnosis of focal nodular hyperplasia in a patient with cirrhosis should be made with great caution (Figures 20.1 and 20.2) [6].

Importance

Hepatocellular carcinoma has a wide spectrum of findings including an appearance indistinguishable from focal nodular hyperplasia. As such, hepatocellular carcinoma should be the primary consideration for any solid lesion in a cirrhotic liver that is not a hemangioma [2]. This approach may result in earlier diagnosis and treatment of hepatocellular carcinoma and prevent inappropriate management resulting from a potentially mistaken diagnosis of focal nodular hyperplasia.

Typical clinical scenario

The possibility of hepatocellular carcinoma mimicking focal nodular hyperplasia arises primarily in patients with cirrhosis, of whatever etiology. The diagnosis of cirrhosis is usually obvious due to the presence of diffuse hepatic nodularity and signs of portal hypertension.

Differential diagnosis

There is some controversy as to whether focal nodular hyperplasia occurs in cirrhosis. From a pathological perspective, it has been argued that focal nodular hyperplasia is unlikely to occur in cirrhosis, and even if it did, it would be practically impossible to diagnose since the disorganized liver tissue of focal nodular hyperplasia would be indistinguishable from the disorganized liver tissue of cirrhosis [1]. Conversely, several recent reports have described nodules that were pathologically and radiologically virtually indistinguishable from focal nodular hyperplasia in cirrhotic patients [7–13]. Reasonable conclusions from these data are that focal nodular hyperplasia is very rare in cirrhosis and should only be diagnosed after imaging confirmation of long-term stability, supplemented by biopsy where appropriate. Increased FDG uptake at PET may help point to a diagnosis of hepatocellular carcinoma [6].

Teaching point

Hepatocellular carcinoma is the primary consideration for any solid mass in a cirrhotic liver [2], even if the appearances suggest focal nodular hyperplasia. The diagnosis of focal nodular hyperplasia should rarely, if ever, be made in a cirrhotic liver and certainly only after demonstration of long-term stability, supplemented by biopsy or PET as appropriate.

REFERENCES

1 Hytiroglou P, Theise ND. Differential diagnosis of hepatocellular nodular lesions. *Semin Diagn Pathol* 1998; **15**: 285–299.

2 Coakley FV, Schwartz LH. Imaging of hepatocellular carcinoma: a practical approach. *Semin Oncol* 2001; **28**: 460–473.

3 Craig JR, Peters RL, Edmondson HA. Tumors of the liver and intrahepatic bile ducts. *In: Atlas of tumor pathology* (Series 2, Fascicle 26). Washington, DC: Armed Forces Institute of Pathology, 1989.

4 Jeong YY, Yim NY, Kang HK. Hepatocellular carcinoma in the cirrhotic liver with helical CT and MRI: imaging spectrum and pitfalls of cirrhosis-related nodules. *Am J Roentgenol* 2005; **185**: 1024–1032.

5 Vilgraine V, Flejou JF, Arrive L, et al. Focal nodular hyperplasia of the liver: MR imaging and pathologic correlation in 37 patients. *Radiology* 1992; **184**: 699–703.

6 Rafaelian O, Coakley FV, Qayyum A, et al. Magnetic resonance imaging of hepatocellular carcinoma mimicking focal nodular hyperplasia: a potential pitfall in patients with cirrhosis? *J Comput Assist Tomogr* 2006; **30**: 796–799.

7 Sugihara S, Nakashima O, Kiyomatsu K, et al. A case of liver cirrhosis with a hyperplastic nodular lesion. *Acta Pathol Jpn* 1990; **40**: 699–703.

8 Terada T, Kitani S, Ueda K, et al. Adenomatous hyperplasia of the liver resembling focal nodular hyperplasia in patients with chronic liver disease. *Virchows Arch A Pathol Anat Histopathol* 1993; **422**: 247–252.

9 Quaglia A, Tibballs J, Grasso A, et al. Focal nodular hyperplasia-like areas in cirrhosis. *Histopathology* 2003; **42**: 14–21.

10 An HJ, Illei P, Diflo T, et al. Scirrhous changes in dysplastic nodules do not indicate high-grade status. *J Gastroenterol Hepatol* 2003; **18**: 660–665.

11 Libbrecht L, Bielen D, Verslype C, et al. Focal lesions in cirrhotic explant livers: pathological evaluation and accuracy of pretransplantation imaging examination. *Liver Transpl* 2002; **8**: 749–761.

12 Nakashima O, Kurogi M, Yamaguchi R, et al. Unique hypervascular nodules in alcoholic liver cirrhosis: identical to focal nodular hyperplasia-like nodules? *J Hepatol* 2004; **41**: 992–998.

13 Lee YH, Kim SH, Cho MY, Shim KY, Kim MS. Focal nodular hyperplasia-like nodules in alcoholic liver cirrhosis: radiologic-pathologic correlation. *Am J Roentgenol* 2007; **188**: W459–W463.

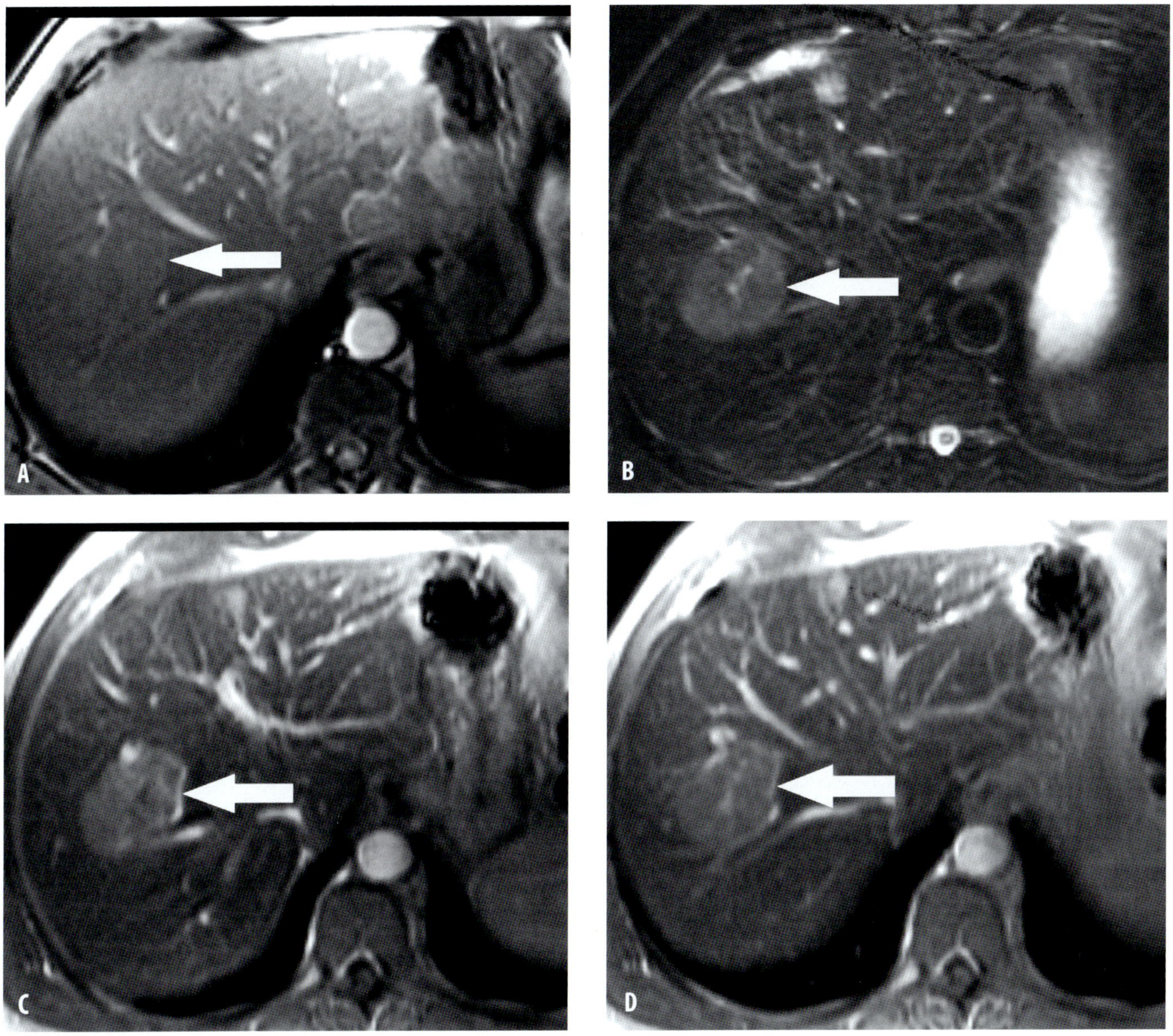

Figure 20.1 **A.** Axial spoiled gradient-echo T1-weighted MR image of the liver in a 56 year old man with hepatitis C and an elevated alpha-fetoprotein of 146 IU/mL shows a subtle mass (arrow) of isointense signal. **B.** Axial fat-suppressed fast spin-echo T2-weighted MR image of the liver shows the mass (arrow) is of mildly increased signal and has a bright central scar. **C.** Axial spoiled gradient-echo T1-weighted MR image of the liver in the arterial phase of enhancement after dynamic intravenous gadolinium administration demonstrates the mass (arrow) to be hypervascular. **D.** Axial spoiled gradient-echo T1-weighted MR image of the liver in the delayed phase of enhancement after dynamic intravenous gadolinium administration demonstrates delayed enhancement of the central scar within the mass (arrow). A diagnosis of hepatocellular carcinoma was demonstrated at surgery.

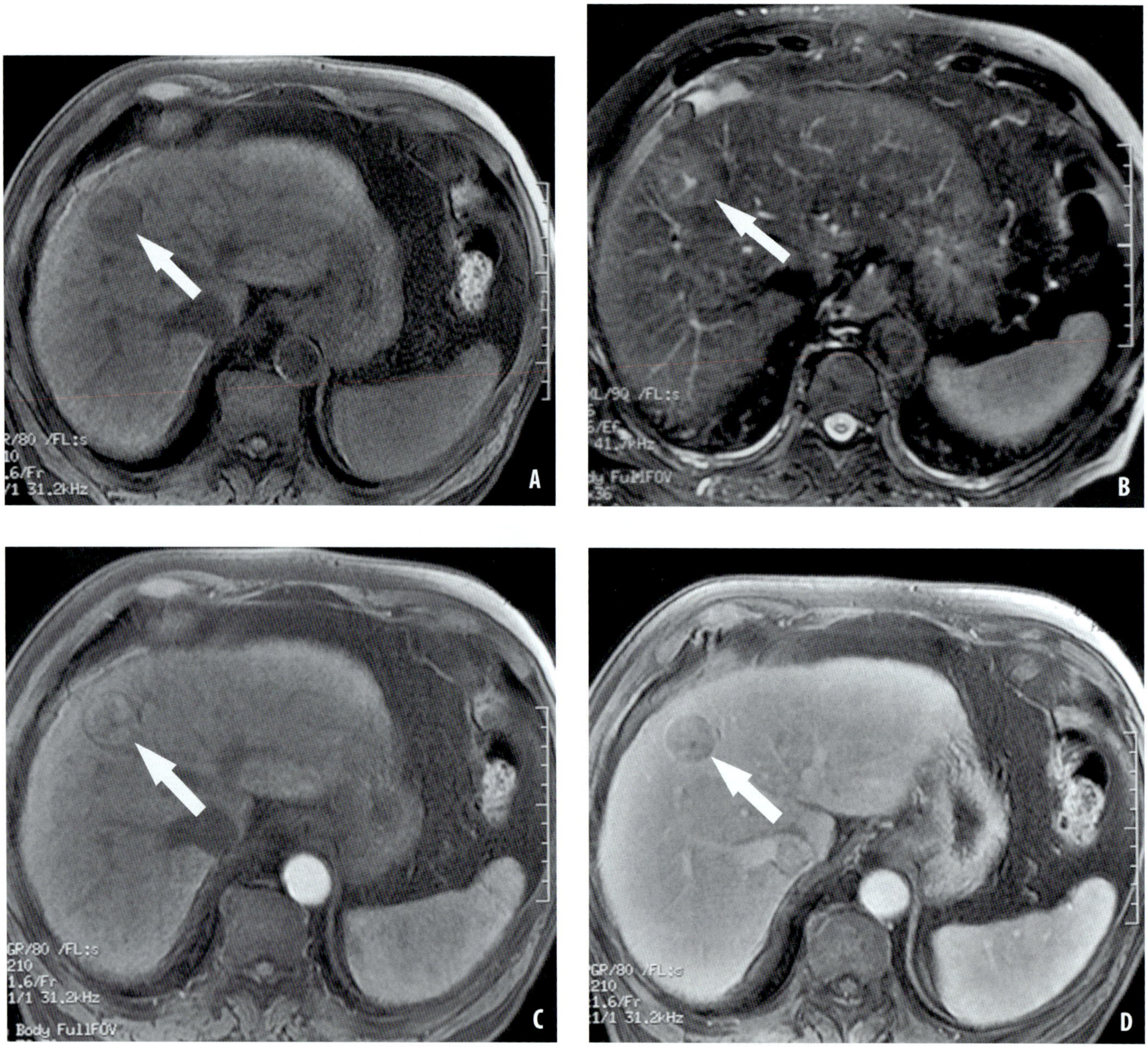

Figure 20.2 A. Axial fat-suppressed spoiled gradient-echo T1-weighted MR image of the liver performed for surveillance in a 70 year old man with long-standing hepatitis B and a normal alpha-fetoprotein shows a mass (arrow) of mildly hypointense signal intensity. **B.** Axial fat-suppressed fast spin-echo T2-weighted MR image of the liver shows the mass (arrow) is of mildly increased signal intensity with a bright central scar. **C.** Axial fat-suppressed spoiled gradient-echo T1-weighted MR image of the liver in the arterial phase of enhancement after dynamic intravenous gadolinium administration demonstrates the mass (arrow) to be subtly hypervascular. **D.** Axial fat-suppressed spoiled gradient-echo T1-weighted MR image of the liver in the portal venous phase of enhancement after dynamic intravenous gadolinium administration demonstrates the mass (arrow) to be predominantly hypovascular. **E.** Axial fat-suppressed spoiled gradient-echo T1-weighted MR image of the liver in the delayed phase of enhancement after dynamic intravenous gadolinium administration demonstrates contrast washout in most of the mass (arrow), with delayed enhancement of the central scar. **F.** Axial image from a whole body FDG PET shows the mass (arrow) to be hypermetabolic (maximum SUV of 3.3). A diagnosis of hepatocellular carcinoma was confirmed at surgery.

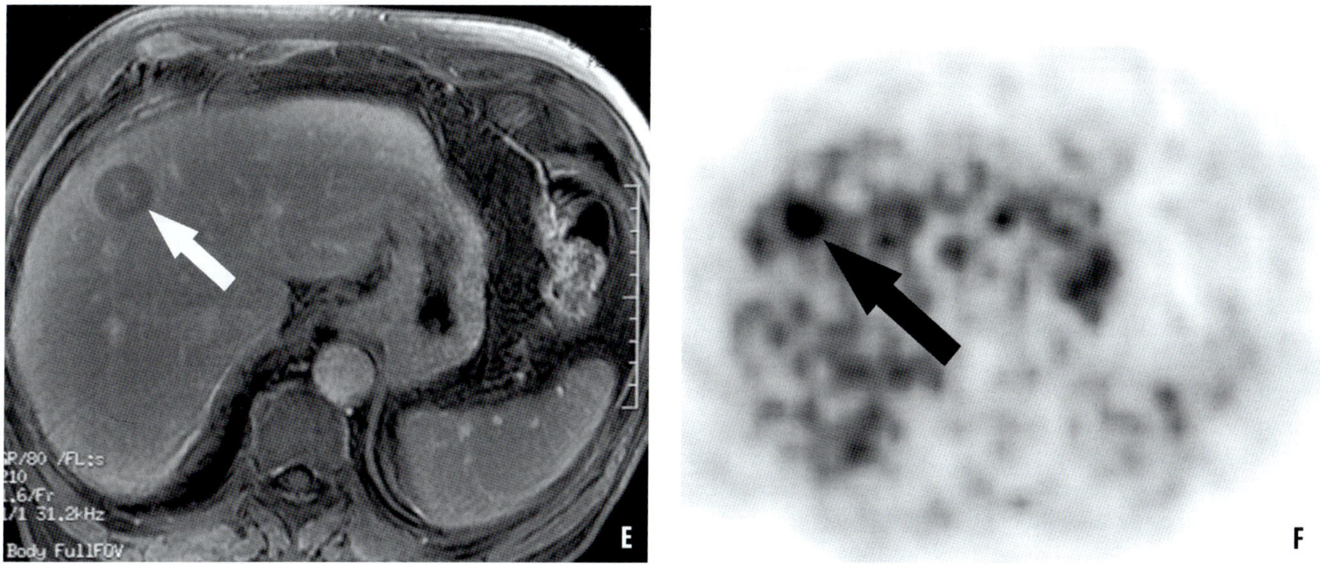

Figure 20.2 (cont.)

CASE 21

Paradoxical signal gain in the liver

Imaging description

In and out of phase T1-weighted gradient-echo MRI of the liver is mainly used to evaluate diffusc fatty infiltration. Normal liver has the same signal intensity on in and out of phase images (Figure 21.1). Fatty liver shows a loss of signal on out of phase versus in phase images (Figure 21.2), due to chemical-shift-related signal cancellation between fat and water protons. Occasionally, the liver is brighter on out of phase versus in phase images (Figure 21.3). This is known as paradoxical signal gain and is due to hepatic iron overload causing T2* related signal loss on in phase imaging [1], since in phase images are typically acquired with a longer echo time (in a 1.5T scanner, out of phase images are usually acquired at a TE of 2.1 milliseconds with in phase images acquired at a TE of 4.2 milliseconds). With a longer echo time, iron-induced signal loss becomes more pronounced (such T2* effects are negligible in livers unaffected by iron overload).

Importance

Paradoxical signal gain on out of phase MRI can be seen with liver iron concentrations of 80 μmol iron per gram and above [2]. This threshold is about twice the upper limit of normal liver iron concentration (36 μmol per gram) and is considered clinically significant [3], so that the detection of this phenomenon merits recognition and reporting. Liver iron overload also confounds the detection of diffuse fatty infiltration on out of phase MRI [4], because iron and fat have the opposite effect on signal dropout.

Typical clinical scenario

Hepatic iron overload (hemosiderosis) may be primary (due to hemochromatosis or other rare genetic or congenital disorders such as neonatal hemochromatosis) or secondary (due to chronic liver disease or repeated transfusions for hematological disorders) [5, 6]. In practice, hemosiderosis is probably most commonly seen in patients who have had repeated transfusions for sickle cell disease, thalassemia, or bone marrow transplantation [7].

Differential diagnosis

The pattern of organ involvement on out of phase MRI can help distinguish the type of iron overload. Primary hemochromatosis generally affects the liver and pancreas and spares the spleen and bone marrow. In contrast, secondary hemosiderosis affects the liver, bone marrow, and spleen but not the pancreas [1]. In brief, signal loss (due to iron overload) in the pancreas but not the spleen indicates primary hemochromatosis (Figure 21.4), while the opposite pattern suggests secondary hemochromatosis (Figure 21.5). I remember this as Primary affects Pancreas and Secondary affects Spleen. It has been postulated that reduced signal intensity in the liver and spleen after chemotherapy might also reflect a decrease in lymphoid tissue [8], but such a mechanism would presumably affect both organs similarly on out of and in phase images, and not cause the reversal of relative liver to spleen signal intensity seen with iron overload.

Teaching point

Paradoxical signal gain on out of phase versus in phase gradient-echo MRI indicates hepatic iron overload, and in practice is often due to transfusional hemosiderosis.

REFERENCES

1 Taouli B, Ehman RL, Reeder SB. Advanced MRI methods for assessment of chronic liver disease. *Am J Roentgenol* 2009; **193**: 14–27.

2 Alustiza JM, Castiella A. Liver fat and iron at in-phase and opposed-phase MR imaging. *Radiology* 2008; **246**: 641.

3 Gandon Y, Olivié D, Guyader D, et al. Non-invasive assessment of hepatic iron stores by MRI. *Lancet* 2004; **363**: 357–362.

4 Westphalen AC, Qayyum A, Yeh BM, et al. Liver fat: effect of hepatic iron deposition on evaluation with opposed-phase MR imaging. *Radiology* 2007; **242**: 450–455.

5 Deugnier Y, Brissot P, Loréal O. Iron and the liver: update 2008. *J Hepatol* 2008; **48**(Suppl 1): S113–S123.

6 Coakley FV, Hricak H, Filly RA, Barkovich AJ, Harrison MR. Complex fetal disorders: effect of MR imaging on management–preliminary clinical experience. *Radiology* 1999; **213**: 691–696.

7 Strasser SI, Kowdley KV, Sale GE, McDonald GB. Iron overload in bone marrow transplant recipients. *Bone Marrow Transplant* 1998; **22**: 167–173.

8 Donnelly LF, Bisset GS 3rd. Dark spleens and livers on MRI after chemotherapy: is it really iron overload? *Pediatr Radiol* 1998; **28**: 486.

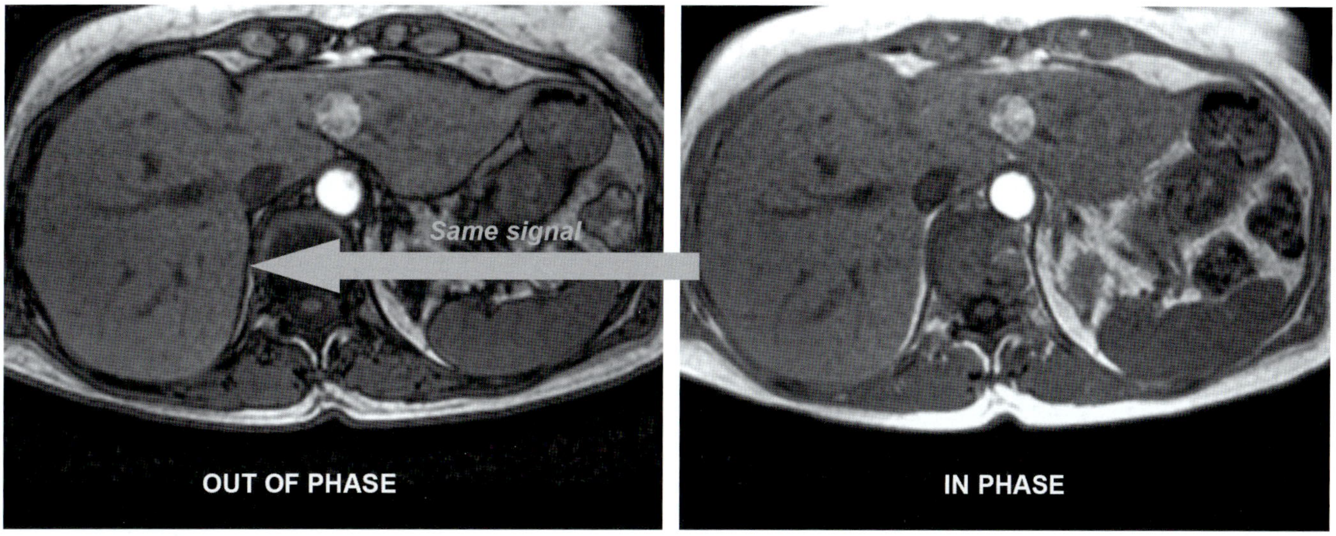

Figure 21.1 Photomontage of spoiled gradient-echo out of and in phase axial T1-weighted MR images in a 55 year old woman with breast cancer and a normal liver shows the hepatic signal intensity does not change on out of versus in phase images.

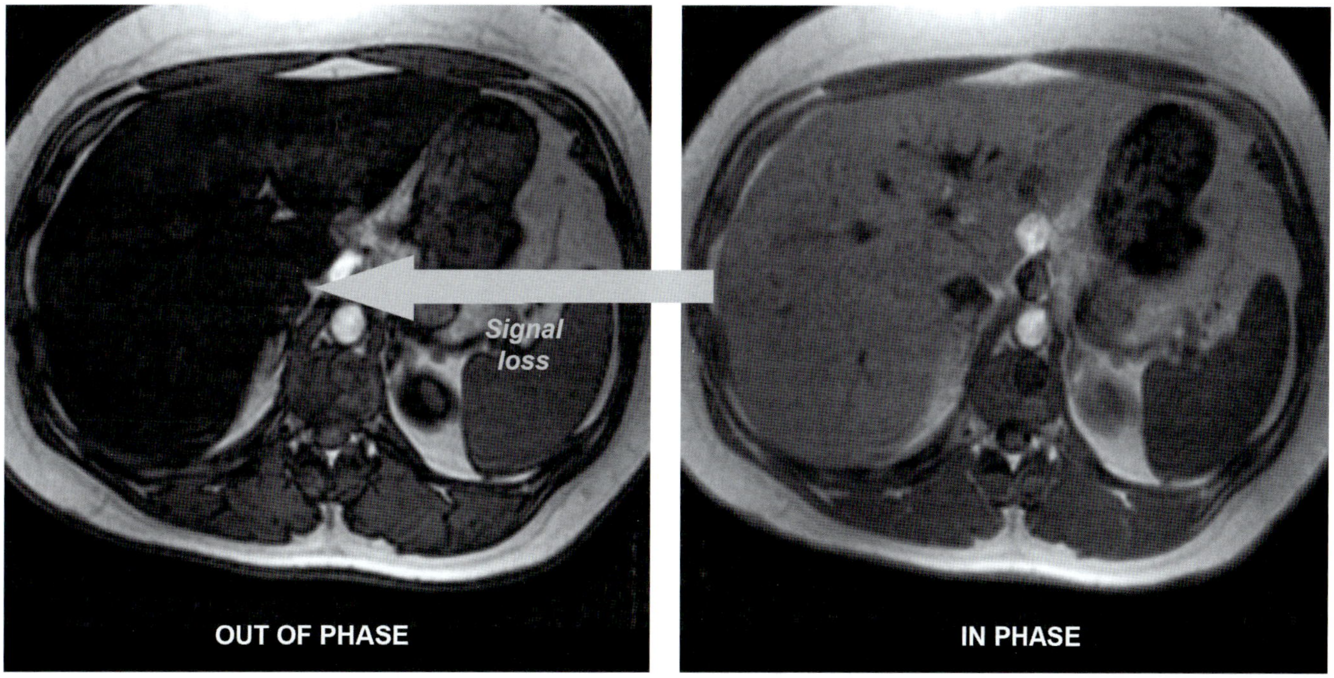

Figure 21.2 Photomontage of spoiled gradient-echo out of and in phase axial T1-weighted MR images in a 31 year old woman with suspected renal artery stenosis and diffuse fatty infiltration of the liver shows the hepatic signal intensity drops on out of versus in phase images.

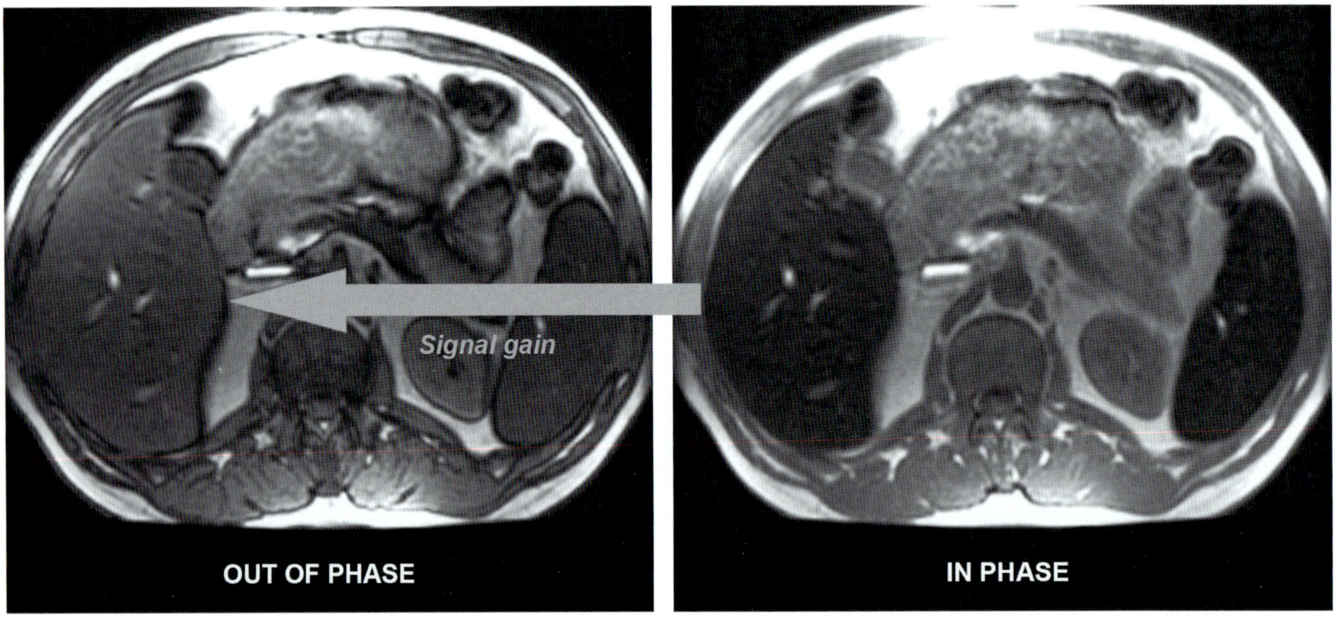

Figure 21.3 Photomontage of spoiled gradient-echo out of and in phase axial T1-weighted MR images in a 51 year old man with non-Hodgkin's lymphoma treated by autologous stem cell bone marrow transplantation three months earlier shows paradoxical hepatic signal intensity gain on out of versus in phase images, consistent with hemosiderosis. The patient had received multiple blood transfusions during his treatment by chemotherapy and bone marrow transplantation.

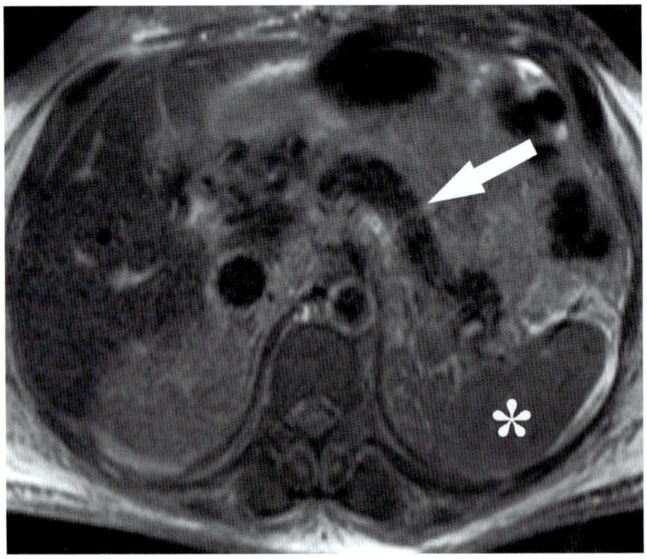

Figure 21.4 Axial T2-weighted MR image in a 46 year old man three months after liver transplantation for primary hemochromatosis shows diffusely reduced signal intensity in the pancreas (arrow) with normal signal intensity in the spleen (asterisk).

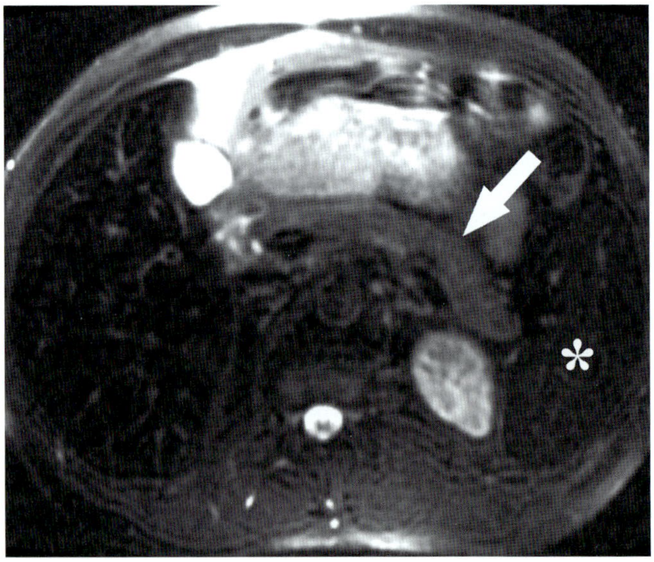

Figure 21.5 Axial T2-weighted MR image in the patient shown in Figure 21.3 with transfusional hemosiderosis. Note that in contrast to primary hemochromatosis, the pancreas (arrow) appears of normal signal intensity while the spleen (asterisk) is of reduced signal intensity.

CASE 22 Peribiliary cysts

Imaging description

Peribiliary cysts are small (2–20 mm) thin-walled non-communicating retention cysts of the serous glands adjacent to the intrahepatic bile ducts [1, 2]. If numerous and contiguous, these periductal cysts result in an appearance of periportal fluid-filled tubes that may mimic biliary dilatation (including Caroli's disease) or periportal edema (Figures 22.1 and 22.2) [3–6].

Importance

Peribilary cysts are usually inconsequential, and only matter insofar as they mimic more serious pathology or may be markers of associated underlying disease. Rarely, peribiliary cysts may cause biliary obstruction [7, 8].

Typical clinical scenario

Peribiliary cysts usually occur in patients over 50 and are reportedly commoner in men [3]. They may be idiopathic or secondary to adult polycystic disease (Figures 22.3 and 22.4), cirrhosis, idiopathic portal hypertension, extrahepatic biliary obstruction, systemic infection, or liver metastases [9].

Differential diagnosis

While peribiliary cysts can closely mimic biliary dilatation, close inspection may show that the periportal findings are on both sides of the portal vein branches (Figures 22.1 and 22.4), in contrast to dilated bile ducts which are only visible on one side of the portal vein branches. CT cholangiography can elegantly confirm the diagnosis by opacifying the bile ducts and documenting that they are distinct to the peribiliary cysts (Figures 22.1 and 22.2) [3]. Periportal edema, which appears as circumferential thin halos of low attenuation around the portal vein branches, likely represents dilatation of the lymphatics in the loose areolar tissues of the portal triads and can occur with trauma, hepatitis, portal lymphadenopathy, tumors in the porta hepatis, cardiac failure, liver transplantation, and bone marrow transplantation [10, 11]. Periductal spread of pancreatic carcinoma, cholangiocarcinoma, post-transplantation lymphoproliferative disorder, lymphoma, Kaposi sarcoma, extramedullary hematopoiesis, schistosomiasis, and hepatic necrosis due to hepatic artery occlusion after liver transplantation have been reported as rare causes of linear periportal low attenuation [12–19].

Teaching point

Peribiliary cysts should be considered as an alternative diagnosis for apparent biliary dilatation at cross-sectional imaging, particularly when seen in a patient without jaundice or other features of biliary obstruction. CT cholangiography can be a helpful confirmatory test in difficult cases by opacifying the biliary tract separate to the non-communicating peribiliary cysts.

REFERENCES

1 Nakanuma Y, Kurumaya H, Ohta C. Multiple cysts in the hepatic hilum and their pathogenesis: a suggestion of periductal gland origin. *Virchows Arch A Pathol Anat Histopathol* 1984; **404**: 341–350.

2 Terada T, Minato H, Nakanuma Y, et al. Ultrasound visualization of hepatic peribiliary cysts: a comparison with morphology. *Am J Gastroenterol* 1992; **87**: 1499–1502.

3 Itai Y, Ebihara R, Tohno E, et al. Hepatic peribiliary cysts: multiple tiny cysts within the larger portal tract, hepatic hilum, or both. *Radiology* 1994; **191**: 107–110.

4 Baron RL, Campbell WL, Dodd GD 3rd. Peribiliary cysts associated with severe liver disease: imaging-pathologic correlation. *Am J Roentgenol* 1994; **162**: 631–636.

5 Terayama N, Matsui O, Hoshiba K, et al. Peribiliary cysts in liver cirrhosis: US, CT, and MR findings. *J Comput Assist Tomogr* 1995; **19**: 419–423.

6 Fusai G, Tucker O, Nik Sulaiman NM, et al. Peribiliary cysts can mimic Caroli's disease: a case report. *Int J Surg Pathol* 2005; **13**: 379–382.

7 Johnson MA, Ravichandran P, Surendran R. Solitary extra-hepatic hilar peribiliary cyst presenting with obstructive jaundice: a case report. *Acta Chir Belg* 2007; **107**: 716–719.

8 Kai K, Eguchi Y, Kumagai T, Sugita Y, Tokunaga O. An autopsy case of obstructive jaundice due to hepatic multiple peribiliary cysts accompanying hepatolithiasis. *Hepatol Res* 2008; **38**: 211–216.

9 Terada T, Nakanuma Y. Pathological observations of intrahepatic peribiliary glands in 1,000 consecutive autopsy livers. III. Survey of necroinflammation and cystic dilatation. *Hepatology* 1990; **12**: 1229–1233.

10 Morehouse H, Leibman AJ, Biempica L, Hoffman J. Infiltrating periductal neoplasm mimicking biliary dilatation on computed tomography. *J Comput Assist Tomogr* 1983; **7**: 721–723.

11 Lawson TL, Thorsen MK, Erickson SJ, et al. Periportal halo: a CT sign of liver disease. *Abdom Imag* 1993; **18**: 42–46.

12 Donnelly LF. CT imaging of immunocompromised children with acute abdominal symptoms. *Am J Roentgenol* 1996; **167**: 909–913.

13 Itai Y, Araki T, Furui S, et al. Computed tomography of primary intrahepatic biliary malignancy. *Radiology* 1983; **147**: 485–490.

14 Strouse PJ, Platt JF, Francis IR, Bree RL. Tumorous intrahepatic lymphoproliferative disorder in transplanted livers. *Am J Roentgenol* 1996; **167**: 1159–1162.

15 Luburich P, Bru C, Ayuso MC, Azon A, Condom E. Hepatic Kaposi sarcoma in AIDS: US and CT findings. *Radiology* 1990; **175**: 172–174.

16 Kobayashi A, Sugihara M, Kurosaki M, et al. CT characteristics of intrahepatic periportal extramedullary hematopoiesis. *J Comput Assist Tomogr* 1989; **13**: 354–356.

17 Fataar S, Bassiony H, Satyanath S, et al. CT of hepatic schistosomiasis mansoni. *Am J Roentgenol* 1985; **145**: 63–66.

18 Meyer JI, Khalil R, Remer EM. Periportal low attenuation due to hepatic necrosis in a patient after liver transplant. *Am J Roentgenol* 1996; **167**: 98–99.

19 Coakley FV, O'Reilly EM, Schwartz LH, Panicek DM, Castellino RA. Non-Hodgkin's lymphoma as a cause of intrahepatic periportal low attenuation on computed tomography. *J Comput Assist Tomogr* 1997; **21**: 726–728.

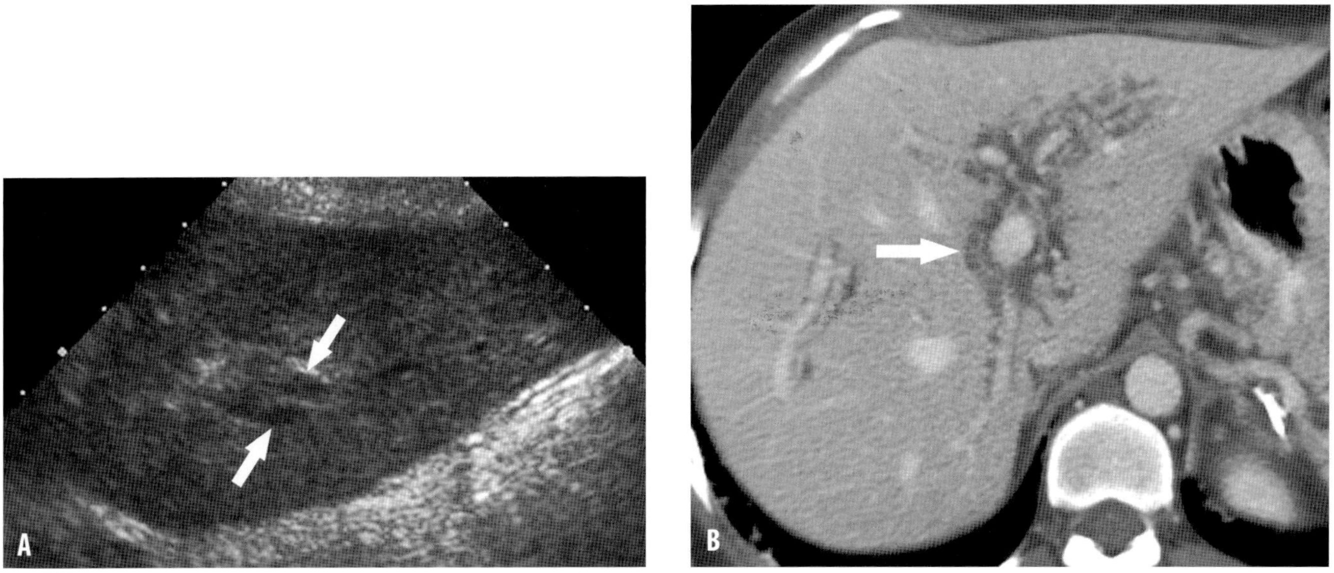

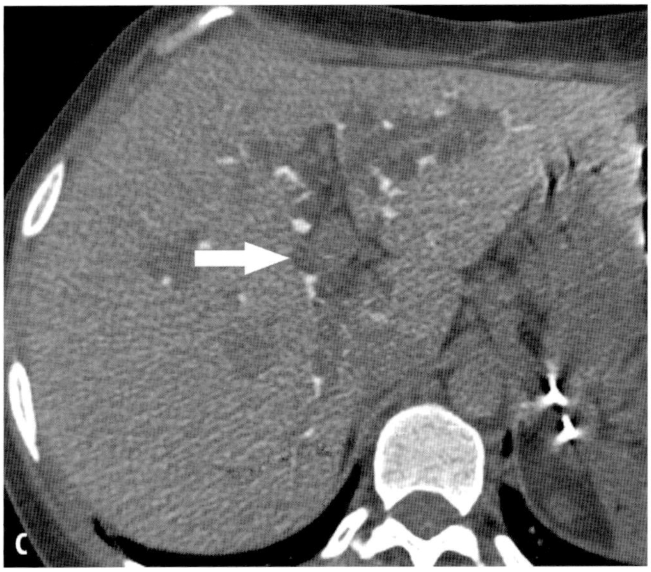

Figure 22.1 **A.** Ultrasound image in a 34 year old woman with right upper quadrant pain and a history of primary pulmonary hypertension shows parallel tubular structures (arrows), suggestive of intrahepatic biliary dilatation. **B.** Axial contrast-enhanced CT image of the liver shows periportal low-density tubular structures (arrow) which are somewhat suggestive of intrahepatic biliary dilatation. However, note that the structures are visible on both sides of the portal veins, raising the possibility of peribiliary cysts. **C.** Axial CT cholangiogram image shows that the low-density structures (arrow) are distinct from the opacified and normal caliber intrahepatic bile ducts, confirming the diagnosis of peribiliary cysts.

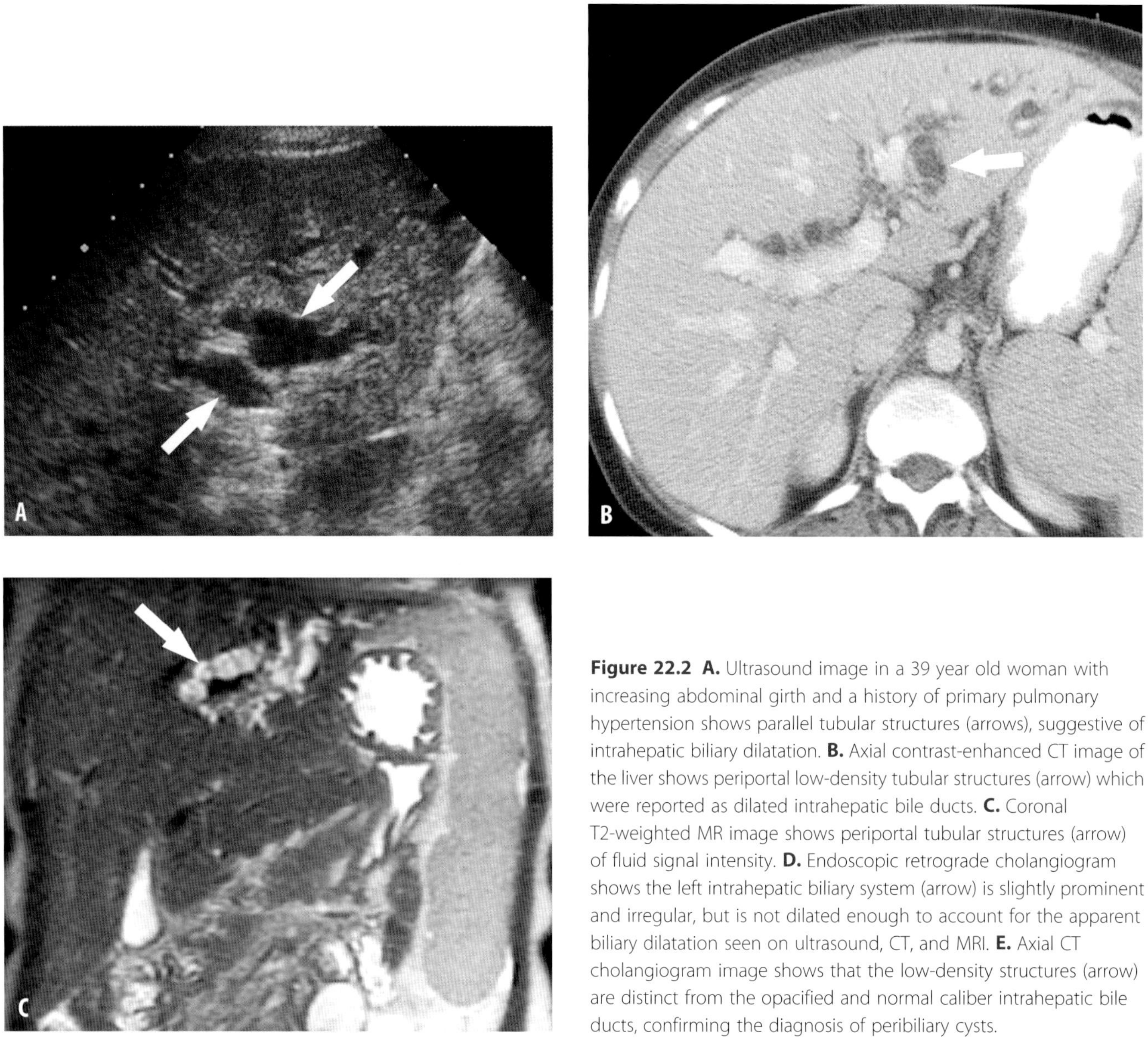

Figure 22.2 **A.** Ultrasound image in a 39 year old woman with increasing abdominal girth and a history of primary pulmonary hypertension shows parallel tubular structures (arrows), suggestive of intrahepatic biliary dilatation. **B.** Axial contrast-enhanced CT image of the liver shows periportal low-density tubular structures (arrow) which were reported as dilated intrahepatic bile ducts. **C.** Coronal T2-weighted MR image shows periportal tubular structures (arrow) of fluid signal intensity. **D.** Endoscopic retrograde cholangiogram shows the left intrahepatic biliary system (arrow) is slightly prominent and irregular, but is not dilated enough to account for the apparent biliary dilatation seen on ultrasound, CT, and MRI. **E.** Axial CT cholangiogram image shows that the low-density structures (arrow) are distinct from the opacified and normal caliber intrahepatic bile ducts, confirming the diagnosis of peribiliary cysts.

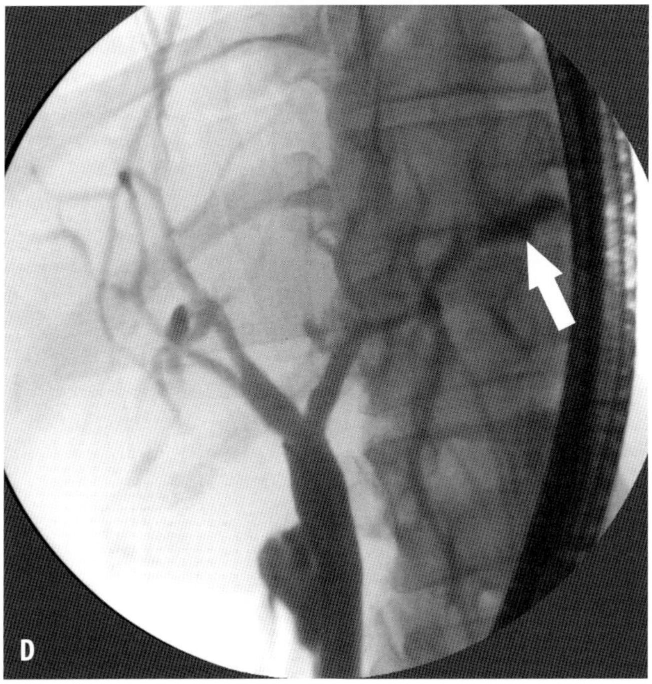

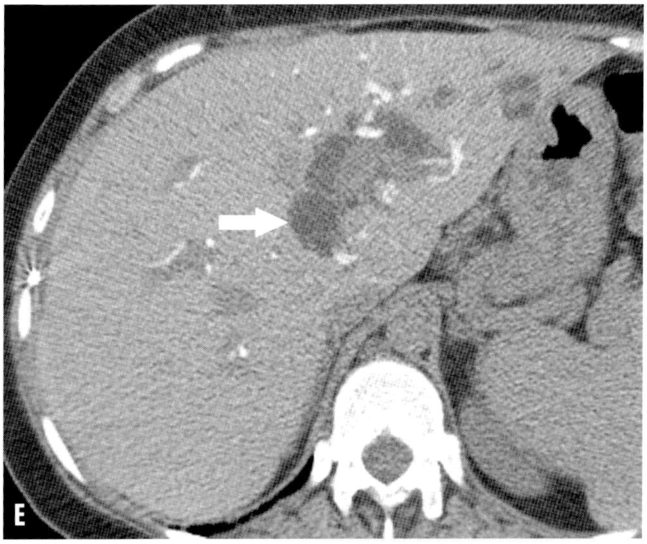

Figure 22.2 (cont.)

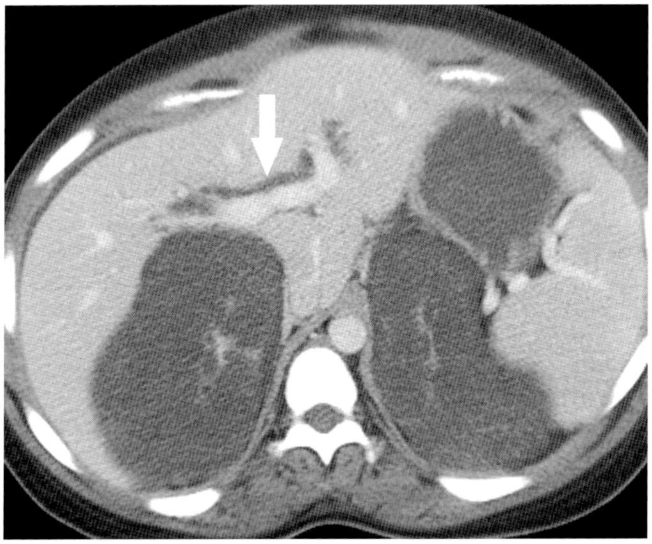

Figure 22.3 Axial contrast-enhanced CT image in a 27 year old woman with generalized abdominal pain and a history of autosomal dominant polycystic kidney disease (note the kidneys are enlarged and mostly replaced by multiple cysts) shows apparent biliary dilatation (arrow) due to peribiliary cysts.

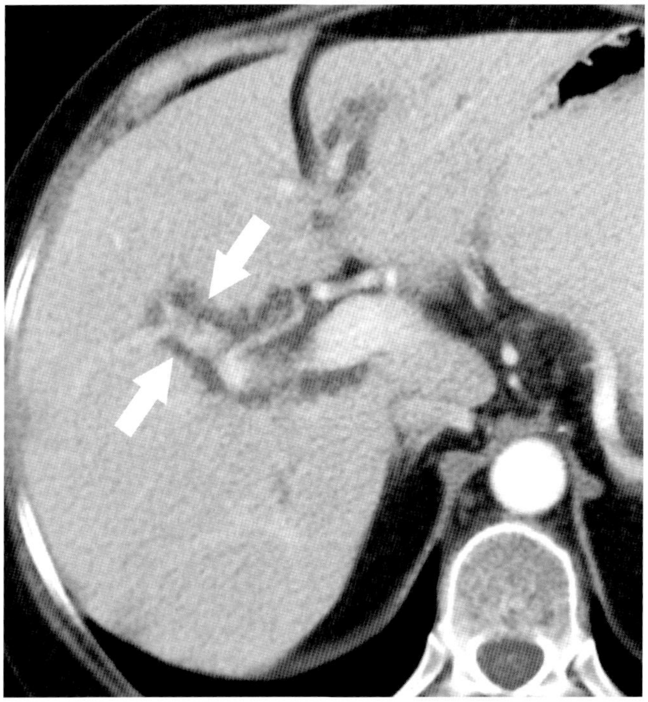

Figure 22.4 Axial contrast-enhanced CT image in a 52 year old man with autosomal dominant polycystic kidney disease. Peribiliary cysts are present and partially mimic biliary dilatation, but can be distinguished because they are arranged on both sides (arrows) of the portal veins.

CASE 23 Pseudo-Klatskin tumor due to malignant masquerade

Imaging description

The term malignant masquerade [1] refers to benign idiopathic fibroinflammatory stricturing of the common hepatic duct confluence that is clinically and radiologically (Figure 23.1) indistinguishable from hilar cholangiocarcinoma (Klatskin tumor).

Importance

While, virtually by definition, malignant masquerade cannot be diagnosed by imaging alone, recognition of this pseudotumor emphasizes the importance of offering surgery to all patients with what appears to be a resectable hilar cholangiocarcinoma. Palliative treatment of these patients, particularly with metallic endobiliary stenting, is rarely appropriate without a histological diagnosis because the assumption of malignancy may be incorrect and because such treatment may complicate or preclude subsequent surgery.

Typical clinical scenario

Studies have consistently shown that 5 to 10% of patients with a presumptive preoperative diagnosis of hilar cholangiocarcinoma are ultimately found to have an idiopathic benign stricture at final histopathological review [2].

Differential diagnosis

Benign idiopathic fibroinflammatory strictures of the common hepatic duct confluence appear to represent several different histopathological entities, including lymphoplasmacytic sclerosing pancreatitis and cholangitis, primary sclerosing cholangitis, granulomatous disease, non-specific fibrosis and inflammation, and occult stone disease [2]. Lymphoplasmacytic sclerosing pancreatitis and cholangitis is a distinct and relatively recently recognized autoimmune disorder [3] which may primarily manifest in the pancreas (where it is also known as lymphoplasmacytic sclerosing pancreatitis or autoimmune pancreatitis) but can also primarily affect the biliary system [4]. With respect to occult stones, in one series, 6 of 22 patients with malignant masquerade were found to have stones that were undetected at preoperative imaging [2]. It is unclear whether such occult stones were the cause or effect of the associated stricture, but in either case the preoperative clinical and radiological impression was of malignancy. Some patients with benign idiopathic fibroinflammatory stricture of the common hepatic duct confluence have associated autoimmune, connective tissue, or fibrotic disorders, and it may be that in such cases the condition represents a localized proliferative condition in the porta hepatis, analogous to primary retroperitoneal fibrosis in the retroperitoneum [5]. Other conditions that only partially or rarely mimic Klatskin tumor include recurrent pyogenic cholangitis, acquired immunodeficiency syndrome cholangiopathy, inflammatory pseudotumor, Mirizzi syndrome, xanthogranulomatous cholangitis, sarcoidosis, hepatic arterial chemotherapy-induced sclerosis (Figure 23.2), hepatocellular carcinoma, metastases, lymphoma, and leukemia [6].

Teaching point

The diagnosis of biliary hilar cholangiocarcinoma (Klatskin tumor) cannot be made on radiological features alone, because benign idiopathic strictures with a variety of histopathological etiologies may produce identical clinical and imaging findings.

REFERENCES

1 Hadjis NS, Collier NA, Blumgart LH. Malignant masquerade at the hilum of the liver. *Br J Surg* 1985; **72**: 659–661.

2 Corvera CU, Blumgart LH, Darvishian F, et al. Clinical and pathologic features of proximal biliary strictures masquerading as hilar cholangiocarcinoma. *J Am Coll Surg* 2005; **201**: 862–869.

3 Hamano H, Kawa S, Horiuchi A, et al. High serum IgG4 concentrations in patients with sclerosing pancreatitis. *N Engl J Med* 2001; **344**: 732–738.

4 Abraham SC, Cruz-Correa M, Argani P, et al. Lymphoplasmacytic chronic cholecystitis and biliary tract disease in patients with lymphoplasmacytic sclerosing pancreatitis. *Am J Surg Pathol* 2003; **27**: 441–451.

5 Binkley CE, Eckhauser FE, Colletti LM. Unusual causes of benign biliary strictures with cholangiographic features of cholangiocarcinoma. *J Gastrointest Surg* 2002; **6**: 676–681.

6 Menias CO, Surabhi VR, Prasad SR, et al. Mimics of cholangiocarcinoma: spectrum of disease. *Radiographics* 2008; **28**: 1115–1129.

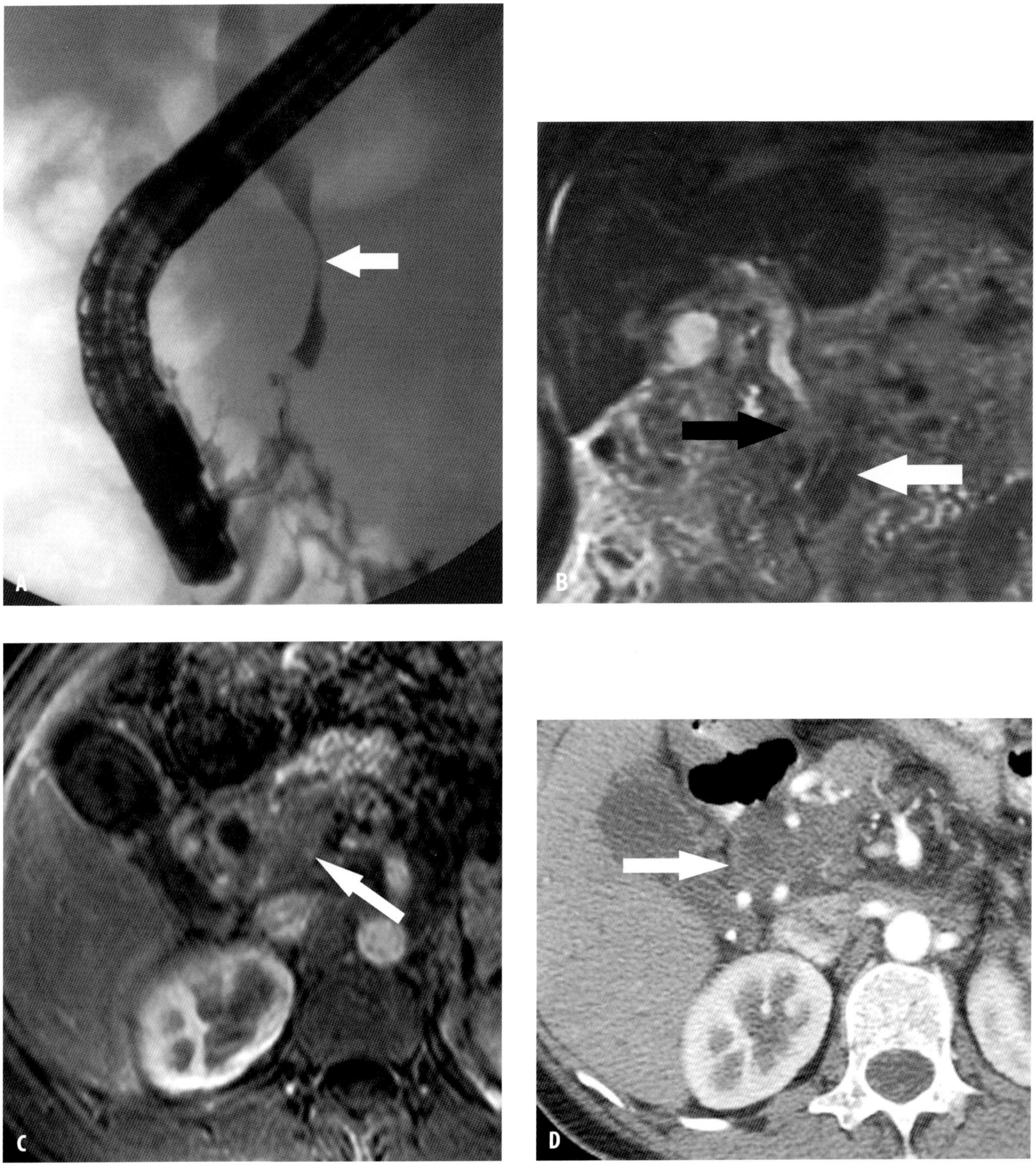

Figure 23.1 **A.** ERCP image from a 54 year old Vietnamese woman with a long history of abdominal pain, showing a tight stricture (arrow) in the common duct. Endoscopy revealed esophageal varices, and subsequent liver biopsy showed only mild periportal fibrosis. **B.** Coronal T2-weighted MR image shows an ill-defined and relatively hyperintense mass (black arrow) superior to the pancreatic head (white arrow). **C.** Axial gradient-echo fat-saturated T1-weighted MR image obtained in the late arterial phase of enhancement after intravenous gadolinium administration shows a hypovascular mass (arrow) adjacent to the pancreatic head. **D.** Axial contrast-enhanced CT image obtained in the late arterial phase of enhancement confirms the presence of a hypovascular mass (arrow) extending towards the porta hepatis.

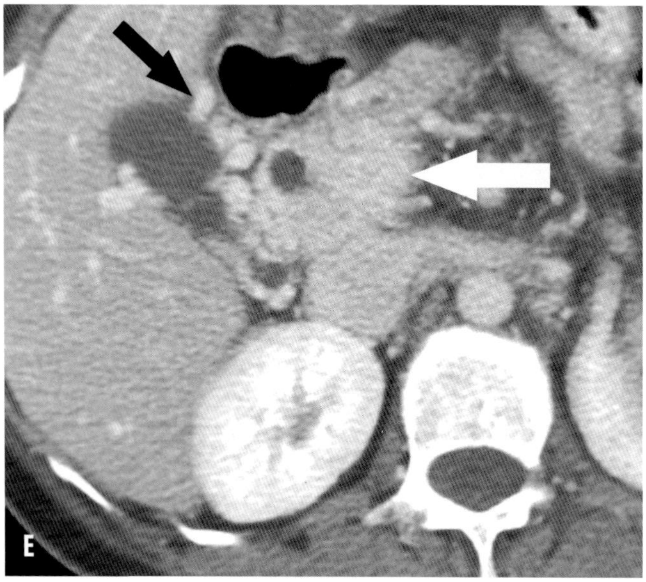

Figure 23.1 (cont.)
E. Axial contrast-enhanced CT image obtained in the delayed phase of enhancement shows delayed enhancement in the mass (arrow). Varices (black arrow) around the gallbladder were due to associated cavernous transformation of the portal vein. A CT-guided fine needle aspiration biopsy revealed only fibrous tissue. Surgical exploration confirmed the mass was benign, although resection was not possible because of extensive varices.

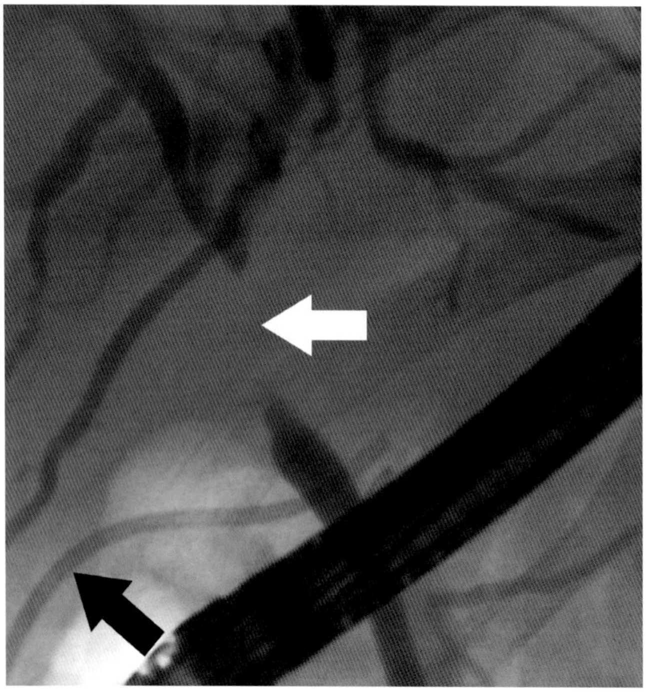

Figure 23.2 ERCP image in a 65 year old man with jaundice and a history of hepatic arterial chemotherapy infusion for colorectal metastases to the liver. A short segment stricture (white arrow) is seen at the common hepatic duct confluence due to hepatic arterial chemotherapy-induced sclerosis. Note the presence of a hepatic arterial catheter (black arrow).

CASE 24 Adenomyomatosis of the gallbladder

Imaging description

Adenomyomatosis (or diverticular disease of the gallbladder) is an acquired hyperplastic condition characterized by excessive proliferation of surface epithelium with deepened invaginations or diverticula (so-called Rokitansky-Aschoff sinuses) extending into the thickened muscular layer of the gallbladder wall [1]. Adenomyomatosis of the gallbladder results in focal or diffuse wall thickening that contains small cyst-like spaces at cross-sectional imaging (Figure 24.1). These cyst-like spaces may give rise to the "pearl necklace" sign at T2-weighted MRI (Figure 24.2) [2]. The condition has a predilection for the gallbladder fundus (Figure 24.3). The central gallbladder may also be affected, resulting in a relatively typical "hourglass" configuration (Figure 24.4).

Importance

Gallbladder wall thickening may suggest the diagnosis of gallbladder cancer, resulting in unnecessary workup or even surgery. The "compartmentalization" of the gallbladder in the hourglass type of adenomyomatosis may result in failure to identify the distal compartment at ultrasound (Figure 24.5) or contribute to incomplete cholecystectomy when only the distal half of the gallbladder is removed at surgery (Figure 24.6).

Typical clinical scenario

Depending on the series, adenomyomatosis is seen in 1 to 8.7% of cholecystectomy specimens [3–5]; this wide range may reflect population differences or variability in the rigorousness of pathological diagnostic criteria and assessment. Adenomyomatosis is usually associated with gallstones [4], in which case treatment of the gallstones determines clinical management. The appropriate treatment strategy for acalculous adenomyomatosis is not well established.

Differential diagnosis

Focal or diffuse gallbladder wall thickening is a relatively frequent observation at cross-sectional imaging and is often incidental. When seen incidentally, the major diagnostic considerations are adenomyomatosis and gallbladder cancer (cholecystitis being unlikely in an asymptomatic patient). The detection of intramural diverticula appears reliable for making the diagnosis of adenomyomatosis [6], although adenomyomatosis can occasionally appear solid and resemble malignancy (Figures 24.7 and 24.8). Other causes of focal and diffuse gallbladder wall thickening include polyps (of varying etiology), cholecystitis, hepatitis, cirrhosis, and ascites [7, 8].

Teaching point

Adenomyomatosis is the likely cause of focal or diffuse gallbladder wall thickening if small cyst-like spaces are seen within the affected portion of the gallbladder.

REFERENCES

1 Bilhartz LE. Acalculous cholecystitis, cholesterolosis, adenomyomatosis, and polyps of the gallbladder. In: Feldman M, Friedman LS, Sleisenger MH, eds. *Sleisenger & Fordtran's gastrointestinal and liver disease: pathology/diagnosis/ management*, 7th edition. Philadelphia, PA: Saunders, 2002; 1116–1130.

2 Haradome H, Ichikawa T, Sou H, et al. The pearl necklace sign: an imaging sign of adenomyomatosis of the gallbladder at MR cholangiopancreatography. *Radiology* 2003; **227**: 80–88.

3 Shepard VD, Walters W, Dockerty MB. Benign neoplasms of the gallbladder. *Arch Surg* 1942; **45**: 1–18.

4 Williams I, Slavin G, Cox A, Simpson P, de Lacey G. Diverticular disease (adenomyomatosis) of the gallbladder: a radiological-pathological survey. *Br J Radiol* 1986; **59**: 29–34.

5 Ootani T, Shirai Y, Tsukada K, Muto T. Relationship between gallbladder carcinoma and the segmental type of adenomyomatosis of the gallbladder. *Cancer* 1992; **69**: 2647–2652.

6 Ching BH, Yeh BM, Westphalen AC, et al. CT differentiation of adenomyomatosis and gallbladder cancer. *Am J Roentgenol* 2007; **189**: 62–66.

7 Huang CS, Lien HH, Jeng JY, Huang SH. Role of laparoscopic cholecystectomy in the management of polypoid lesions of the gallbladder. *Surg Laparosc Endosc Percutan Tech* 2001; **11**: 242–247.

8 Grand D, Horton KM, Fishman EK. CT of the gallbladder: spectrum of disease. *Am J Roentgenol* 2004; **183**: 163–170.

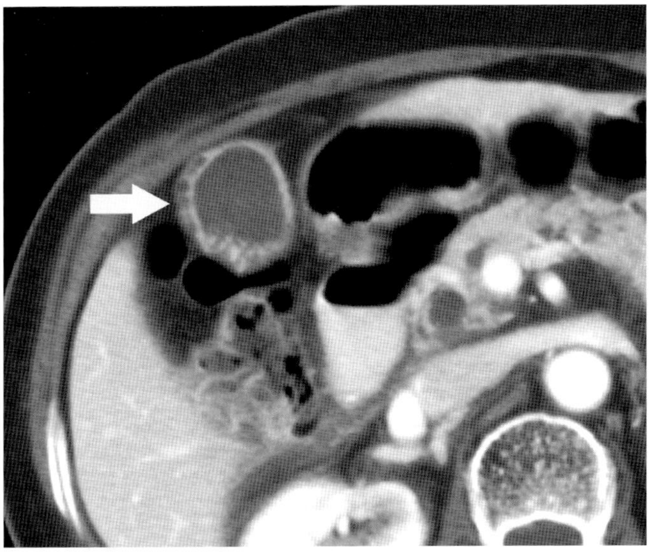

Figure 24.1 Axial contrast-enhanced CT image in a 65 year old woman with breast cancer. Focal thickening of the gallbladder wall (arrow) is seen to consist of small cyst-like spaces, indicative of adenomyomatosis.

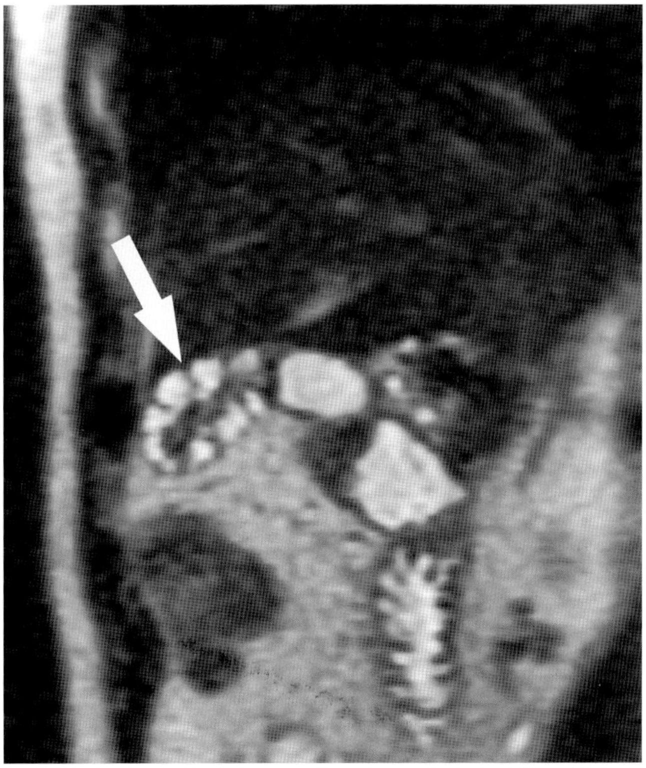

Figure 24.2 Sagittal T2-weighted MR image in a 72 year old man with vague upper abdominal pain. Adenomyomatosis of the distal portion of the gallbladder (arrow) results in the so-called pearl necklace sign.

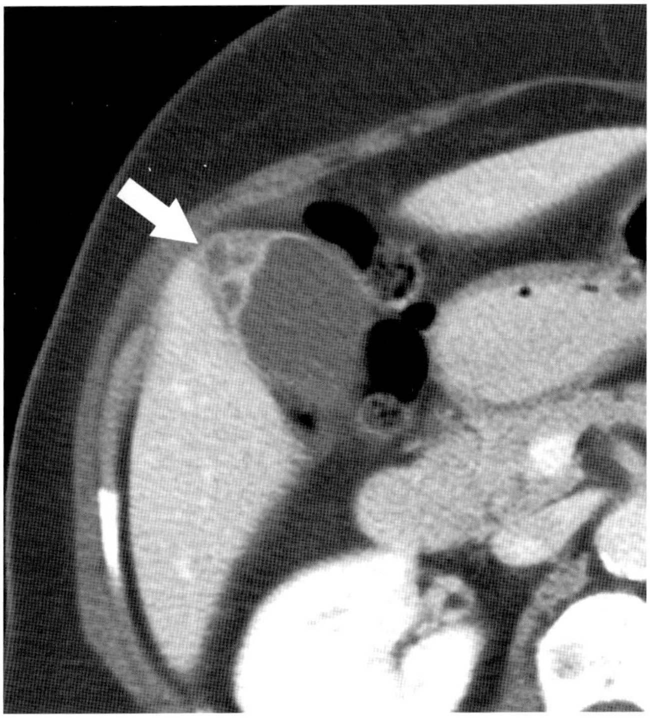

Figure 24.3 Axial contrast-enhanced CT image in a 53 year old woman with right upper quadrant pain. Focal gallbladder wall thickening at the gallbladder fundus (arrow) contains multiple small cystic spaces, typical of adenomyomatosis, which has a predilection for the fundus.

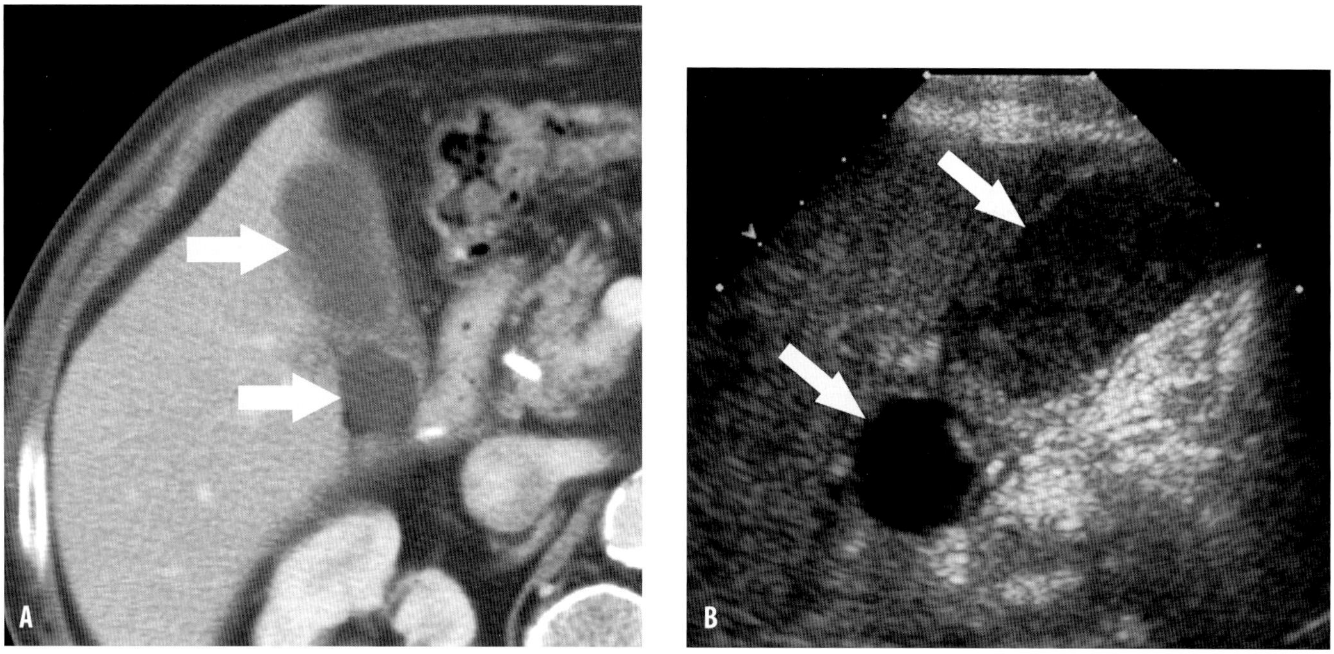

Figure 24.4 A. Axial contrast-enhanced CT image in a 72 year old woman with colorectal cancer. The gallbladder is separated into two compartments (arrows) by focal thickening in the mid-gallbladder. Such an "hourglass" configuration is typical of adenomyomatosis.
B. Ultrasound image of the gallbladder also shows the hourglass configuration, with two distinct compartments (arrows). Note that the bile in the more distal compartment is more echogenic, suggesting functional separation between the compartments. On occasion, stones trapped in the distal compartment of a gallbladder affected by adenomyomatosis can be missed by ultrasound.

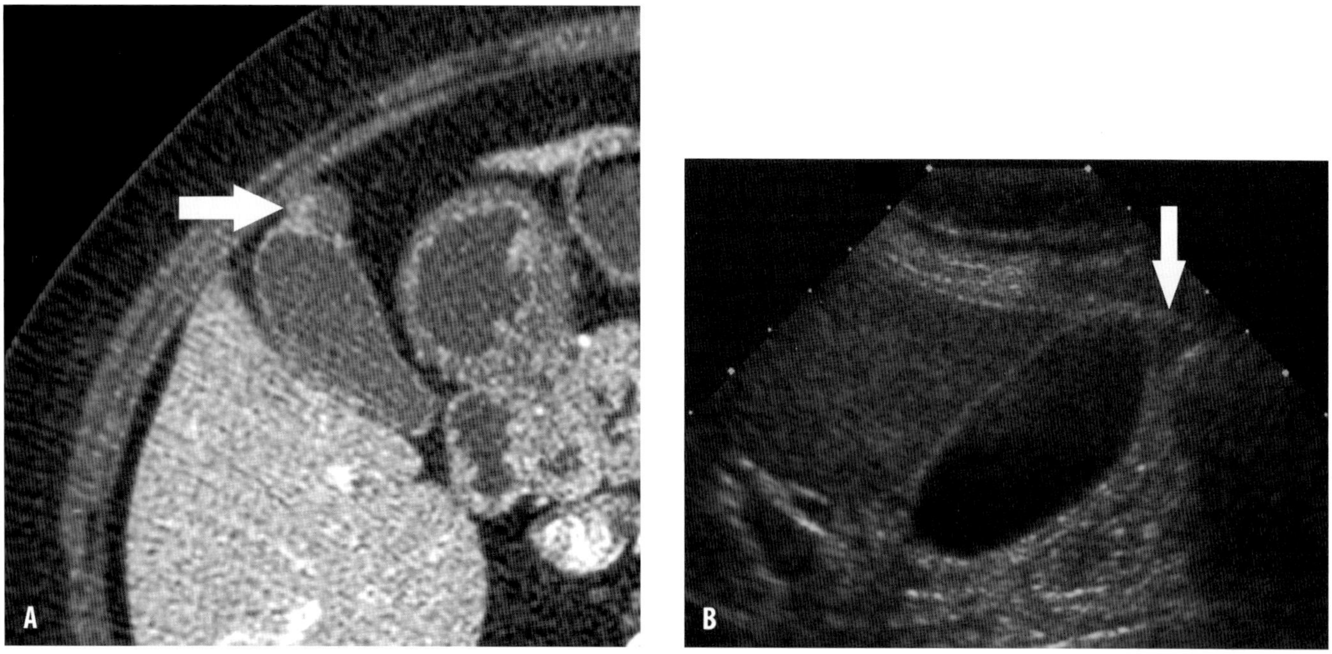

Figure 24.5 A. Axial contrast-enhanced CT image in a 54 year old woman with abdominal pain. The gallbladder has an hourglass configuration, with the fundus (arrow) being "pinched off" from the rest of the gallbladder. **B.** Ultrasound image of the gallbladder was prospectively interpreted as normal, and the fundal compartment (arrow) was not identified.

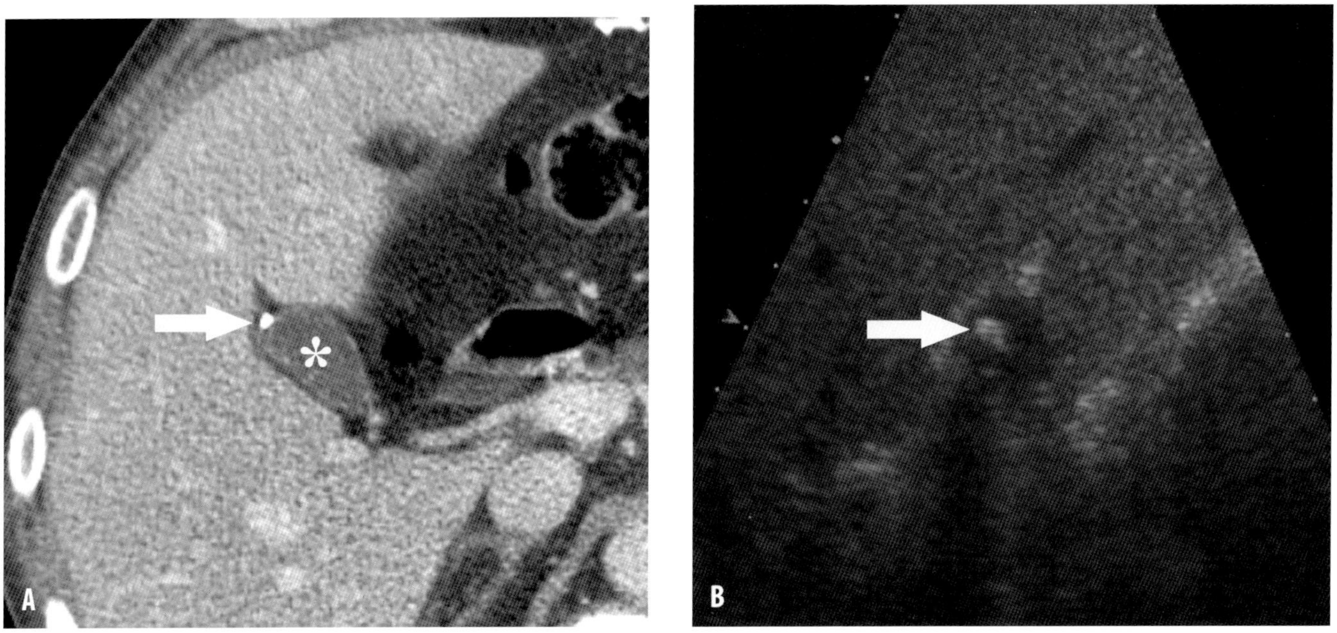

Figure 24.6 A. Axial contrast-enhanced CT image in a 65 year old man with recurrent right upper quadrant pain 4 months after laparoscopic cholecystectomy. A portion of the gallbladder (asterisk) remains in the gallbladder fossa, and contains a small calcified stone (arrow). **B.** Ultrasound image confirms the presence of a small stone (arrow) in a gallbladder remnant. Such an incomplete cholecystectomy may occur when only the distal half of a gallbladder with an hourglass configuration due to adenomyomatosis is removed at surgery.

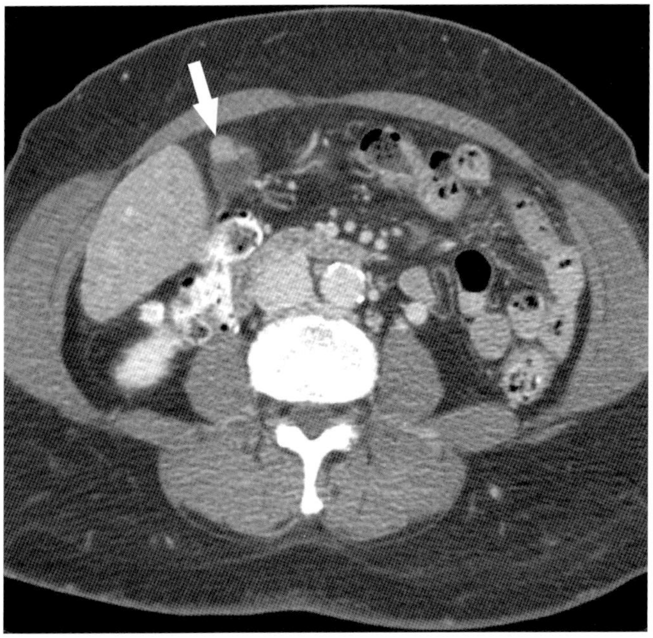

Figure 24.7 Axial contrast-enhanced CT image in an 80 year old woman with microhematuria. A 1 cm solid-appearing nodule at the gallbladder fundus (arrow) is concerning for possible gallbladder cancer, but only adenomyomatosis was found at surgery. Adenomyomatosis can occasionally appear solid and resemble malignancy.

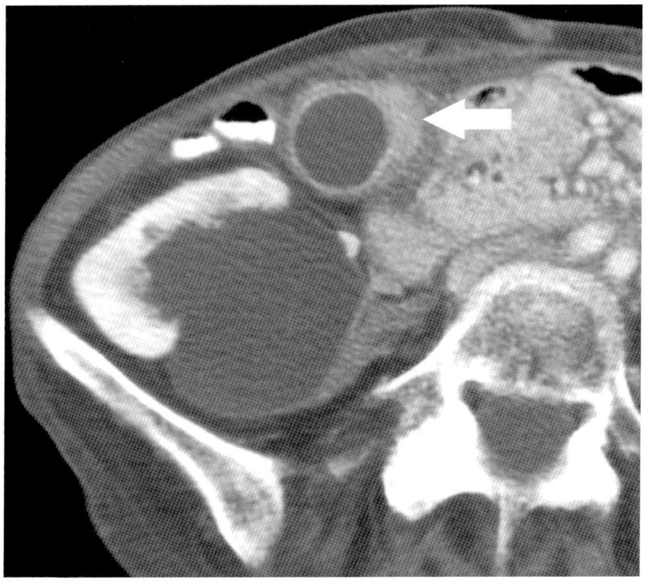

Figure 24.8 Axial contrast-enhanced CT image in a 73 year old woman with abdominal pain. Solid-appearing thickening of the gallbladder wall at the fundus was considered concerning for malignancy, but only adenomyomatosis was found at surgery.

CASE **25**

Pseudotumor of the distal common bile duct

Imaging description

Tumor is the primary consideration when an eccentric focal narrowing or mural-based filling defect is seen at endoscopic retrograde cholangiopancreatography (ERCP) in the distal common bile duct, but this appearance has also been described as a transient and presumably physiological phenomenon that likely reflects transient duct contraction or peristalsis (Figures 25.1–25.4).

Importance

Recognition of this pseudotumor may help avoid unnecessary testing or surgery.

Typical clinical scenario

In one series of eight patients with the appearance of a pseudotumor in the distal common bile duct [1], only one patient went to surgery (and no pathological abnormality was found). As such the anatomic basis of this pseudotumor appearance is largely speculative and it is difficult to draw general conclusions as to the typical clinical scenario. Many of the patients in this report had prior hepatobiliary pathology or intervention; the extent to which this may have altered motility or contraction dynamics in the distal common duct is unknown.

Differential diagnosis

Recent reporting of the pseudotumor of the distal common bile duct [1] likely reflects a contemporary update to earlier studies describing the so-called "pseudocalculus sign" in the distal common duct that may be seen on endoscopic, percutaneous, intra-operative, and MR cholangiopancreatography [2–5]. These studies also stated that the finding was due to contraction of the bile duct, and suggested that the pseudolesion could be distinguished from a true calculus by dynamic ERCP [6]. More recently, dynamic MR cholangiopancreatography has been used to visualize real-time common duct contractile activity [7]. Other features that favor the diagnosis of a pseudotumor include resolution on subsequent images, or an absent inferior border [3]. Other causes of a pseudocalculus at cholangiography include insertion of the cystic duct or posterior compression from the right hepatic artery [8, 9], though these etiologies will usually result in a pseudocalculus in a more proximal segment of the common duct rather than in the distal common duct. Of course, the diagnosis of a pseudotumor should only be made when the possibility of a true tumor or stone can be confidently excluded.

Teaching point

An eccentric focal narrowing or mural-based filling defect in the distal common bile duct can be artifactual in nature and likely reflects duct contraction or peristalsis. This highlights the need to interpret abnormalities of the distal common bile duct seen on static ERCP images with caution, particularly if only a limited number of images are available. If the pathological or artifactual etiology of a distal common bile duct finding is uncertain, close inspection of remaining images, correlation with cross-sectional imaging such as CT or MRI, or even repeat ERCP, may be helpful in making this determination.

REFERENCES

1 Tan JH, Coakley FV, Wang ZJ, et al. Pseudotumor of the distal common bile duct at endoscopic retrograde cholangiopancreatography. *Clin Imaging (in press).*

2 Samardar P. The pseudocalculus sign. *Radiology* 2002; **223**: 239–240.

3 Mujahed Z, Evans JA. Pseudocalculus defect in cholangiography. *Am J Roentgenol Radium Ther Nucl Med* 1972; **116**: 337–341.

4 Martin MB, Kon ND, Ott DJ, Sterchi JM. Pseudocalculus sign. A pitfall of static cholangiography. *Am Surg* 1986; **52**: 197–200.

5 Lautatzis M, Shoenut JP, Scurrah J, Micflikier AB. Pseudocalculus of the common bile duct. *Can J Surg* 1988; **31**: 37–38.

6 Wertheimer M, Brooke WS, Koehler PR, Nelson JA. Pseudocalculus of the common bile duct. A dynamic radiographic differentiation from true retained stone. *Am J Surg* 1975; **130**: 742–745.

7 Van Hoe L, Mermuys K, Vanhoenacker P. MRCP pitfalls. *Abdom Imaging* 2004; **29**: 360–387.

8 Baer JW, Abiri M. Right hepatic artery as a cause of pseudocalculus in the biliary tree. *Gastrointest Radiol* 1982; **7**: 269–273.

9 Berkowitz I, Bornman PC, Kottler RE. Cystic duct entry–another cause of pseudocalculus. *Endoscopy* 1990; **22**: 85–87.

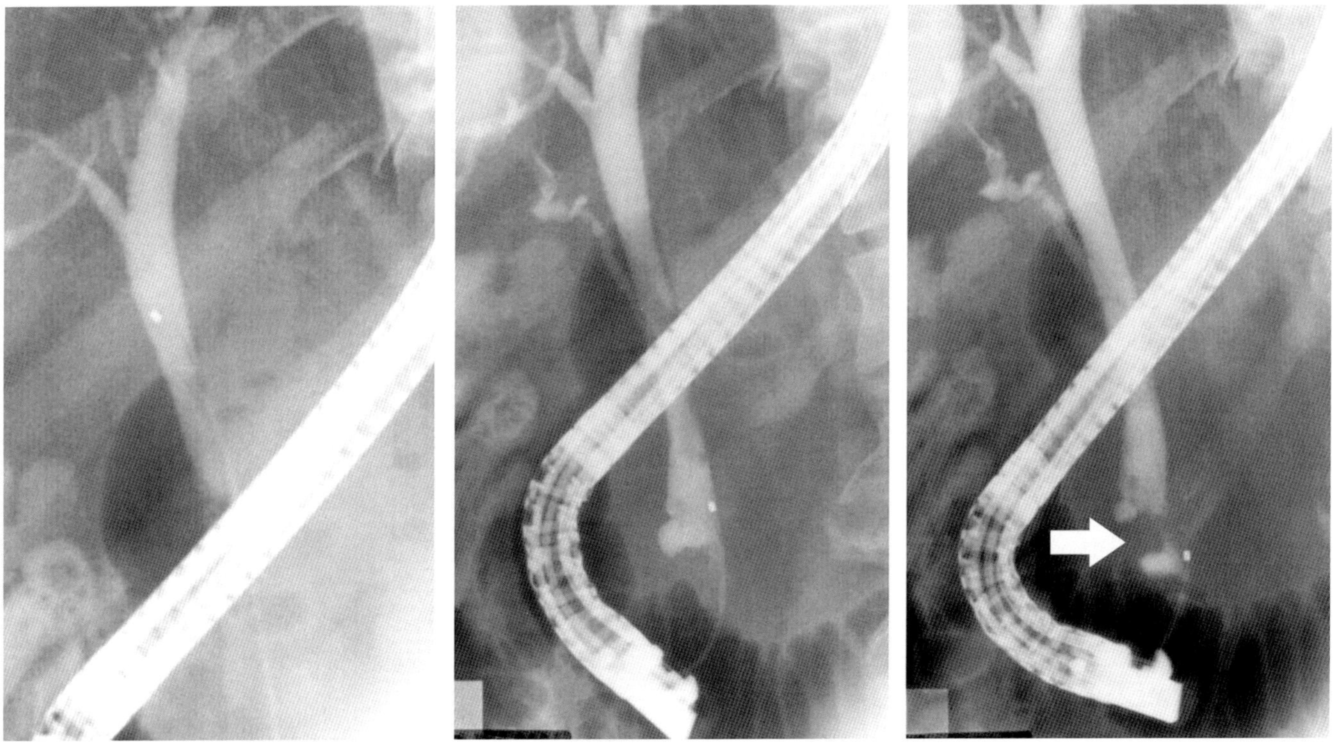

Figure 25.1 Photomontage of three successive images from an ERCP in a 66 year old man with abdominal pain and jaundice. An apparent eccentric filling defect (arrow) is seen in the distal common duct on the final image. Pancreatoduodenectomy was performed for suspected intraductal cholangiocarcinoma. No tumor was found in the specimen.

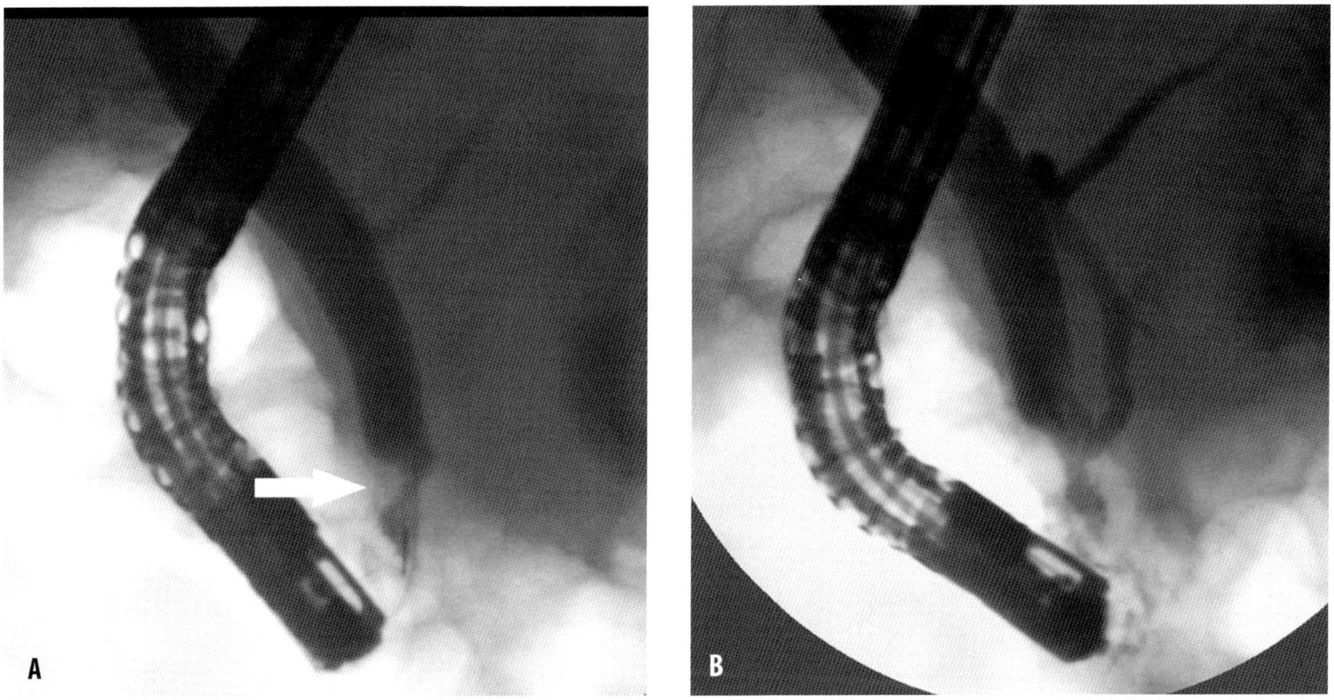

Figure 25.2 A. ERCP image from a 60 year old woman with abdominal pain showing eccentric narrowing (arrow) of the distal common bile duct. **B.** Subsequent image during the same study shows resolution of the finding. The patient later underwent a laparoscopic cholecystectomy for acute cholecystitis, with a normal intra-operative cholangiogram.

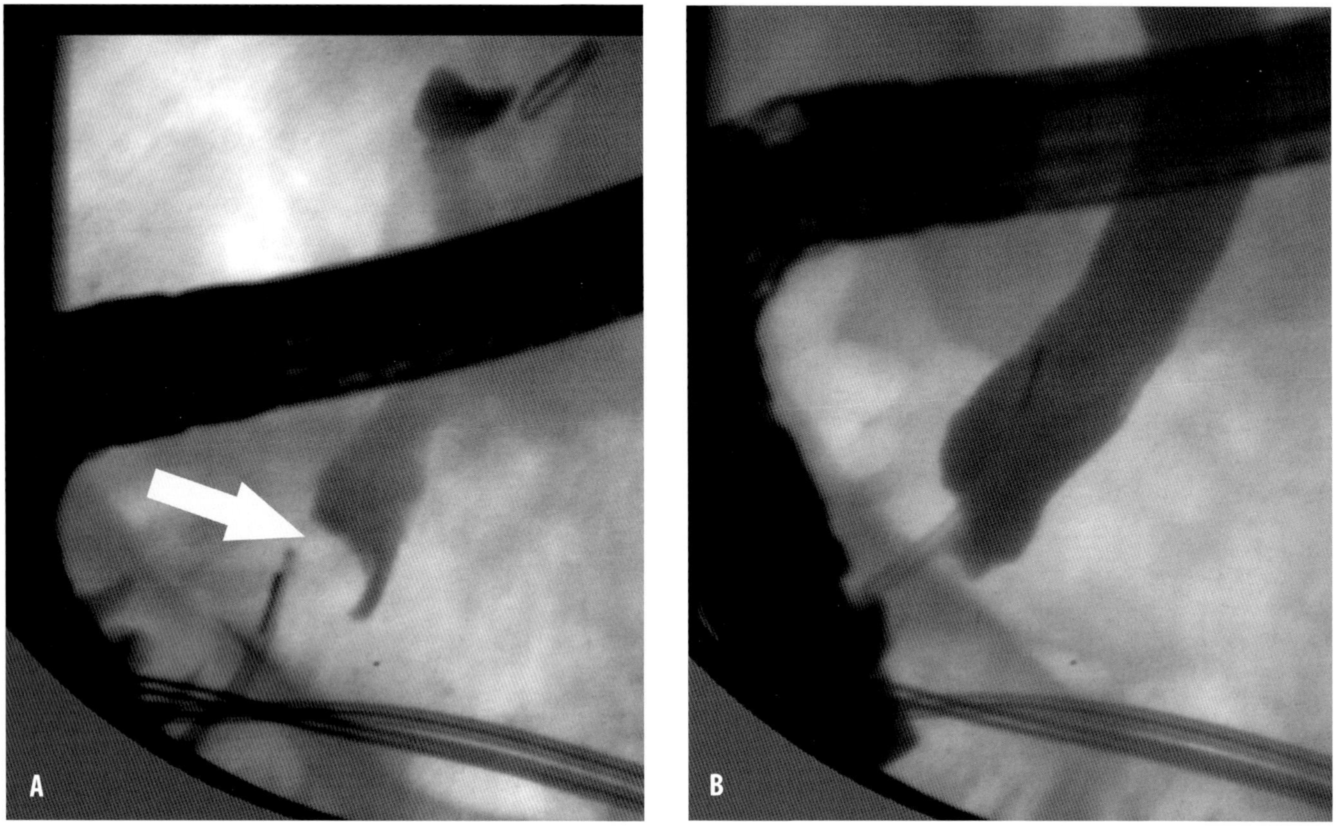

Figure 25.3 **A.** ERCP image in a 59 year old man with a history of liver transplantation. A filling defect of the distal common bile duct (arrow) is seen without visualization of the inferior border. **B.** Subsequent image during the same study shows resolution of the finding.

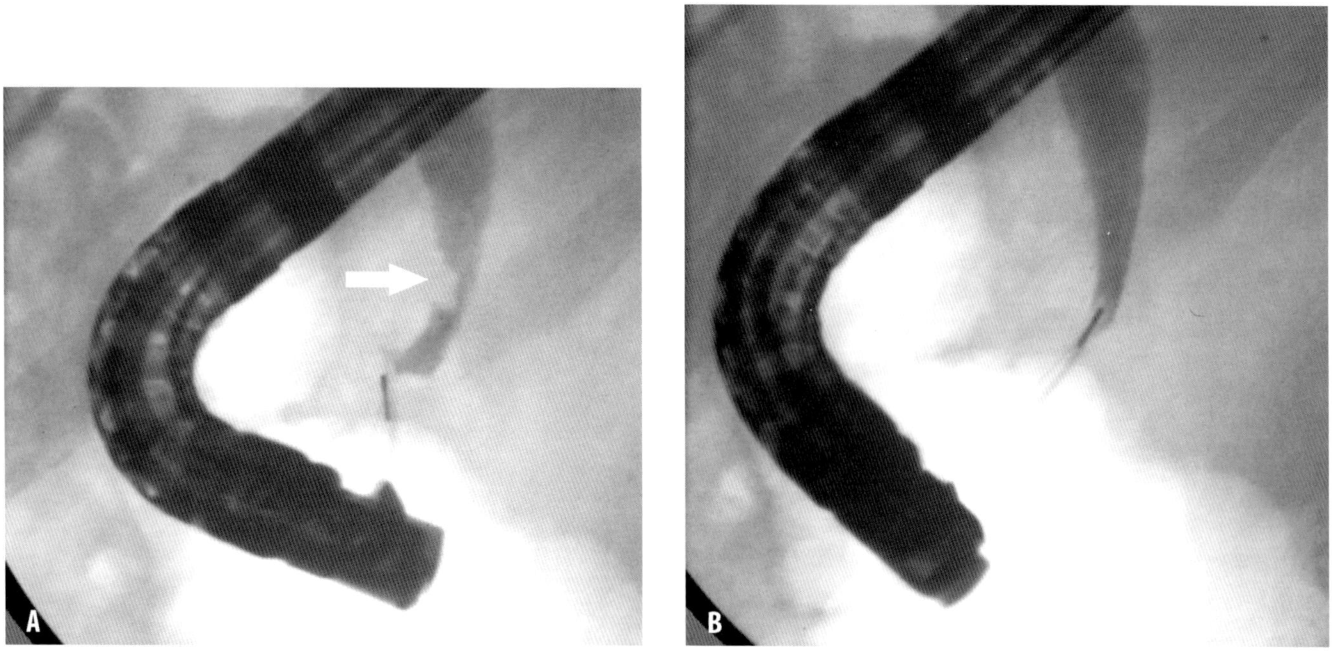

Figure 25.4 **A.** ERCP image in a 66 year old woman with abdominal pain and a history of cholecystectomy and sphincterotomy. A filling defect of the distal common bile duct (arrow) is seen. **B.** Subsequent image during the same study shows resolution of the finding.

CASE **26**

Pancreaticobiliary maljunction

Imaging description

Pancreaticobiliary maljunction (also known as anomalous pancreaticobiliary ductal union or common channel syndrome) is a congenital anomaly in which the pancreatic and biliary ducts join prematurely outside the duodenal wall, so that two-way reflux can occur between the pancreatic and biliary systems [1]. The diagnosis can be made at ERCP or MRI when a long common channel is seen between the distal pancreatic and biliary ducts (Figures 26.1–26.4). Threshold lengths of 12 to 15 mm have been suggested to make the diagnosis [1, 2], although in my experience the diagnosis can generally be made on visual inspection alone. Because hydrostatic pressures are usually higher in the pancreatic duct, pancreaticobiliary maljunction often results in reflux of pancreatic secretions into the bile duct, and so can be diagnosed biochemically when high levels of amylase are found in bile obtained from the bile duct or gallbladder, either percutaneously or at laparotomy [1]. Visualization of pancreaticobiliary reflux in common channel syndrome has also been reported using secretin-stimulated dynamic MRCP [3].

Importance

Pancreaticobiliary maljunction has been linked with a wide variety of pancreatic and biliary diseases, including gallbladder cancer, cholangiocarcinoma, pancreatic cancer, adenomyomatosis, pancreatitis, and cholelithiasis [4]. It should be noted that these associations have been extensively described in reports from Japanese and Taiwanese investigators, with far fewer reports from Europe and the United States. As such, the global relevance of pancreaticobiliary maljunction is unclear, although conceptually it is possible that mixing of pancreatic and biliary secretions might be pathogenic or even carcinogenic.

Typical clinical scenario

In Japanese and Taiwanese studies, the prevalence of pancreaticobiliary maljunction ranges from 1.5 to 8.7% [3, 5]. In a North American study, a prevalence of 3.5% was reported in a series of 2847 patients [6]. All of these frequencies seem higher than I would have expected based on my own clinical experience. The appropriate management of isolated pancreaticobiliary maljunction is controversial. Suggested strategies include endoscopic "deroofing" of the common channel or prophylactic cholecystectomy, since gallbladder cancer appears to be the primary malignancy associated with pancreaticobiliary maljunction [1, 6].

Differential diagnosis

No strict criteria exist to differentiate pancreaticobiliary maljunction from choledochocele, and the distinction may be largely arbitrary since it is reasonable to regard pancreaticobiliary maljunction as an incomplete form of choledochocele [6]. Note that pancreaticobiliary maljunction is present in most patients with unequivocal choledochal cysts (Figures 26.5 and 26.6) [7]. Close apposition of the distal common bile and pancreatic ducts has been reported to falsely simulate pancreaticobiliary maljunction at MRCP [8].

Teaching point

The presence of a long common channel between the distal common bile duct and distal pancreatic duct is indicative of pancreaticobiliary maljunction, and is associated with a variety of pancreatic and biliary diseases and choledochal cysts.

REFERENCES

1 Kamisawa T, Anjiki H, Egawa N, et al. Diagnosis and clinical implications of pancreatobiliary reflux. *World J Gastroenterol* 2008; **14**: 6622–6626.

2 The Japanese Study Group on Pancreaticobiliary Maljunction. Diagnostic criteria of pancreaticobiliary maljunction. *J Hep Bil Pancr Surg* 1994; **1**: 219–221.

3 Hosoki T, Hasuike Y, Takeda Y, et al. Visualization of pancreaticobiliary reflux in anomalous pancreaticobiliary junction by secretin-stimulated dynamic magnetic resonance cholangiopancreatography. *Acta Radiol* 2004; **45**: 375–382.

4 Wang HP, Wu MS, Lin CC, et al. Pancreaticobiliary diseases associated with anomalous pancreaticobiliary ductal union. *Gastrointest Endosc* 1998; **48**: 184–189.

5 Yamauchi S, Koga A, Matsumoto S, Tanaka M, Nakayama F. Anomalous junction of pancreaticobiliary duct without congenital choledochal cyst: a possible risk factor for gallbladder cancer. *Am J Gastroenterol* 1987; **82**: 20–24.

6 Elton E, Hanson BL, Biber BP, Howell DA. Dilated common channel syndrome: endoscopic diagnosis, treatment, and relationship to choledochocele formation. *Gastrointest Endosc* 1998; **47**: 471–478.

7 Irie H, Honda H, Jimi M, et al. Value of MR cholangiopancreatography in evaluating choledochal cysts. *Am J Roentgenol* 1998; **171**: 1381–1385.

8 Kamisawa T, Okamoto T. Pitfalls of MRCP in the diagnosis of pancreaticobiliary maljunction. *JOP* 2004; **5**: 488–490.

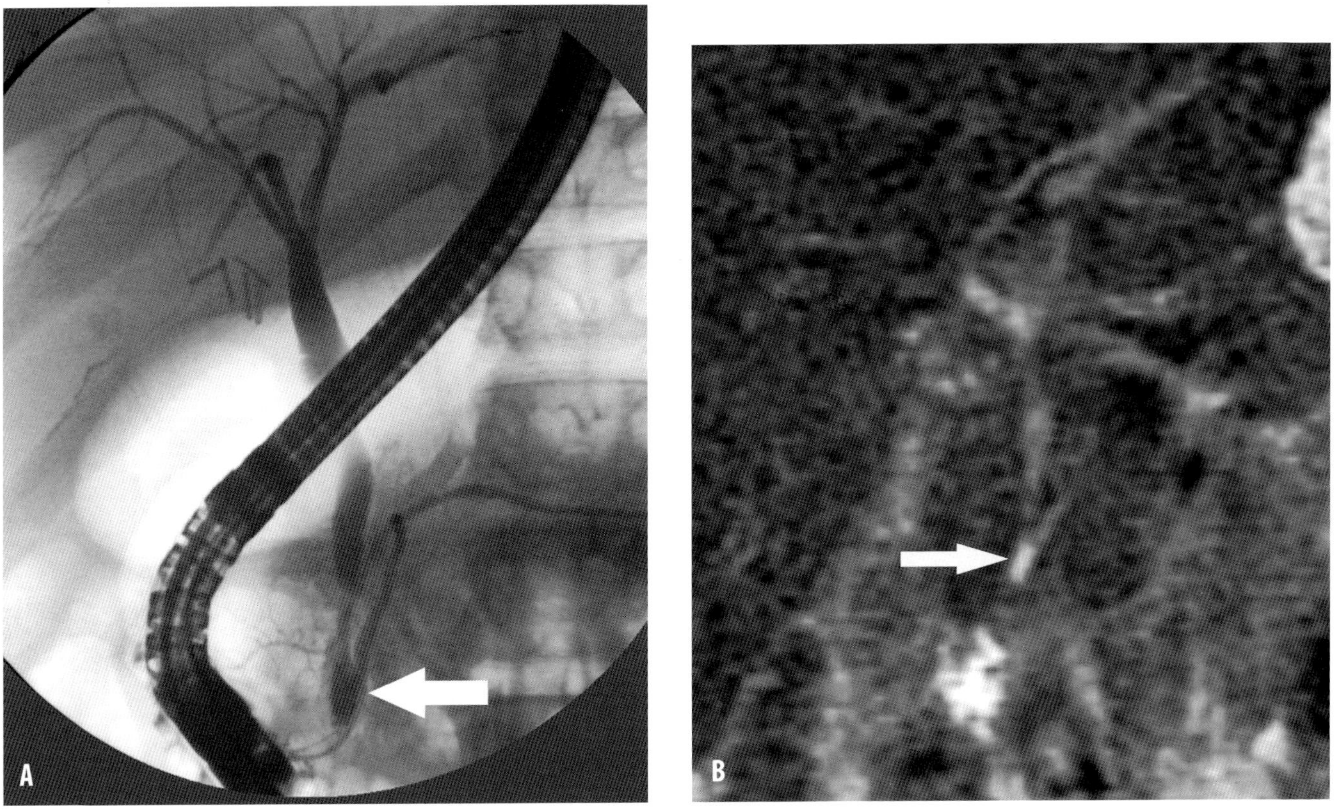

Figure 26.1 A. ERCP image from a 13 year old girl with a history of recurrent pancreatitis and prior cholecystectomy for gallstones shows a common channel (arrow) between the distal common bile duct and distal pancreatic duct. **B.** Corresponding coronal T2-weighted MR image also shows a common channel (arrow).

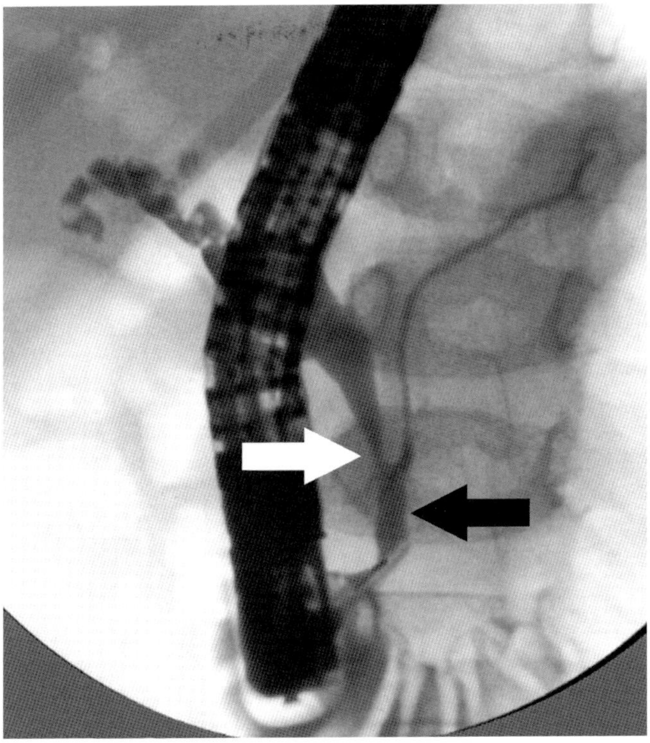

Figure 26.2 ERCP image from a 4 year old girl with recurrent pancreatitis shows a common channel (black arrow) between the distal common bile duct and distal pancreatic duct, with a narrowing (white arrow) at the junction of the common bile duct and the common channel. The patient subsequently underwent cholecystectomy and choledochoduodenostomy.

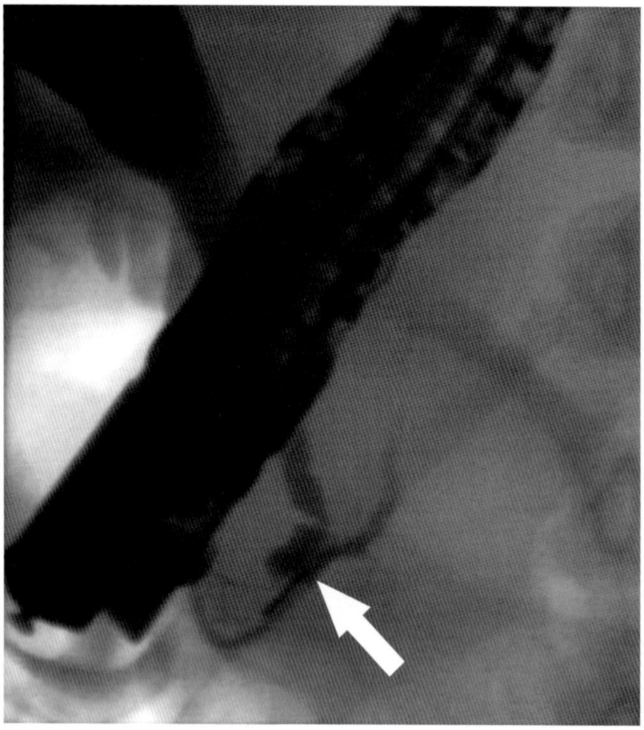

Figure 26.3 ERCP image from a 2 year old girl with recurrent pancreatitis shows a common channel (arrow) between the distal common bile duct and distal pancreatic duct.

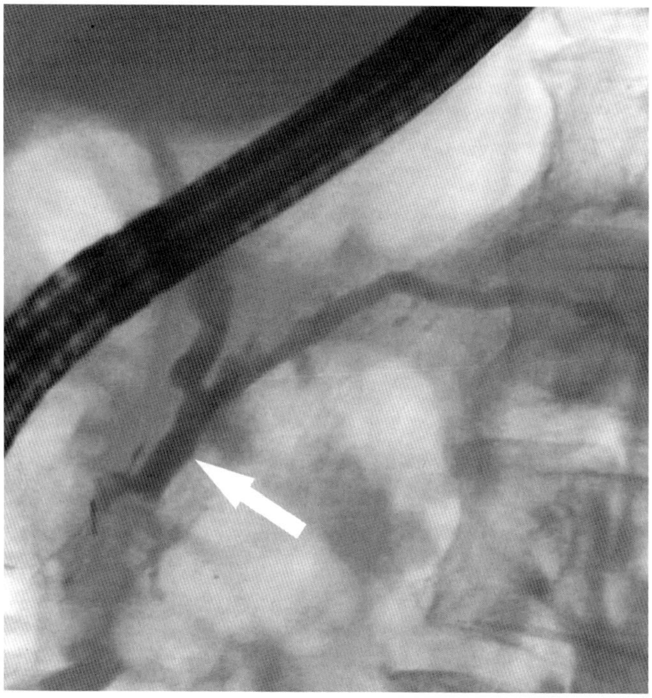

Figure 26.4 ERCP image from a 58 year old man with chronic pancreatitis and prior cholecystectomy for gallstones shows a common channel (arrow) between the distal common bile duct and pancreatic duct.

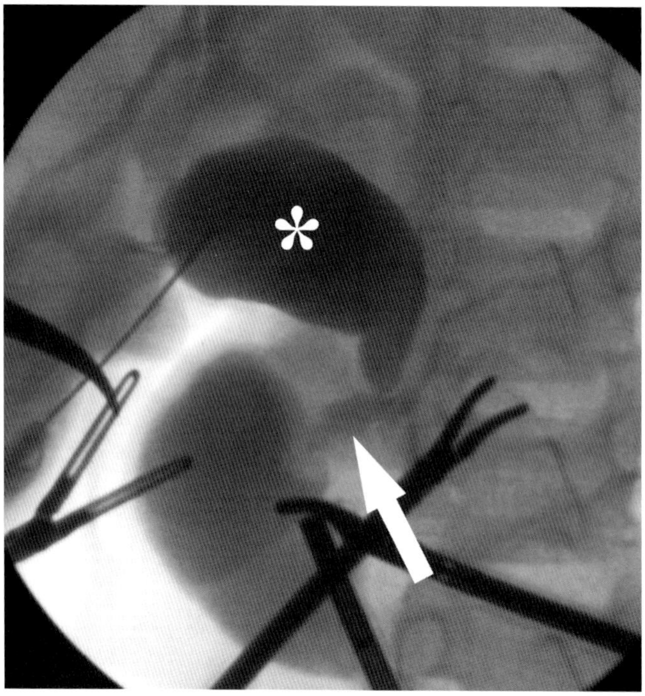

Figure 26.5 Intraoperative cholangiogram during laparoscopic choledochal cyst excision and choledochojejunostomy in a 4 year old girl with recurrent jaundice due to a choledochal cyst (asterisk) shows a common channel (arrow) between the distal common bile duct and distal pancreatic duct.

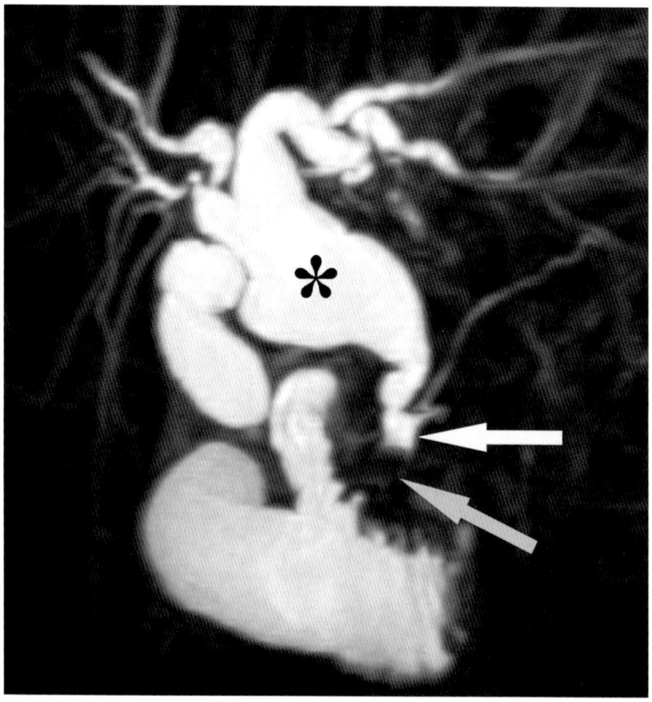

Figure 26.6 MRCP image in a 6 year old girl with obstructive jaundice due to a choledochal cyst (asterisk) shows a common channel (white arrow) between the distal common bile duct and distal pancreatic duct. A stone (grey arrow) is visible in the common channel.

CASE

27 Pseudofluid due to complete splenic infarction

Imaging description

Complete splenic infarction can result in a crescentic or ovoid low-density structure in the left upper quadrant that can mimic a fluid collection or fluid-filled colon on contrast-enhanced CT (Figures 27.1–27.3) [1].

Importance

Misidentification of the infarcted spleen as a fluid collection might result in inappropriate attempts at drainage, while failure to recognize splenic infarction may result in a missed opportunity to expeditiously diagnose the underlying cause.

Typical clinical scenario

Splenic infarction is commonly due to thromboembolism in conditions such as atrial fibrillation, left ventricular thrombus, or endocarditis. Other causes include local thromboses (particularly in aortic disease, hematologic diseases such as sickle cell disease, leukemia, and myelodysplastic syndrome), vasculitis, and iatrogenic injury [2–6].

Differential diagnosis

The key to the correct diagnosis of complete splenic infarction is failure to identify a normal spleen during what should be the routine "checklisting" of the major abdominal viscera (liver, gallbladder, *spleen*, pancreas, kidneys, and adrenal glands) when reviewing an abdominal CT. Once the absence of a normal spleen is recognized, the diagnosis is straightforward and no real differential possibilities arise.

Teaching point

The possibility of complete splenic infarction should be considered when the spleen cannot be separately identified from an apparent fluid collection or fluid-filled colon in the left upper quadrant.

REFERENCES

1 Wetton CW, Tran TL. Case report: splenic infarct in sickle cell disease. *Clin Radiol* 1995; **50**: 573–574.

2 Nores M, Phillips EH, Morgenstern L, Hiatt JR. The clinical spectrum of splenic infarction. *Am Surg* 1998; **64**: 182–188.

3 Jaroch MT, Broughan TA, Hermann RE. The natural history of splenic infarction. *Surgery* 1986; **100**: 743–749.

4 Cassar K, Munro A. Iatrogenic splenic injury. *J R Coll Surg Edinb* 2002; **47**: 731–741.

5 Takayasu K, Moniyama N, Muramatsu Y, et al. Splenic infarction: a complication of transcatheter hepatic arterial embolization. *Radiology* 1984; **5**: 371–375.

6 Peters TG, Steinmetz SR, Cowan GS Jr. Splenic injury and repair during bariatric surgical procedures. *South Med J* 1990; **83**: 166–169.

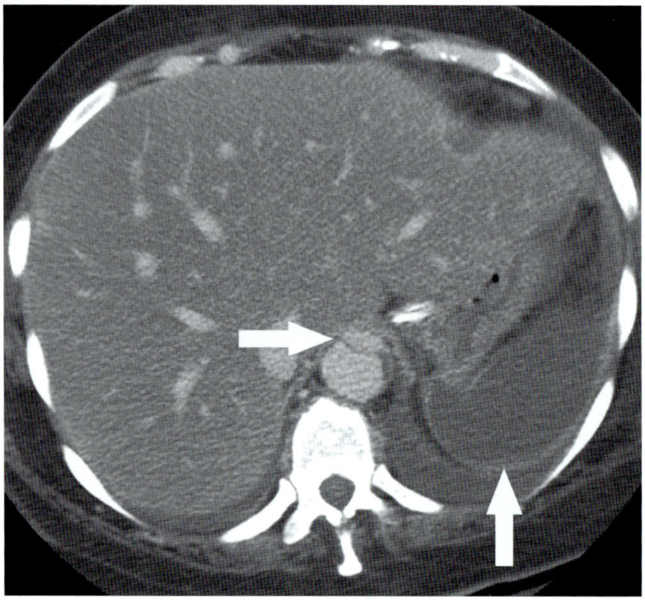

Figure 27.1 Axial contrast-enhanced CT in a 39 year old woman with a type B aortic dissection resulting in complete infarction of the spleen (vertical arrow). Note the dissection flap (horizontal arrow) in the aorta. The density of the infarcted spleen is such that it resembles a fluid collection.

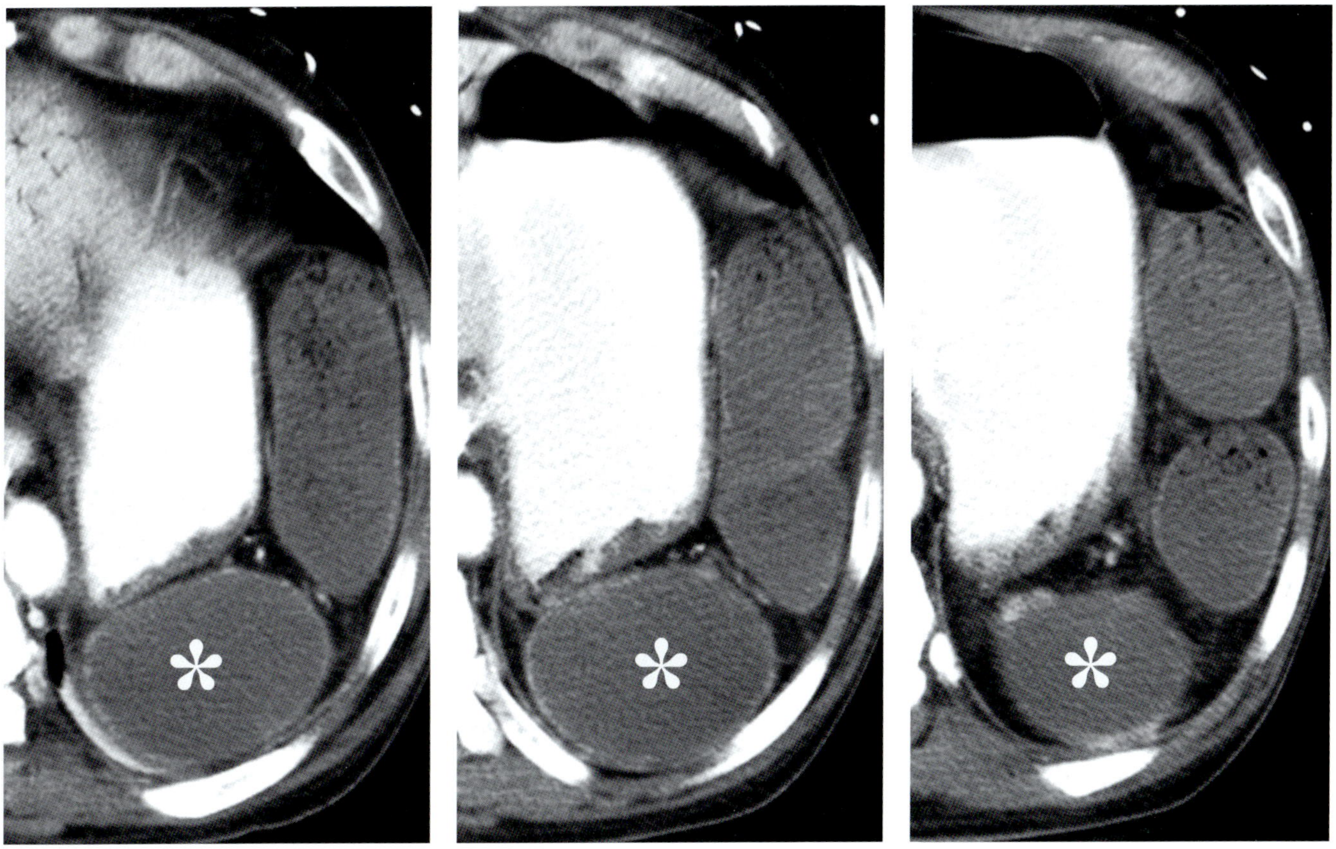

Figure 27.2 Photomontage of axial contrast-enhanced CT images arranged from superior (left image) to inferior (right image) in a 75 year old woman with acute left upper quadrant pain and atrial fibrillation showing complete infarction of the spleen (asterisks). The non-enhancing spleen resembles the adjacent fluid-filled left colon.

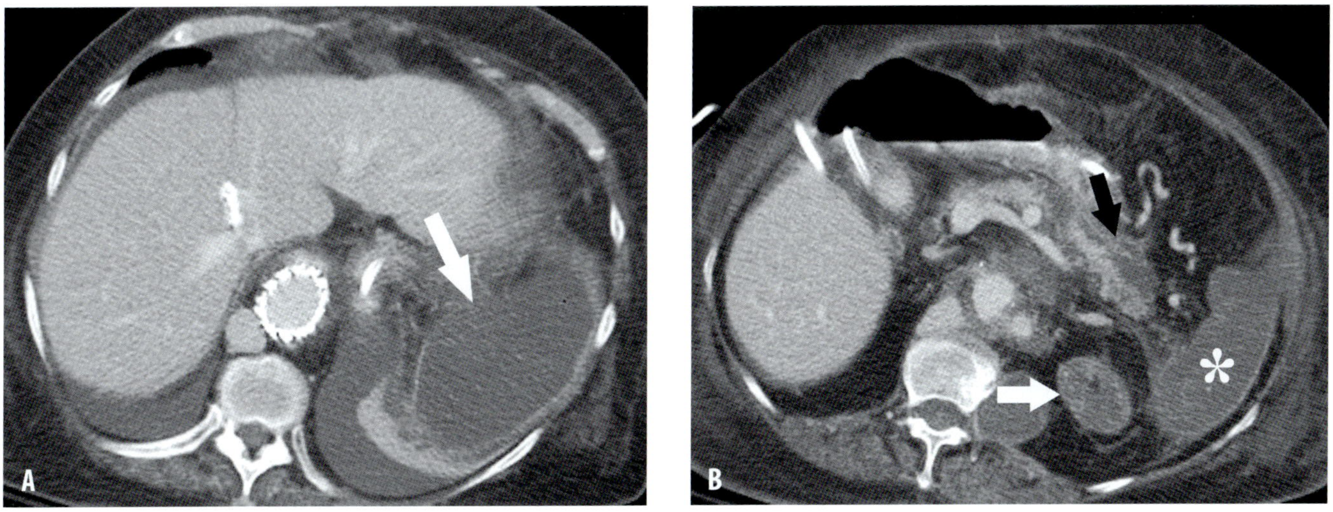

Figure 27.3 A. Axial contrast-enhanced CT image obtained after endovascular aortic stent graft of a leaking thoracoabdominal aneurysm in an 80 year old woman showing complete infarction of the spleen (arrow). The density of the infarcted spleen is such that it resembles a fluid collection. B. Axial contrast-enhanced CT image obtained at a more inferior level again shows the infarcted spleen (asterisk) and also shows concomitant infarction of the left kidney (white arrow) and inflammatory changes (black arrow) around the distal pancreas due to presumed ischemic pancreatitis.

CASE 28 Pseudosubcapsular hematoma

Imaging description

The lateral segment of the left hepatic lobe occasionally extends laterally and wraps around the spleen, where it can be mistaken for subcapsular or perisplenic hematoma at ultrasound or CT (Figures 28.1–28.3) [1–3].

Importance

An erroneous diagnosis of subcapsular or perisplenic hematoma may result in unnecessary additional tests, procedures, or surgery [1].

Typical clinical scenario

Lateral extension of the left hepatic lobe is usually an anatomic variant, and as such may be seen incidentally in any patient. Occasionally, it may be related to compensatory left lobe hypertrophy, as in cirrhosis (Figure 28.1). From a clinical perspective, given the increasing use of ultrasound in the emergency room [4], recognition of this pitfall is particularly important in patients with blunt abdominal trauma since it is then more likely to be confused with subcapsular hematoma.

Differential diagnosis

Careful ultrasound scanning from lateral to medial or medial to lateral will show continuity between the extended lateral segment of the left hepatic lobe and the rest of the liver, preventing misintcrpretation of the anatomic variant as a subcapsular hematoma. In difficult cases or if acoustic access is limited, the true nature of the finding can be easily established by contrast-enhanced CT.

Teaching point

The possibility of an unusually extensive lateral segment of the left hepatic lobe should be considered when an apparent subcapsular or perisplenic hematoma is seen at ultrasound.

REFERENCES

1 Cholankeril JV, Zamora BO, Ketyer S. Left lobe of the liver draping around the spleen: a pitfall in computed tomography diagnosis of perisplenic hematoma. *J Comput Tomogr* 1984; **8**: 261–267.

2 Li DK, Cooperberg PL, Graham MF, Callen P. Pseudo perisplenic "fluid collections": a clue to normal liver and spleen echogenic texture. *J Ultrasound Med* 1986; **5**: 397–400.

3 Crivello MS, Peterson IM, Austin RM. Left lobe of the liver mimicking perisplenic collections. *J Clin Ultrasound* 1986; **14**: 697–701.

4 McKenney KL, Pellei DD. Perisplenic pseudolesion: a case report and literature review. *Emergency Radiology* 2001; **8**: 159–161.

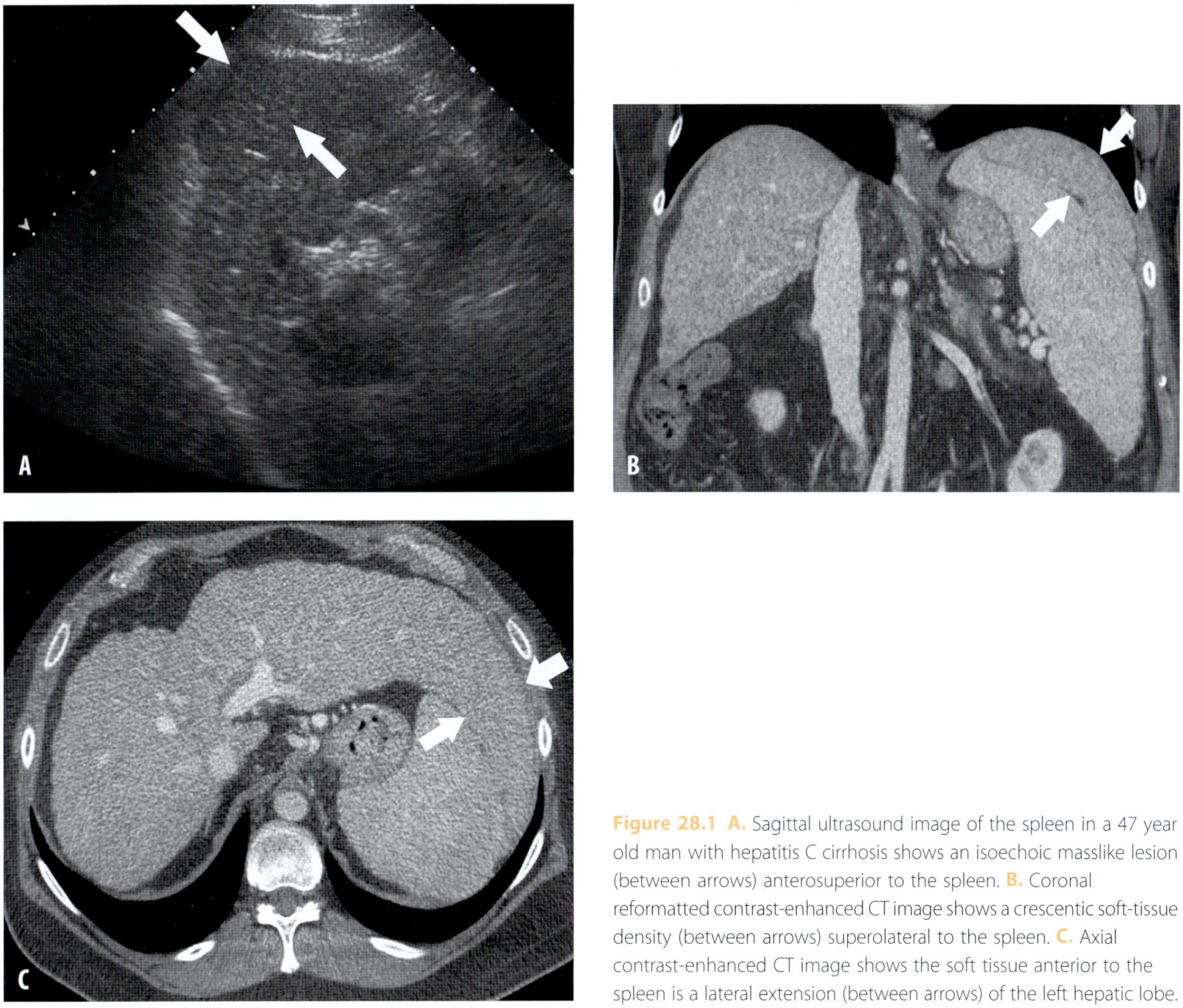

Figure 28.1 **A.** Sagittal ultrasound image of the spleen in a 47 year old man with hepatitis C cirrhosis shows an isoechoic masslike lesion (between arrows) anterosuperior to the spleen. **B.** Coronal reformatted contrast-enhanced CT image shows a crescentic soft-tissue density (between arrows) superolateral to the spleen. **C.** Axial contrast-enhanced CT image shows the soft tissue anterior to the spleen is a lateral extension (between arrows) of the left hepatic lobe.

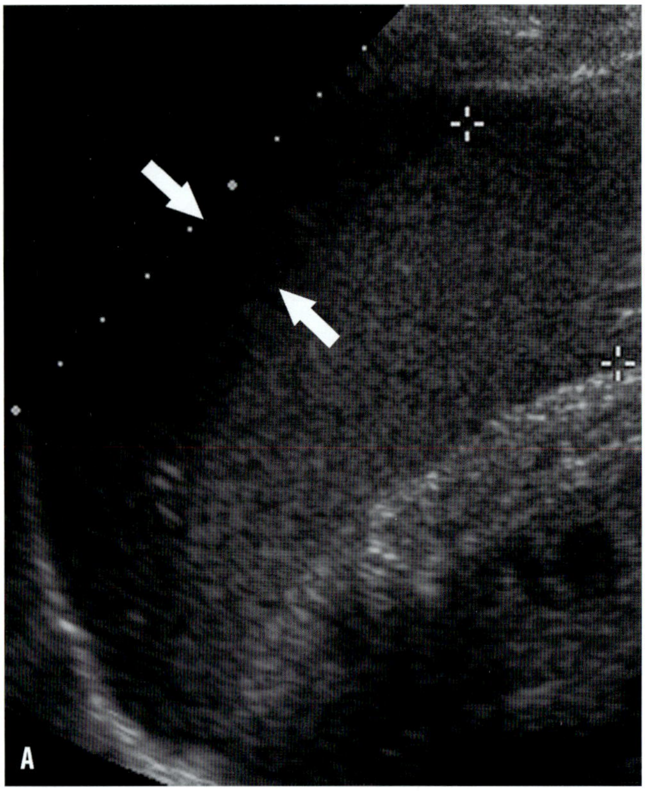

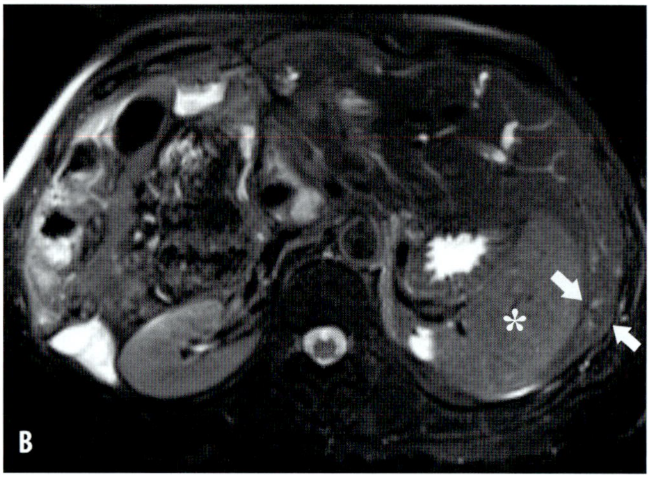

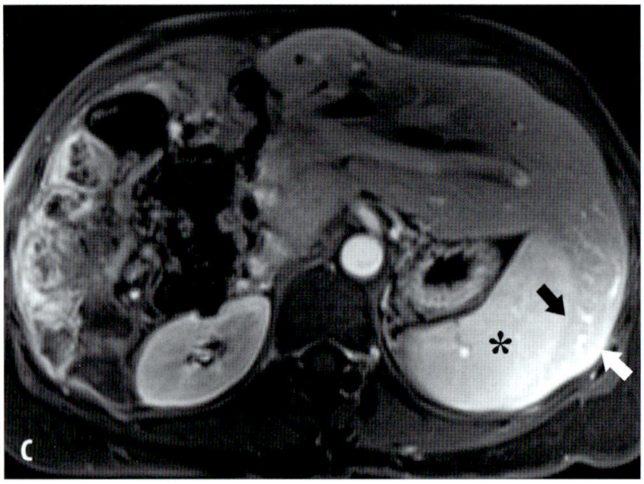

Figure 28.2 A. Sagittal ultrasound image of the spleen in a 52 year old man with fever and a prior right hepatectomy for metastatic neuroendocrine cancer shows an apparent crescentic hypoechoic lesion (between arrows) superior to the spleen (between calipers). B. Axial T2-weighted MR image shows the lateral segment of the left hepatic lobe (between arrows) is wrapped around the spleen (asterisk), and accounts for the abnormality seen above the spleen on ultrasound. C. Axial T1-weighted post-gadolinium MR image also shows the lateral segment of the left hepatic lobe (between arrows) is wrapped around the spleen (asterisk).

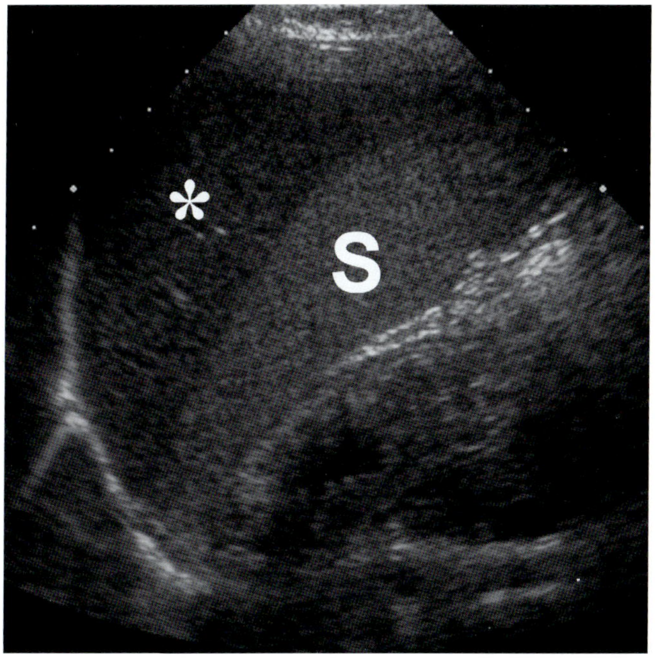

Figure 28.3 Sagittal ultrasound image of the spleen in a 47 year old man with hepatitis C cirrhosis shows a hypoechoic masslike lesion (asterisk) superior to the spleen (S) that could be interpreted as a subcapsular or perisplenic hematoma. Contrast-enhanced CT (not shown) confirmed this was a lateral extension of the left hepatic lobe. *Image for Figure 28.3 graciously provided by Dr Peter Callen, UCSF.*

CASE 29 Splenic hemangioma

Imaging description

Splenic hemangiomas are non-encapsulated benign proliferations of vascular channels that range from capillary to cavernous in size [1]. Smaller hemangiomas are usually solid while larger hemangiomas may be partially cystic or calcified, presumably secondary to thrombosis and infarction (Figures 29.1 and 29.2). At ultrasound, splenic hemangiomas typically appear as well-defined solid echogenic or complex partially cystic masses. At CT, hemangiomas appear as hypodense well-circumscribed masses with marked homogeneous enhancement of the solid components. Delayed enhancement has been described for splenic hemangiomas at CT, but they reportedly have a mottled heterogeneous appearance in contrast to the typical centripetal enhancement of hepatic hemangiomas [2]. At MRI, smaller splenic hemangiomas resemble hepatic hemangiomas with T2 hyperintensity and delayed centripetal enhancement [3], but larger lesions are more variable [1]. There is some discrepancy between the reported CT versus MRI delayed enhancement patterns [2, 3], but this may partially reflect the fact that routine MRI includes more delayed acquisition times than routine CT. While splenic hemangiomas may demonstrate centripetal enhancement (Figures 29.3 and 29.4), they lack the typical nodular globular pattern of peripheral enhancement seen in hepatic hemangiomas [1, 4]. This may reflect differences in vascular supply (dual blood supply to the liver versus single blood supply to the spleen), but this is speculative.

Importance

Splenic hemangiomas are usually small and incidental, but may raise concern of malignancy in the general population or of metastasis in the oncology population.

Typical clinical scenario

Splenic hemangiomas have an autopsy prevalence of 0.3 to 14%, and are the commonest benign tumors in the spleen [1]. Hemangiomas can grow slowly and present as palpable nontender masses in the left upper quadrant. Diffuse splenic hemangiomas may occur as part of the Klippel-Trenaunay or Proteus syndromes [5, 6]. Diffuse splenic hemangiomas can be complicated by rupture or hypersplenism [1]. Kasabach-Merritt syndrome refers to the combination of diffuse splenic hemangiomas and hypersplenism [7]. Such larger or symptomatic lesions may require splenectomy. Malignant degeneration has been reported [8], but such cases may represent primary angiosarcoma.

Differential diagnosis

Differential considerations for a solid splenic mass include hamartoma, angiosarcoma, lymphoma, metastases, and granulomatous disease. Most of these entities are hypovascular. Splenic angiosarcoma is very rare and usually large and advanced at diagnosis, with disseminated metastases in the liver and elsewhere [1]. Hamartomas may be hypervascular and T2 hyperintense, but the T2 hyperintensity is heterogeneous rather than homogeneous as is typically the case for splenic hemangiomas [3]. In any case, confusing one incidental benign tumor for another would usually be of little consequence. Given these considerations, it has been suggested that a T2 hyperintense mass under 4 cm with centripetal enhancement in the spleen can be considered a hemangioma without tissue verification [3], and this seems a reasonable approach to masses with the typical imaging findings (Figures 29.3 and 29.4). Biopsy may be considered for larger or otherwise problematic lesions. Traditionally, splenic mass biopsy has been avoided because of concerns related to non-diagnostic specimens and the risk of hemorrhage [9], but more recent series suggest this can be a diagnostic and safe technique [10, 11].

Teaching point

A solid incidental hypervascular splenic mass is likely a hemangioma, especially if under 4 cm, uniformly T2 hyperintense, and centripetally enhancing. Larger splenic hemangiomas are more variable in appearance and are occasionally associated with hypersplenism.

REFERENCES

1 Abbott RM, Levy AD, Aguilera NS, Gorospe L, Thompson WM. From the archives of the AFIP: primary vascular neoplasms of the spleen: radiologic-pathologic correlation. *Radiographics* 2004; **24**: 1137–1163.

2 Ferrozzi F, Bova D, Draghi F, Garlaschi G. CT findings in primary vascular tumors of the spleen. *Am J Roentgenol* 1996; **166**: 1097–1101.

3 Ramani M, Reinhold C, Semelka R, et al. Splenic hemangiomas and hamartomas: MR imaging characteristics of 28 lesions. *Radiology* 1997; **202**: 166–172.

4 Semelka RC, Brown ED, Ascher SM, et al. Hepatic hemangiomas: a multi-institutional study of appearance on T2-weighted and serial gadolinium-enhanced gradient-echo MR images. *Radiology* 1994; **192**: 401–406.

5 Kocaman O, Alponat A, Aygün C, et al. Lower gastrointestinal bleeding, hematuria and splenic hemangiomas in Klippel-Trenaunay syndrome: a case report and literature review. *Turk J Gastroenterol* 2009; **20**: 62–66.

6 Wang Z, Yu Z, Su Y, et al. Kasabach-Merritt syndrome caused by giant hemangiomas of the spleen in patients with Proteus syndrome. *Blood Coagul Fibrinolysis* 2007; **18**: 505–508.

7 Dufau JP, le Tourneau A, Audouin J, Delmer A, Diebold J. Isolated diffuse hemangiomatosis of the spleen with Kasabach-Merritt-like syndrome. *Histopathology* 1999; **35**: 337–344.

8 Husni EA. The clinical course of splenic hemangioma. *Arch Surg* 1961; **83**: 681–688.

9 Ros PR, Moser RP Jr, Dachman AH, Murari PJ, Olmsted WW. Hemangioma of the spleen: radiologic-pathologic correlation in ten cases. *Radiology* 1987; **162**: 73–77.

10 Keogan MT, Freed KS, Paulson EK, Nelson RC, Dodd LG. Imaging-guided percutaneous biopsy of focal splenic lesions: update on safety and effectiveness. *Am J Roentgenol* 1999; **172**: 933–937.

11 Tam A, Krishnamurthy S, Pillsbury EP, et al. Percutaneous image-guided splenic biopsy in the oncology patient: an audit of 156 consecutive cases. *J Vasc Interv Radiol* 2008; **19**: 80–87.

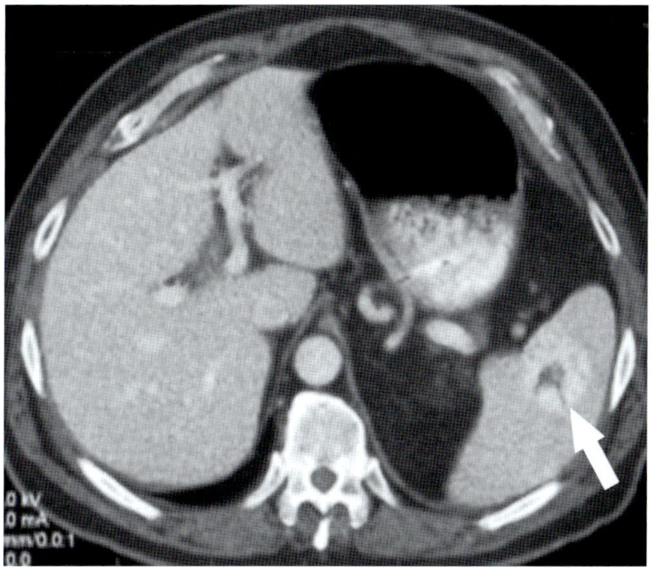

Figure 29.1 Axial contrast-enhanced CT image in a 61 year old man with two left renal masses (not shown) discovered during evaluation of a self-limiting episode of small bowel obstruction shows a peripherally enhancing 5 cm mass (arrow) in the spleen. The patient underwent open partial nephrectomy and splenectomy, and surgical pathology established a diagnosis of two renal cell carcinomas in the kidney and a capillary hemangioma in the spleen.

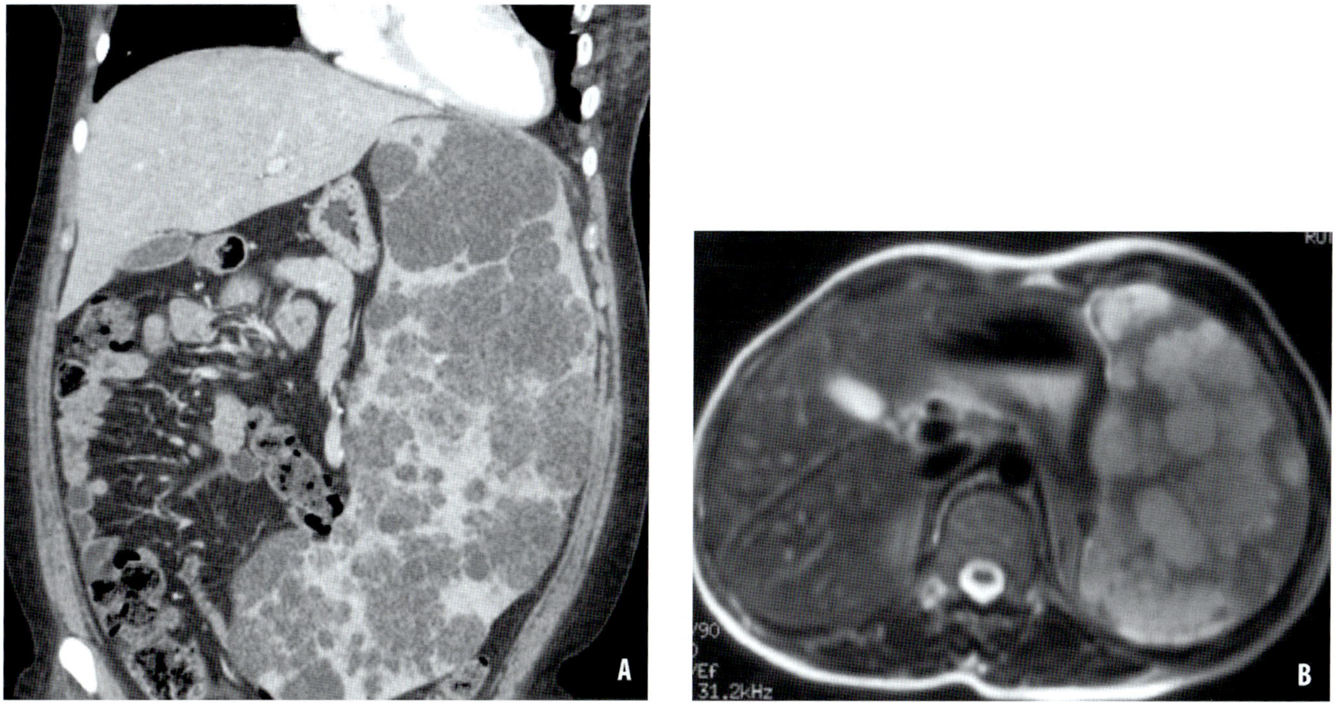

Figure 29.2 **A.** Coronal reformatted contrast-enhanced CT image in a 22 year old woman with Proteus syndrome and slowly progressive splenomegaly over the previous 10 years causing early satiety and pain with eating. The spleen is markedly enlarged by innumerable cyst-like lesions. Elective splenectomy showed multicystic hemangioma. Larger hemangiomas can be variably cystic. **B.** Axial T2-weighted MR image shows the hemangiomas are of markedly increased signal, consistent with their cystic nature.

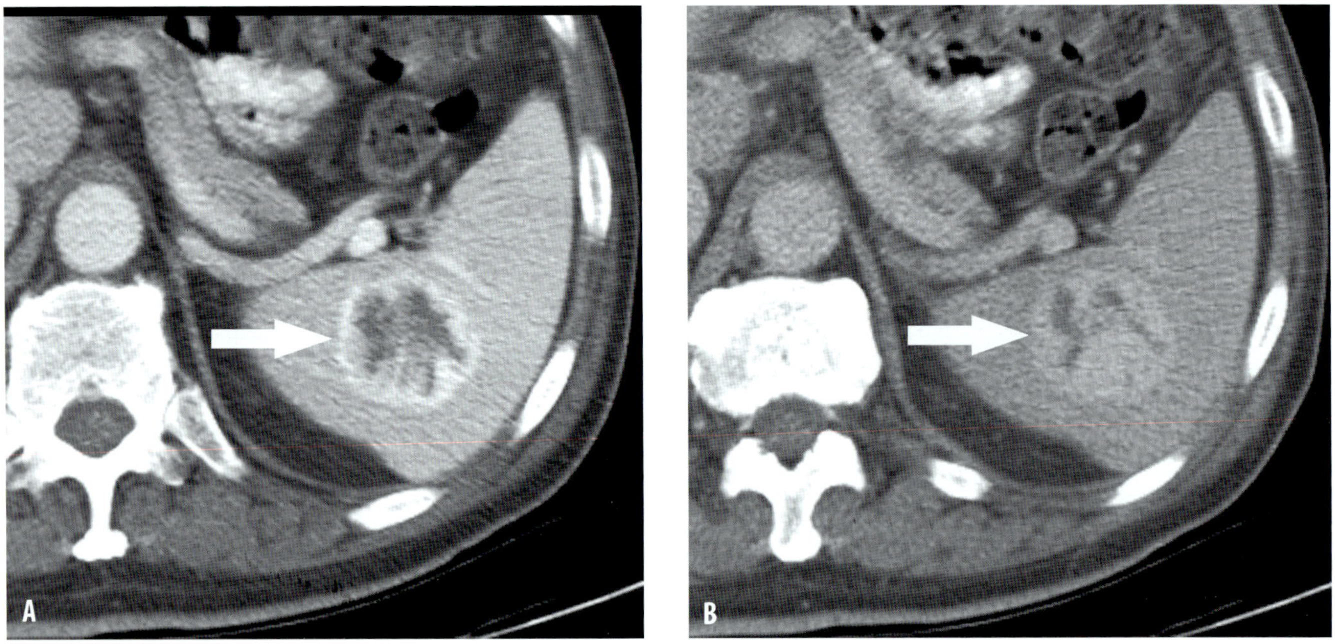

Figure 29.3 **A.** Axial contrast-enhanced CT image in a 70 year old man with locally recurrent rectal cancer shows a hypervascular peripherally enhancing mass (arrow) in the spleen. **B.** Axial delayed phase contrast-enhanced CT image at the same level shows centripetal enhancement within the mass (arrow). The mass was presumed to be a hemangioma given the characteristic imaging findings, and remained stable for over two years of follow-up, despite progressive disease in the pelvis.

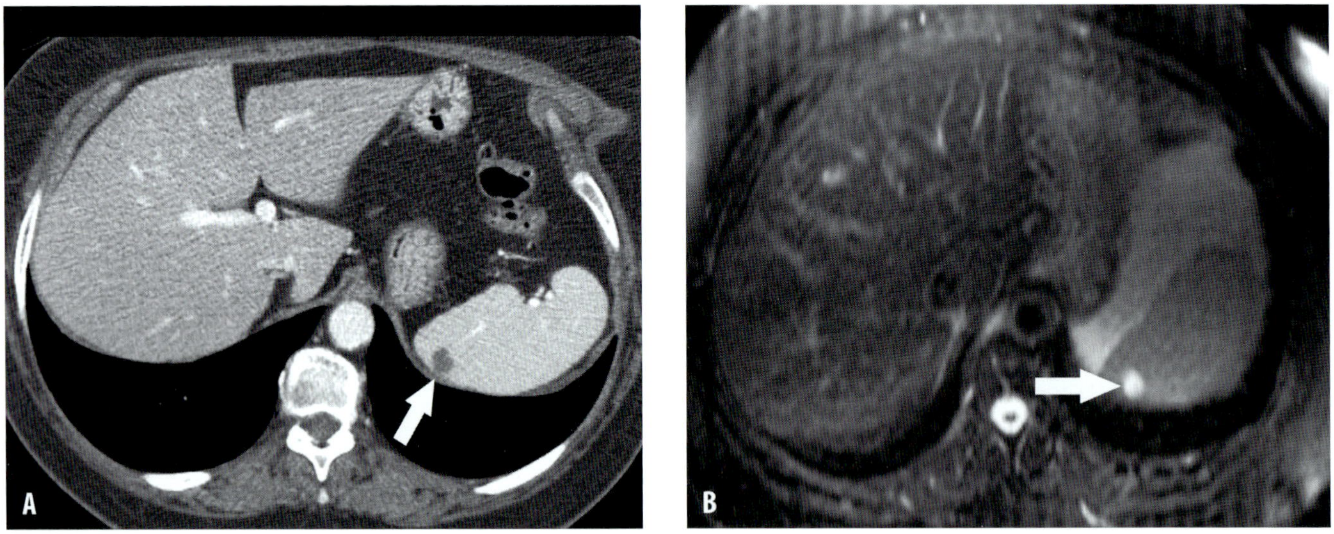

Figure 29.4 **A.** Axial contrast-enhanced CT image in a 63 year old woman with abdominal pain shows a non-specific hypovascular mass (arrow) in the spleen. **B.** Axial T2-weighted MR image shows the mass (arrow) is of markedly increased signal. **C.** Axial T1-weighted post-gadolinium MR image in the early phase shows the mass (arrow) is hypovascular, but difficult to discern against the background of heterogeneous splenic enhancement. **D.** Axial T1-weighted post-gadolinium MR image in the intermediate phase shows peripheral enhancement in the mass (arrow). **E.** Axial T1-weighted post-gadolinium MR image in the delayed phase shows uniform centripetal enhancement in the mass (arrow). The mass was presumed to be a hemangioma given the characteristic imaging findings, and remained stable for over two years of follow-up.

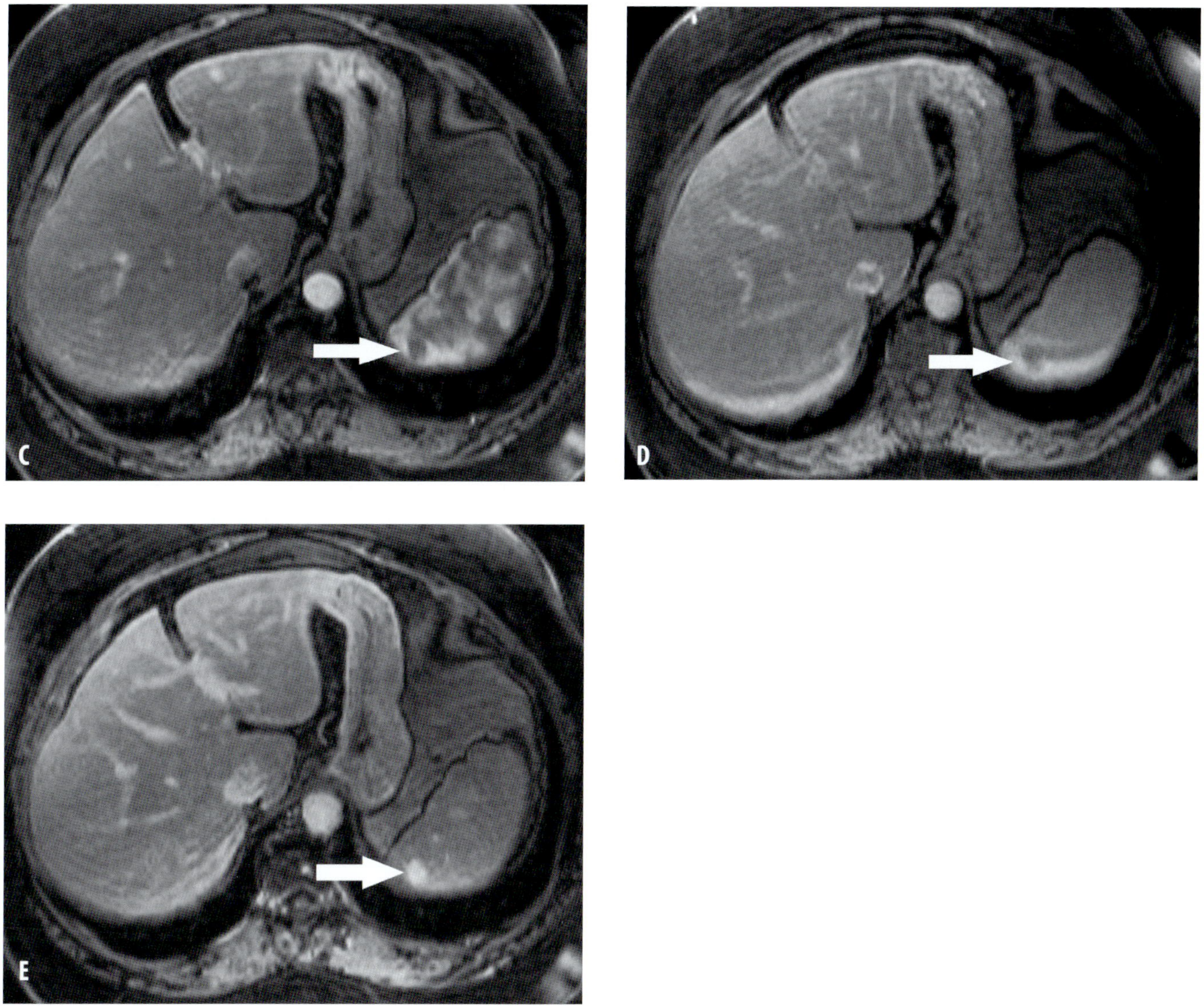

Figure 29.4 (cont.)

CASE **30**

Littoral cell angioma

Imaging description

Littoral cell angioma of the spleen is a rare vascular tumor first described in 1991 [1] that is thought to arise from the littoral cells which normally line the splenic sinuses of the red pulp. These lining cells have dual endothelial/vascular and macrophage/histiocytic potential, and this duality is a distinctive morphologic and immunophenotypic feature of littoral cell angioma. Littoral cell angioma typically occurs as multiple, similarly sized, well-circumscribed but non-encapsulated nodules of spongelike vascular spaces within a variably enlarged spleen [2]. At ultrasound, littoral cell angioma may manifest as diffuse heterogeneity or multiple nodules that are hypoechoic, isoechoic, or hyperechoic [3–6]. At non-enhanced or enhanced CT, the tumor is seen as multiple hypodense nodules. This finding is non-specific, but homogeneous enhancement of the lesions such that the nodules become isoattenuating and virtually invisible on delayed contrast-enhanced CT images appears to be a relatively distinctive diagnostic feature (Figure 30.1) [2, 6, 7]. At MRI, the nodules of littoral cell angioma are typically of low T1 and low T2 signal intensity, probably due to hemosiderin deposition secondary to phagocytosis of red blood cells [2].

Importance

The finding of multiple nodules in the spleen generally suggests serious pathology, such as metastases, lymphoma, abscesses, or granulomatous disease such as sarcoidosis or tuberculosis. This differential and the failure to consider littoral cell angioma may even result in splenectomy [6]. Interestingly, this may not be as bad as it sounds – while these tumors were originally considered benign, their natural history is not well established and malignant behavior has been reported [8, 9].

Typical clinical scenario

Littoral cell angiomas may occur at any age and have no gender predilection. The tumor may be discovered incidentally, but may also cause hypersplenism and so the splenic abnormality is often found during investigation of anemia, thrombocytopenia, or systemic ill health [1–3, 7]. In patients with hypersplenism, splenectomy may be appropriate for both definitive evaluation and treatment.

Differential diagnosis

Splenic metastases are rare in the absence of widely disseminated malignancy, lymphoma is usually associated with adenopathy elsewhere, and splenic abscesses are usually associated with abscesses in the liver. The role of fine-needle aspiration biopsy is not well established [10], but this might obviate splenectomy in those without clinically important hypersplenism [2]. Splenic biopsy is often considered risky because the spleen is a vascular organ, but the reported complication rate with image-guided percutaneous biopsy is 0–2% [11]. A complication rate of 10.3% has been reported for patients with refractory thrombocytopenia or vascular splenic neoplasms [12], so a careful evaluation of clotting function and risk to benefit ratio should occur before attempting to biopsy suspected littoral cell angioma.

Teaching point

Littoral cell angioma should be considered when multiple splenic nodules are found as an isolated imaging finding, particularly if associated with splenomegaly, hypersplenism, or isodense enhancement on delayed contrast-enhanced images. Biopsy may be appropriate in patients with good clotting function who do not require splenectomy.

REFERENCES

1 Falk S, Stutte HJ, Frizzera G. Littoral cell angioma: a novel splenic vascular lesion demonstrating histiocytic differentiation. *Am J Surg Pathol* 1991; **15**: 1023–1033.

2 Abbott RM, Levy AD, Aguilera NS, Gorospe L, Thompson WM. From the archives of the AFIP: primary vascular neoplasms of the spleen: radiologic-pathologic correlation. *Radiographics* 2004; **24**: 1137–1163.

3 Ziske C, Meybehm M, Sauerbruch T, Schmidt-Wolf IG. Littoral cell angioma as a rare cause of splenomegaly. *Ann Hematol* 2001; **80**: 45–48.

4 Goldfeld M, Cohen I, Loberant N, et al. Littoral cell angioma of the spleen: appearance on sonography and CT. *J Clin Ultrasound* 2002; **30**: 510–513.

5 Oliver-Goldaracena JM, Blanco A, Miralles M, Martin-Gonzalez MA. Littoral cell angioma of the spleen: US and MR imaging findings. *Abdom Imaging* 1998; **23**: 636–639.

6 Kinoshita LL, Yee J, Nash SR. Littoral cell angioma of the spleen: imaging features. *Am J Roentgenol* 2000; **174**: 467–469.

7 Levy AD, Abbott RM, Abbondanzo SL. Littoral cell angioma of the spleen: CT features with clinicopathologic comparison. *Radiology* 2004; **230**: 485–490.

8 Ben-Izhak O, Bejar J, Ben-Eliezer S, Vlodavsky E. Splenic littoral cell haemangioendothelioma: a new low-grade variant of malignant littoral cell tumour. *Histopathology* 2001; **39**: 469–475.

9 Meybehm M, Fischer HP. Littoral cell angiosarcoma of the spleen: morphologic, immunohistochemical findings and consideration of histogenesis of a rare splenic tumor. *Pathologie* 1997; **18**: 401–405.

10 Heese J, Bocklage T. Specimen fine-needle aspiration cytology of littoral cell angioma with histologic and immunohistochemical confirmation. *Diagn Cytopathol* 2000; **22**: 39–44.

11 Keogane MT, Freed KS, Paulson EK, Nelson RC, Dodd LG. Imaging-guided percutaneous biopsy of focal splenic lesions: update on safety and effectiveness. *Am J Roentgenol* 1999; **172**: 933–937.

12 Lucey BC, Boland GW, Maher MM, et al. Percutaneous nonvascular splenic intervention: a 10-year review. *Am J Roentgenol* 2002; **179**: 1591–1596.

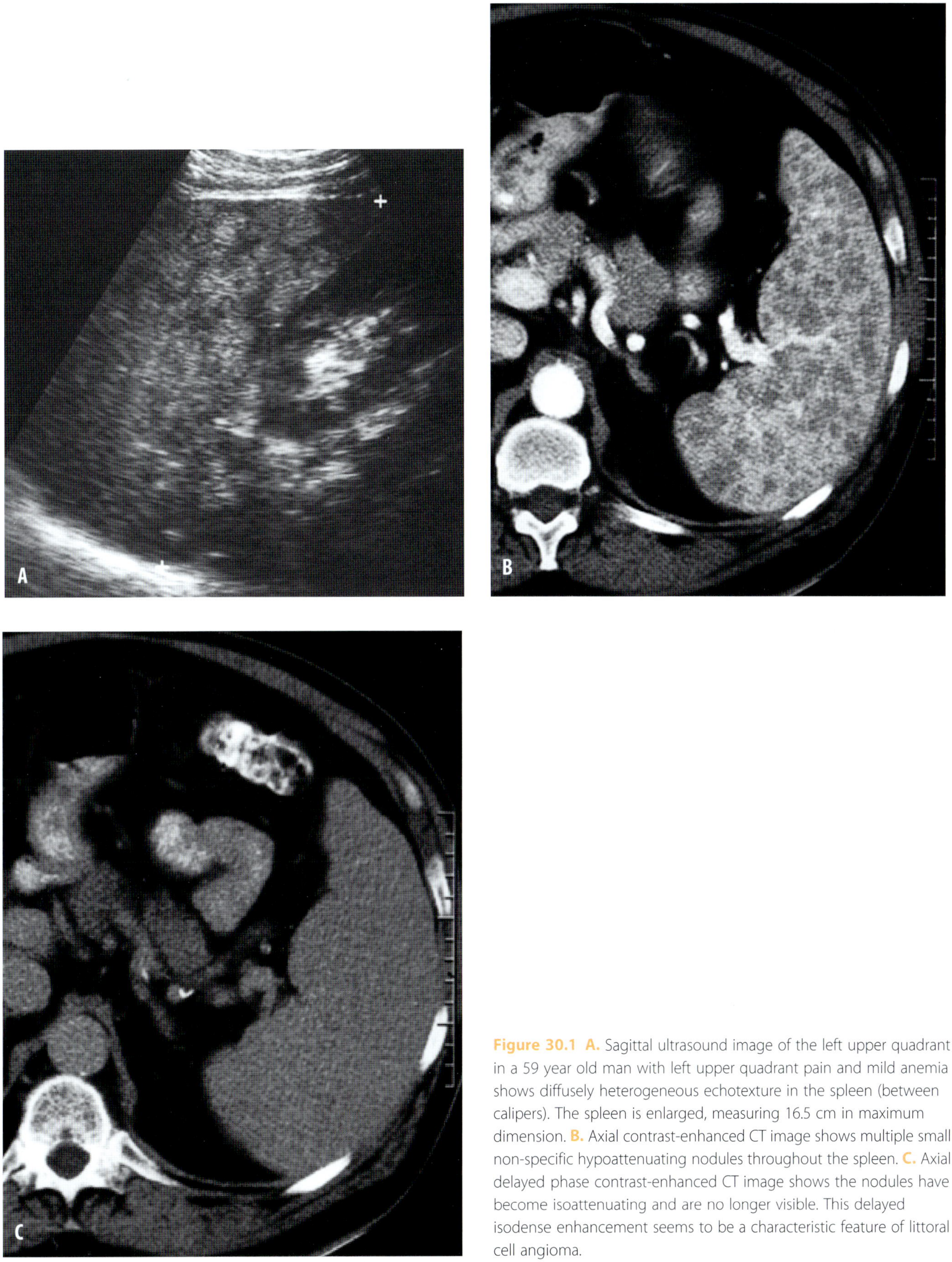

Figure 30.1 A. Sagittal ultrasound image of the left upper quadrant in a 59 year old man with left upper quadrant pain and mild anemia shows diffusely heterogeneous echotexture in the spleen (between calipers). The spleen is enlarged, measuring 16.5 cm in maximum dimension. B. Axial contrast-enhanced CT image shows multiple small non-specific hypoattenuating nodules throughout the spleen. C. Axial delayed phase contrast-enhanced CT image shows the nodules have become isoattenuating and are no longer visible. This delayed isodense enhancement seems to be a characteristic feature of littoral cell angioma.

CASE 31 Groove pancreatitis

Imaging description

Groove pancreatitis, first described in 1973, is a rare form of focal chronic pancreatitis that occurs in the "groove" between the pancreatic head, duodenum, and common bile duct [1–3]. Cross-sectional imaging typically demonstrates a mass in the pancreatic head adjacent to the duodenum that is hypoechoic at ultrasound, poorly enhancing at CT, and iso- or slightly hyperintense on T2-weighed MRI with delayed enhancement after intravenous gadolinium on T1-weighted images (Figures 31.1 and 31.2) [4–6]. The pancreatic and common bile duct may show smooth tapering but are usually not markedly dilated and jaundice is rare. Intralesional cysts and duodenal narrowing or wall thickening may also be seen [6, 7].

Importance

Groove pancreatitis is frequently misdiagnosed as pancreatic adenocarcinoma. Many cases go to surgery with malignancy as the preoperative diagnosis [8].

Typical clinical scenario

Groove pancreatitis has been primarily reported in middle-aged alcoholic men [2, 3]. The etiology of the condition is unknown, but suggestions include inflammation secondary to blockage of the duct of Santorini or inflammation of heterotopic pancreatic tissue [8].

Differential diagnosis

The confident distinction between groove pancreatitis and adenocarcinoma of the pancreatic head is often difficult on imaging. The lack of marked biliary dilatation in the presence of an ill-defined pancreatic head mass should suggest a diagnosis other than primary pancreatic adenocarcinoma, and raise consideration of groove pancreatitis. Lymphoplasmacytic (sclerosing) pancreatitis, also known as autoimmune pancreatitis, is a rare form of chronic pancreatitis characterized pathologically by a mixed inflammatory infiltrate that centers on the pancreatic ducts [9]. Autoimmune pancreatitis can be focal and masslike at imaging and resemble groove pancreatitis (Figure 31.3). Unlike groove pancreatitis, it is not associated with alcohol use, is commonly associated with other autoimmune conditions, and responds well to steroids. Neuroendocrine tumor and metastasis to the pancreas can also result in a pancreatic head mass with little or no biliary dilatation, but are usually better defined (Figures 31.4 and 31.5). In addition, neuroendocrine tumor is often of hyperintense T2 signal intensity on MRI (Figure 31.5).

Teaching point

Groove pancreatitis should be considered when an ill-defined pancreatic head mass with little or no biliary dilatation is seen at imaging in a middle-aged alcoholic man.

REFERENCES

1 Becker V. Bauchepeicheldrüse. In: Doerr W, Seifert G, Ühlinger E, eds. *Spezielle Pathologische Anatomie.* Berlin: Springer, 1973; 252–445.

2 Stole M, Weiss W, Volkholz H, Rosch W. A special type of segmental pancreatitis: "groove pancreatitis". *Hepatogastroenterology* 1982; **29**: 198–208.

3 Becker V, Mischke U. Groove pancreatitis. *Int J Pancreatol* 1991; **10**: 173–182.

4 Balakrishnan V, Chatni S, Radhakrishnan L, Narayanan VA, Nair P. Groove pancreatitis: a case report and review of literature. *JOP* 2007; **8**: 592–597.

5 Itoh S, Yamakawa K, Shimamoto K, Endo T, Ishigaki T. CT findings in groove pancreatitis: correlation with histopathological findings. *J Comput Assist Tomogr* 1994; **18**: 911–915.

6 Irie H, Honda H, Kuroiwa T, et al. MRI of groove pancreatitis. *J Comput Assist Tomogr* 1998; **22**: 651–665.

7 Blasbalg R, Baroni RH, Costa DN, Machado MC. MRI features of groove pancreatitis. *Am J Roentgenol* 2007; **189**: 73–80.

8 Shudo R, Obara T, Tanno S, et al. Segmental groove pancreatitis accompanied by protein plugs in Santorini's duct. *J Gastroenterol* 1998; **33**: 289–294.

9 Kawamoto S, Siegelman SS, Hruban RH, Fishman EK. Lymphoplasmacytic sclerosing pancreatitis (autoimmune pancreatitis): evaluation with multidetector CT. *Radiographics* 2008; **28**: 157–170.

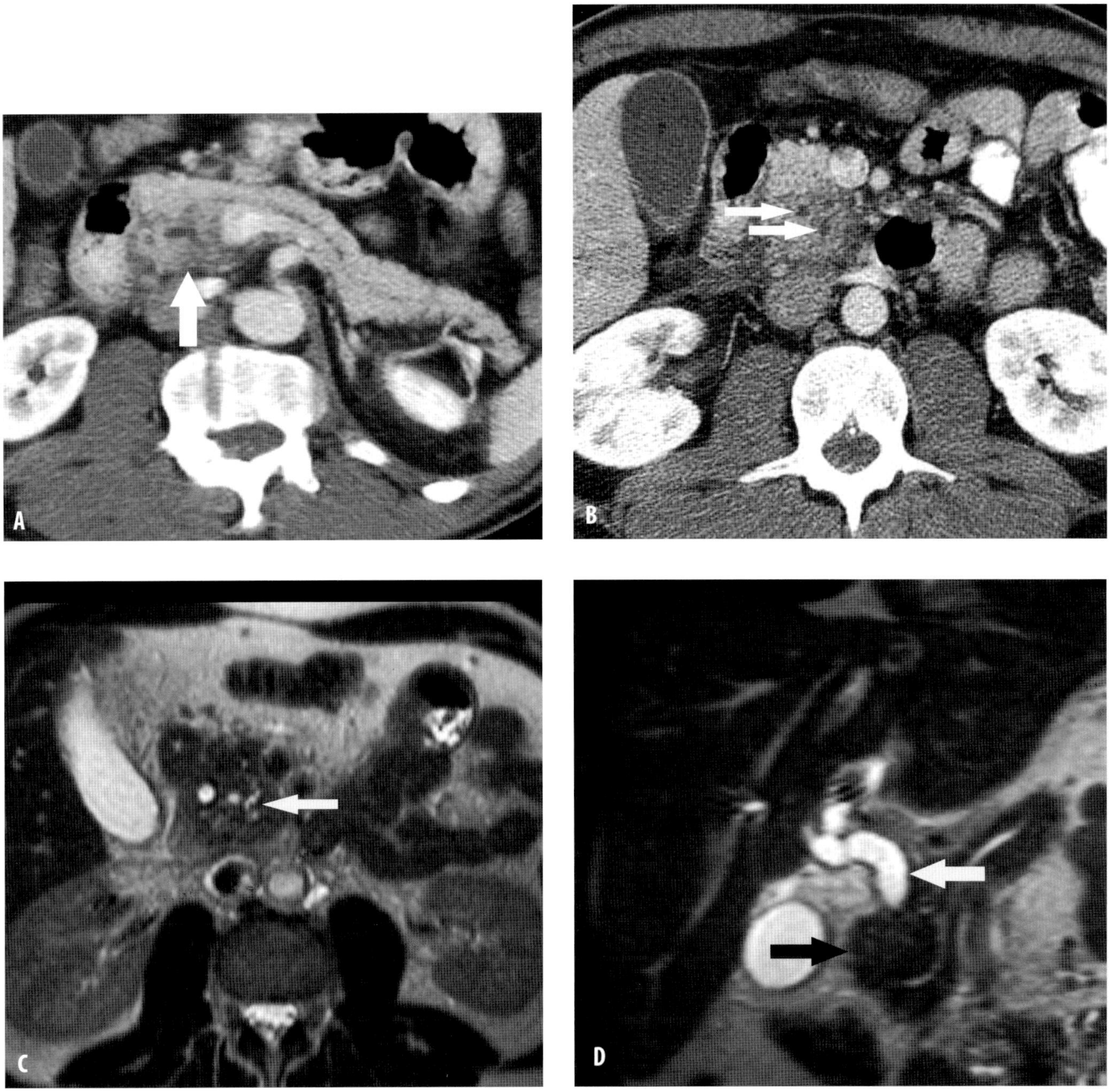

Figure 31.1 **A.** Curved planar axial contrast-enhanced CT image in a 45 year old man with several months of increasing right upper quadrant pain and a history of heavy alcohol use. An ill-defined mass (arrows) is seen between the pancreatic head and duodenum. **B.** Axial contrast-enhanced CT image at another level through the mass shows several small cyst-like spaces (arrows). **C.** Axial T2-weighted MR image shows the cyst-like lesions are branching tubular structures (arrow), presumably small dilated ducts. Such a finding would not be expected within a pancreatic adenocarcinoma. **D.** Coronal T2-weighted MR image shows the mass (black arrow) is causing only minimal fullness of the common bile duct (white arrow). Note that a pancreatic adenocarcinoma of this size in this location would typically cause marked biliary obstruction and jaundice. **E.** Axial gadolinium-enhanced T1-weighted MR image shows the mass (white arrow) is poorly enhancing when compared to the normal adjacent pancreatic parenchyma (grey arrow). Pancreatoduodenectomy (Whipple's procedure) was performed for presumed adenocarcinoma. Histopathological analysis revealed groove pancreatitis.

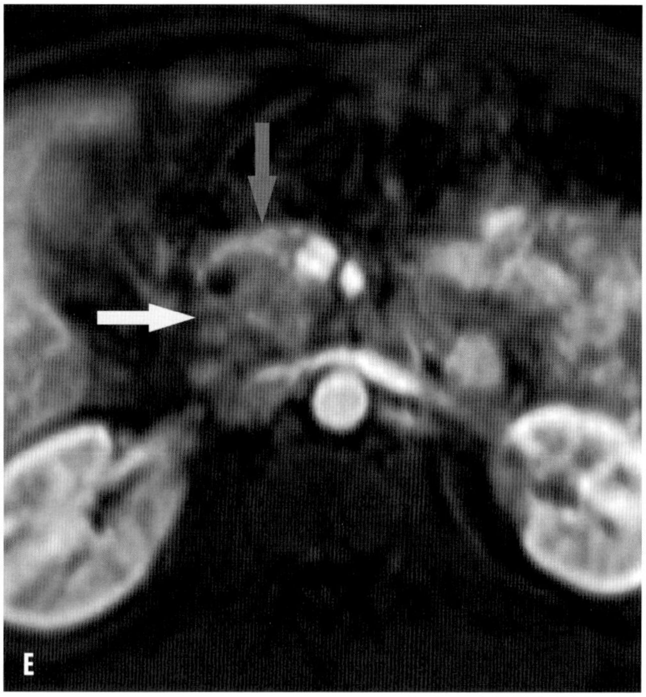

Figure 31.1 (cont.)

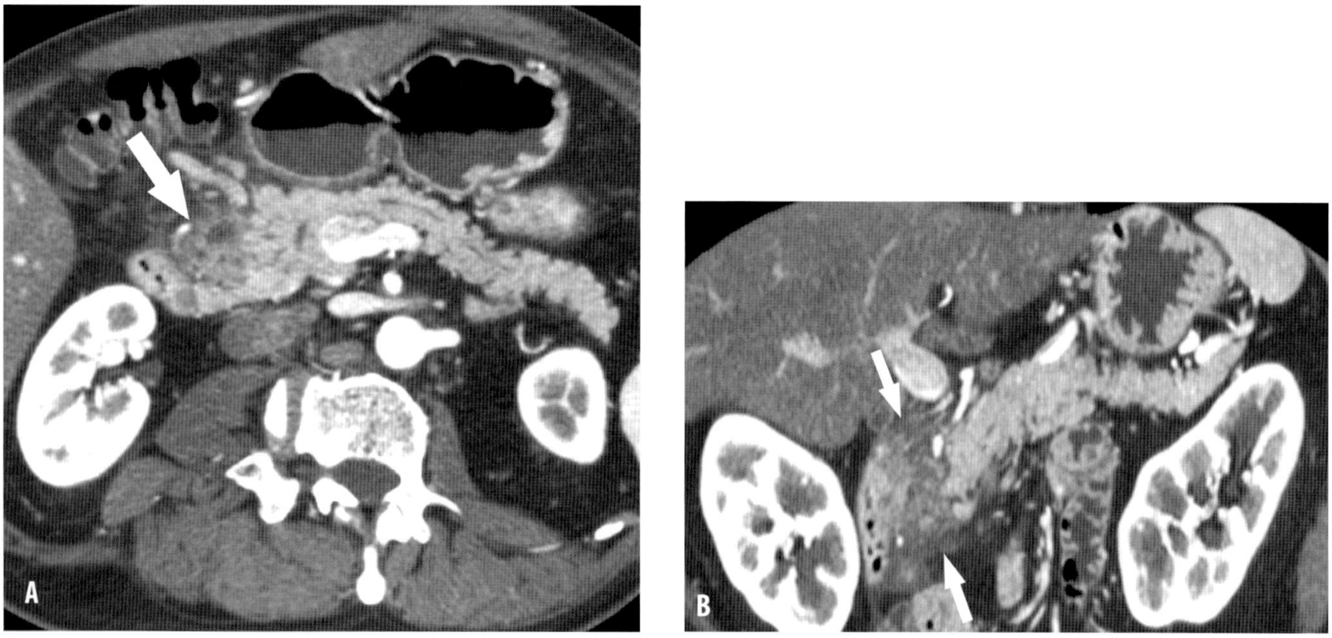

Figure 31.2 A. Curved planar axial contrast-enhanced CT image in a 55 year old man with a one-year history of abdominal pain and weight loss and a prior history of heavy alcohol use. A hypodense mass (arrow) is seen in the groove between the pancreatic head and duodenum. **B.** Curved planar coronal contrast-enhanced CT image shows the mass (between arrows) lies in the groove between the pancreas and duodenum and is not causing biliary obstruction (note absence of dilated bile ducts in the liver), which would be atypical for a pancreatic adenocarcinoma of this size and location. Biopsy confirmed the diagnosis of groove pancreatitis.

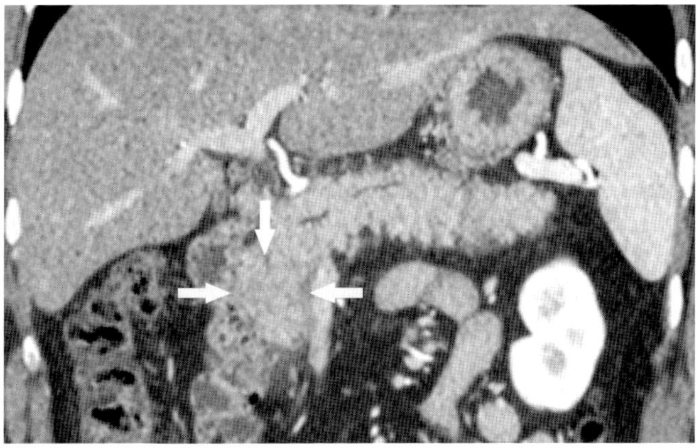

Figure 31.3 Axial contrast-enhanced CT image in a 49 year old man with several months of abdominal pain and bloating. An ill-defined subtle mass (arrows) is seen in the pancreatic head. Pancreatoduodenectomy (Whipple's procedure) was performed for presumed adenocarcinoma. Histopathological analysis revealed autoimmune pancreatitis.

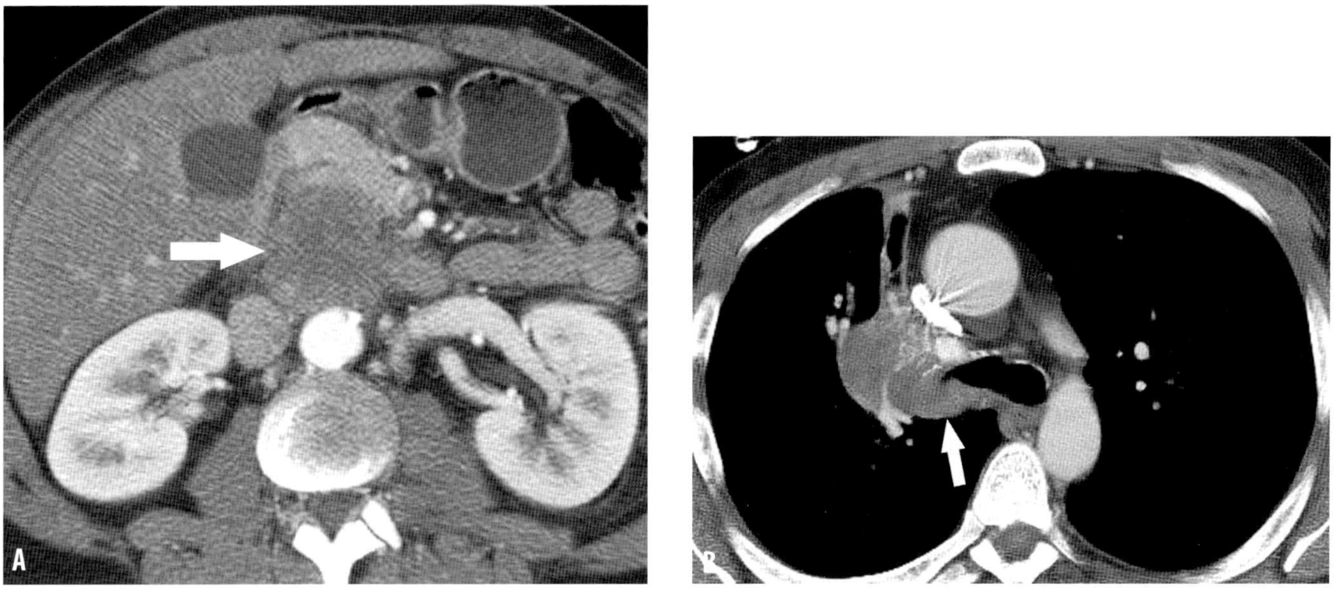

Figure 31.4 **A.** Axial contrast-enhanced CT image in a 50 year old man with abdominal pain and weight loss and history of smoking and heavy alcohol use. A hypodense mass (arrow) adjacent to the pancreatic head is rather too well defined to represent groove pancreatitis. The lack of biliary dilatation (note absence of dilated bile ducts in the liver) argues against pancreatic adenocarcinoma. **B.** Axial contrast-enhanced CT image through the chest shows a mass (arrow) related to the right upper lobe bronchus. Biopsy confirmed a primary non-small cell lung cancer with metastasis to the pancreas.

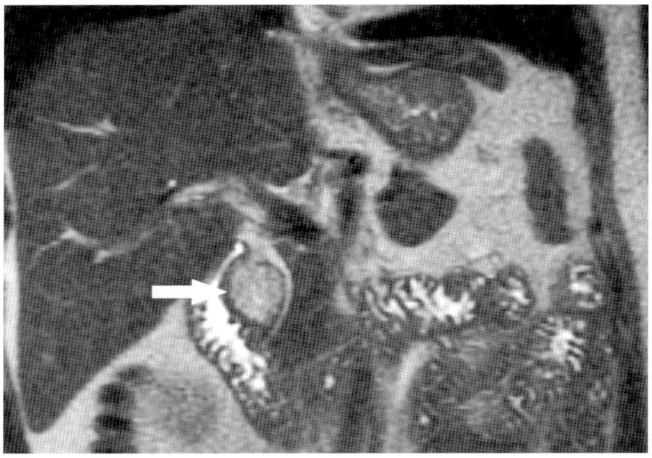

Figure 31.5 Coronal T2-weighted MR image in 57 year old woman with an asymptomatic pancreatic mass found incidentally at screening abdominal ultrasound for chronic hepatitis B. A mass (arrow) between the pancreatic duct and duodenum is too well defined and too bright to be groove pancreatitis. A diagnosis of non-functional neuroendocrine tumor was established by resection.

CASE 32 Intrapancreatic accessory spleen

Imaging description

At cross-sectional imaging, intrapancreatic accessory spleens appear as incidental and relatively non-specific solid smooth enhancing masses of variable size in the pancreatic tail (Figures 32.1–32.3). On ultrasound, accessory spleens are usually mildly and homogeneously echogenic with posterior acoustic enhancement, and may demonstrate vascular communication with the splenic vessels [1]. At dynamic CT, accessory spleens remain isointense to the spleen across all phases [2, 3]. On MRI, intrapancreatic spleens show low T1 signal intensity, high T2 signal intensity, and isointensity to the spleen on post-gadolinium dynamic sequences [2–4]. The signal on T2-weighted sequences can be slightly brighter than that of the spleen [2–3]. At contrast-enhanced CT or MRI, an intrapancreatic accessory spleen may demonstrate lesser or comparable enhancement relative to the pancreas in the arterial phase and lesser, comparable or higher enhancement on later phases [4], though in my experience an intrapancreatic accessory spleen enhances in parallel with the native spleen and so is usually slightly hypervascular relative to the pancreas.

Importance

An intrapancreatic accessory spleen may mimic more sinister pancreatic pathology including malignancy, resulting in unnecessary surgery unless a benign diagnosis can be confidently established [2, 5].

Typical clinical scenario

An intrapancreatic accessory spleen was found in 61 of 3000 autopsies (2%), as an incidental congenital variant [6]. Given that the diagnosis is rarely made in radiological practice, it is likely most intrapancreatic spleens are occult at cross-sectional imaging. Occasionally, an intrapancreatic accessory spleen may be seen after splenectomy (Figure 32.3), presumably reflecting compensatory hypertrophy.

Differential diagnosis

Differential diagnoses for a solid enhancing mass detected in the pancreas at ultrasound, CT, or MRI include neuroendocrine tumor, adenocarcinoma, metastases, or lymphoma [7]. Neuroendocrine tumors can be radiologically indistinguishable from intrapancreatic accessory spleen because they can also be solid, hypervascular, and incidental (Figure 32.4) [8]. Adenocarcinoma is usually hypovascular, irregular, and associated with ductal obstruction. Metastases typically occur in the setting of a known primary malignancy, particularly lung cancer, breast cancer, melanoma, or renal cell carcinoma. Of these, renal cell cancer metastases can be hypervascular and could resemble an intrapancreatic accessory spleen (Figure 32.5). In cases where additional characterization is required, advanced imaging techniques such as superparamagnetic iron oxide-enhanced MRI, sulfur colloid scintigraphy, or heat-damaged red blood cell scintigraphy may be helpful [2, 4, 9]. Accessory spleens contain Kuppfer cells which take up superparamagnetic iron oxide particles and will therefore demonstrate signal loss after the administration of this contrast agent. Nuclear medicine tests for characterization of splenic tissue also rely on processes specific to the spleen, such as the uptake of sulfur colloid by Kuppfer cells or the recycling of damaged red blood cells (Figures 32.1 and 32.2). Finally, it has been shown that splenic tissue is typically of high signal on diffusion-weighted MRI [10] and it is possible that this may be a relatively straightforward method of making the diagnosis of intrapancreatic accessory spleen, based on the strikingly similar high signal intensity between the pancreatic mass and the native spleen on diffusion-weighted MRI (Figure 32.6). Contrast-enhanced ultrasound has been reported to show similar enhancement patterns between the accessory and native spleen [11], but this technology is not widely available. Finally, it should be remembered that accessory spleens are prone to the same pathological processes as the native spleen, including infarction and neoplasia [12, 13], and such processes may add complexity to the imaging findings and evaluation.

Teaching point

Intrapancreatic accessory spleen should be included in the differential considerations for an incidental solid smooth enhancing mass in the pancreatic tail. Imaging options for specific characterization include sulfur colloid or heat-damaged red blood cell scintigraphy, superparamagnetic iron oxide-enhanced MRI, and possibly diffusion-weighted MRI.

REFERENCES

1 Subramanyam BR, Balthazar EJ, Horii SC. Sonography of the accessory spleen. *Am J Roentgenol* 1984; **143**: 47–49.

2 Kim SH, Lee JM, Han JK, et al. MDCT and superparamagnetic iron oxide (SPIO)-enhanced MR findings of intrapancreatic accessory spleen in seven patients. *Eur Radiol* 2006; **16**: 1887–1897.

3 Kim SH, Lee JM, Han JK, et al. Intrapancreatic accessory spleen: findings on MR imaging, CT, US and scintigraphy, and the pathologic analysis. *Korean J Radiol* 2008; **9**: 162–174.

4 Herédia V, Altun E, Bilaj F, et al. Gadolinium- and superparamagnetic-iron-oxide enhanced MR findings of intrapancreatic accessory spleen in five patients. *Magn Reson Imaging* 2008; **26**: 1273–1278.

5 Uchiyama S, Chijiiwa K, Hiyoshi M, et al. Intrapancreatic accessory spleen mimicking endocrine tumor of the pancreas: case report and review of the literature. *J Gastrointest Surg* 2008; **12**: 1471–1473.

6 Halpert B, Gyorkey F. Lesions observed in accessory spleens of 311 patients. *Am J Clin Pathol* 1959; **32**: 165–168.

7 Kalra MK, Maher MM, Mueller PR, Saini S. State-of-the-art imaging of pancreatic neoplasms. *Br J Radiol* 2003; **76**: 857–865.

8 Fidler JL, Johnson CD. Imaging of neuroendocrine tumors of the pancreas. *Int J Gastrointest Cancer* 2001; **30**: 73–85.
9 Ota T, Tei M, Yoshioka A, et al. Intrapancreatic accessory spleen diagnosed by technetium-99m heat-damaged red blood cell SPECT. *J Nucl Med* 1997; **38**: 494–495.
10 Yoshikawa T, Kawamitsu H, Mitchell DG, et al. ADC measurement of abdominal organs and lesions using parallel imaging technique. *Am J Roentgenol* 2006; **187**: 1521–1530.
11 Kim SH, Lee JM, Lee JY, et al. Contrast-enhanced sonography of intrapancreatic accessory spleen in six patients. *Am J Roentgenol* 2007; **188**: 422–428.
12 Itano O, Shiraga N, Kouta E, et al. Epidermoid cyst originating from an intrapancreatic accessory spleen. *J Hepatobiliary Pancreat Surg* 2008; **15**: 436–439.
13 Lee S, Hwang HI, Yum SM, et al. Accessory splenic infarction presenting as a hemorrhagic tumor in the pancreas. *Korean J Gastroenterol* 2008; **52**: 48–51.

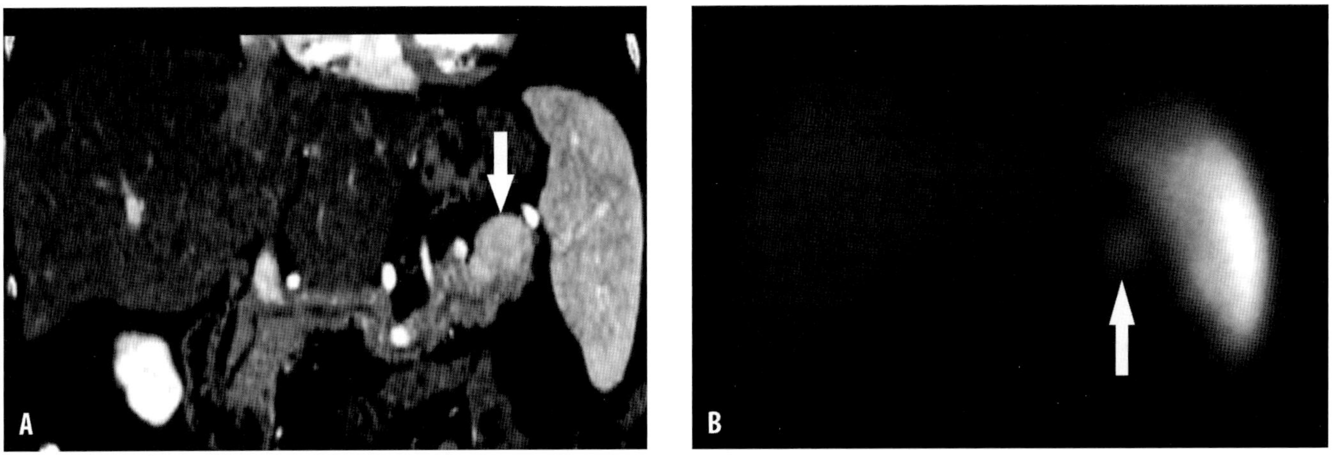

Figure 32.1 **A.** Coronal reformatted contrast-enhanced CT image in a 74 year old woman with chronic abdominal pain showing a smooth enhancing solid mass (arrow) in the pancreatic tail. The patient was referred for possible surgery, but the possibility of an intrapancreatic accessory spleen was suggested after further review of the images. **B.** Coronal image from heat-damaged red blood cell scintigraphy demonstrates uptake in the mass (arrow), confirming the diagnosis of an intrapancreatic accessory spleen.

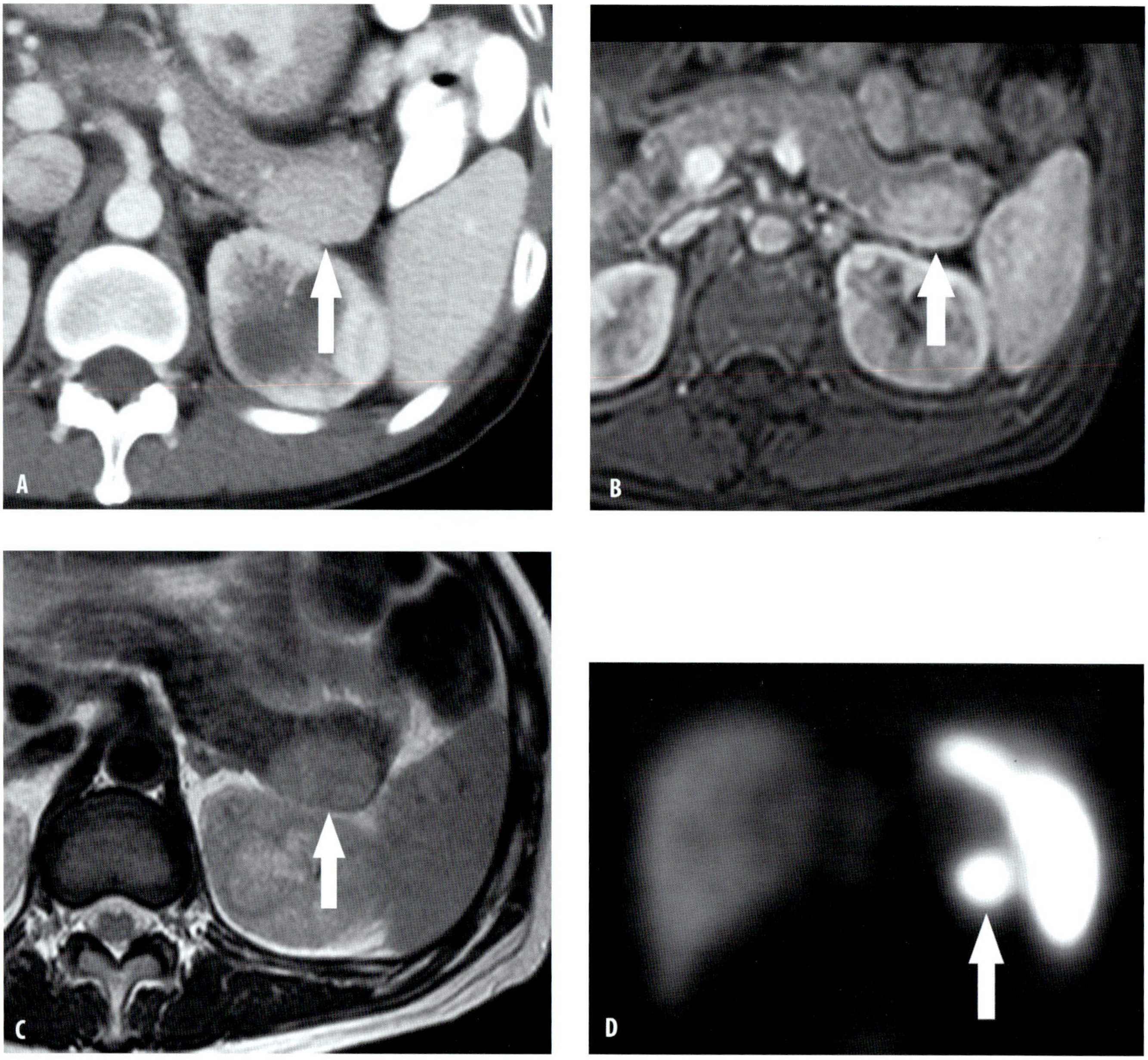

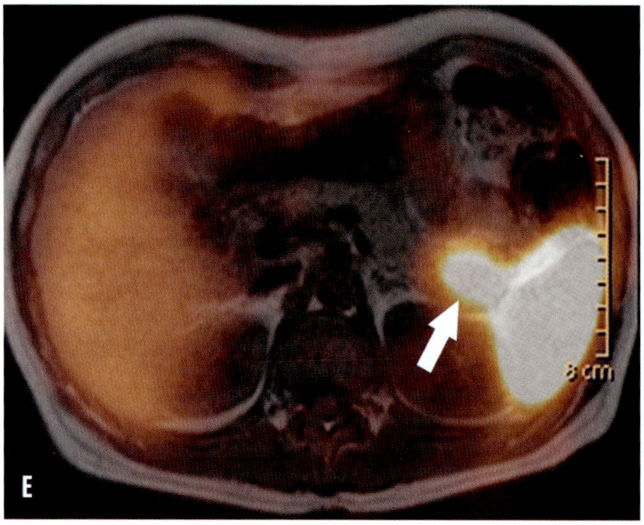

Figure 32.2 **A.** Axial contrast-enhanced CT image in a 44 year old man being staged for lymphoma showing a smooth enhancing solid mass (arrow) in the pancreatic tail. **B.** Axial T1-weighted gadolinium-enhanced MR image with fat saturation confirms the presence of a solid enhancing mass (arrow) in the pancreatic tail. **C.** Axial T2-weighted MR image shows the mass (arrow) is of moderately high signal intensity that is similar to the signal intensity of the native spleen. **D.** Coronal image from heat-damaged red blood cell scintigraphy demonstrates uptake in the mass confirming the diagnosis of an intrapancreatic accessory spleen (arrow). **E.** Fused axial T1-weighted unenhanced MR and heat-damaged red blood cell scintigraphic images showing the increased radiotracer uptake corresponds to the pancreatic tail mass.

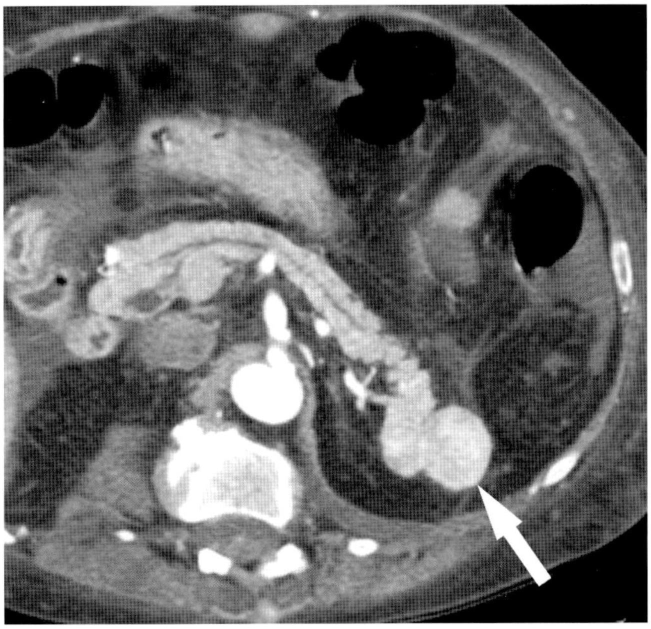

Figure 32.3 Axial curved planar reformatted contrast-enhanced CT image in a 59 year old woman with chronic abdominal pain, known right heart failure, and a remote history of splenectomy after trauma. A lobulated hypervascular intrapancreatic accessory spleen (arrow) is relatively large, presumably reflecting post-splenectomy hypertrophy.

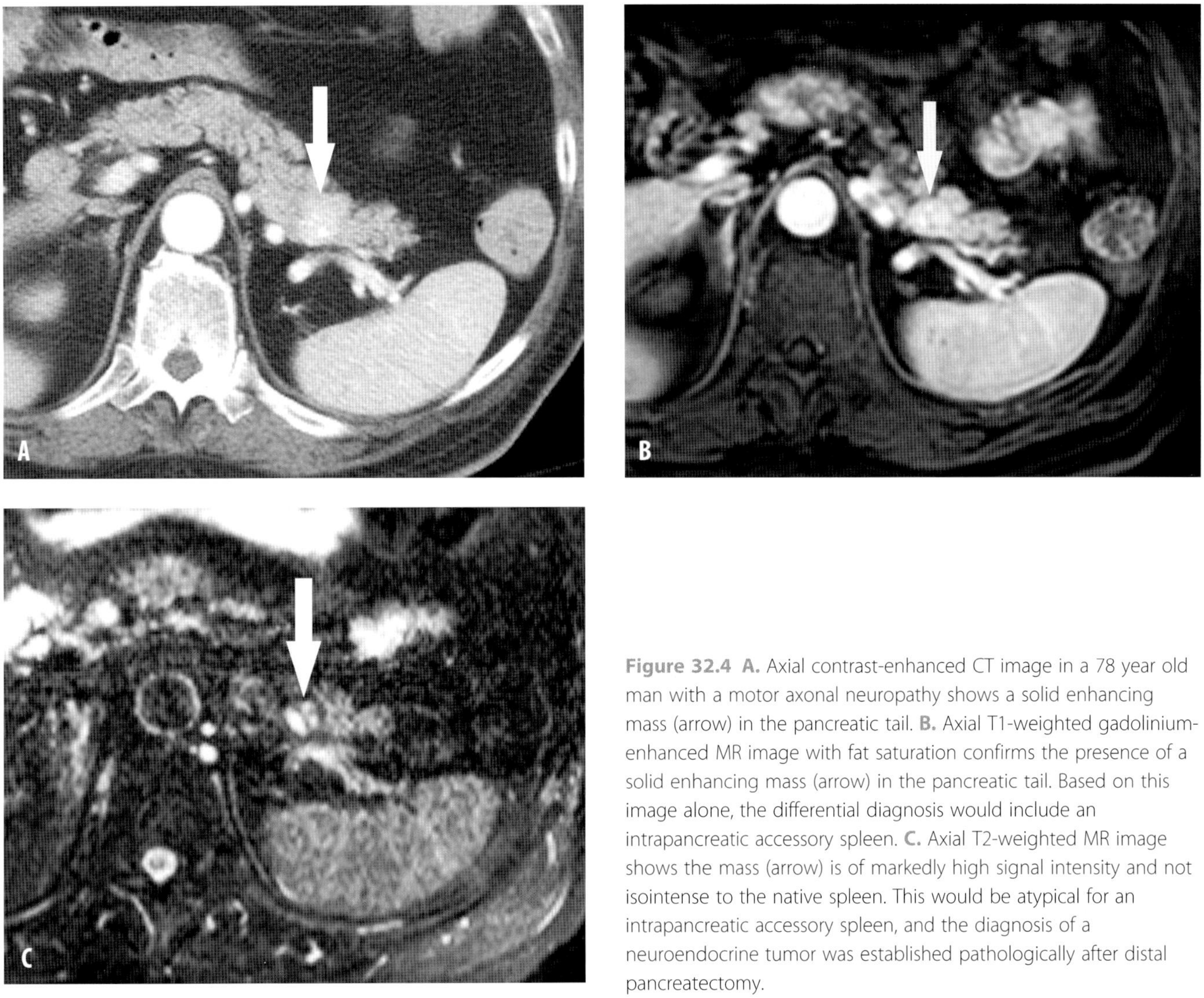

Figure 32.4 **A.** Axial contrast-enhanced CT image in a 78 year old man with a motor axonal neuropathy shows a solid enhancing mass (arrow) in the pancreatic tail. **B.** Axial T1-weighted gadolinium-enhanced MR image with fat saturation confirms the presence of a solid enhancing mass (arrow) in the pancreatic tail. Based on this image alone, the differential diagnosis would include an intrapancreatic accessory spleen. **C.** Axial T2-weighted MR image shows the mass (arrow) is of markedly high signal intensity and not isointense to the native spleen. This would be atypical for an intrapancreatic accessory spleen, and the diagnosis of a neuroendocrine tumor was established pathologically after distal pancreatectomy.

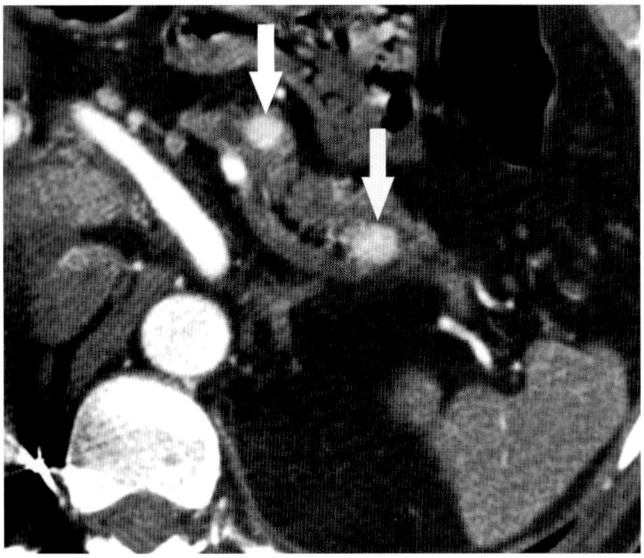

Figure 32.5 Axial contrast-enhanced arterial phase CT image in a 64 year old woman with a history of right nephrectomy for renal cell carcinoma shows two hypervascular metastases (arrows) in the pancreas. Metastases to the pancreas from renal cell carcinoma may enhance and resemble intrapancreatic accessory spleens, but note in this case the masses are enhancing more than the native spleen, suggesting they do not represent accessory splenic tissue.

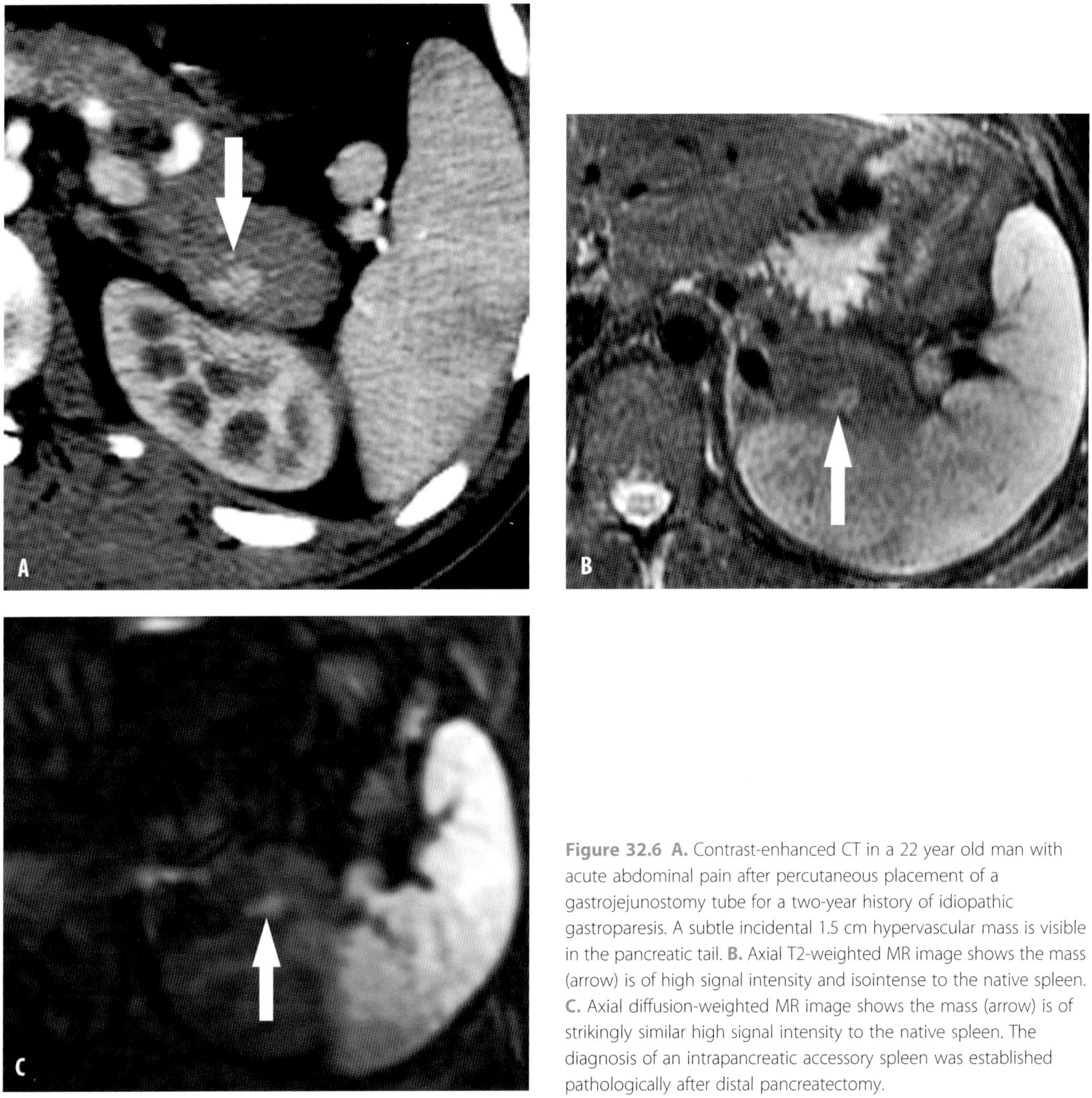

Figure 32.6 A. Contrast-enhanced CT in a 22 year old man with acute abdominal pain after percutaneous placement of a gastrojejunostomy tube for a two-year history of idiopathic gastroparesis. A subtle incidental 1.5 cm hypervascular mass is visible in the pancreatic tail. **B.** Axial T2-weighted MR image shows the mass (arrow) is of high signal intensity and isointense to the native spleen. **C.** Axial diffusion-weighted MR image shows the mass (arrow) is of strikingly similar high signal intensity to the native spleen. The diagnosis of an intrapancreatic accessory spleen was established pathologically after distal pancreatectomy.

CASE 33 # Pancreatic cleft

Imaging description

The normal pancreas has a lobulated surface. Occasionally peripancreatic fat "trapped" within a deep cleft or fold between these lobulations may mimic a mass or fracture at cross-sectional imaging (Figures 33.1 and 33.2) [1–3].

Importance

An erroneous diagnosis of a pancreatic mass or fracture may result in unnecessary additional testing and patient anxiety.

Typical clinical scenario

Pancreatic clefts are anatomic variants that are likely to cause most diagnostic confusion when seen incidentally in patients being imaged for unrelated reasons (may be mistaken for a mass) or after trauma (may be mistaken for a fracture).

Differential diagnosis

The diagnosis of a pancreatic cleft should be considered for any apparent small pancreatic lesion that abuts the gland surface, and can usually be confirmed by examining non-enhanced or thin section CT images [1]. MRI sequences with and without fat saturation may also be helpful.

Teaching point

The possibility of a pancreatic cleft should be considered for an apparent small pancreatic lesion that abuts the gland surface.

REFERENCES

1 Freeny PC. Cross-sectional imaging of the pancreas. *Bailliére's Clin Gastroenterol* 1995; **9**: 135–151.

2 Brandon JC, Izenberg SD, Fields PA, et al. Pancreatic clefts caused by penetrating vessels: a potential diagnostic pitfall for pancreatic fracture on CT. *Emerg Radiol* 2000; **7**: 283–286.

3 Cirillo RL Jr, Koniaris LG. Detecting blunt pancreatic injuries. *J Gastrointest Surg* 2002; **6**: 587–598.

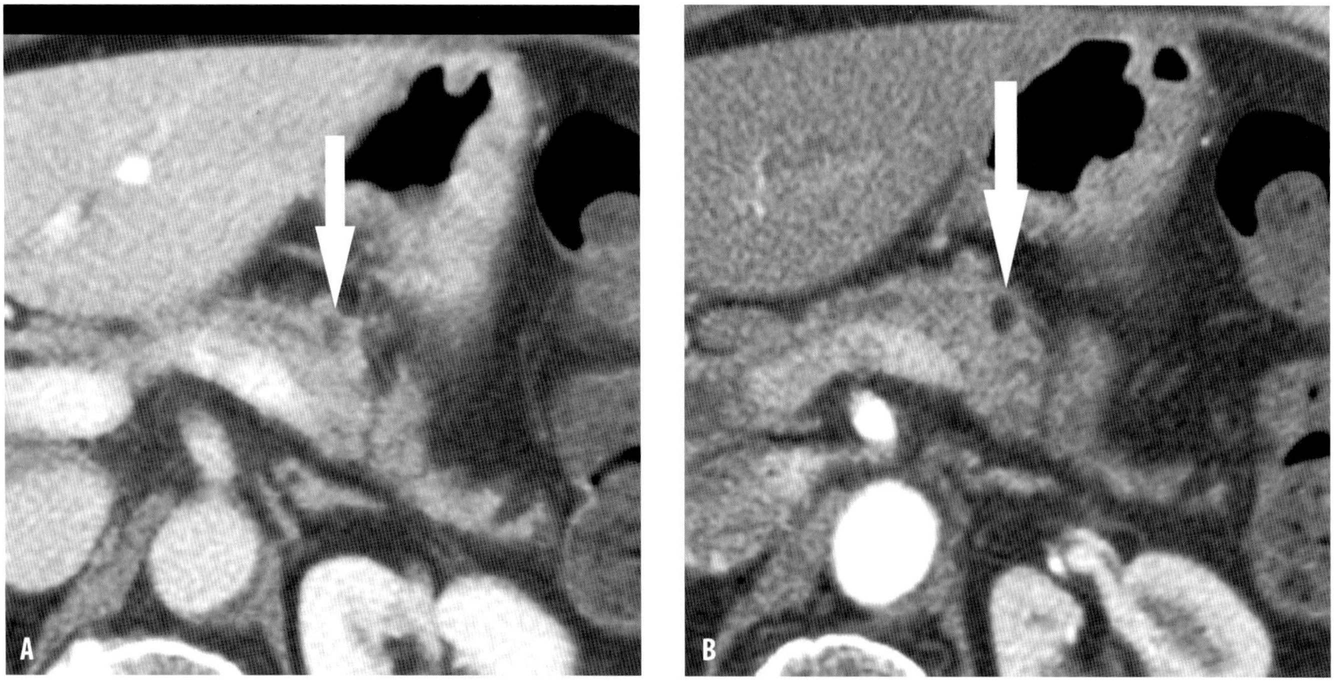

Figure 33.1 A. Axial contrast-enhanced portal venous phase CT image obtained at 5 mm collimation in a 64 year old woman with chronic hepatitis C undergoing surveillance for hepatocellular carcinoma shows an apparent small hypodense mass (arrow) in the pancreas. **B.** Axial contrast-enhanced arterial phase CT image obtained at 2.5 mm collimation shows the apparent mass (arrow) is of fat density, consistent with fat within a pancreatic cleft.

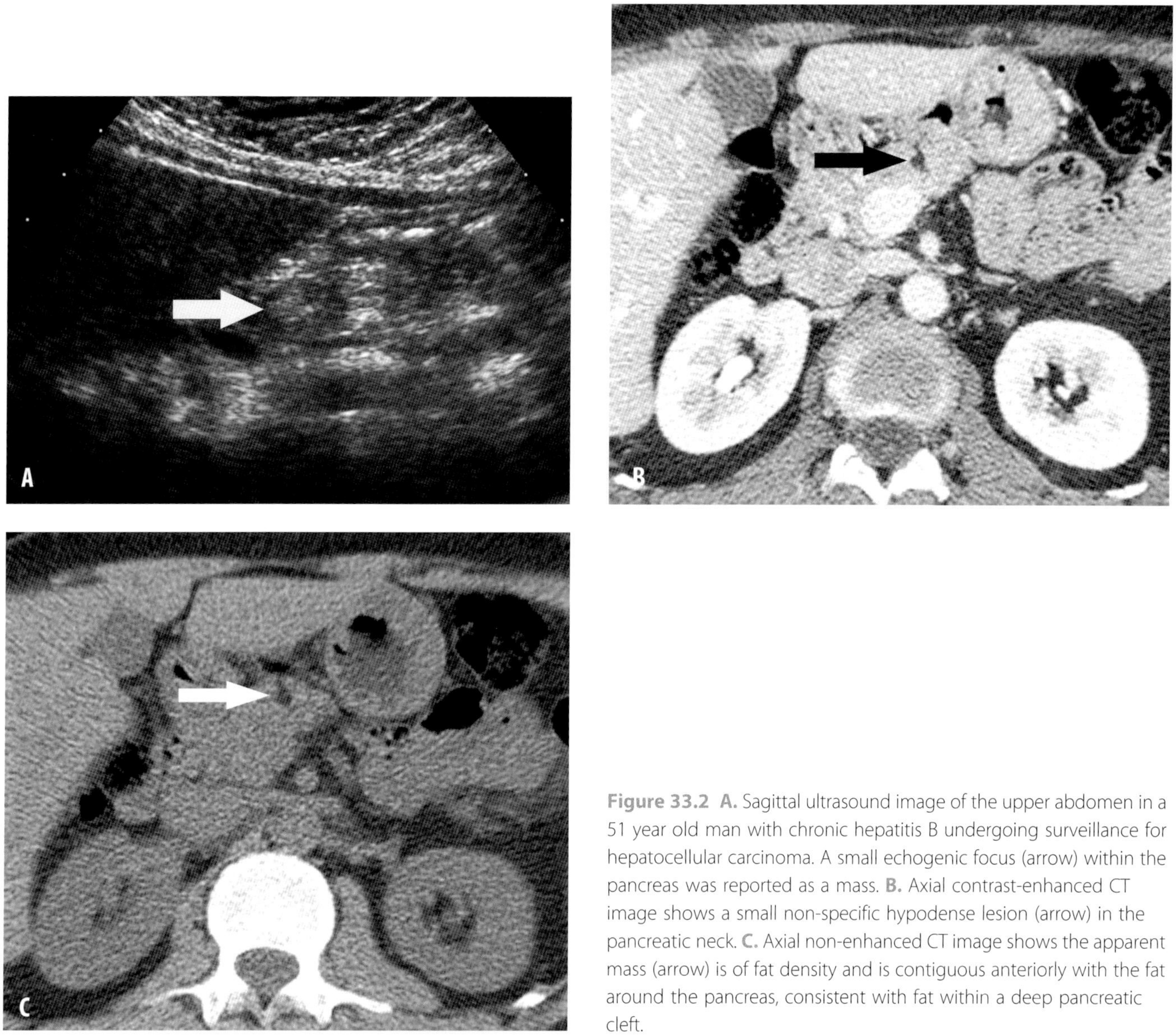

Figure 33.2 A. Sagittal ultrasound image of the upper abdomen in a 51 year old man with chronic hepatitis B undergoing surveillance for hepatocellular carcinoma. A small echogenic focus (arrow) within the pancreas was reported as a mass. B. Axial contrast-enhanced CT image shows a small non-specific hypodense lesion (arrow) in the pancreatic neck. C. Axial non-enhanced CT image shows the apparent mass (arrow) is of fat density and is contiguous anteriorly with the fat around the pancreas, consistent with fat within a deep pancreatic cleft.

CASE 34 Colloid carcinoma of the pancreas

Imaging description

Colloid (or mucinous) carcinoma of the pancreas is a subtype of pancreatic adenocarcinoma characterized histologically by large pools of extracellular mucin which comprise at least 50% of the tumor volume and which surround central balls of malignant epithelium [1, 2]. Colloid carcinoma accounts for 1–3% of invasive pancreatic adenocarcinomas and is commonly found in association with intraductal papillary mucinous neoplasms [3, 4]. At CT, colloid carcinomas may have a well-circumscribed border and low density (Figures 34.1 and 34.2), potentially suggesting a benign or less aggressive cystic process (these findings are not invariable and colloid carcinoma can also be indistinguishable from non-mucinous adenocarcinoma).

Importance

Colloid carcinoma of the pancreas, particularly when it is low density, well circumscribed, and associated with pancreatic ductal dilatation, may be mistaken for benign pathology such as pancreatitis with pseudocyst (Figure 34.2) or intraductal papillary mucinous neoplasm.

Typical clinical scenario

Colloid carcinoma of the pancreas should be considered when a low-density well-circumscribed pancreatic mass with ductal dilatation is seen in a middle-aged or elderly patient, particularly if the patient is known to have intraductal papillary mucinous neoplasm.

Differential diagnosis

The CT appearance of colloid carcinoma may suggest a cystic lesion, and therefore the imaging differential considerations include pseudocyst, mucinous cystic neoplasm, serous cystadenoma, and intraductal papillary mucinous neoplasm. While various "typical" features of these specific cystic lesions have been described, imaging remains of limited accuracy in the characterization of cystic pancreatic masses [5], and colloid carcinoma should be added to the list of potential considerations.

> **Teaching point**
>
> Colloid carcinoma is a subtype of pancreatic adenocarcinoma that may be well circumscribed and low density at CT, falsely suggesting a cystic lesion; this possibility should be added to the usual differential considerations for an apparent pancreatic cyst seen at imaging.

REFERENCES

1 Molavi D, Argani P. Distinguishing benign dissecting mucin (stromal mucin pools) from invasive mucinous carcinoma. *Adv Anat Pathol* 2008; **15**: 1–17.

2 Hamilton SR. *Pathology and genetics of tumours of the digestive system: WHO classification of tumours*, volume 2. Lyon: World Health Organization, 2000.

3 Adsay NV, Pierson C, Sarkar F, et al. Colloid (mucinous noncystic) carcinoma of the pancreas. *Am J Surg Pathol* 2001; **25**: 26–42.

4 Seidel G, Zahurak M, Iacobuzio-Donahue C, et al. Almost all infiltrating colloid carcinomas of the pancreas and periampullary region arise from in situ papillary neoplasms: a study of 39 cases. *Am J Surg Pathol* 2002; **26**: 56–63.

5 Visser BC, Yeh BM, Qayyum A, et al. Characterization of cystic pancreatic masses: relative accuracy of CT and MRI. *Am J Roentgenol* 2007; **189**: 648–656.

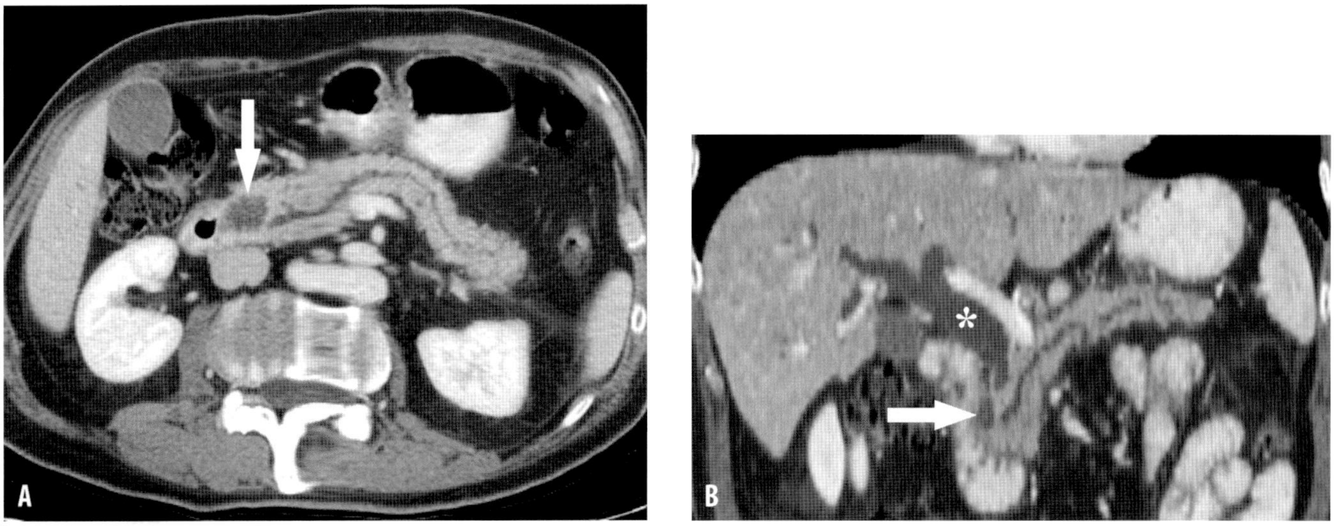

Figure 34.1 A. Axial curved planar CT image through the pancreas in a 62 year old man with obstructive jaundice due to a colloid carcinoma of the pancreatic head. The tumor is visible as a well-circumscribed low-density mass in the pancreatic head, with mild upstream dilatation of the pancreatic duct. **B.** Coronal curved planar CT image through the pancreas demonstrated the well-circumscribed low-density tumor (arrow) with upstream biliary (asterisk) and pancreatic ductal dilatation.

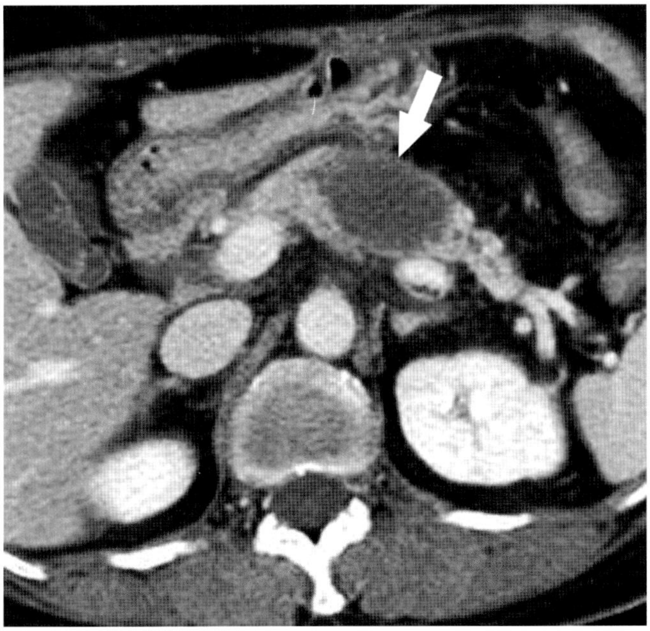

Figure 34.2 Axial contrast-enhanced CT image in a 64 year old woman with abdominal pain and jaundice due to a colloid carcinoma of the pancreas. The tumor is visible as a well-circumscribed hypodense mass (arrow) with mild upstream pancreatic ductal dilatation. The findings were initially interpreted as indicating pancreatitis and pseudocyst.

CASE

35 Minor adrenal nodularity or thickening

Imaging description

Minor adrenal nodularity or thickening is a common finding at CT or MRI [1], and represents a diagnostic problem that is distinct from and commoner than the more well-described dilemma of an incidental adrenal mass [1, 2]. Smooth enlargement has been defined as diffuse thickening of the adrenal glands with a limb thickness of over 6 to 8 mm (Figure 35.1) [3], while nodularity has been defined as multifocal surface irregularity without a unifocal or dominant nodule over 1 cm in diameter or unifocal surface irregularity under 1 cm in diameter (Figure 35.2) [4].

Importance

Minor adrenal nodularity or thickening is most problematic when seen in a patient with a known primary malignancy, because of the concern that such abnormalities are an early sign of metastases. This concern is particularly acute in patients with primary lung cancer, where the frequency of metastases to the adrenal glands at autopsy may be as high as 35% [5].

Typical clinical scenario

Minor morphologic abnormalities of the adrenal glands are common. For example, in one series of 197 patients with lung cancer, two independent readers reported smooth enlargement in 11 to 18% of adrenal glands and minor nodularity in 18 to 23% [4].

Differential diagnosis

In a study of 197 patients with lung cancer and no obvious adrenal metastases at baseline CT [4], adrenal metastases subsequently developed over a mean follow-up period of 1.3 years in 13 adrenal glands of 11 patients. Kaplan-Meier analysis showed no association between baseline adrenal morphology and the subsequent development of adrenal metastases. This lack of association between minor morphological abnormality and the subsequent development of metastases strongly suggests that minor adrenal nodularity or thickening is not an early manifestation of metastatic disease.

Teaching point

Smooth enlargement or nodularity of an adrenal gland should not be interpreted as a potential early sign of metastatic disease in patients with a known primary malignancy.

REFERENCES

1 Grossrubatscher E, Vignati F, Possa M, Lohi P. The natural history of incidentally discovered adrenocortical adenomas: a retrospective evaluation. *J Endocrinol Invest* 2001; **24**: 846–855.

2 Grumbach MM, Biller BM, Braunstein GD, et al. Management of the clinically inapparent adrenal mass ("incidentaloma"). *Ann Intern Med* 2003; **138**: 424–429.

3 Vincent JM, Morrison ID, Armstrong P, Reznek RH. Computed tomography of diffuse, non-metastatic enlargement of the adrenal glands in patients with malignant disease. *Clin Radiol* 1994; **49**: 456–460.

4 Benitah N, Yeh BM, Qayyum A, et al. Minor morphologic abnormalities of adrenal glands at CT: Prognostic importance in patients with lung cancer. *Radiology* 2005; **235**: 517–522.

5 Abrams HL, Spiro R, Goldstein N. Metastases in carcinoma analysis of 1000 autopsied cases. *Cancer* 1950; **3**: 74.

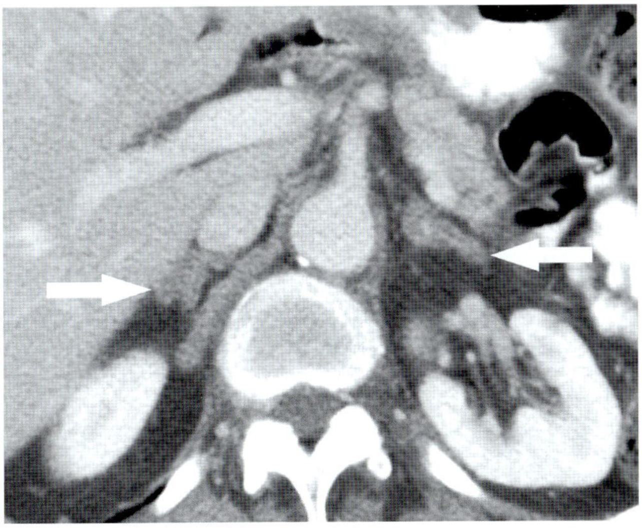

Figure 35.1 Axial contrast-enhanced CT image in an 84 year old woman with unexplained weight loss. Both adrenal glands (arrows) are thickened and nodular, but without a discrete unifocal lesion.

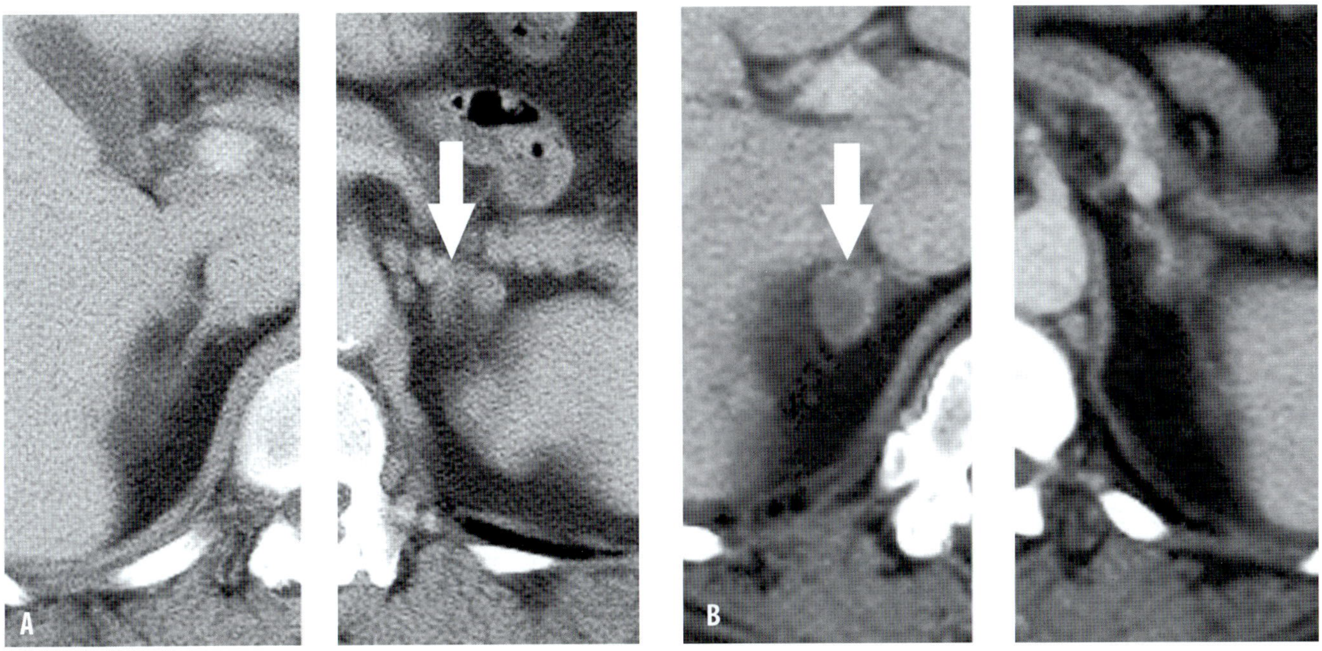

Figure 35.2 **A.** Photomontage of contrast-enhanced CT images in a 57 year old woman with stage IIIB lung cancer. The right adrenal gland is normal while the left adrenal gland (arrow) is diffusely nodular. **B.** Photomontage of contrast-enhanced CT images obtained two years later shows a metastasis has developed in the right adrenal gland, while the left adrenal gland remains diffusely nodular.

CASE 36 Adrenal pseudotumor due to gastric fundal diverticulum

Imaging description

Gastric diverticula are rare, with an approximate incidence of one in every 2400 contrast studies of the upper gastrointestinal tract [1]. Most are posterior and near the gastroesophageal junction, which likely reflects a congenital origin [2]. In this location, a gastric diverticulum typically appears as a thin-walled gas and fluid-filled mass measuring 1–5 cm in diameter above the left kidney (Figures 36.1 and 36.2).

Importance

A gastric fundal diverticulum, especially if filled only with fluid and not air and fluid, may mimic a cystic retroperitoneal or adrenal mass at cross-sectional imaging (Figures 36.3 and 36.4) [2, 3].

Typical clinical scenario

While gastric diverticula are usually incidental and asymptomatic, complications such as bleeding, polyp formation, and malignancy (Figure 36.5) have been reported [4–7].

Differential diagnosis

An air and fluid-filled gastric diverticulum has a very limited differential diagnosis. Conceivably a retroperitoneal abscess or super-infected cystic tumor might be considerations, though it is most likely these entities would be thick-walled and the patient would be septic – strong clues to the correct diagnosis. A purely fluid-filled gastric diverticulum can be difficult to distinguish from benign or malignant retroperitoneal tumors such as lymphangioma or sarcoma. CT with positive oral contrast can then be helpful in establishing the correct diagnosis (Figures 36.3 and 36.4). Careful scrutiny for the presence of a separate left adrenal gland can usually prevent confusion with an adrenal mass.

Teaching point

The possibility of a gastric fundal diverticulum should be considered when an apparent cystic mass or tumor is seen above the left kidney.

REFERENCES

1 Palmer ED. Collective review: gastric diverticula. *Int Abstr Surg* 1951; **92**: 417–423.

2 Schwartz AN, Goiney RC, Graney DO. Gastric diverticulum simulating an adrenal mass: CT appearance and embryogenesis. *Am J Roentgenol* 1986; **146**: 553–554.

3 Silverman PM. Gastric diverticulum mimicking adrenal mass: CT demonstration. *J Comput Assist Tomogr* 1986; **10**: 709–710.

4 Gibbons CP, Harvey L. An ulcerated gastric diverticulum–a rare cause of haematemesis and melaena. *Postgrad Med J* 1984; **60**: 693–695.

5 Chatni S, Nair SP, Nandakumar R, et al. Polyp within gastric diverticulum. *Indian J Gastroenterol* 2006; **25**: 205.

6 Fork FT, Tóth E, Lindström C. Early gastric cancer in a fundic diverticulum. *Endoscopy* 1998; **30**: S2.

7 Adachi Y, Mori M, Haraguchi Y, Sugimachi K. Gastric diverticulum invaded by gastric adenocarcinoma. *Am J Gastroenterol* 1987; **82**: 807.

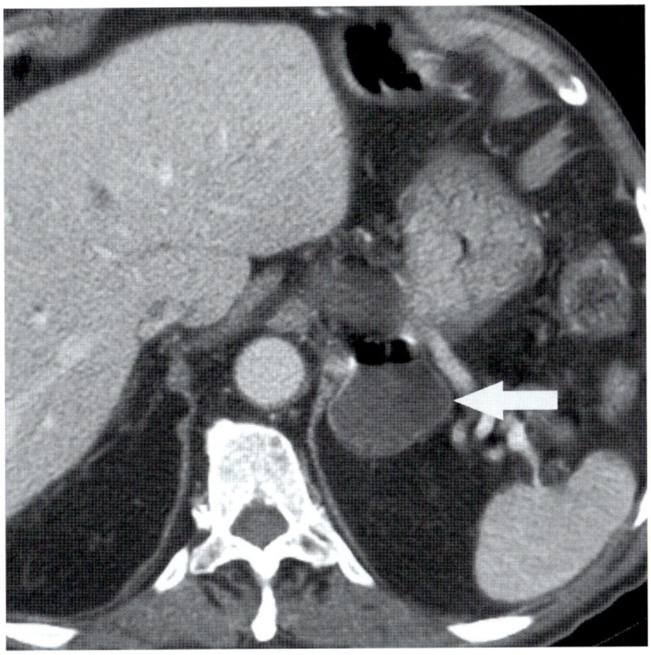

Figure 36.1 Axial contrast-enhanced CT image in a patient with newly diagnosed lung cancer showing an incidental gastric fundal diverticulum (arrow). The typical appearance of a thin-walled gas and fluid-filled mass posterior to the gastric fundus is evident.

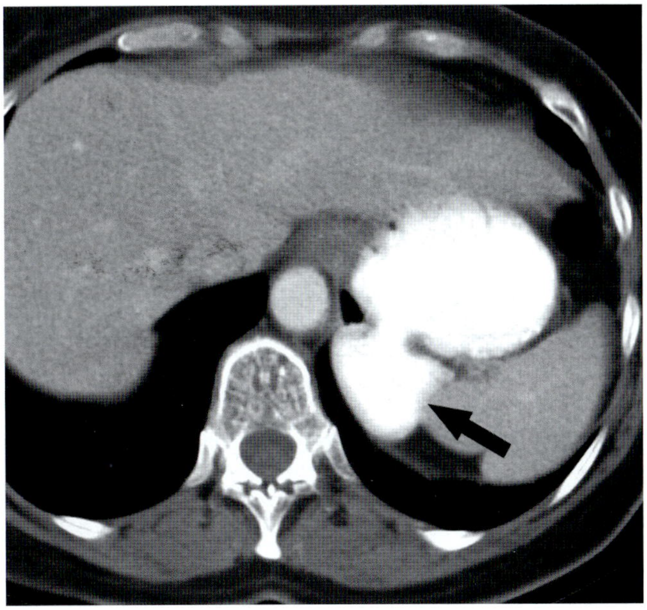

Figure 36.2 Axial contrast-enhanced CT image in a patient with newly diagnosed colon cancer showing an incidental gastric fundal diverticulum (arrow). In this case oral contrast has been administered, so the communication between the diverticulum and the stomach is readily appreciated.

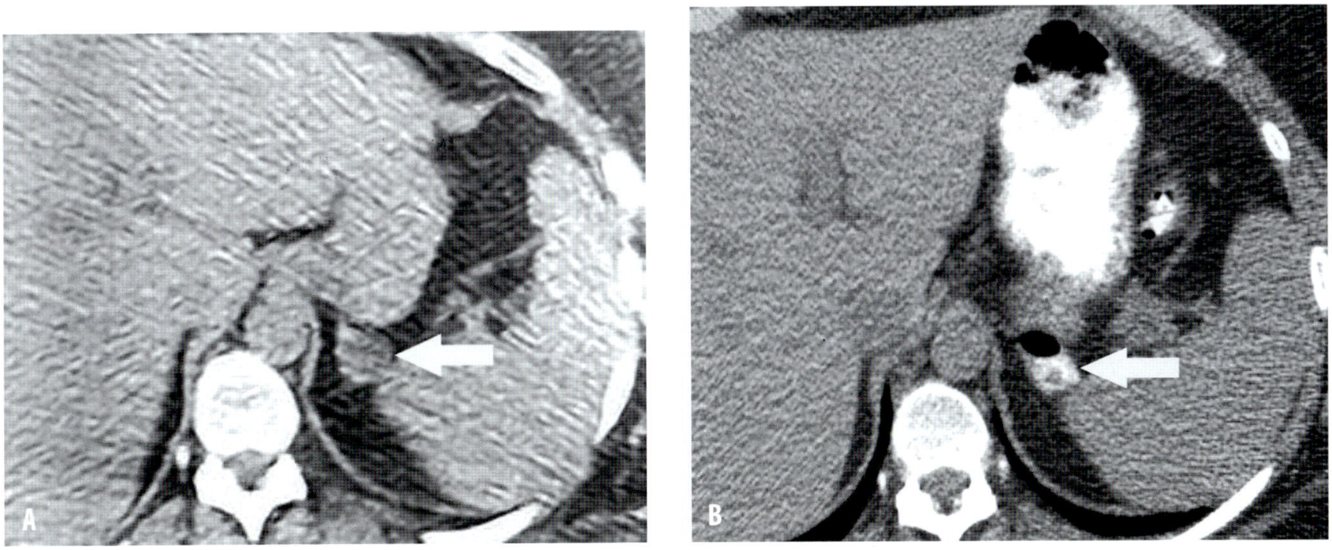

Figure 36.3 **A.** Axial non-enhanced CT image in a patient with newly diagnosed breast cancer. The study was performed for planning of radiation therapy, and was reported as showing a left adrenal mass (arrow), although the left adrenal gland was visible separately on other images (not shown). **B.** Axial non-enhanced CT image from a subsequent study performed with oral contrast. The "mass" (arrow) is seen to contain both gas and oral contrast, confirming the diagnosis of a gastric fundal diverticulum.

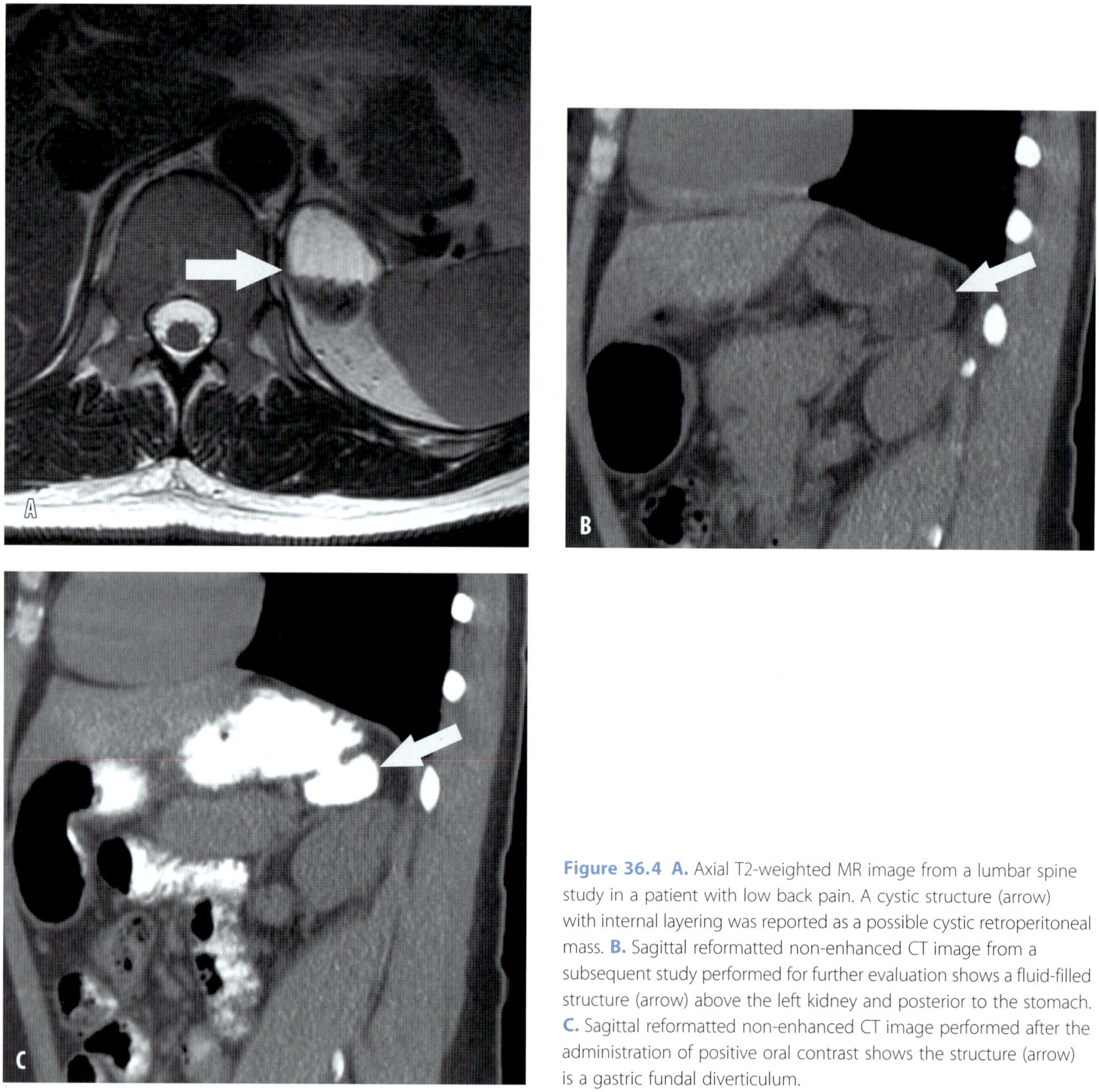

Figure 36.4 **A.** Axial T2-weighted MR image from a lumbar spine study in a patient with low back pain. A cystic structure (arrow) with internal layering was reported as a possible cystic retroperitoneal mass. **B.** Sagittal reformatted non-enhanced CT image from a subsequent study performed for further evaluation shows a fluid-filled structure (arrow) above the left kidney and posterior to the stomach. **C.** Sagittal reformatted non-enhanced CT image performed after the administration of positive oral contrast shows the structure (arrow) is a gastric fundal diverticulum.

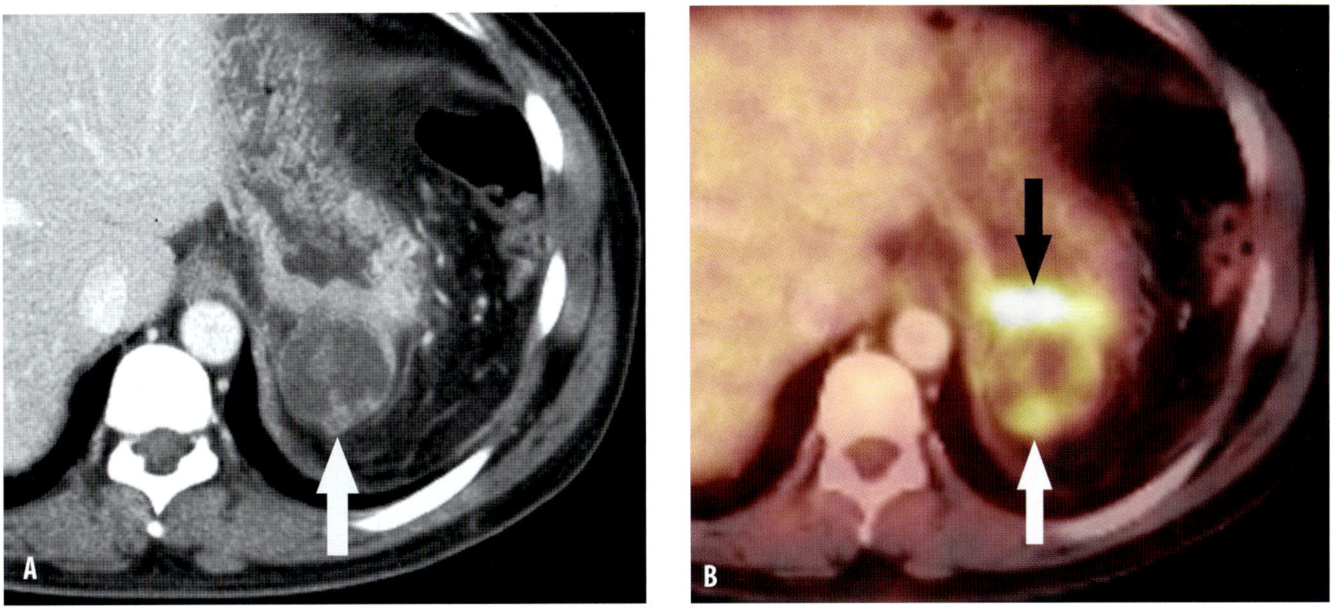

Figure 36.5 A. Axial contrast-enhanced CT image in a patient with newly diagnosed gastric cancer showing solid nodules (arrow) within a gastric fundal diverticulum. **B.** Fused axial PET/CT image shows increased FDG uptake in the tumor nodules within the diverticulum (white arrow) and also in pathologically confirmed infiltrative cancer in the gastric wall around the mouth of the diverticulum (black arrow). Malignancy is a rare but reported complication of gastric fundal diverticula.

CASE 37 Adrenal pseudotumor due to horizontal lie

Imaging description

When a limb or bifurcation of the adrenal gland lies in the horizontal plane, this orientation can result in the artifactual appearance of a mass in the adrenal gland on axial images. This artifact can be appreciated by correlation with coronal or sagittal reformatted images (Figures 37.1–37.3).

Importance

An adrenal pseudotumor due to horizontal lie may mimic a true tumor of the adrenal gland, provoking unnecessary additional investigations and patient anxiety.

Typical clinical scenario

This pseudotumor occurs primarily as an incidental finding on axial CT images that are of sufficient thickness (5 mm or so) to result in partial voluming of the horizontally oriented portion of the adrenal gland, creating an apparent mass. Thinner CT sections are less likely to result in this artifact.

Differential diagnosis

The artifactual nature of such a pseudotumor can be appreciated by correlation with coronal or sagittal reformatted images. In addition such pseudotumors are typically seen only on one axial image, while a true globular mass might be expected to have roughly similar dimensions in all directions, and to therefore be seen on more than one axial slice.

Teaching point

The possibility of a pseudotumor due to horizontal lie should be considered for an apparent mass in the adrenal gland that is seen only on one axial CT image, and correlation with coronal or sagittal reformatted images should confirm this diagnosis.

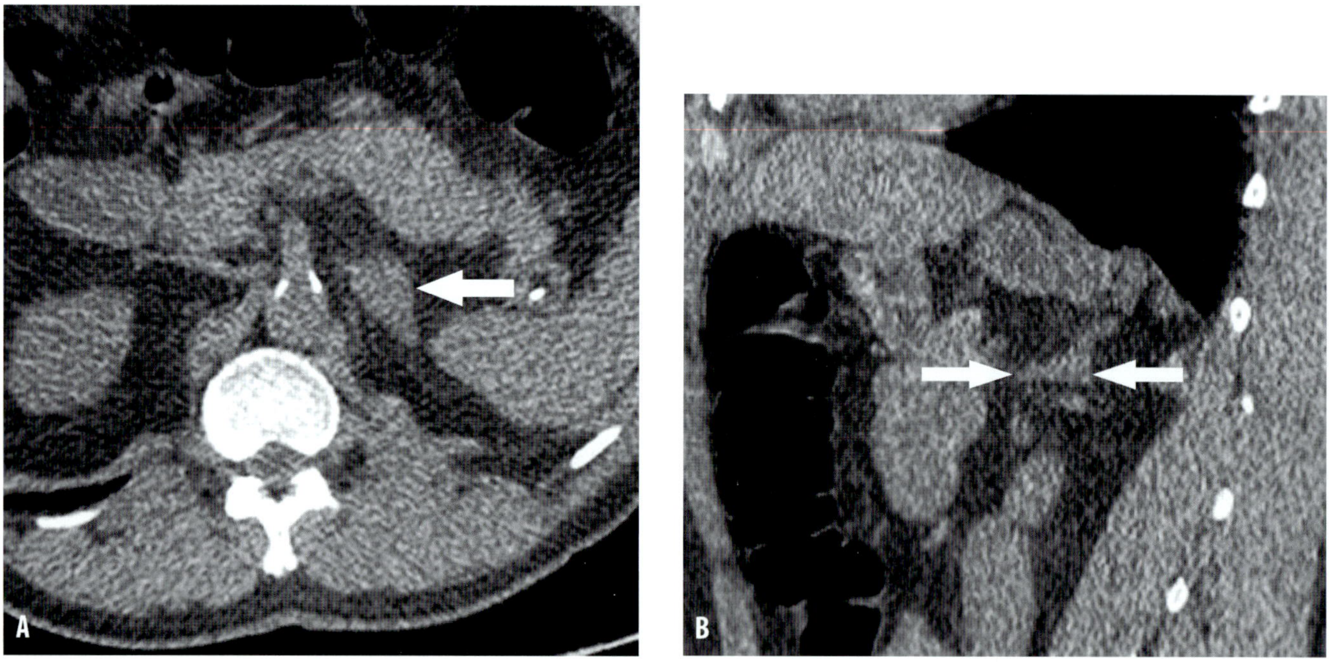

Figure 37.1 A. Axial non-enhanced CT image in a 70 year old man undergoing virtual colonoscopy. An apparent nodule (arrow) is seen in the left adrenal gland. **B.** Sagittal reformatted CT image shows the nodule is due to axial imaging through a horizontally oriented portion of the adrenal gland (between arrows) and no true left adrenal mass is present.

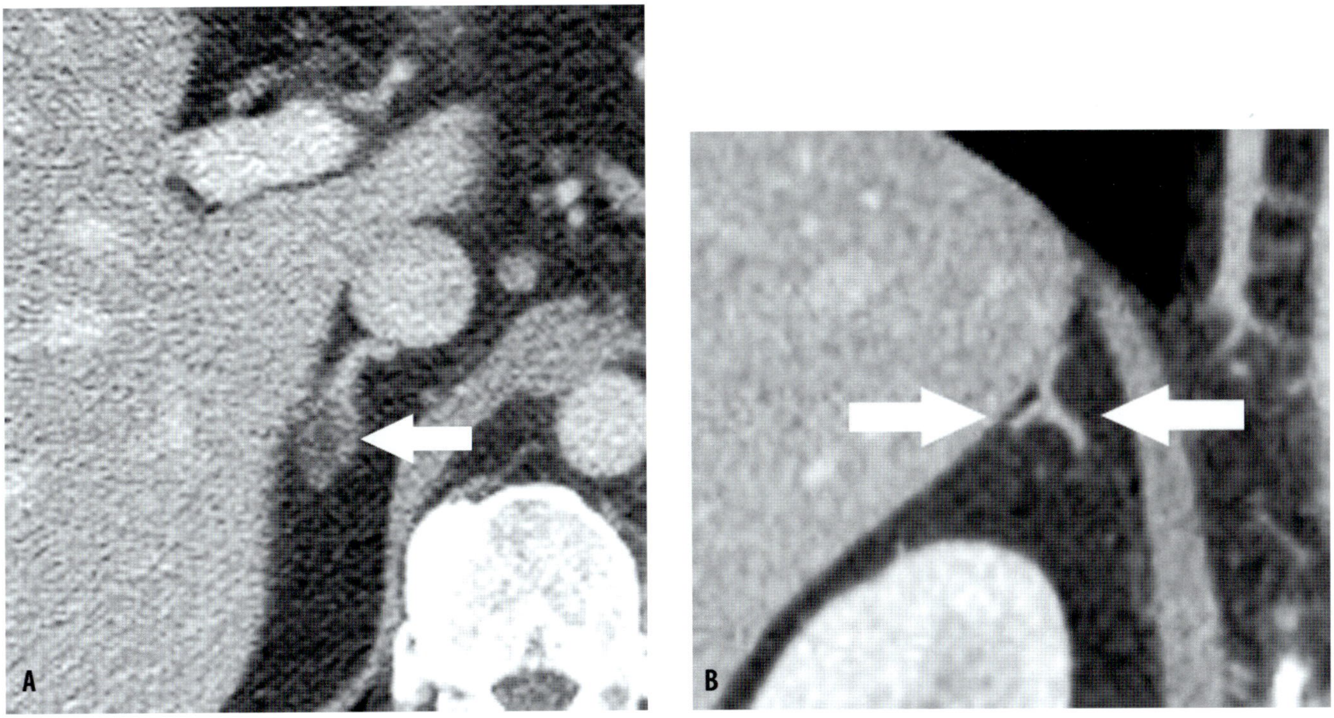

Figure 37.2 A. Axial contrast-enhanced CT image in a 56 year old man with penile cancer. An apparent nodule (arrow) is seen in the right adrenal gland. **B.** Coronal reformatted CT image shows the nodule is due to axial imaging through a horizontally oriented bifurcation of the adrenal gland (between arrows) and no true right adrenal mass is present.

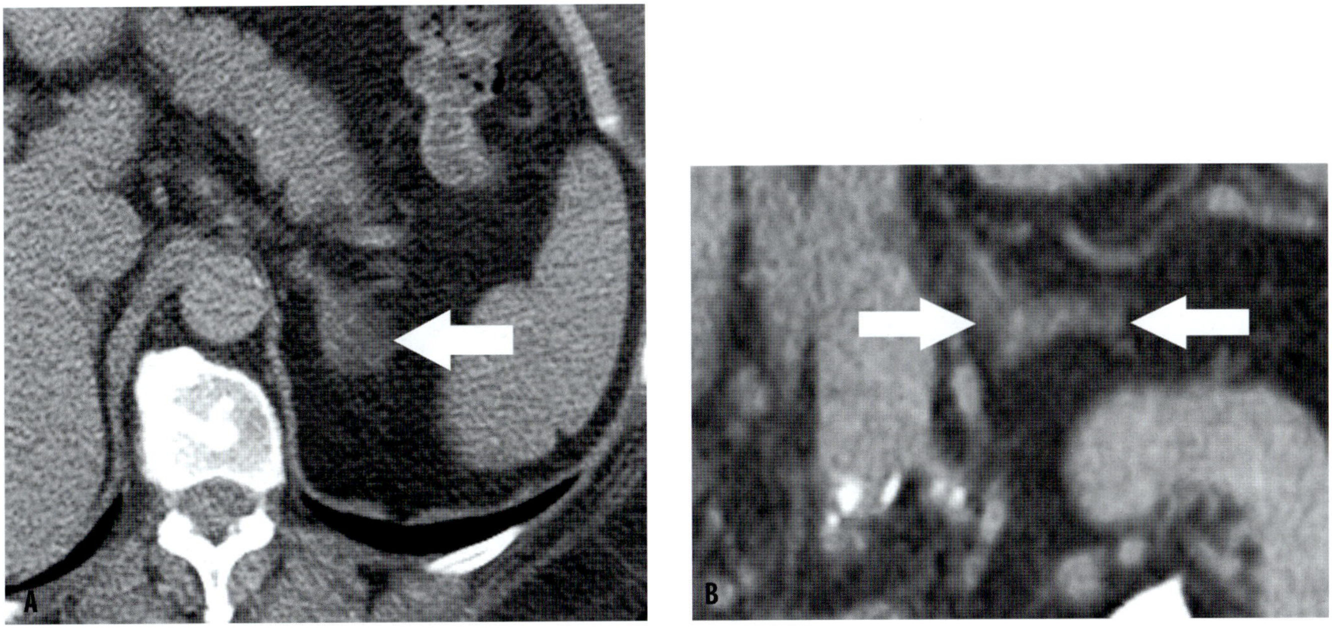

Figure 37.3 A. Axial non-enhanced CT image in a 68 year old woman with previously ablated renal cell cancer. An apparent nodule (arrow) is seen in the left adrenal gland. **B.** Coronal reformatted delayed contrast-enhanced CT image shows the nodule is due to axial imaging through a horizontally oriented portion of the adrenal gland (between arrows) and no true left adrenal mass is present.

CASE 38 Adrenal pseudotumor due to varices

Imaging description

Tortuous and dilated periadrenal portosystemic collateral veins in patients with portal hypertension may simulate an adrenal mass at CT or MRI (Figures 38.1 and 38.2) [1–4].

Importance

The misidentification of periadrenal varices as an adrenal mass may result in unnecessary workup that may be compounded by endocrine disturbances related to cirrhosis [1]. Attempted biopsy of such a pseudotumor could be potentially catastrophic.

Typical clinical scenario

An adrenal pseudotumor due to varices occurs in patients with portal hypertension, and mainly on the left side where the left inferior phrenic vein acts as a recognized portosystemic collateral pathway from the splenic to the left renal vein (Figure 38.1) [2, 4].

Differential diagnosis

Critical clues that usually allow correct identification of this vascular pseudotumor are [2]:

- Recognition that the patient has portal hypertension.
- Identification of the adrenal gland separate to the suspected tumor (not always possible).
- Isodensity to the adjacent veins on sequential phases.
- Tubular or tortuous "wormlike" configuration.
- Continuity of the pseudotumor with venous structures over serial slices or on multiplanar reformatted images.

Teaching point

In a patient with portal hypertension, the possibility of a pseudotumor due to varices should be considered for an apparent adrenal mass, especially if left-sided.

REFERENCES

1 Paramesh AS, Meiners R, Fairchild RB, et al. Retroperitoneal varices presenting as an adrenal pseudotumor in a cirrhotic patient. *World Med J* 2006; **105**: 44–46.

2 Brady TM, Gross BH, Glazer GM, Williams DM. Adrenal pseudomasses due to varices: angiographic-CT-MRI-pathologic correlations. *Am J Roentgenol* 1985; **145**: 301–304.

3 Gokan T, Ohgiya Y, Nobusawa H, Munechika H. Commonly encountered adrenal pseudotumours on CT. *Br J Radiol* 2005; **78**: 170–174.

4 Mitty HA, Cohen BA, Sprayregen S, Schwartz K. Adrenal pseudotumors on CT due to dilated portosystemic veins. *Am J Roentgenol* 1983; **141**: 727–730.

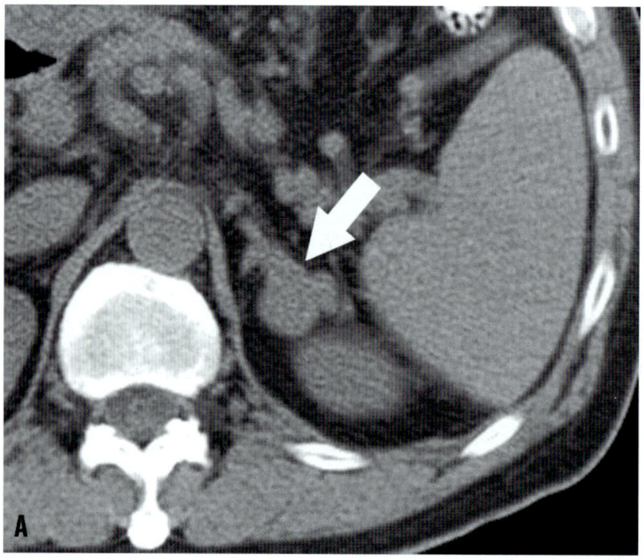

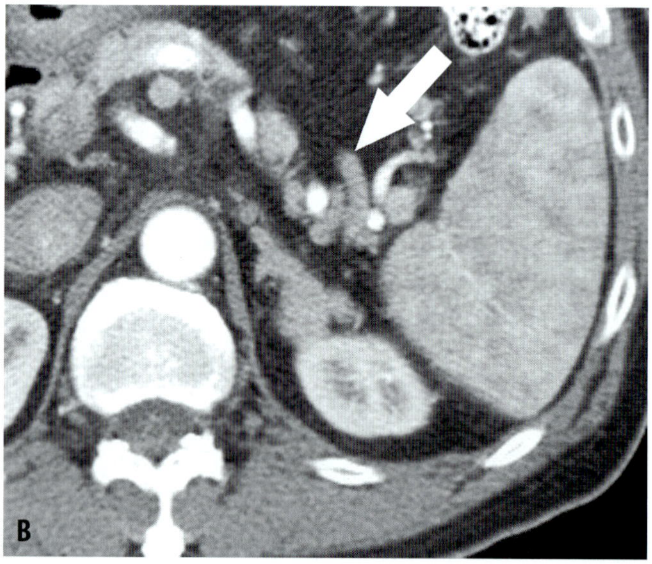

Figure 38.1 **A.** Axial non-enhanced CT image in a 61 year-old man awaiting liver transplantation for end-stage cirrhosis shows an apparent left adrenal mass (arrow). **B.** Axial arterial phase contrast-enhanced CT image shows the mass is enhancing. Note that the enhancement is similar to the enhancement in the splenic vein (arrow). **C.** Axial portal venous phase contrast-enhanced CT image again shows isodense enhancement of the mass and the splenic vein (arrow). **D.** On close inspection, the left adrenal gland can be tentatively identified as a separate structure (shaded yellow). **E.** Coronal curved planar reformatted CT image confirms the mass is a pseudolesion due to a periadrenal portosystemic collateral vein (arrow) draining to the left renal vein (asterisk). **F.** Color Doppler ultrasound image demonstrates prominent perisplenic varices, but it is difficult to confirm that these varices correspond to the finding on CT.

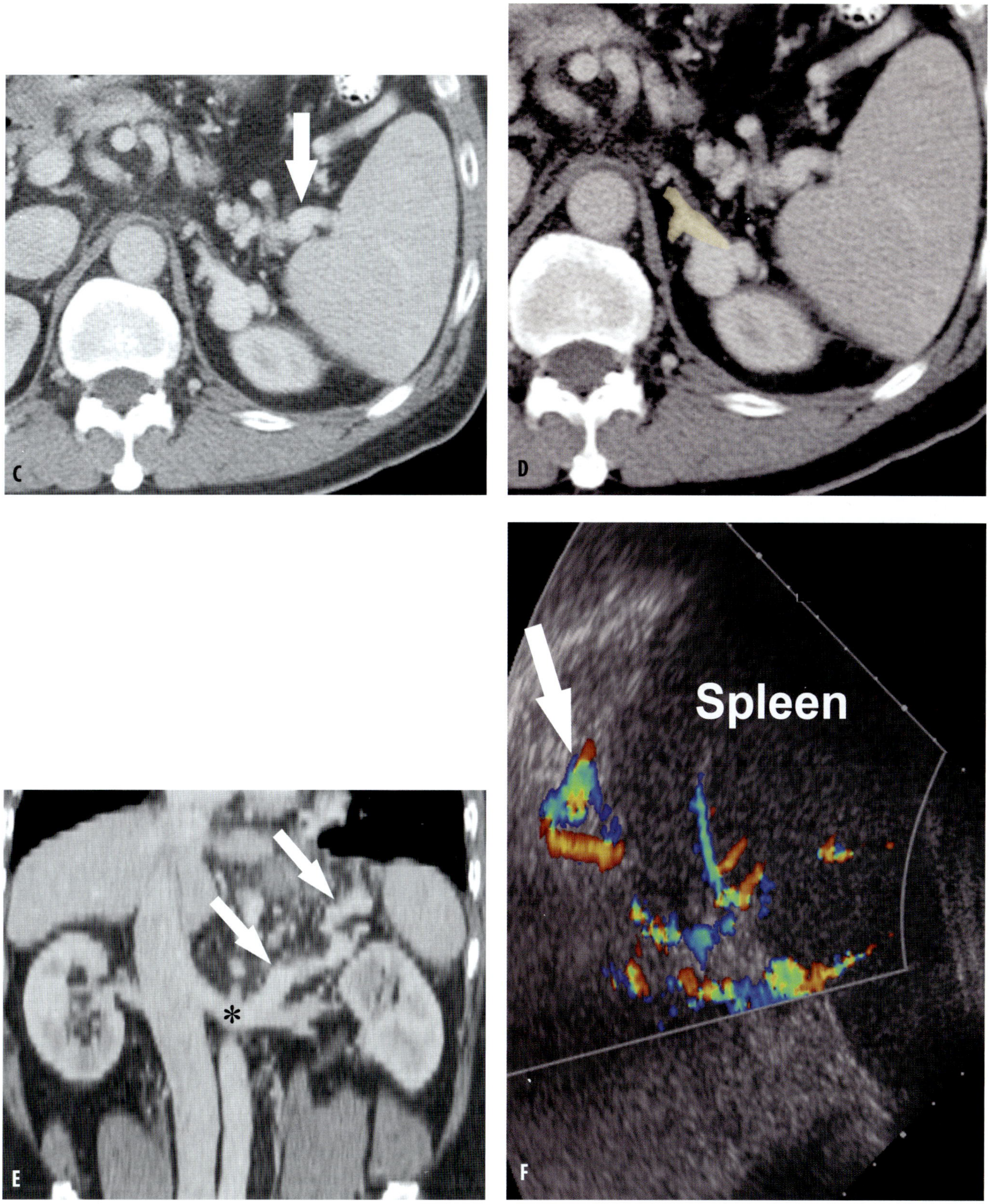

Figure 38.1 (cont.)

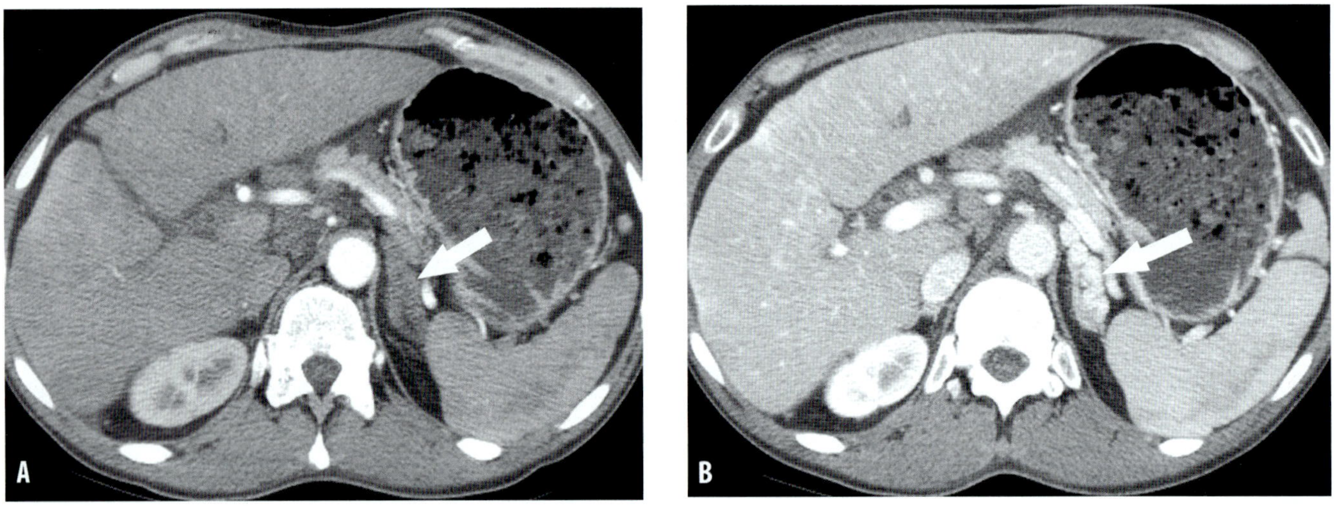

Figure 38.2 A. Axial arterial phase contrast-enhanced CT image in a 43 year old man with recurrent hepatitis C cirrhosis three years after liver transplantation. An apparent mass (arrow) is seen in the left adrenal gland. **B.** Axial portal venous phase contrast-enhanced CT image shows that the apparent adrenal mass is due to a tortuous and dilated periadrenal portosystemic collateral vein (arrow). Part of the normal left adrenal gland is visible medial to the varix.

CASE **39**

Adrenal pseudoadenoma

Imaging description

The normal steroid-producing cells of the adrenal cortex are rich in intracellular lipid because steroids are based on the cholesterol molecule. Adrenal adenomas are benign tumors derived from the adrenal cortex and also contain substantial amounts of intracellular lipid. Two of the critical imaging findings used to characterize adenomas are based on this fact; namely, adenomas are typically of low density (less than 10–18 Hounsfield Units) at non-enhanced CT and lose signal on opposed phase gradient-echo MRI when compared to in phase gradient-echo MRI [1, 2]. While both these signs are highly specific, there are some exceptions:

- Fluid-filled cystic or necrotic masses can be of low density on non-enhanced CT (Figures 39.1 and 39.2) [3].
- Other adrenal tumors can occasionally be rich in intracellular lipid, and demonstrate signal loss on opposed phase gradient-echo MRI, including adrenocortical carcinoma, metastases from primary tumors that contain intracellular lipid such as clear cell renal cell carcinoma or hepatocellular carcinoma (Figure 39.3), and pheochromocytoma [4–8].

Importance

The non-invasive characterization of adrenal adenomas by non-enhanced CT or opposed phase gradient-echo MRI is one of the major advances of modern genitourinary radiology, but erroneous diagnosis of an adenoma due to one of the rare exceptions detailed above could theoretically result in a missed opportunity for treatment of clinically important pathology, such as adrenocortical carcinoma or pheochromocytoma, or result in inappropriate treatment of the primary malignancy in the setting of unrecognized metastatic spread of clear cell renal cell carcinoma or hepatocellular carcinoma.

Typical clinical scenario

In practice, the vast majority of adrenal nodules that meet non-enhanced CT density or opposed phase MRI signal intensity criteria for adenomas are indeed adenomas. The rare exceptions listed above should be considered when:

- Systemic symptoms suggest the possibility of adrenocortical carcinoma or pheochromocytoma.
- The apparent adenoma shows progression on serial imaging or increased uptake on FDG PET scanning.
- The patient has a primary malignancy that is capable of intracellular lipid production, such as clear cell renal cell carcinoma or hepatocellular carcinoma.

Differential diagnosis

Cystic masses mimicking an adrenal adenoma on non-enhanced CT can be easily identified by the administration of contrast, and the main importance of this pitfall is as a reminder to check contrast-enhanced images to confirm an adrenal lesion is solid before making a diagnosis of adenoma based on low density on non-enhanced CT images. Fat-containing masses such as adrenocortical carcinoma, metastases from clear cell renal cell carcinoma or hepatocellular carcinoma, and pheochromocytoma can simulate an adrenal adenoma on both CT and MRI. Identification of these mimics may require correlation with clinical or biochemical findings (e.g., systemic ill health in the case of adrenocortical carcinoma and symptomatic or biochemical findings related to excess catecholamine production in pheochromocytoma), comparison with prior studies (adrenal adenomas should be essentially stable over time), or additional investigation with PET or biopsy.

Teaching point

Contrast-enhanced CT images should be checked to confirm an adrenal lesion is solid before making a diagnosis of adenoma based on low density on non-enhanced CT images, since adrenal cysts can also be of low density on non-contrast CT. For a solid lesion that appears to be an adenoma on CT or MRI, the possibility of an adenoma mimic should be considered if systemic symptoms suggest adrenocortical carcinoma or pheochromocytoma, the apparent adenoma shows progression on serial imaging or increased uptake on FDG PET scanning, or the patient has a primary malignancy that is capable of intracellular lipid production, such as clear cell renal cell carcinoma or hepatocellular carcinoma.

REFERENCES

1 Lee MJ, Hahn PF, Papanicolau N, et al. Benign and malignant adrenal masses: CT distinction with attenuation coefficients, size, and observer analysis. *Radiology* 1991; **179**: 415–418.

2 Mitchell DG, Crovello M, Matteucci T, Petersen RO, Miettinen MM. Benign adrenocortical masses: diagnosis with chemical shift MR imaging. *Radiology* 1992; **185**: 345–351.

3 Tung GA, Pfister RC, Papanicolaou N, Yoder IC. Adrenal cysts: imaging and percutaneous aspiration. *Radiology* 1989; **173**: 107–111.

4 Yamada T, Saito H, Moriya et al. Adrenal carcinoma with a signal loss on chemical shift magnetic resonance imaging. *J Comput Assist Tomogr* 2003; **27**: 606–608.

5 Nakayama T, Yoshimitsu K, Irie H, et al. Fat in liver metastasis from renal cell carcinoma detected by chemical shift MR imaging. *Abdom Imaging* 2003; **28**: 657–659.

6 Shinozaki K, Yoshimitsu K, Honda H, et al. Metastatic adrenal tumor from clear-cell renal cell carcinoma: a pitfall of chemical shift MR imaging. *Abdom Imaging* 2001; **26**: 439–442.

7 Sydow BD, Rosen MA, Siegelman ES. Intracellular lipid within metastatic hepatocellular carcinoma of the adrenal gland: a potential diagnostic pitfall of chemical shift imaging of the adrenal gland. *Am J Roentgenol* 2006; **187**: W550–W551.

8 Ramsay JA, Asa SL, van Nostrand AW, Hassaram ST, de Harven EP. Lipid degeneration in pheochromocytomas mimicking adrenal cortical tumors. *Am J Surg Pathol* 1987; **11**: 480–486.

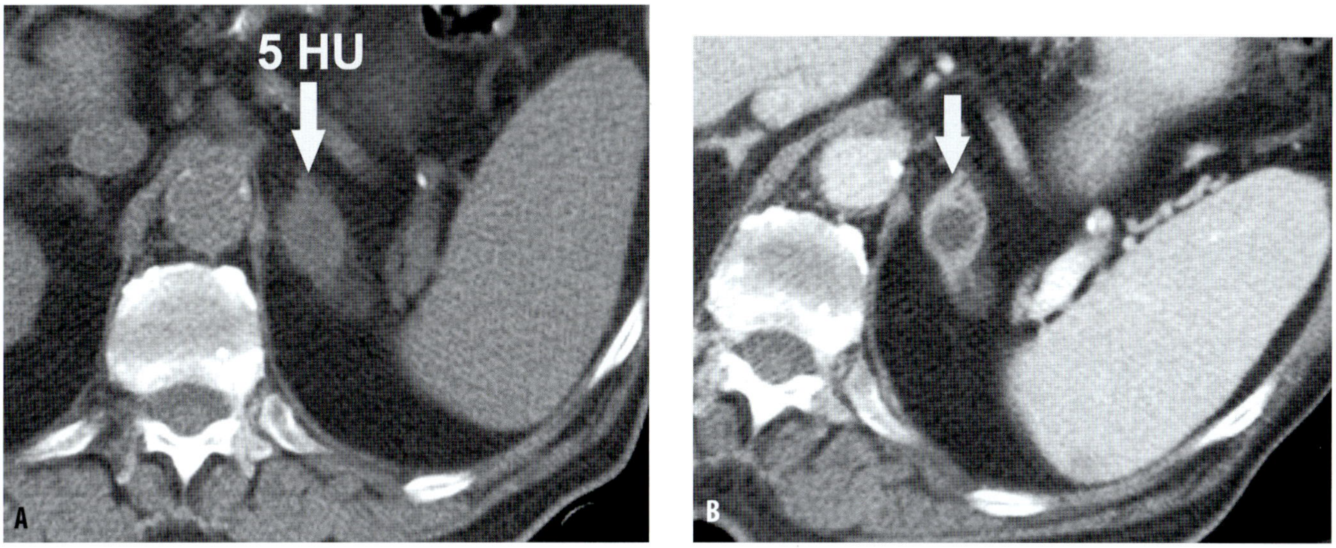

Figure 39.1 A. Axial non-enhanced CT image in an 80 year old man with newly diagnosed transitional cell carcinoma of the bladder shows a low-density (5 Hounsfield Units) mass (arrow) in the left adrenal gland. On non-contrast images alone, this lesion could reasonably be interpreted as a probable adrenal adenoma. **B.** Corresponding axial contrast-enhanced CT image shows the adrenal mass is actually an adrenal cyst with an enhancing peripheral wall and a fluid-filled central cavity. The mass was unchanged from a prior study performed two years earlier, and therefore was considered benign and a tissue diagnosis was not obtained.

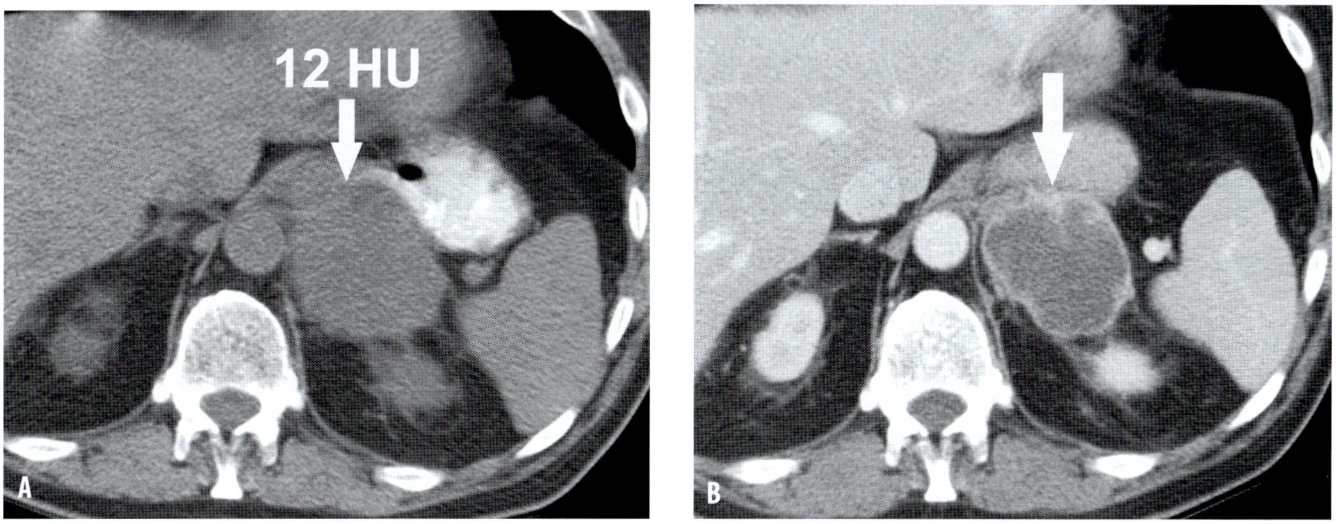

Figure 39.2 A. Axial non-enhanced CT image in a 65 year old man with a history of transitional cell carcinoma of the bladder treated by radical cystectomy and ileal conduit formation. A low-density (12 Hounsfield Units) mass (arrow) in the left adrenal gland could reasonably be interpreted as a probable adrenal adenoma. **B.** Corresponding axial contrast-enhanced CT image shows the adrenal mass is actually a cystic mass with an enhancing irregular peripheral wall and a fluid-filled central cavity. The mass was not present on a CT performed one year earlier and was also associated with new retroperitoneal adenopathy, leading to a a presumptive diagnosis of adrenal metastasis.

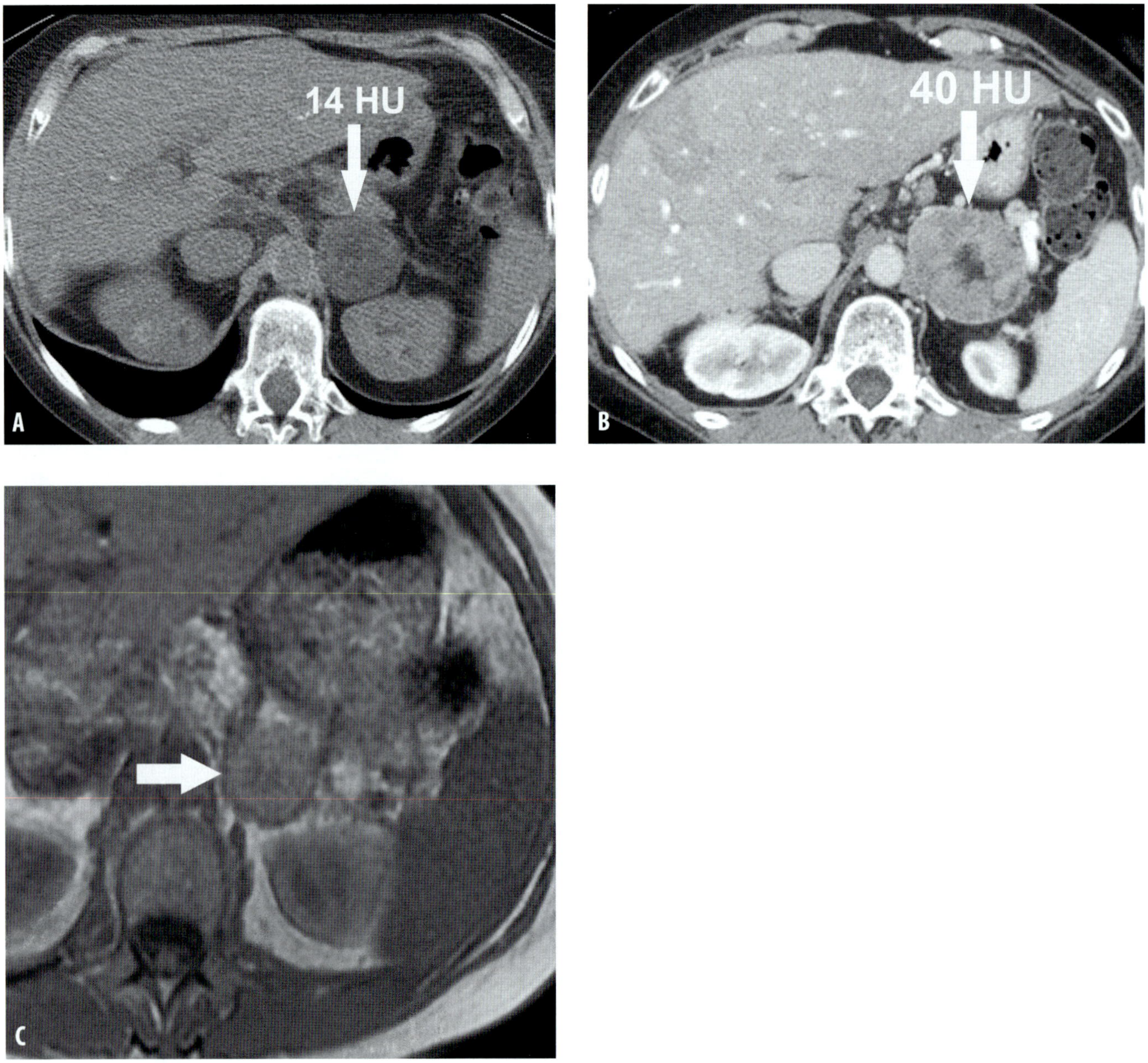

Figure 39.3 **A.** Axial non-enhanced CT image in a 62 year old woman three months after transarterial chemoembolization of a hepatocellular carcinoma shows a low-density (14 Hounsfield Units) mass (arrow) in the left adrenal gland. **B.** Corresponding axial contrast-enhanced CT image shows the adrenal mass is solid and enhancing. **C.** Axial in phase gradient-echo T1-weighted MR image shows the adrenal mass (arrow) is of intermediate signal intensity. **D.** Axial opposed phase gradient-echo T1-weighted MR image shows unequivocal signal loss in the mass (arrow). The CT and MRI findings shown above are consistent with an adrenal adenoma, but the mass was not present on a CT scan performed before chemoembolization of the hepatocellular carcinoma. Because of this, the patient proceeded to additional investigation of the adrenal mass. **E.** Axial PET image shows moderately increased FDG uptake within the mass (arrow), with a standardized uptake value of 4.6. Based on this finding, the possibility of a fat-containing metastasis was raised, and confirmed on CT-guided percutaneous biopsy.

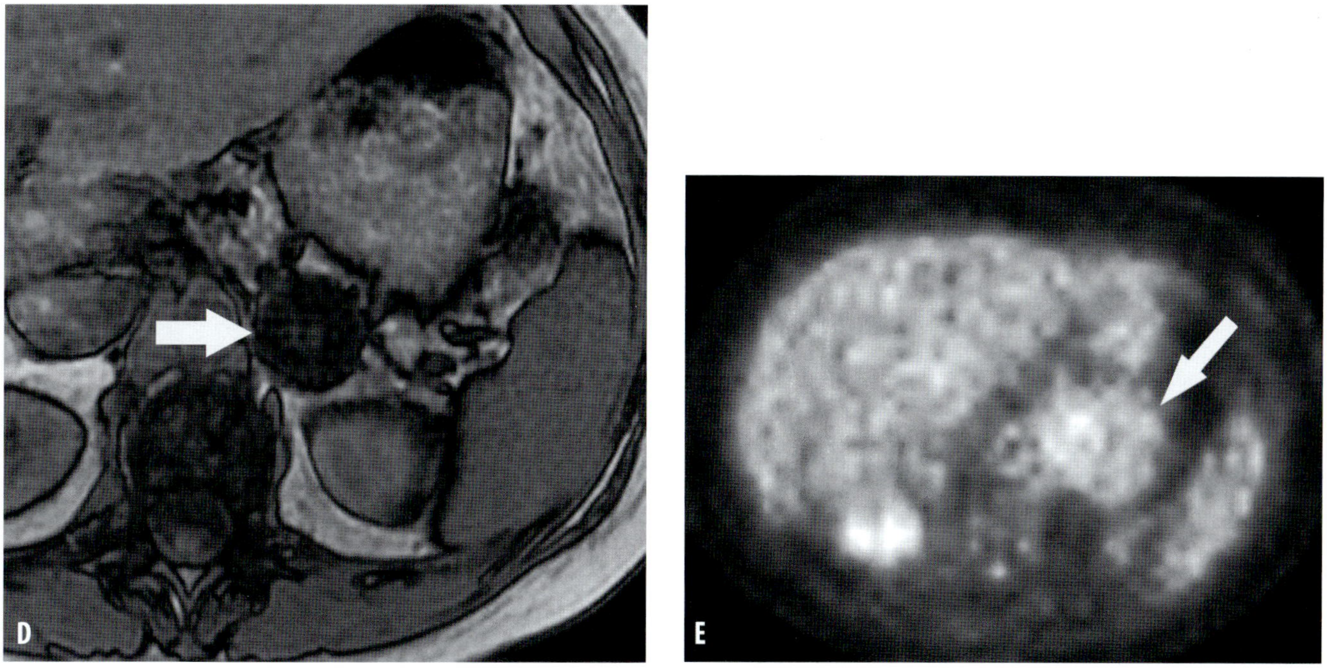

Figure 39.3 (cont.)

CASE **40**

Radiation nephropathy

Imaging description

Radiation nephropathy (or nephritis) refers to the renal damage that may develop months to years after therapeutic radiation that includes all or part of the kidneys. At CT or MRI, radiation nephropathy initially appears as delayed or persistent enhancement in the irradiated portion of the kidney with later development of renal atrophy or caliectasis [1–4]. These changes may be seen at CT or MRI (Figures 40.1–40.4), and are most frequently seen in the medial aspect of the upper poles of both kidneys after radiation therapy of the retroperitoneum or spine (the upper poles are preferentially affected because they lie closer to the midline). Other patterns of involvement may be seen depending on the radiation field.

Importance

It could be argued that radiation nephropathy is of little clinical relevance because it is unlikely to be mistaken for serious pathology and often goes unrecognized. That said, recognition of radiation nephropathy remains of importance because it indicates a history of malignancy and may prompt closer scrutiny of the images for recurrent disease or other therapy-induced complications.

Typical clinical scenario

Radiation nephropathy is seen only in patients who have been irradiated for malignant disease, and may develop in the months or years after upper abdominal radiation that includes the kidneys in the radiation field, and persist indefinitely.

Differential diagnosis

Reduced enhancement with linear borders that somewhat resembles radiation nephropathy may seen with infection (Figure 40.5) or infarction (Figure 40.6), but in both conditions the clinical scenario will be distinctive and the renal abnormalities are typically wedge-shaped and do not have the true non-anatomic straight margins of radiation change.

Teaching point

Radiation nephropathy is the likely diagnosis when focal reduced enhancement or atrophy in the kidneys at CT or MRI has straight borders corresponding to a radiation field.

REFERENCES

1 Bluemke DA, Fishman EK, Kuhlman JE, Zinreich ES. Complications of radiation therapy: CT evaluation. *Radiographics* 1991; **11**: 581–600.

2 Kwek JW, Iyer RB, Dunnington J, Faria S, Silverman PM. Spectrum of imaging findings in the abdomen after radiotherapy. *Am J Roentgenol* 2006; **187**: 1204–1211.

3 Moore L, Curry NS, Jenrette JM. Computed tomography of acute radiation nephritis. *Urol Radiol* 1986; **8**: 89–91.

4 Anderson BI, Lauver JW, Ross P, Fitzgerald RH. Demonstration of radiation nephritis by computed tomography. *Comput Radiol* 1982; **6**: 187–191.

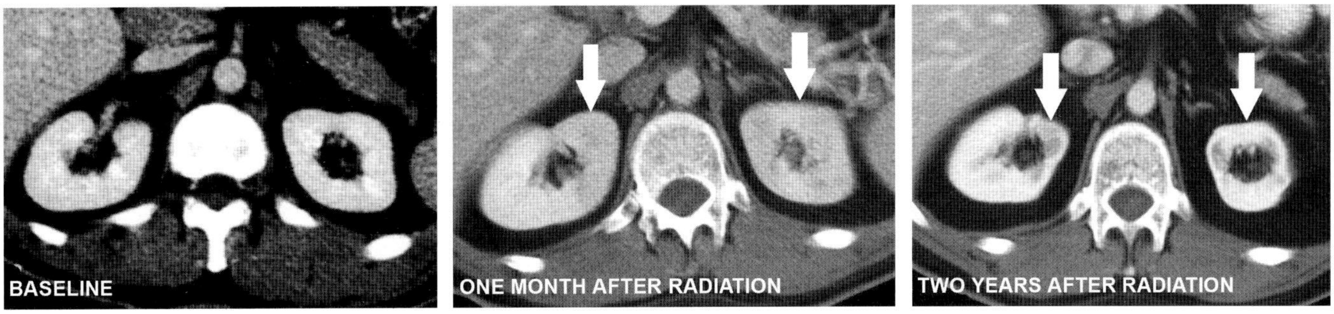

Figure 40.1 Photomontage of serial contrast-enhanced CT images performed before and after external beam radiation of the retroperitoneum and left upper quadrant for treatment of gastric cancer in a 76 year old man. The kidneys are normal in appearance prior to radiation. One month after treatment, the irradiated portions of the kidneys, specifically the upper pole of the left kidney and the medial aspect of the upper pole of the right kidney (arrows), demonstrate reduced enhancement. Two years later, the irradiated portions of the kidneys (arrows) demonstrate both reduced enhancement and atrophy.

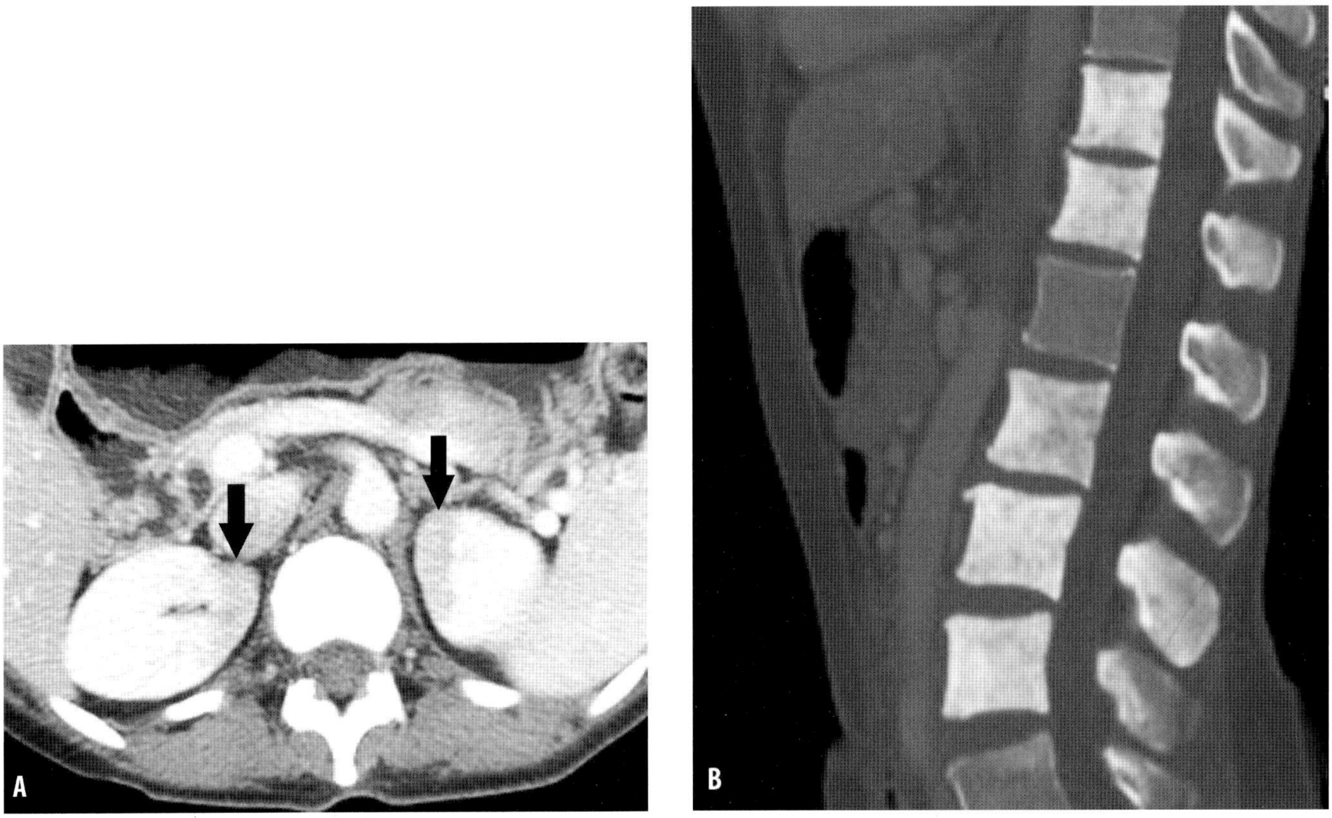

Figure 40.2 **A.** Axial contrast-enhanced CT image in a 50 year old woman 6 months after radiation of the spinal metastases from breast cancer, demonstrating reduced enhancement in the medial aspects of the upper poles of both kidneys. This distribution, and the straight lateral margin of the affected portion of the kidneys, is typical of radiation nephropathy after spinal radiation. **B.** Sagittal reformatted CT image of the lumbar spine, demonstrating multiple osteoblastic bony metastases.

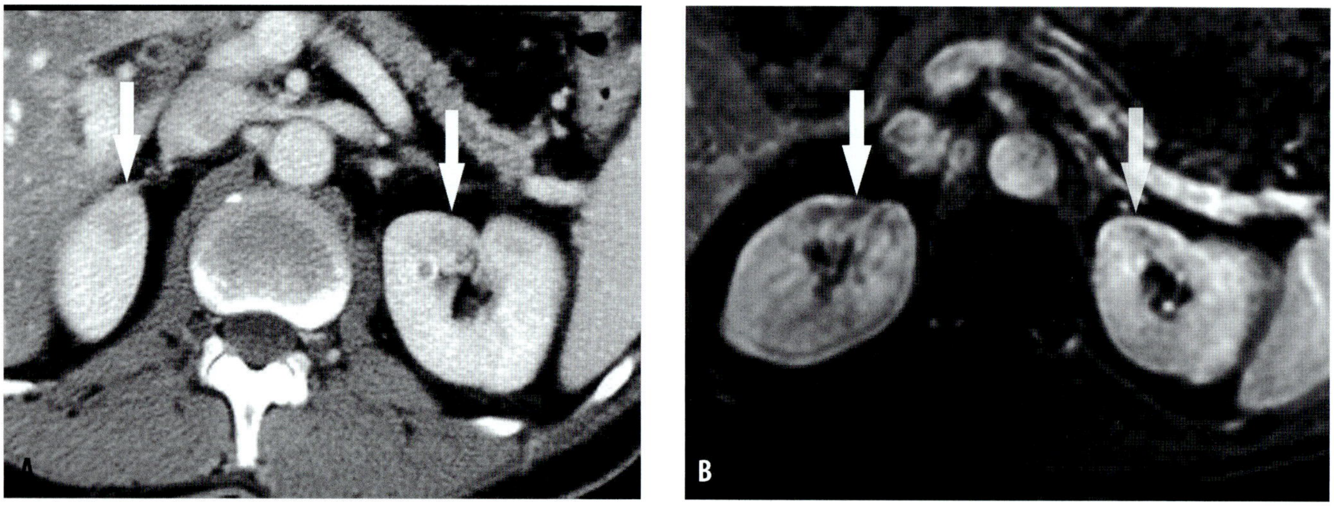

Figure 40.3 A. Axial contrast-enhanced CT image in a 55 year old man 4 months after upper abdominal radiation for pancreatic cancer, demonstrating reduced enhancement in the anterior aspects of both kidneys (arrows). The straight posterior margin of the affected portions of the kidneys is typical of radiation nephropathy. This case illustrates how the distribution pattern of radiation nephropathy varies with the radiation field. **B.** Axial gadolinium-enhanced T1-weighted MR image with fat saturation also shows reduced enhancement in the anterior aspects of both kidneys (arrows).

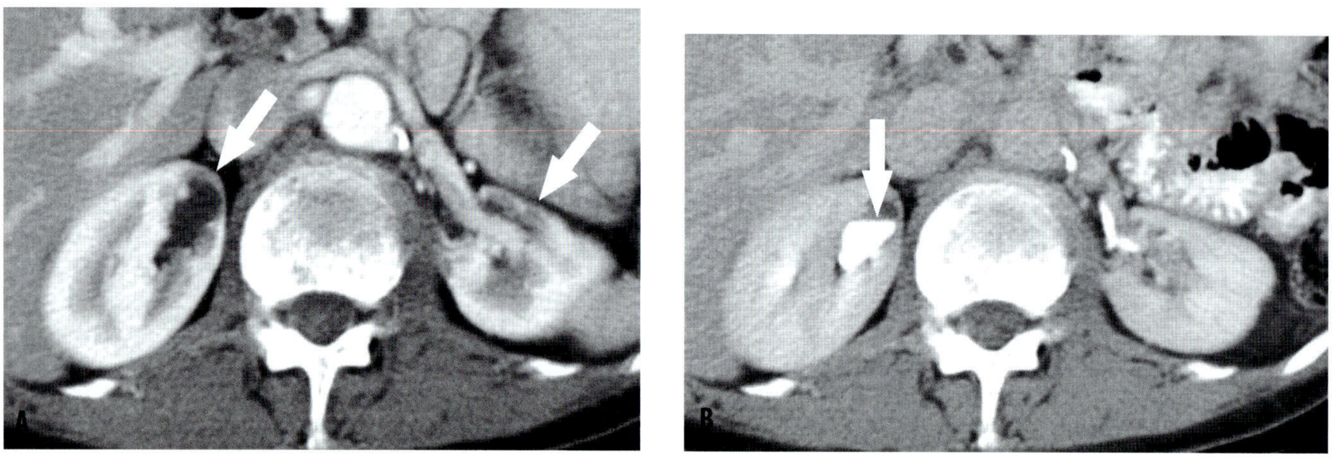

Figure 40.4 A. Axial contrast-enhanced CT image in a 69 year old man with a remote history of upper abdominal radiation for lymphoma, demonstrating reduced enhancement and atrophy in the upper poles of both kidneys (arrows). **B.** Axial contrast-enhanced CT image in the delayed phase of enhancement showing delayed opacification of a focally dilated calyx (arrow) in the right upper pole.

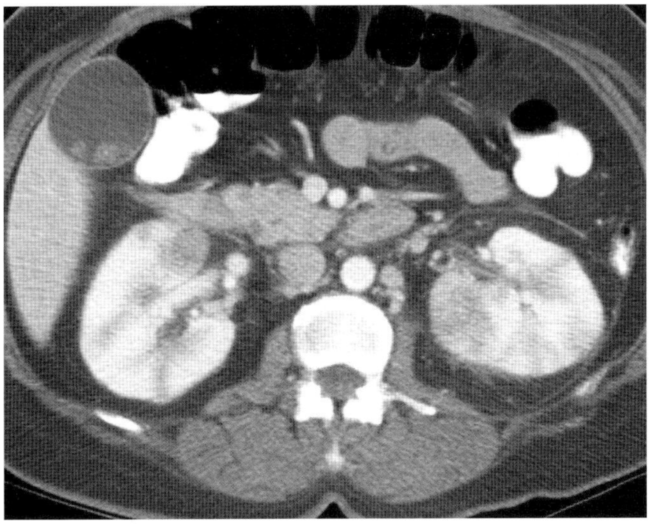

Figure 40.5 Axial contrast-enhanced CT image in a 48 year old diabetic woman with back pain, fever, urinary frequency, and a history of recurrent urinary tract infections shows multiple foci of wedge-shaped reduced perfusion due to bilateral pyelonephritis. Symptoms and CT changes resolved after antibiotic treatment. The CT findings somewhat resemble radiation nephropathy.

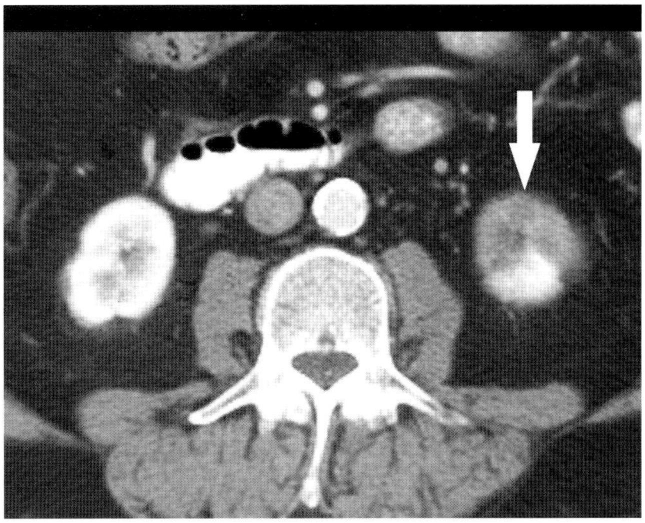

Figure 40.6 Axial contrast-enhanced CT image in a 71 year old man with acute abdominal pain and a history of paroxysmal atrial fibrillation. A renal infarct due to an angiographically proven embolus appears as a focus of reduced enhancement (arrow) with a linear border in the upper pole of the left kidney.

CASE

41 Lithium nephropathy

Imaging description

Long-term lithium therapy (primarily used for treatment of bipolar disorder) commonly results in impaired renal concentrating ability (leading to nephrogenic diabetes insipidus) and occasionally chronic kidney disease due to tubulointerstitial nephropathy. The latter constitutes lithium nephropathy, which is characterized clinically by decreased glomerular filtration rate and pathologically by chronic focal interstitial nephritis with tubular atrophy, parenchymal fibrosis, sclerotic glomeruli, tubular dilatation, and cyst formation [1–4]. The latter can be seen at imaging. The cysts of lithium nephropathy are typically small (1–2 mm), variable in number, and either randomly or primarily cortical in location (Figures 41.1–41.3) [5]. The kidneys may be normal in size or slightly atrophic.

Importance

The finding of abundant, uniform, and symmetrically distributed renal microcysts in normal-sized kidneys in a patient on long-term lithium therapy with renal insufficiency is strongly suggestive of lithium nephropathy and may eliminate the need for diagnostic confirmation by renal biopsy [5].

Typical clinical scenario

The multiple small scattered renal microcysts of lithium nephropathy are usually detected incidentally when a patient on lithium is imaged for unrelated reasons. The appearance is relatively characteristic, and should be correlated with lithium usage. In the largest published series to date, all of 16 patients on long-term lithium therapy with renal impairment demonstrated at least some renal microcysts at MRI [5]. In this series, the degree of renal impairment was variable (creatinine clearances of 20 to 70 mL/min) and all but three of the patients had clinical features of nephrogenic diabetes insipidus.

Differential diagnosis

Simple renal cysts increase in number and diameter with age, but even in those over 45 years of age the average number of simple cysts detected at MRI is only 1.9 [6], so that confusion with lithium nephropathy is unlikely. Autosomal dominant polycystic kidney disease is characterized by renal enlargement due to innumerable cysts of varying sizes, including larger and hemorrhagic cysts (Figure 41.4) [7]. Autosomal recessive polycystic kidney disease is usually diagnosed in childhood, and is characterized by innumerable small cysts that may completely replace the renal parenchyma, unlike lithium nephropathy (Figure 41.5). Glomerulocystic kidney disease is a rare renal cystic disease with sporadic or familial occurrence, characterized histologically by multiple small subcapsular or cortical cysts [8]. Medullary cystic kidney disease is a hereditary disease characterized clinically by chronic renal failure and radiologically by cortical thinning and multiple cysts in the medulla and corticomedullary junction [9]. Acquired cystic kidney disease refers to the development of renal cysts in patients who have advanced chronic renal failure or who are on dialysis (Figure 41.6). Unlike lithium nephropathy, these patients have end-stage renal failure with small kidneys [10].

Teaching point

The detection of multiple scattered renal microcysts in a patient with mild to moderate renal impairment and normal or slightly atrophic kidneys should suggest the diagnosis of lithium nephropathy, and correlation with lithium usage is usually sufficient to confirm the diagnosis.

REFERENCES

1 Grünfeld JP, Rossier BC. Lithium nephrotoxicity revisited. *Nat Rev Nephrol* 2009; **5**: 270–276.

2 Markovitz GS, Radhakrishnan J, Kammbham N, et al. Lithium nephrotoxicity: a progressive combined glomerular and tubulointerstitial nephropathy. *J Am Soc Nephrol* 2000; **11**: 1439–1448.

3 Walker RG. Lithium nephrotoxicity. *Kidney Int Suppl* 1993; **42**: S93–S98.

4 Aurell M, Svalander C, Wallin L, Alling C. Renal function and biopsy findings in patients on long-term lithium treatment. *Kidney Int* 1981; **20**: 663–670.

5 Farres MT, Ronco P, Saadoun D, et al. Chronic lithium nephropathy: MR imaging for diagnosis. *Radiology* 2003; **229**: 570–574.

6 Nascimento AB, Mitchell DG, Zhang X, et al. Rapid MR imaging detection of renal cysts: age-based standards. *Radiology* 2001; **221**: 628–632.

7 Choyke PL. Inherited cystic diseases of the kidney. *Radiol Clin North Am* 1996; **34**: 925–946.

8 Egashira K, Nakata H, Hashimoto O, Kaizu K. MR imaging of adult glomerulocystic kidney disease. *Acta Radiol* 1991; **32**: 251–253.

9 Levine E, Hartman DS, Meilstrup JW, et al. Current concepts and controversies in imaging of renal cystic diseases. *Urol Clin North Am* 1997; **24**: 523–543.

10 Levine E. Acquired cystic kidney disease. *Radiologic Clinics of North America* 1996; **34**: 947–964.

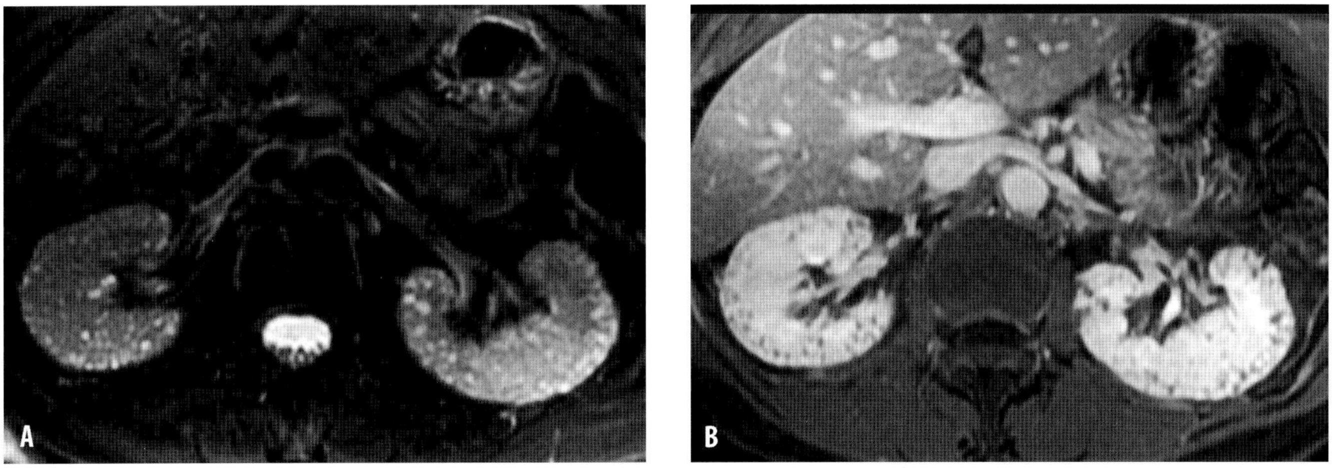

Figure 41.1 A. Axial T2-weighted MR image in a 65 year old woman with renal insufficiency, hypertension, and long-term lithium usage for bipolar disorder. The patient was referred for evaluation of possible renal artery stenosis. Multiple tiny T2 hyperintense lesions are visible scattered throughout both kidneys, with a slight cortical predominance. **B.** Axial post-gadolinium T1-weighted image shows the lesions are non-enhancing and consistent with cysts. The appearances are typical of microcysts due to lithium nephropathy.

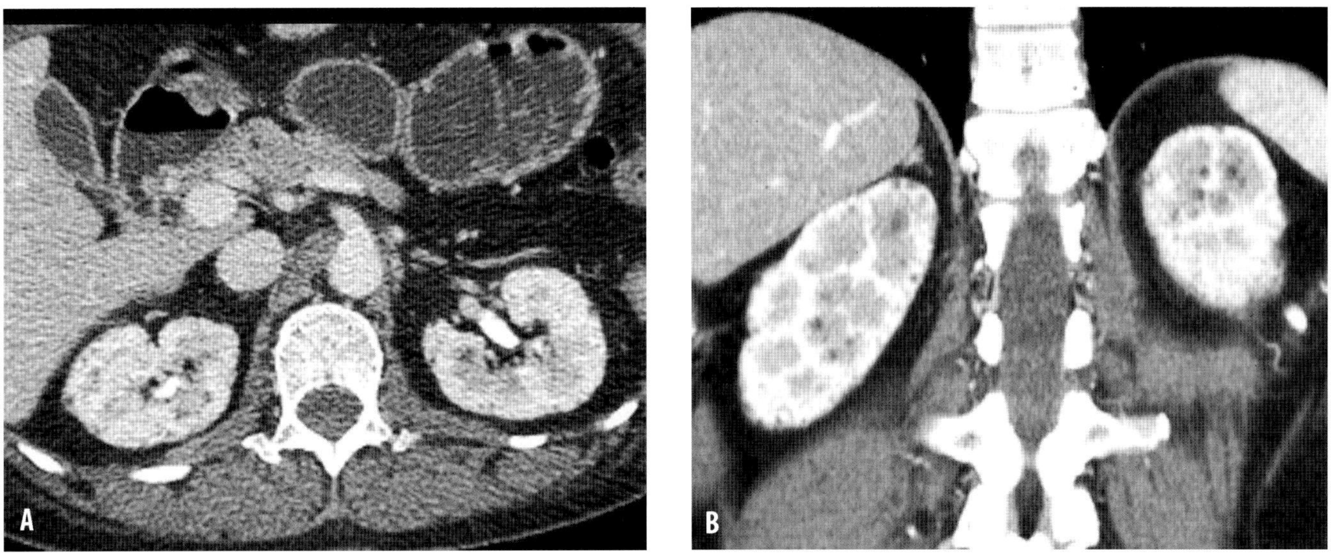

Figure 41.2 A. Axial contrast-enhanced CT image in a 55 year old woman with abdominal discomfort, mild renal impairment, and a history of bipolar disorder previously treated with long-term lithium showing multiple hypodense lesions scattered throughout both kidneys. **B.** Coronal plane reformatted CT image showing the hypodense lesions are scattered throughout the cortex and medulla. The findings are consistent with microcysts due to lithium nephropathy.

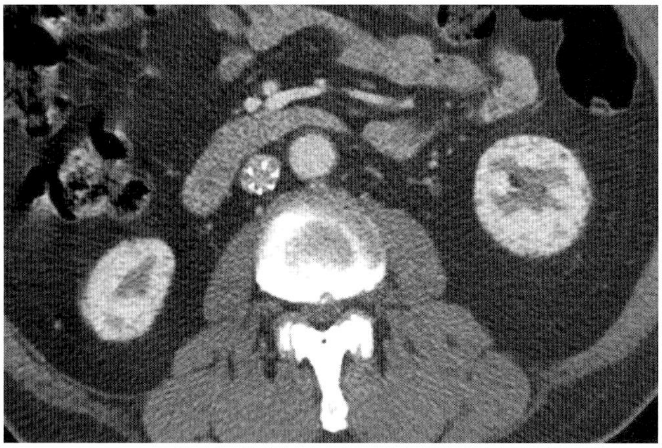

Figure 41.3 Axial contrast-enhanced CT image in a 56 year old man with acute appendicitis (not shown), renal impairment, and a history of bipolar disorder previously treated with long-term lithium showing multiple hypodense lesions scattered throughout both kidneys, consistent with microcysts due to lithium nephropathy.

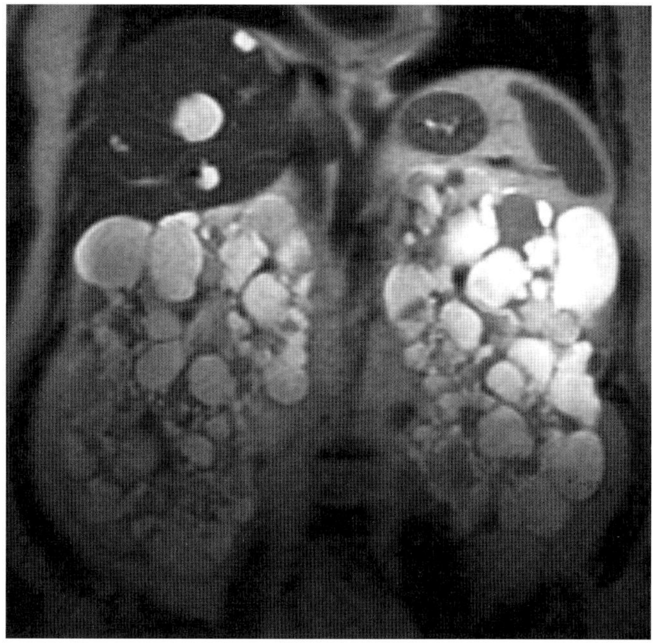

Figure 41.4 Coronal T2-weighted MR image obtained for renal assessment in a 42 year old man with autosomal dominant polycystic renal disease. The kidneys are markedly enlarged by innumerable cysts of varying sizes, with no normal parenchyma visible.

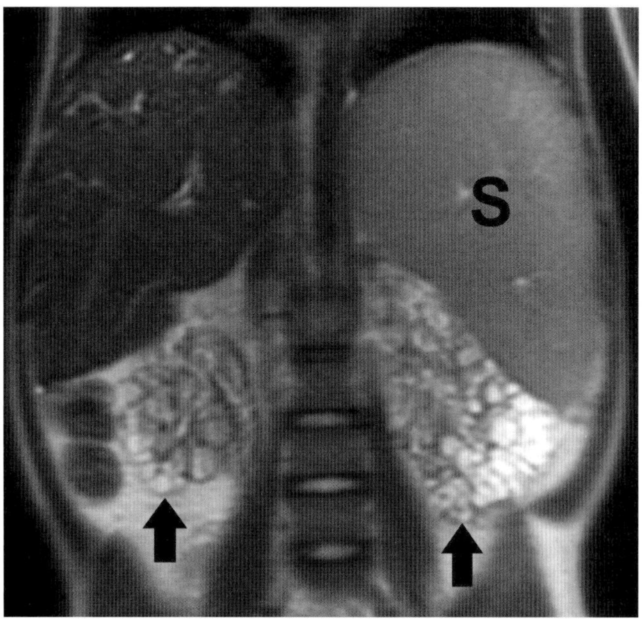

Figure 41.5 Coronal T2-weighted MR image obtained in a 12 year old girl because of fever. The patient had a history of renal transplantation for renal failure due to autosomal recessive polycystic kidney disease. Both kidneys (arrows) are replaced by innumerable small cysts with no normal parenchyma visible. Splenomegaly (S) is due to co-existent congenital hepatic fibrosis with portal hypertension.

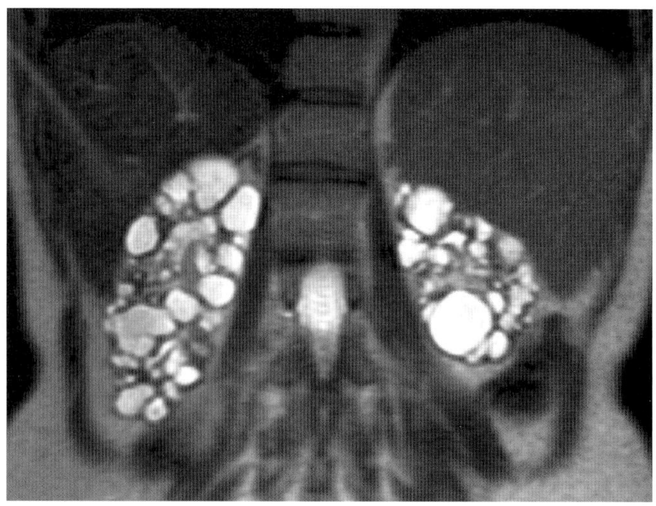

Figure 41.6 Coronal T2-weighted MR image obtained in a 42 year old woman because of chronic hepatitis B showing slightly atrophic kidneys containing multiple cysts of varying size. The patient also had a history of long-term hemodialysis for chronic renal failure due to polyarteritis nodosa, and the appearance of the kidneys is consistent with acquired cystic disease.

CASE 42 Pseudoenhancement of small renal cysts

Imaging description

Renal cyst pseudoenhancement is the artifactual increase in CT density that can be seen in small (less than 15 mm in diameter) simple renal cysts after the intravenous administration of iodinated contrast material (Figure 42.1) [1–4]. Pseudoenhancement is not simply due to partial volume averaging, because the artifact occurs even in experimental phantom studies where the effects of partial volume averaging have been removed. Pseudoenhancement is believed to be a consequence of beam-hardening effects of the enhanced renal parenchyma combined with artifact introduced by the CT image reconstruction algorithm [1].

Importance

Solid renal masses are typically distinguished from renal cysts at CT by the presence of enhancement after the intravenous administration of iodinated contrast material, and an increase in density of 10 Hounsfield Units has generally been regarded as the threshold for enhancement [5]. Pseudoenhancement may therefore lead to mischaracterization of a small renal cyst as an enhancing neoplasm and unnecessary intervention.

Typical clinical scenario

Renal cysts are present in 20 to 40% of the population [6]. Small renal cysts are particularly common. In one study, 67 of 258 patients (26%) undergoing CT had one or more renal cysts (as proven by MRI) of 10 mm or less in diameter [7]. In another study, 38% of simulated cysts measuring up to 15 mm in diameter demonstrated pseudoenhancement with an increase in density that ranged from 10.3 to 28.3 Hounsfield Units [8]. Factors that increase the likelihood of pseudoenhancement are smaller cyst size, higher background renal parenchymal enhancement, a higher number of CT detector rows, and higher kVp [1–3, 9–11].

Differential diagnosis

Pseudoenhancement makes it difficult to characterize small hypodense renal lesions by density measurements. Another practical problem is that these lesions are often not well seen on non-enhanced images, precluding confident placement of a region of interest measurement to determine pre-contrast density. However, such lesions can still be characterized with reasonable accuracy by subjective visual impression, because experienced radiologists are generally able to distinguish the vast majority of small hypodense renal lesions that are simple cysts from the small minority that are solid masses, especially small renal cell carcinomas. In one study, three independent readers demonstrated accuracies of 90 to 100% in the subjective visual characterization of 14 simple cysts and 6 renal cell carcinomas that were all less than or equal to 15 mm in diameter (Figure 42.2) [12].

Teaching point

CT density measurements are unreliable in determining if a small (15 mm or less) hypodense renal lesion enhances, because an increase in density may be artifactual due to pseudoenhancement. Instead, visual inspection by an experienced radiologist allows for accurate discrimination of the vast majority of small hypodense renal lesions that are simple cysts from the small minority that are really small renal cell carcinomas.

REFERENCES

1 Maki DD, Birnbaum BA, Chakraborty DP, et al. Renal cyst pseudoenhancement: beam-hardening effects on CT numbers. *Radiology* 1999; **213**: 468–472.

2 Coulam CH, Sheafor DH, Leder RA, et al. Evaluation of pseudoenhancement of renal cysts during contrast-enhanced CT. *Am J Roentgenol* 2000; **174**: 493–498.

3 Bae KT, Heiken JP, Siegel CL, Bennett HF. Renal cysts: is attenuation artifactually increased on contrast-enhanced CT images? *Radiology* 2000; **216**: 792–796.

4 Birnbaum BA, Jacobs JE, Ramchandani P. Multiphasic renal CT: comparison of renal mass enhancement during the corticomedullary and nephrographic phases. *Radiology* 1996; **200**: 753–758.

5 Bosniak MA, Rofsky NM. Problems in the detection and characterization of small renal masses. *Radiology* 1996; **200**: 286–287.

6 Tada S, Yamagishi J, Kobayashi H, Hata Y, Kobari T. The incidence of simple renal cyst by computed tomography. *Clin Radiol* 1983; **34**: 437–439.

7 Chithriki M, JoeBN, Yeh BM, et al. Incidental small hypoattenuating renal lesions: to follow or not to follow. *Am J Roentgenol* 2004; **182**: S4.

8 Birnbaum BA, Maki DD, Chakraborty DP, Jacobs JE, Babb JS. Renal cyst pseudoenhancement: evaluation with an anthropomorphic body CT phantom. *Radiology* 2002; **225**: 83–90.

9 Heneghan JP, Spielmann AL, Sheafor DH, et al. Pseudoenhancement of simple renal cysts: a comparison of single and multidetector helical CT. *J Comput Assist Tomogr* 2002; **26**: 90–94.

10 Abdulla C, Kalra MK, Saini S, et al. Pseudoenhancement of simulated renal cysts in a phantom using different multidetector CT scanners. *Am J Roentgenol* 2002; **179**: 1473–1476.

11 Wang ZJ, Coakley FV, Fu Y, et al. Renal cyst pseudoenhancement at multidetector CT: what are the effects of number of detectors and peak tube voltage? *Radiology* 2008; **248**: 910–916.

12 Patel NS, Yeh BM, Qayyum A, et al. The characterization of small hypoattenuating renal masses on contrast-enhanced CT. *Am J Roentgenol* 2004; **182**: S5.

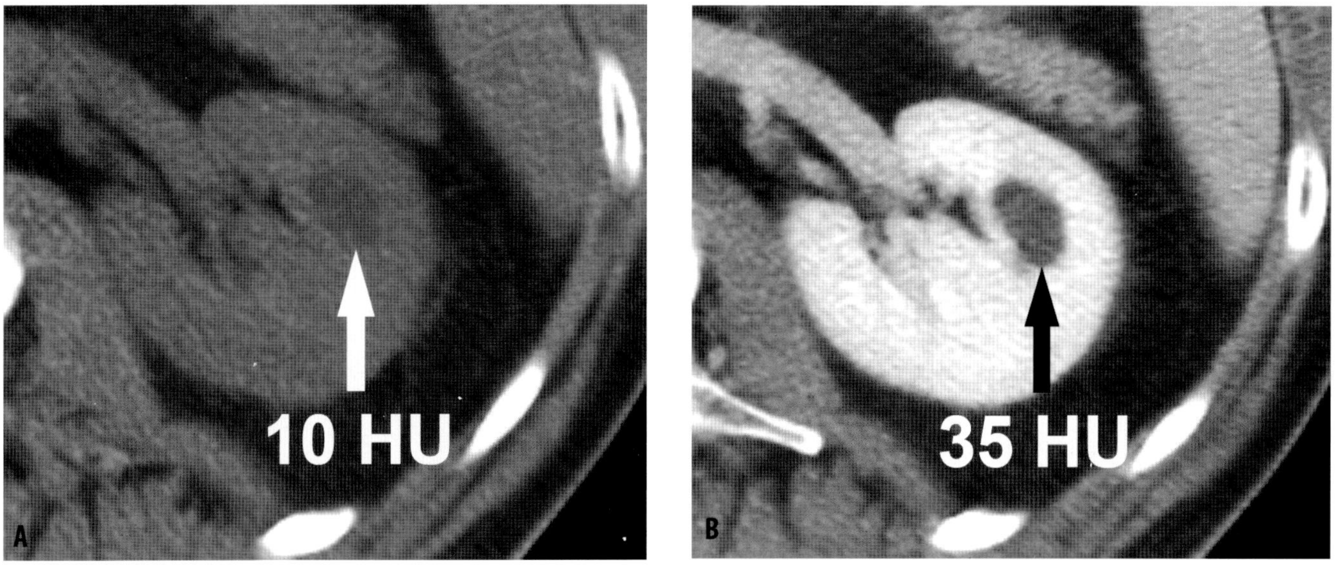

Figure 42.1 A. Axial non-enhanced CT image in a 52 year old man with chronic hepatitis shows a low-density (10 Hounsfield Units) lesion (arrow) in the left kidney. MRI (not shown) confirmed the lesion was a small cyst (based on non-enhancement after intravenous gadolinium and fluid signal intensity on T2-weighted imaging). **B.** Axial contrast-enhanced CT image shows an apparent increase in density in the cyst (arrow) to 35 HU. This artifactual increase in density that may be seen in small renal cysts surrounded by enhancing parenchyma is known as pseudoenhancement.

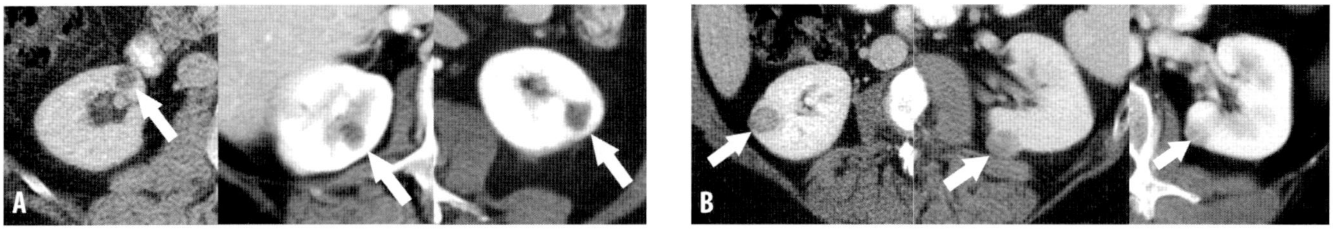

Figure 42.2 A. Photomontage of three small hypoattenuating renal lesions (arrows) at contrast-enhanced CT in three different patients. While density measurements are of limited utility in the evaluation of these lesions, all three appear visually cystic, and were proven to be small cysts at MRI. **B.** Photomontage of three small hypoattenuating renal lesions (arrows) at contrast-enhanced CT in three different patients, all of which were surgically proven to be renal cell carcinomas. Note that these lesions all appear visually distinct from the three small cysts shown in Figure 42.2A.

CASE

43 Pseudotumor due to focal masslike parenchyma

Imaging description

Historically, prominent masslike foci of renal parenchyma simulating tumors were a well-recognized pitfall at intravenous pyelography, but such pseudotumors can also be seen at ultrasound and sometimes CT. Masslike foci of renal parenchyma may be congenital or acquired, and causes include a hypertrophied column of Bertin, lobar dysmorphism, persistent fetal lobation, dromedary or splenic hump, and scarring with focal sparing or compensatory hypertrophy (Figures 43.1–43.4) [1–9]. In 1744, the French anatomist Exupere Joseph Bertin noted that the renal cortex not only envelops the kidney, but also extends in radial bands between the medullary pyramids. These radial bands are now known as the septa or columns of Bertin [1]. When unusually thickened, they are said to be hypertrophied and may mimic a mass. As insightfully noted by Yeh et al., the tissue is not actually hypertrophied, but rather reflects incomplete fusion of the fetal lobes or sub-kidneys that normally join together to form the kidney during gestation and arguably would be more correctly termed "junctional parenchyma" [3]. Renal lobar dysmorphism is a related congenital anomaly in which a diminutive lobe develops between two normal lobes. It resembles a hypertrophied column of Bertin, except that it also contains medullary tissue [4]. Fetal lobation is another form of incomplete lobar fusion in which persistent grooves in the renal outline demarcate the junction between fetal lobes [5]. The tissue between these grooves may sometimes appear masslike. The splenic or dromedary hump is a prominent bulge on the superolateral border of the left kidney that is believed to arise secondary to molding of the upper pole of the left kidney by the spleen during development.

Importance

Misdiagnosis of a renal mass may result in unnecessary procedures such as biopsy or surgery, exposing the patient to needless risk and anxiety.

Typical clinical scenario

Many of these pseudotumors are congenital anatomic variants, and so may be seen in any patient undergoing imaging. For example, a hypertrophied column of Bertin measuring more than 1 cm in diameter was seen in 22 of 200 kidneys (11%) studied with ultrasound [2]. Most were located in the mid-kidney, and they were more common on the left side. Focal compensatory hypertrophy should be considered when an apparent renal mass is seen in a patient with a potential cause of renal cortical loss and scarring, such as reflux nephropathy or recurrent urinary tract infections [8, 9].

Differential diagnosis

The primary consideration is the distinction of a renal pseudotumor from a true solid renal mass. A number of findings and modalities can be helpful. Pseudotumors should be similar to the non-masslike renal parenchyma with respect to echogenicity, CT density, and MRI signal intensity [10]. Hypertrophied columns of Bertin demonstrate a characteristic splaying of the central sinus echoes at ultrasound ("split sinus sign") [1] and the overlying renal contour is smooth and not bulging [2]. The grooves of fetal lobation are sharp and linear, while indentations due to scarring are wide, less well defined, and associated with loss of renal cortex [5]. In problematic cases, other modalities that may help in correct identification of a renal pseudotumor include contrast-enhanced ultrasound or radionuclide imaging [9, 10].

Teaching point

The possibility of a pseudotumor due to focal masslike parenchyma should be considered for an apparent solid renal mass, particularly for a mass seen at ultrasound that is of similar echogenicity to the renal cortex.

REFERENCES

1 Mahony BS, Jeffrey RB, Laing FC. Septa of Bertin: a sonographic pseudotumor. *J Clin Ultrasound* 1983; **11**: 317–319.

2 Lafortune M, Constantin A, Breton G, Vallee C. Sonography of the hypertrophied column of Bertin. *Am J Roentgenol* 1986; **146**: 53–56.

3 Yeh HC, Halton KP, Shapiro RS, Rabinowitz JG, Mitty HA. Junctional parenchyma: revised definition of hypertrophic column of Bertin. *Radiology* 1992; **185**: 725–732.

4 Zwirewich CV, Rowley VA. Lobar dysmorphism of the kidney: report of two cases and review of the literature. *J Comput Assist Tomogr* 1997; **21**: 742–744.

5 Patriquin H, Lefaivre JF, Lafortune M, Russo P, Boisvert J. Fetal lobation. An anatomo-ultrasonographic correlation. *J Ultrasound Med* 1990; **9**: 191–197.

6 Ascenti G, Zimbaro G, Mazziotti S, et al. Contrast-enhanced power Doppler US in the diagnosis of renal pseudotumors. *Eur Radiol* 2001; **11**: 2496–2499.

7 Depner TA, Ryan KG, Yamauchi H. Pseudotumor of the kidney: a sequel to regional glomerulonephritis. *Am J Roentgenol* 1976; **126**: 1197–1202.

8 Damry N, Avni F, Guissard G, et al. Compensatory hypertrophy of renal parenchyma presenting as a mass lesion. *Pediatr Radiol* 2005; **35**: 832–833.

9 Urbania TH, Kammen BF, Nancarrow PA, Morrell RE. Bilateral renal masses in a 10-year-old girl with renal failure and urinary tract infection: the importance of functional imaging. *Pediatr Radiol* 2009; **39**: 172–175.

10 Mazziotti S, Zimbaro F, Pandolfo A, et al. *Usefulness of contrast-enhanced ultrasonography in the diagnosis of renal pseudotumors.* Abdominal Imaging 2009 Feb 5. [Epub ahead of print].

11 Simpson EL, Mintz MC, Pollack HM, Arger PH, Coleman BG. Computed tomography in the diagnosis of renal pseudotumors. *J Comput Tomogr* 1986; **10**: 341–348.

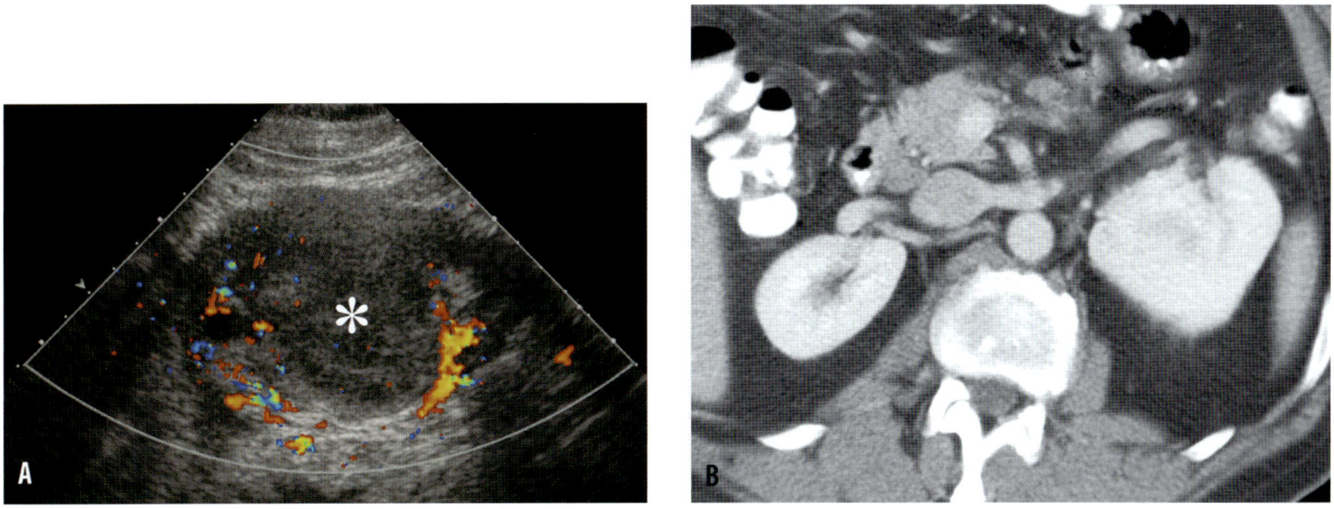

Figure 43.1 A. Longitudinal ultrasound image of the left kidney in a 48 year old man with chronic lymphocytic leukemia shows an apparent mass (asterisk). Note the mass is isoechoic with the renal cortex and splays the central sinus fat. **B.** Axial contrast-enhanced CT image shows the mass is isodense with the rest of the renal parenchyma, suggesting a pseudotumor. Because of concern that the mass might be a leukemic deposit, a CT-guided biopsy (not shown) was performed and demonstrated normal renal parenchyma, confirming a diagnosis of hypertrophied column of Bertin.

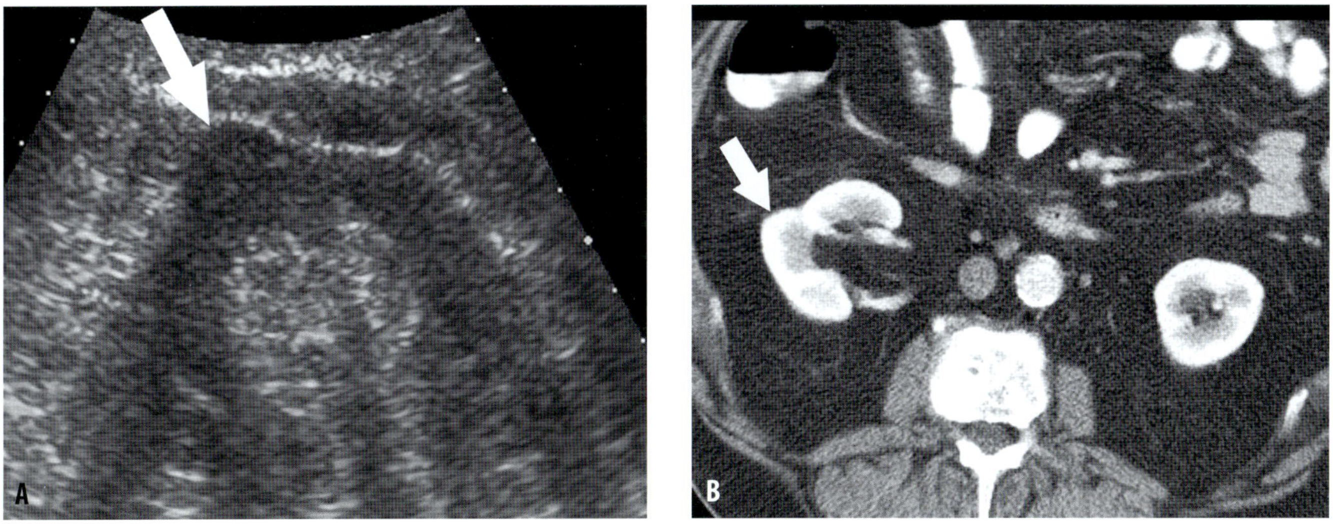

Figure 43.2 A. Longitudinal ultrasound image of the right kidney in a 78 year old man with an abdominal aortic aneurysm shows an apparent exophytic mass (arrow), which was interpreted as either fetal lobation or a renal cell carcinoma. **B.** Axial contrast-enhanced CT image shows the mass (arrow) is due to bulging of the renal outline and not a true tumor, consistent with fetal lobation.

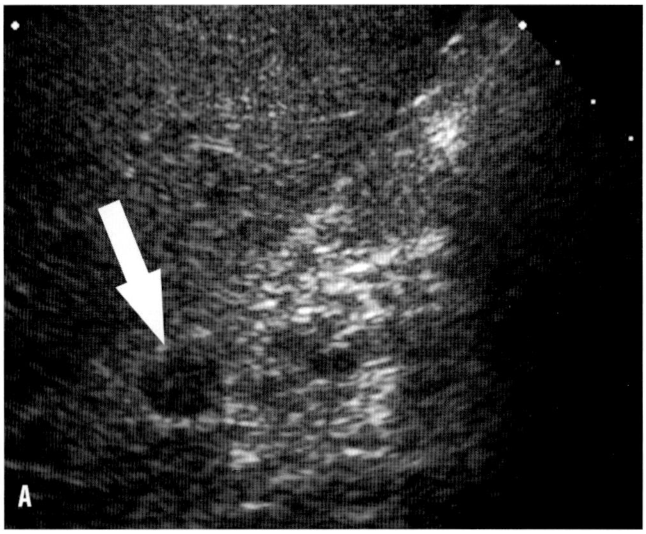

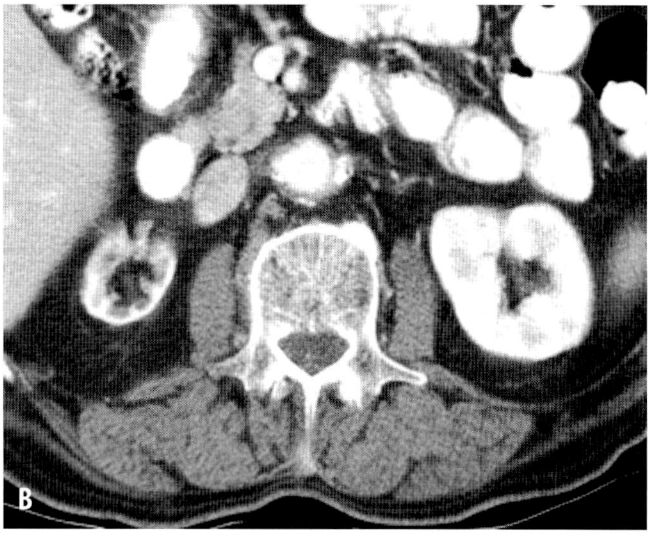

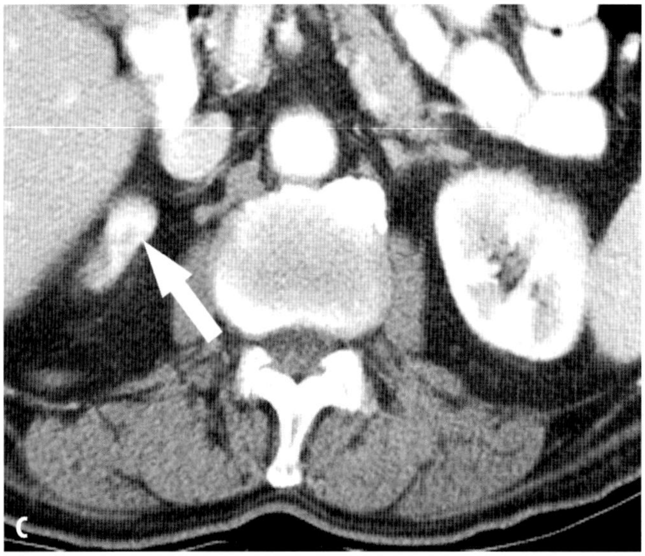

Figure 43.3 **A.** Longitudinal ultrasound image of the right kidney in an 81 year old woman with abdominal pain shows an apparent mass (arrow) arising from the upper pole of an atrophic right kidney. **B.** Axial contrast-enhanced CT at the level of the mid-kidneys confirms marked atrophy of the right kidney. **C.** Axial contrast-enhanced CT at the level of the apparent mass shows it is a focus (arrow) of enhancing renal parenchyma, consistent with relative sparing in an atrophic kidney.

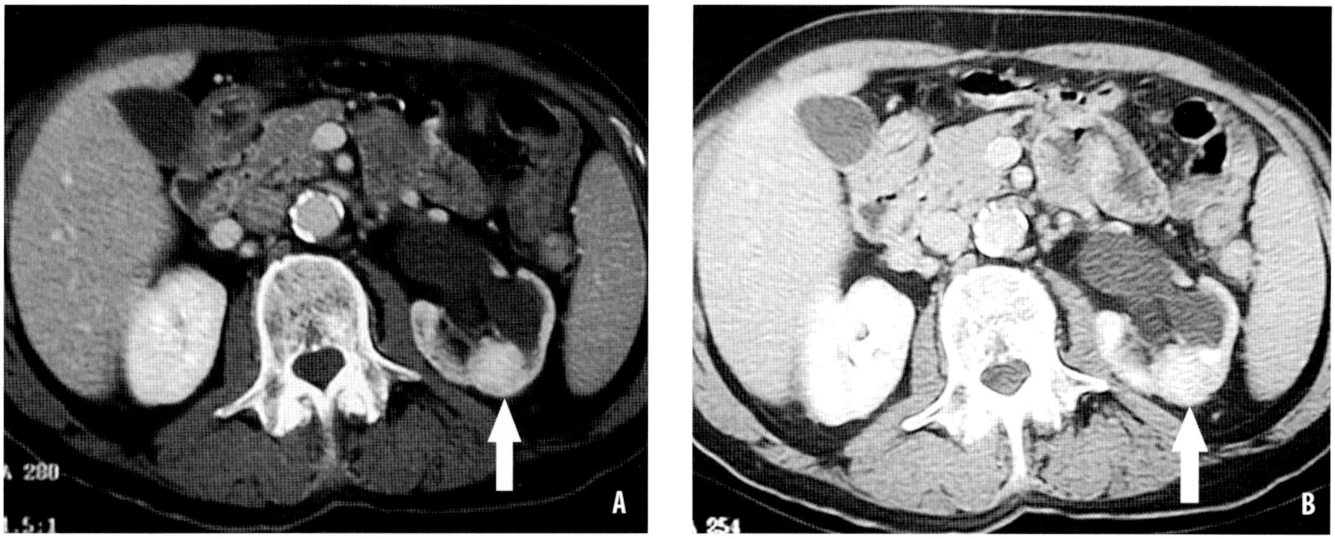

Figure 43.4 A. Axial contrast-enhanced CT image in a 44 year old man with longstanding diminished function in the left kidney due to reflux nephropathy shows an apparent mass (arrow) in the mid-kidney. **B.** Axial delayed phase contrast-enhanced CT image shows the apparent mass (arrow) is isodense with the remaining renal parenchyma. Left nephrectomy confirmed the absence of any tumor in the kidney. The mass was due to an island of relatively spared parenchyma against a background of diffuse atrophy due to reflux nephropathy.

CASE 44 Pseudotumor due to anisotropism

Imaging description

In ultrasound, anisotropism refers to the different echogenicity that can occur within tissues with a directional internal structure depending on the angle of insonation. The term is derived from the Greek *aniso* (meaning not the same) and *tropos* (to turn or reflect). The phenomenon was first described in tendons [1, 2], but can also occur in the kidneys where the radial arrangement of nephrons and intervening tissues results in greater echogenicity from parts of the kidney where the nephrons are perpendicular to the ultrasound beam when compared to parts where the nephrons are parallel to the ultrasound beam [3, 4]. In practice, this can result in an apparent echogenic pseudotumor in the polar parts of the kidneys when the ultrasound beam is centered on the mid-kidney (Figure 44.1).

Importance

Anisotropic renal pseudotumor may be misinterpreted as a true echogenic renal mass, suggestive of either angiomyolipoma or renal cell carcinoma, and result in unnecessary additional workup and patient anxiety.

Typical clinical scenario

This pseudotumor is a technical artifact and so can potentially be seen in any patient undergoing ultrasound of the kidneys.

Differential diagnosis

The key to recognizing anisotropic renal pseudotumor at ultrasound is to compare the image with the apparent mass when the transducer is centered on the mid-kidney to an image obtained when the transducer is closer to a radial alignment with the polar part of the kidney – the anisotropic pseudotumor will not be visible on the latter image, unlike a true mass which should be equally visible on both. In addition, anisotropic renal pseudotumor typically has ill-defined margins and fades gradually into the surrounding tissues, unlike a true renal mass which frequently has well-defined margins.

Teaching point

The possibility of an anisotropic renal pseudotumor should be considered when an apparent echogenic mass is seen at ultrasound in the polar parts of the kidney.

REFERENCES

1 Fornage BD. The hypoechoic normal tendon. A pitfall. *J Ultrasound Med* 1987; **6**: 19–22.

2 Connolly DJ, Berman L, McNally EG. The use of beam angulation to overcome anisotropy when viewing human tendon with high frequency linear array ultrasound. *Br J Radiol* 2001; **74**: 183–185.

3 Rubin JM, Carson PL, Meyer CR. Anisotropic ultrasonic backscatter from the renal cortex. *Ultrasound Med Biol* 1988; **14**: 507–511.

4 Insana MF, Hall TJ, Fishback JL. Identifying acoustic scattering sources in normal renal parenchyma from the anisotropy in acoustic properties. *Ultrasound Med Biol* 1991; **17**: 613–626.

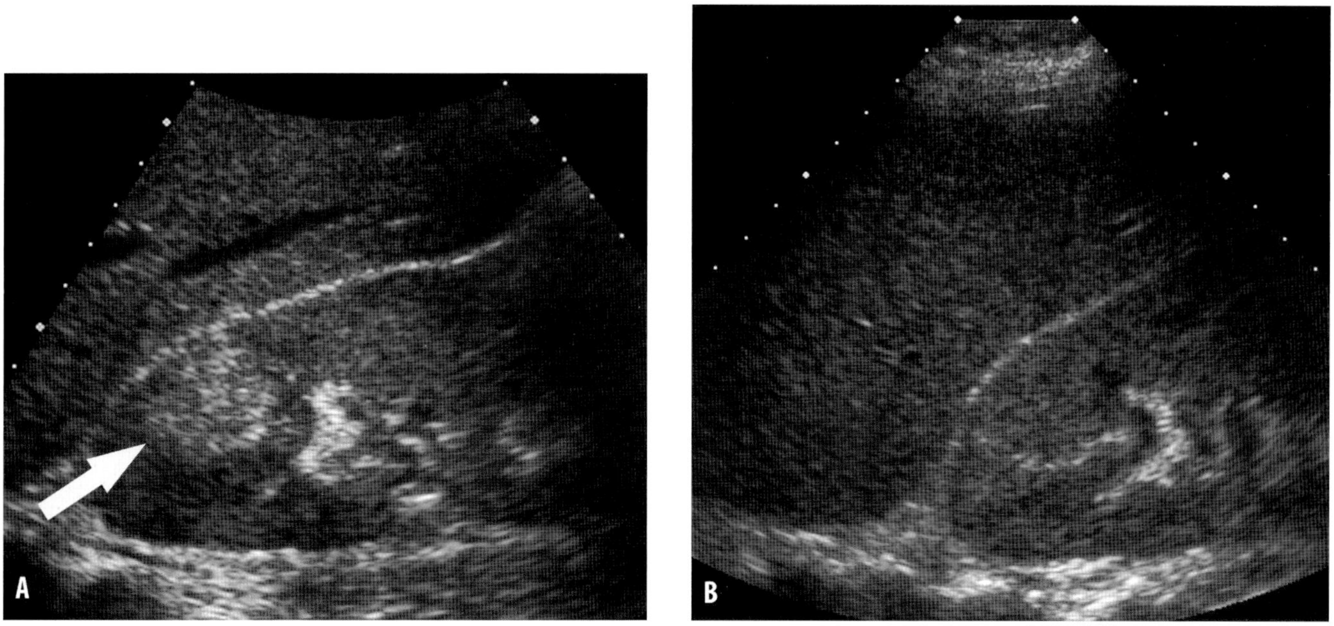

Figure 44.1 A. Longitudinal ultrasound image of the right kidney obtained during routine evaluation of a 21-week gestation pregnancy in a 29 year old woman shows an apparent echogenic mass (arrow) in the upper pole. Note the transducer is centered on the mid-kidney. **B.** Longitudinal ultrasound image of the right kidney obtained during the same study with the transducer centered over the upper pole of the kidney shows the mass is no longer evident. The appearances are typical of an anisotropic renal pseudotumor.
Images graciously provided by Dr Peter Callen, UCSF.

CASE 45

Echogenic renal cell carcinoma mimicking angiomyolipoma

Imaging description

A reported 61% (22 of 36) to 77% (24 of 31) of small renal cell carcinomas are hyperechoic relative to the adjacent renal parenchyma at ultrasound, and 32% (10 of 31) are uniformly and markedly echogenic such that they mimic angiomyolipomas (Figures 45.1 and 45.2) [1, 2]. Larger renal cell carcinomas are usually hypoechoic. Given that there is no particularly plausible reason for echogenicity to depend on tumor size, it is possible that this relationship is artifactual due to diagnostic bias. That is, smaller hypoechoic renal cell carcinomas are less likely to cause contour deformities or other mass effects and may be missed, while small echogenic renal cell carcinomas stand out relative to the renal parenchyma and are more likely to be detected [3].

Importance

The primary concern is that a renal cell cancer misdiagnosed as an angiomyolipoma might progress and become incurable. Based on the available evidence and given that the frequency with which small echogenic renal masses represent renal cell carcinoma rather than angiomyolipoma is unknown, it has been suggested that all non-calcified echogenic renal lesions found on ultrasound need further evaluation with CT [4]. This may be a counsel of perfection, since in practice supplementary CT is inconsistently recommended and often ignored [5]. I have been unable to find any reports of a fatal renal cell carcinoma that was initially diagnosed as an angiomyolipoma on ultrasound. This may mean the majority of small echogenic masses are truly angiomyolipomas, or might just as well reflect the fact that small incidental renal cell carcinomas are often indolent and arguably subclinical [6].

Typical clinical scenario

Echogenic renal masses are usually detected incidentally at ultrasound performed for unrelated reasons, and so may be encountered in any clinical setting.

Differential diagnosis

Several studies have shown that some ultrasound features help in the distinction of angiomyolipoma from echogenic renal cell carcinoma. Specifically, shadowing is seen only with angiomyolipomas (Figure 45.3), while a hypoechoic rim and intratumoral cysts are seen only in renal cell carcinomas (Figure 45.4) [7–9]. Unfortunately, these findings are not present in many cases, limiting their clinical utility.

Teaching point

Most small uniformly and brightly echogenic renal masses seen incidentally at ultrasound are probably angiomyolipomas, but renal cell carcinoma cannot be entirely excluded and confirmation by CT is a reasonable recommendation.

REFERENCES

1 Forman HP, Middleton WD, Melson GL, McClennan BL. Hyperechoic renal cell carcinomas: increase in detection at US. *Radiology* 1993; **188**: 431–434.

2 Yamashita Y, Ueno S, Makita O, et al. Hyperechoic renal tumors: anechoic rim and intratumoral cysts in US differentiation of renal cell carcinoma from angiomyolipoma. *Radiology* 1993; **188**: 179–182.

3 Hélénon O, Correas JM, Balleyguier C, Ghouadni M, Cornud F. Ultrasound of renal tumors. *Eur Radiol* 2001; **11**: 1890–1901.

4 Farrelly C, Delaney H, McDermott R, Malone D. Do all non-calcified echogenic renal lesions found on ultrasound need further evaluation with CT? *Abdom Imaging* 2008; **33**: 44–47.

5 Ikeda AK, Korobkin M, Platt JF, Cohan RH, Ellis JH. Small echogenic renal masses: how often is computed tomography used to confirm the sonographic suspicion of angiomyolipoma? *Urology* 1995; **46**: 311–315.

6 Lee CT, Katz J, Fearn PA, Russo P. Mode of presentation of renal cell carcinoma provides prognostic information. *Urol Oncol* 2002; **7**: 135–140.

7 Yamashita Y, Ueno S, Makita O, et al. Hyperechoic renal tumors: anechoic rim and intratumoral cysts in US differentiation of renal cell carcinoma from angiomyolipoma. *Radiology* 1993; **188**: 179–182.

8 Siegel CL, Middleton WD, Teefey SA, McClennan BL. Angiomyolipoma and renal cell carcinoma: US differentiation. *Radiology* 1996; **198**: 789–793.

9 Zebedin D, Kammerhuber F, Uggowitzer MM, Szolar DH. Criteria for ultrasound differentiation of small angiomyolipomas (< or = 3 cm) and renal cell carcinomas. *Rofo* 1998; **169**: 627–632 [German].

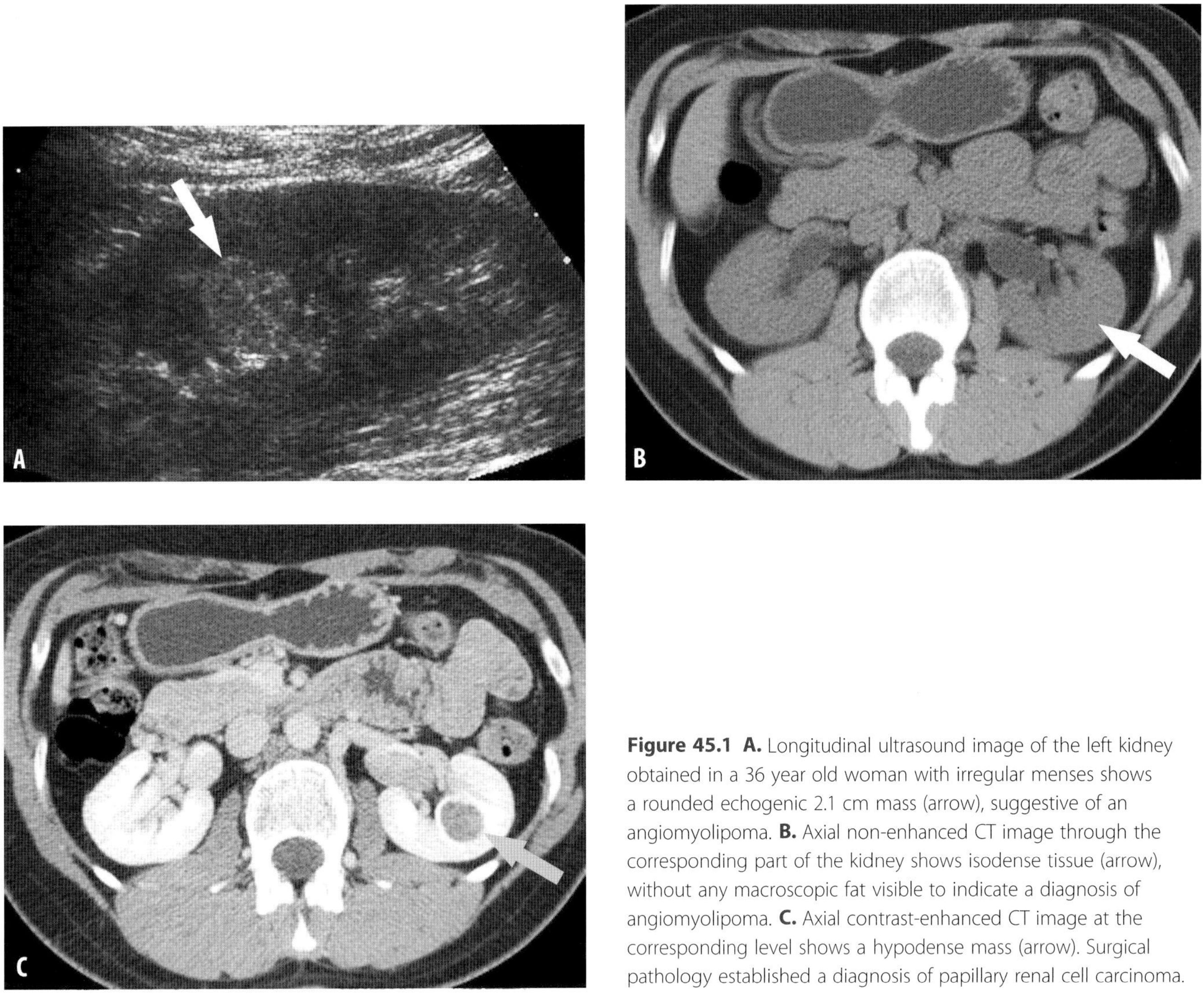

Figure 45.1 **A.** Longitudinal ultrasound image of the left kidney obtained in a 36 year old woman with irregular menses shows a rounded echogenic 2.1 cm mass (arrow), suggestive of an angiomyolipoma. **B.** Axial non-enhanced CT image through the corresponding part of the kidney shows isodense tissue (arrow), without any macroscopic fat visible to indicate a diagnosis of angiomyolipoma. **C.** Axial contrast-enhanced CT image at the corresponding level shows a hypodense mass (arrow). Surgical pathology established a diagnosis of papillary renal cell carcinoma.

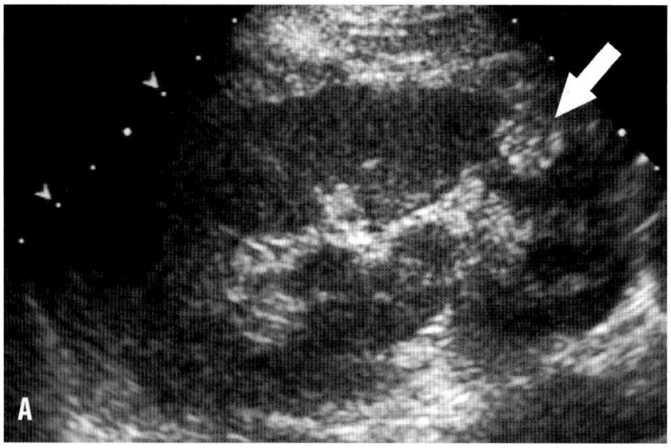

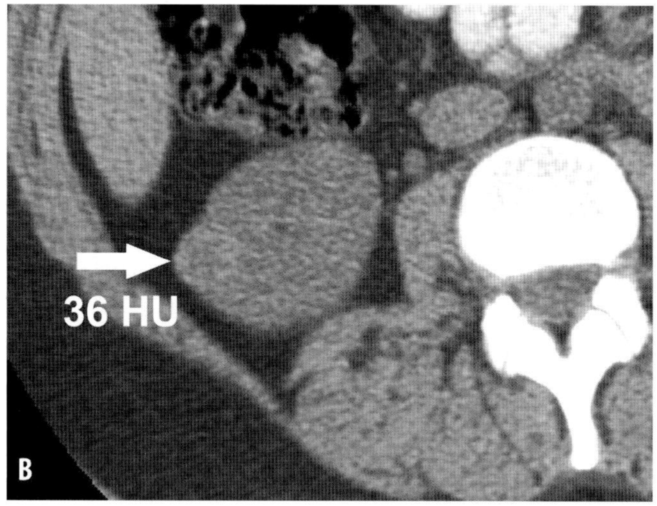

Figure 45.2 A. Longitudinal ultrasound image of the right kidney obtained in a 34 year old woman with gestational trophoblastic disease shows a rounded echogenic 1.4 cm mass (arrow), suggestive of an angiomyolipoma. **B.** Axial non-enhanced CT image through the corresponding part of the kidney shows a subtle mass (arrow), without any macroscopic fat visible to indicate a diagnosis of angiomyolipoma.

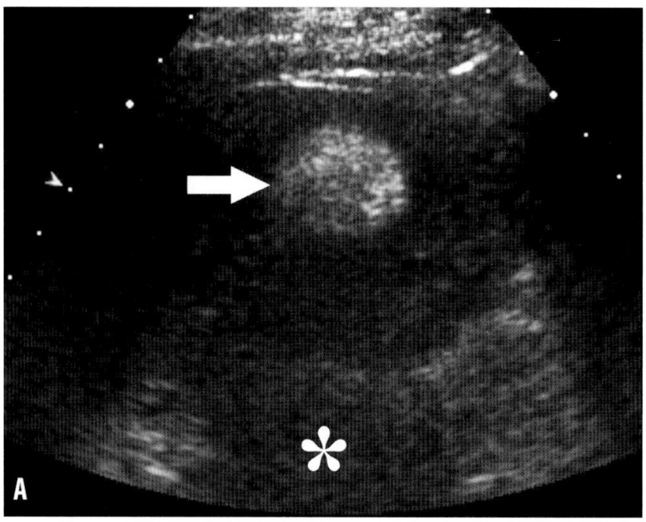

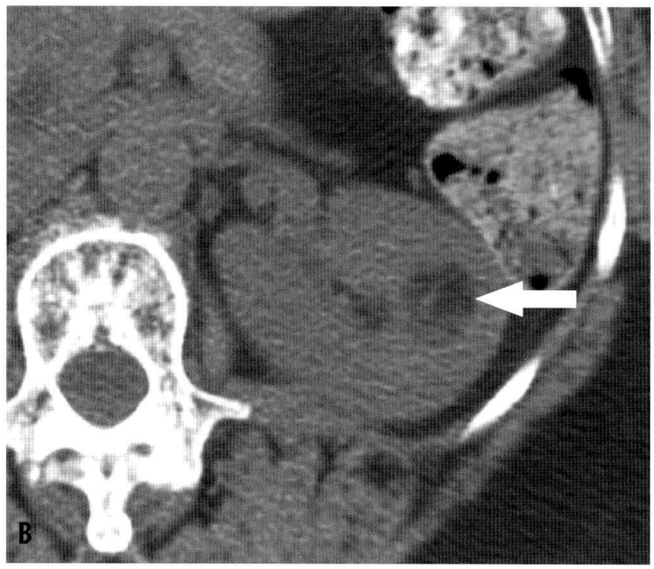

Figure 45.3 A. Longitudinal ultrasound image of the left kidney obtained in a 69 year old woman with locally advanced rectal cancer shows a rounded highly echogenic 2.2 cm mass (arrow), suggestive of an angiomyolipoma. Note the presence of acoustic shadowing (asterisk). **B.** Axial non-enhanced CT image through the corresponding part of the kidney shows a macroscopic fat-containing mass (arrow), confirming the diagnosis of angiomyolipoma. Acoustic shadowing is seen in only a fraction of angiomyolipomas, but seems to be of high positive predictive value.

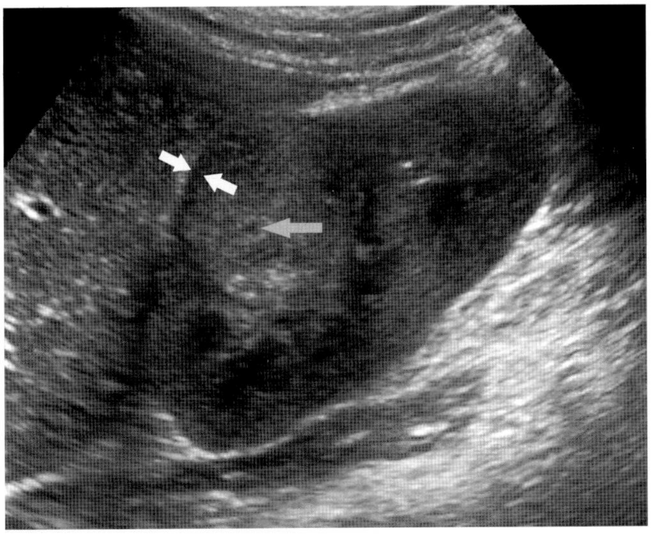

Figure 45.4 Longitudinal ultrasound image of the right kidney obtained in a 31 year old woman with an echogenic 3.5 cm papillary renal cell carcinoma shows the tumor has a hypoechoic rim (between white arrows) and contains an intratumoral cyst (grey arrow).

CASE 46 Pseudohydronephrosis

Imaging description

Fluid-filled structures (e.g., varices or parapelvic cysts) or solid hypoechoic masses (e.g., lymphomas or related conditions) in the renal hilum may simulate a dilated pelvicaliceal system at imaging and result in an erroneous diagnosis of hydronephrosis (Figures 46.1–46.3) [1–9].

Importance

Misidentification of intrarenal varices as hydronephrosis is potentially the most serious error, since attempted percutaneous nephrostomy tube placement could conceivably result in catastrophic bleeding. Misidentification of parapelvic cysts or solid hilar tumors as hydronephrosis could also lead to inappropriate treatment or a missed opportunity for earlier diagnosis and management of malignancy.

Typical clinical scenario

Renal hilar varices are typically manifestations of renal arteriovenous malformations, which may be congenital or acquired due to trauma, surgery, biopsy, malignancy, or inflammation [3]. Parapelvic cysts are found at 1.2 to 1.5% of autopsies, and may be congenital or acquired due to lymphatic blockage [10, 11]. Renal involvement by lymphoma or other malignancies of reduced echogenicity may occur at any age, but is commoner in adults.

Differential diagnosis

Hilar varices are easily recognized at ultrasound, provided Doppler images are acquired, since they contain internal flow. They are also easily recognized as tubular enhancing vascular structures at CT or MRI. Parapelvic cysts can more closely simulate hydronephrosis; pointers to the correct diagnosis include a multilobulated appearance, lack of the typical cauliflower-like intercommunication of dilated calices and pelvis, and the presence of thick septa due to sinus fat or other tissue trapped between the cyst and the pelvicaliceal system. Anechoic or hypoechoic hilar tumor at ultrasound can usually be recognized by masslike morphology or by correlation with CT or MRI.

Teaching point

Apparent pelvicaliceal dilatation can be simulated by renal hilar varices, parapelvic cysts and anechoic or hypoechoic hilar tumor. Close attention to morphology or correlation with appropriately performed CT or MRI usually allows for accurate distinction.

REFERENCES

1 Erden A, Ozcan H, Aytaç S, Sanlidilek U, Cumhur T. Intrarenal varices in portal hypertension: demonstration by color Doppler imaging. *Abdom Imaging* 1996; **21**: 549–550.

2 Kincaid W, Edwards R. Intrarenal varices mimicking hydronephrosis. *Br J Radiol* 1992; **65**: 1038–1039.

3 Kember PG, Peck RJ. Renal arteriovenous malformation mimicking hydronephrosis. *J Clin Ultrasound* 1998; **26**: 95–97.

4 Cronan JJ, Amis ES Jr, Yoder IC, et al. Peripelvic cysts: an impostor of sonographic hydronephrosis. *J Ultrasound Med* 1982; **1**: 229–236.

5 Amis ES Jr, Cronan JJ, Pfister RC. Pseudohydronephrosis on noncontrast computed tomography. *J Comput Assist Tomogr* 1982; **6**: 511–513.

6 Ehrman KO, Kopecky KK, Wass JL, Thomalla JV. Parapelvic lymph cyst in a renal allograft mimicking hydronephrosis: CT diagnosis. *J Comput Assist Tomogr* 1987; **11**: 714–716.

7 Patel U, Huntley L, Kellett MJ. Sonographic features of renal obstruction mimicked by parapelvic cysts. *Clin Radiol* 1994; **49**: 481.

8 Tarzamni MK, Sobhani N, Nezami N, Ghiasi F. Bilateral parapelvic cysts that mimic hydronephrosis in two imaging modalities: a case report. *Cases J* 2008; **1**: 161.

9 Urban BA, Fishman EK. Renal lymphoma: CT patterns with emphasis on helical CT. *Radiographics* 2000; **20**: 197–212.

10 Lee F, Thornbury JR, Juhl JH, Crummy AB, Kuhlman JE. The urinary tract. In: Juhl JH, Crummy AB, Paul LW, eds. *Paul and Juhl's essentials of radiologic imaging*. Philadelphia, PA: Lippincott Williams & Wilkins, 1987; 683.

11 Kabala JE. The kidneys and ureter. In: Sutton D, ed. *Textbook of radiology and imaging*, 7th edition. London: Churchill Livingstone, 2003; 951.

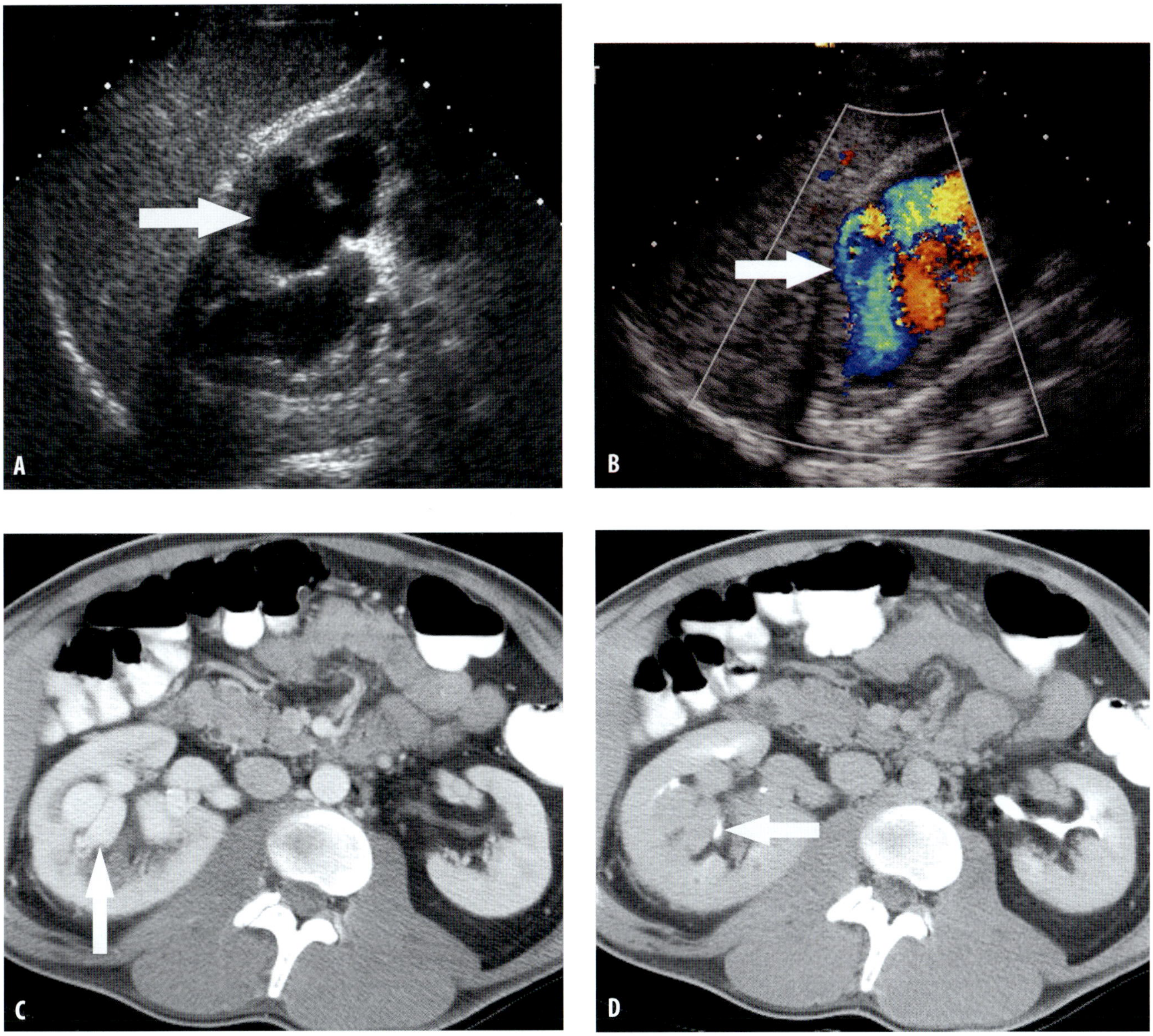

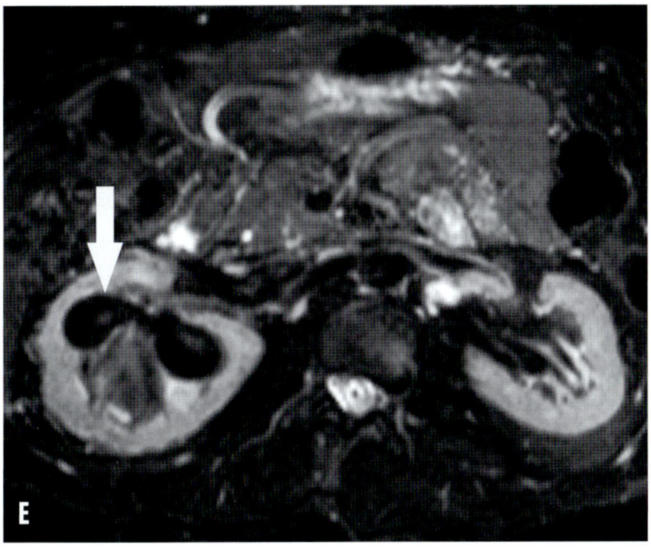

Figure 46.1 A. Longitudinal ultrasound image of the right kidney obtained during evaluation of the liver in a 30 year man with hemophilia and chronic hepatitis (without cirrhosis or portal hypertension) shows apparent pelvicaliceal dilatation (arrow). The study was reported as showing moderate right hydronephrosis. **B.** Doppler ultrasound image shows flow within the apparently dilated pelvicaliceal system. **C.** Axial contrast-enhanced CT image shows a cluster of briskly enhancing tubular structures (arrow) in the renal hilum that appear continuous with the left renal vein. The appearances are consistent with intrarenal varices. **D.** Axial delayed phase contrast-enhanced CT image shows part of the opacified pelvicaliceal system (arrow), which is clearly separate to the hilar varices. **E.** Axial T2-weighted MR image shows the hilar varices as a signal void (arrow) in the renal hilum. The renal abnormality was asymptomatic and has been managed by surveillance, with no change for over five years.

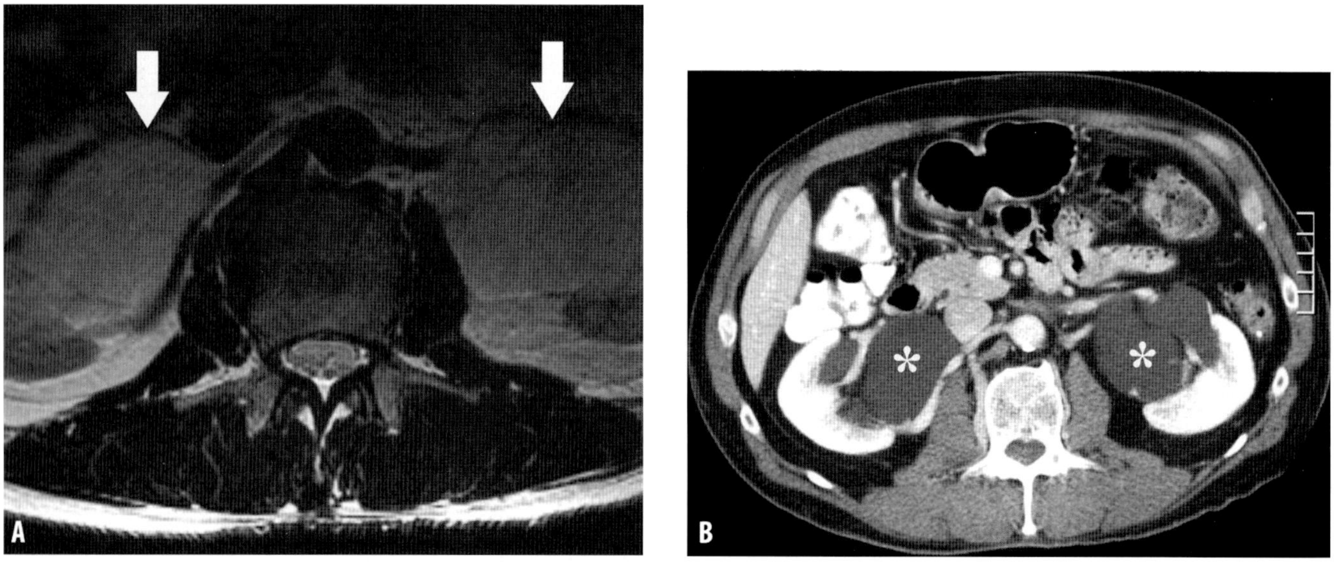

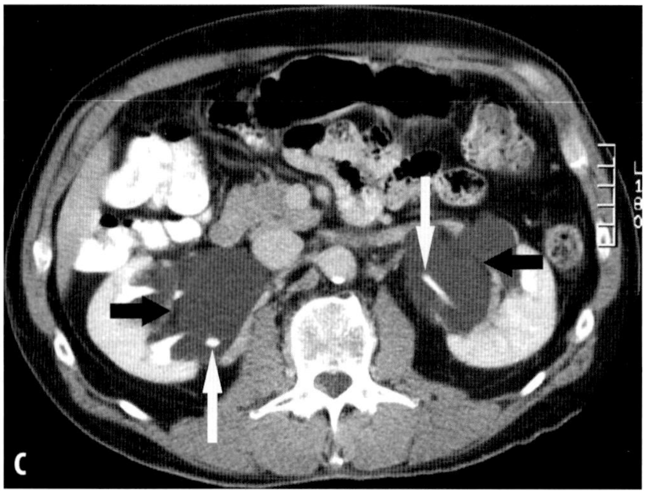

Figure 46.2 A. Axial T2-weighted MR image in a 65 year old man with back pain shows bilateral fluid-filled structures (arrows) in the renal hila. The study was reported as showing marked bilateral hydronephrosis. **B.** Axial contrast-enhanced CT image shows bilateral fluid-filled structures in the renal hila that arguably could reasonably be interpreted as dilated pelvicaliceal systems. **C.** Axial delayed phase contrast-enhanced CT image shows that the fluid-filled structures are actually parapelvic cysts, because the non-dilated pelvicaliceal systems (white arrows) are visualized separately to the fluid-filled structures in the renal hila. Note that fatty septa (black arrows) are visible in the parapelvic cysts. This observation can be an important clue to the diagnosis.

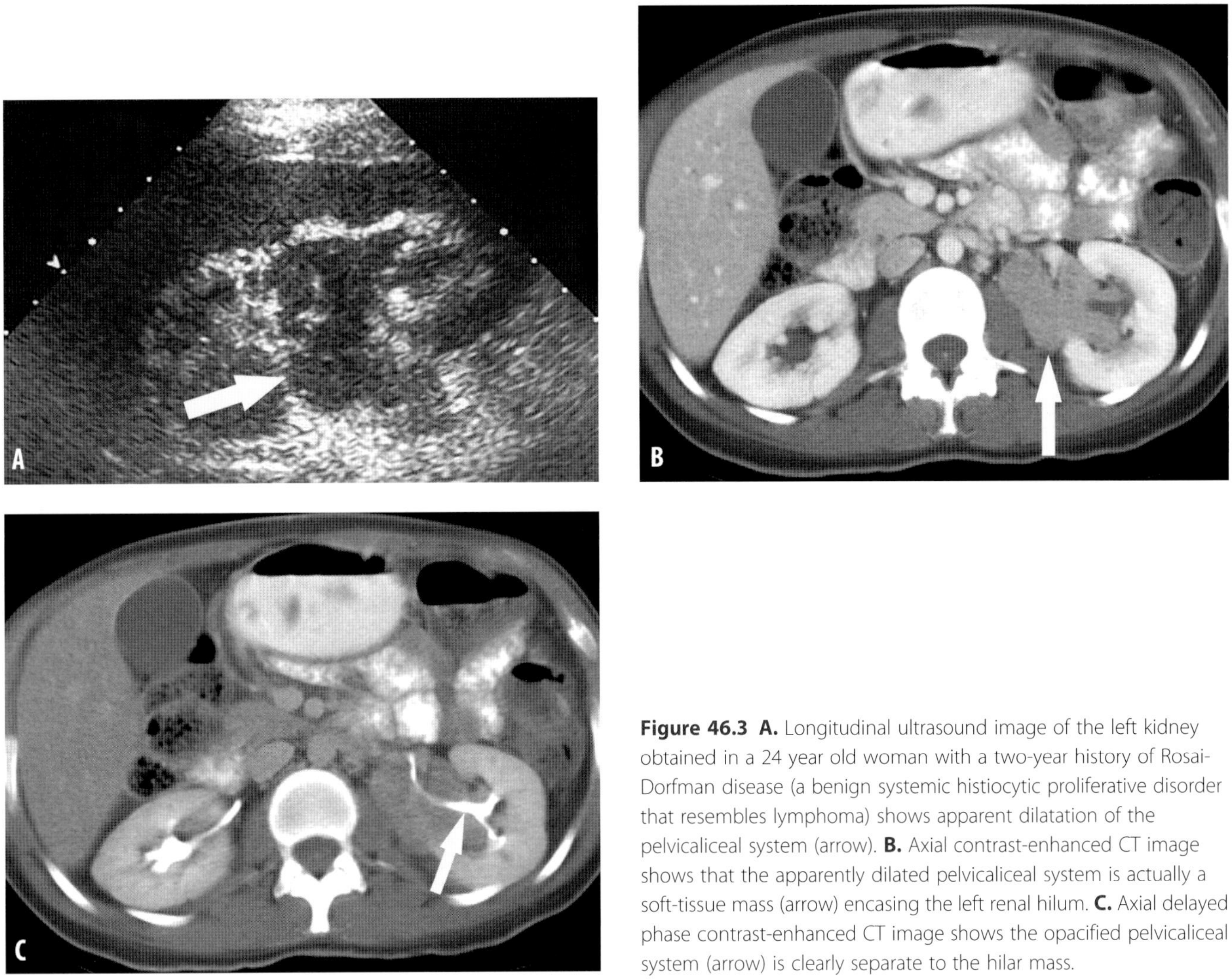

Figure 46.3 A. Longitudinal ultrasound image of the left kidney obtained in a 24 year old woman with a two-year history of Rosai-Dorfman disease (a benign systemic histiocytic proliferative disorder that resembles lymphoma) shows apparent dilatation of the pelvicaliceal system (arrow). **B.** Axial contrast-enhanced CT image shows that the apparently dilated pelvicaliceal system is actually a soft-tissue mass (arrow) encasing the left renal hilum. **C.** Axial delayed phase contrast-enhanced CT image shows the opacified pelvicaliceal system (arrow) is clearly separate to the hilar mass.

CASE

47 Pseudocalculi due to excreted gadolinium

Imaging description

Gadolinium is a rare-earth metal used as an MRI contrast agent because of its paramagnetic properties. Gadolinium has a high atomic number (64, compared to 53 for iodine) and absorbs x-rays, and so functionally can act as a radiographic contrast agent. Before the recognition of nephrogenic systemic fibrosis as a complication of gadolinium administration in patients with renal failure, gadolinium was advocated as an angiographic contrast agent for such patients [1, 2]. Like iodinated contrast, gadolinium is excreted by the kidneys. Concentrated excreted gadolinium is radiodense within the collecting system at CT [3, 4], and this radiodensity can mimic renal calculi when non-enhanced CT is performed within the first few hours after a gadolinium-enhanced MRI study (Figure 47.1) [4, 5]. The phenomenon has not been extensively studied, but limited data suggest the dense appearance of excreted gadolinium at CT in the collecting systems is variable from patient to patient, and cannot be reliably predicted from the time interval since gadolinium administration, patient weight, or simple indices of renal function [4].

Importance

Misdiagnosis of excreted gadolinium as renal calculi can result in unnecessary additional investigations, such as abdominal radiography or intravenous pyelography [5].

Typical clinical scenario

Pseudocalculi due to excreted gadolinium can be seen in any patient who undergoes non-enhanced abdominal CT after a gadolinium-enhanced MRI. In our practice, we see this most often when a patient with cancer is scheduled to have a staging brain MRI and PET/CT without iodinated contrast on the same day.

Differential diagnosis

Increased density in the collecting systems at CT due to excreted gadolinium is less dense than true calcified stones, is non-obstructive, and spread symmetrically and diffusely through the upper tracts (including the ureters). These observations can help suggest the diagnosis and facilitate the distinction from real calculi, but the ultimate confirmation is establishing that the patient had a gadolinium-enhanced MRI scan shortly before the CT scan.

Teaching point

Increased density in the upper urinary tracts seen at non-enhanced CT that is diffuse, non-obstructive, and not as dense as true calcified stones should suggest a diagnosis of pseudocalculi due to excreted gadolinium, and correlation with recent imaging history should help confirm the diagnosis.

REFERENCES

1 Spinosa DJ, Matsumoto AH, Angle JF, et al. Renal insufficiency: usefulness of gadodiamide-enhanced renal angiography to supplement CO_2-enhanced renal angiography for diagnosis and percutaneous treatment. *Radiology* 1999; **210**: 663–672.

2 Slaba SG, El-Hajj LF, Abboud GA, Gebara VA. Selective angiography of cerebral aneurysm using gadodiamide in polycystic kidney disease with renal insufficiency. *Am J Roentgenol* 2000; **175**: 1467–1468.

3 Bloem JL, Wondergem J. Gd-DTPA as a contrast agent in CT. *Radiology* 1989; **171**: 578–579.

4 Gibson RJ, Meanock CI, Torrie EP, Walker TM. An assessment of Gd-DTPA as a CT contrast agent in the renal tract. *Clin Radiol* 1993; **47**: 278–279.

5 Donnelly LF, Nelson RC. Renal excretion of gadolinium mimicking calculi on non-contrast CT. *Pediatr Radiol* 1998; **28**: 417.

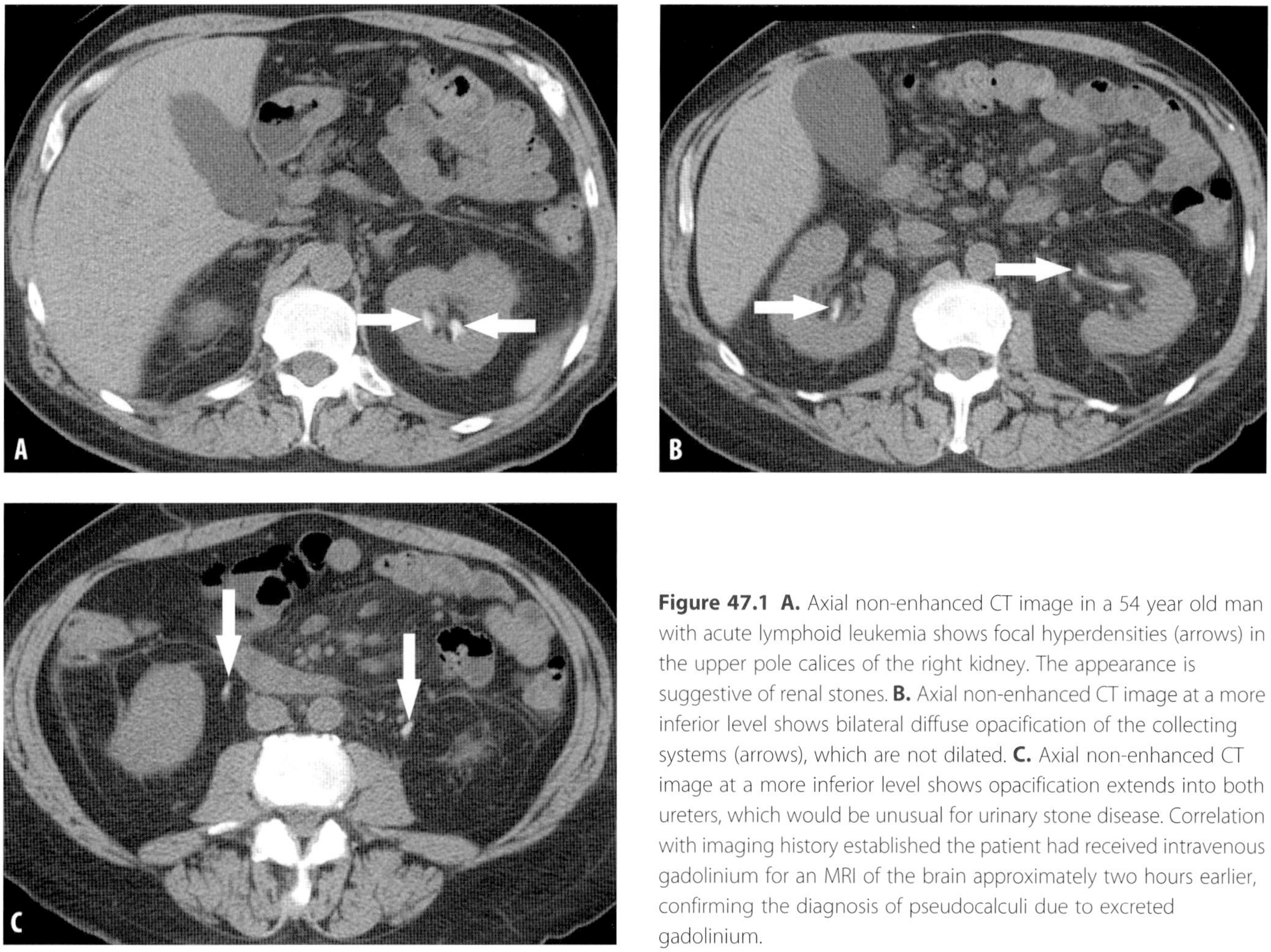

Figure 47.1 **A.** Axial non-enhanced CT image in a 54 year old man with acute lymphoid leukemia shows focal hyperdensities (arrows) in the upper pole calices of the right kidney. The appearance is suggestive of renal stones. **B.** Axial non-enhanced CT image at a more inferior level shows bilateral diffuse opacification of the collecting systems (arrows), which are not dilated. **C.** Axial non-enhanced CT image at a more inferior level shows opacification extends into both ureters, which would be unusual for urinary stone disease. Correlation with imaging history established the patient had received intravenous gadolinium for an MRI of the brain approximately two hours earlier, confirming the diagnosis of pseudocalculi due to excreted gadolinium.

CASE 48 Subtle complete ureteral duplication

Imaging description

Two rules govern the imaging findings of renal duplication with complete ureteral duplication. First, the ureter of the upper renal segment inserts inferiorly and ectopically to the ureter of the lower renal segment (Weigert-Meyer rule) [1], with the upper moiety prone to obstruction and the lower moiety prone to reflux. Second, the appearance of the upper tract predicts the site of insertion, such that a normal pelvicaliceal system and renal segment suggest a normally positioned ureteral orifice, while a hydronephrotic pelvicaliceal system and atrophic renal segment suggest a markedly ectopic ureteral orifice [2]. Accordingly, the diagnosis of complete ureteral duplication is usually radiologically obvious, because the ectopically inserting ureter drains a markedly hydronephrotic moiety (Figure 48.1). However, occasionally the upper pole moiety is small and relatively normal in appearance and then the imaging findings can be subtle and may go unrecognized (Figures 48.2 and 48.3) [3–5]. The term "sub-kidney" has been used to describe the small dysplastic upper moiety of such a duplicated system [6].

Importance

A small subtle upper pole moiety of a duplicated kidney can cause continuous incontinence in girls if there is an associated complete ureteral duplication with an infrasphincteric ectopic ureteral insertion [1–3]. This entity may go unrecognized because the imaging features are relatively inapparent and the ectopic ureter may be invisible even on intravenous urography, presumably due to limited excretion of contrast material from the small dysplastic upper moiety [7]. Correct recognition of the condition allows for relatively straightforward surgical repair, with complete resolution of the distressing symptoms.

Typical clinical scenario

The classic presentation of an upper pole moiety with complete ureteral duplication and an infrasphincteric ectopic ureteral insertion is that of lifelong dribbling of urine or wetness despite successful toilet training [3]. The clinical history is critical to suggesting the possibility of an infrasphincteric ectopic ureteral insertion when a small dysplastic upper pole moiety is demonstrated on imaging, since the ectopic ureter itself may not be directly visualized. MR urography can be helpful in elucidating the abnormality because MR urography can more clearly demonstrate the anatomy of the renal parenchyma, the renal collecting system, the ureter, and the ureteral orifice when compared to visualization on intravenous urography and pelvic ultrasound [8, 9].

Differential diagnosis

The appearance of a small separate pelvicaliceal system in the upper pole is distinctive and is unlikely to be mistaken for anything other than duplication. The real danger is that the finding may be dismissed as inconsequential, because the appearances are not those of a typical "full blown" obstructed upper pole moiety. At ultrasound, the diagnosis may be overlooked or interpreted as insignificant renal duplication, a "hypertrophied column of Bertin", or an adrenal mass [3].

Teaching point

A small subtle upper pole moiety of a duplicated kidney can cause continuous incontinence in girls if there is associated complete ureteral duplication with an infrasphincteric ectopic ureteral insertion; the critical clue is a clinical history of lifelong dribbling of urine or perineal wetness despite successful toilet training.

REFERENCES

1 Berrocal T, López-Pereira P, Arjonilla A, Gutiérrez J. Anomalies of the distal ureter, bladder, and urethra in children: embryologic, radiologic, and pathologic features. *Radiographics* 2002; **22**: 1139–1164.

2 Fernbach SK, Feinstein KA, Spencer K, Lindstrom CA. Ureteral duplication and its complications. *Radiographics* 1997; **17**: 109–127.

3 Carrico C, Lebowitz RL. Incontinence due to an infrasphincteric ectopic ureter: why the delay in diagnosis and what the radiologist can do about it. *Pediatr Radiol* 1998; **28**: 942–949.

4 Braverman RM, Lebowitz RL. Occult ectopic ureter in girls with urinary incontinence: diagnosis by using CT. *Am J Roentgenol* 1991; **156**: 365–366.

5 Gharagozloo AM, Lebowitz RL. Detection of a poorly functioning malpositioned kidney with single ecotopic ureter in girls with urinary dribbling: imaging evaluation in five patients. *Am J Roentgenol* 1995; **164**: 957–961.

6 Yeh HC, Halton KP, Shapiro RS, Rabinowitz JG, Mitty HA. Junctional parenchyma: revised definition of hypertrophic column of Bertin. *Radiology* 1992; **185**: 725–732.

7 Wille S, von Knobloch R, Klose KJ, Heidenreich A, Hofmann R. Magnetic resonance urography in pediatric urology. *Scand J Urol Nephrol* 2002; **37**: 16–21.

8 Riccabona M, Simbrunner J, Ring E, et al. Feasibility of MR urography in neonates and infants with anomalies of the upper urinary tract. *Eur Radiol* 2002; **12**: 1442–1450.

9 Lipson JA, Coakley FV, Baskin LS, Yeh BM. Subtle renal duplication as an unrecognized cause of childhood incontinence: diagnosis by magnetic resonance urography. *J Pediatr Urol* 2008; **4**: 398–400.

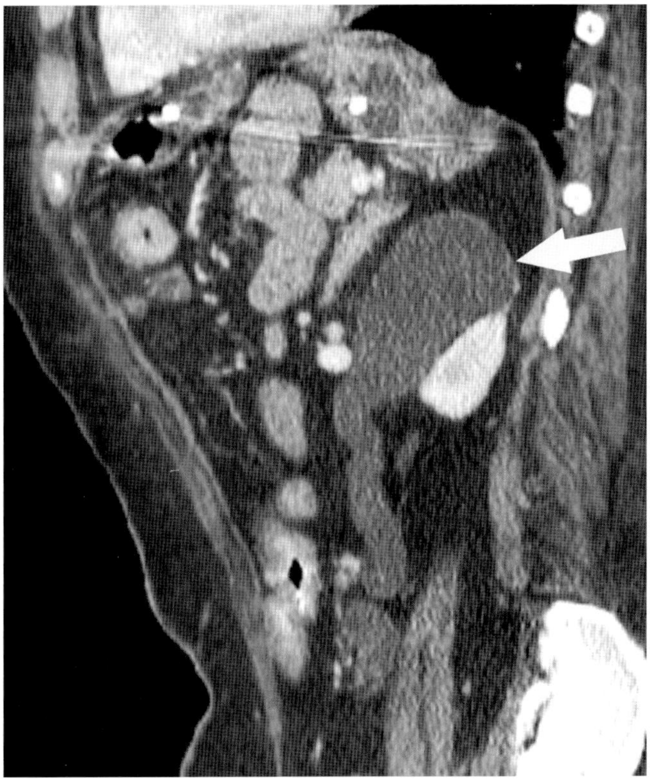

Figure 48.1 Sagittal reformatted contrast-enhanced CT image in a 65 year old woman with a cerebellar syndrome. The study was requested to evaluate for the possibility of a paraneoplastic syndrome secondary to an underlying primary malignancy. A chronically obstructed upper pole moiety (arrow) of a completely duplicated left kidney was discovered incidentally. The upper pole moiety drained to a large ureterocele with an ectopic insertion into the bladder (not shown).

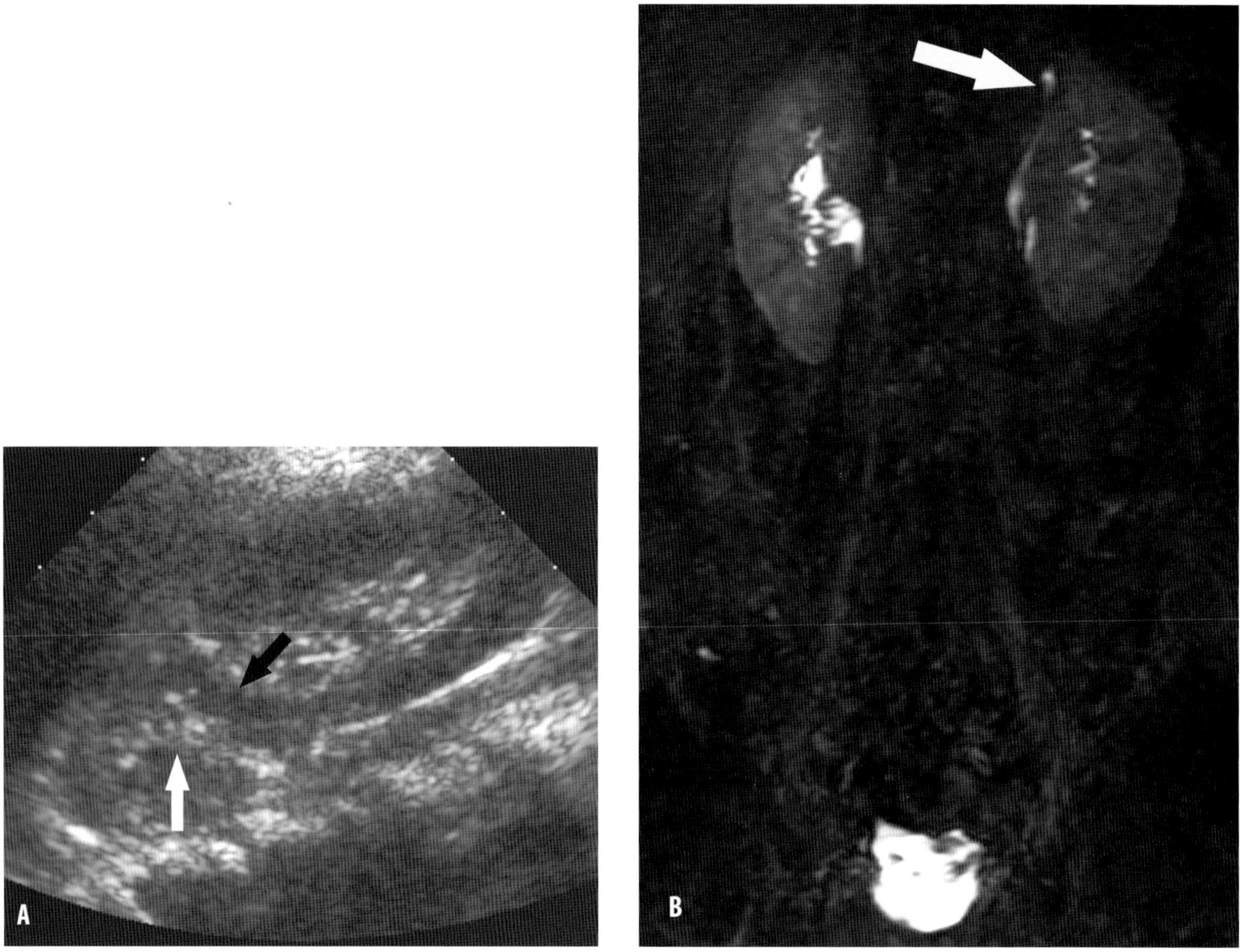

Figure 48.2 A. Left renal ultrasound image in a 3 year old girl with perineal wetness demonstrates a band of renal parenchyma (black arrow) which was misdiagnosed as the superior margin of the kidney. The more superior slightly atrophic renal sub-kidney with associated renal sinus fat (white arrow) was not recognized. **B.** MR urogram shows a duplicated left kidney upper pole ureter (arrow) arising from a poorly enhancing upper pole moiety. Subsequent examination under anesthesia revealed an ectopic ureteral orifice just posterior to the external urethral orifice. Surgical exploration was undertaken and confirmed complete duplication of the left collecting system, with the ureter of the small upper pole moiety draining ectopically. A left ureteroureterostomy of the upper pole moiety ureter into the lower pole moiety ureter and a distal ectopic ureterectomy were performed. The patient recovered uneventfully and is now asymptomatic.

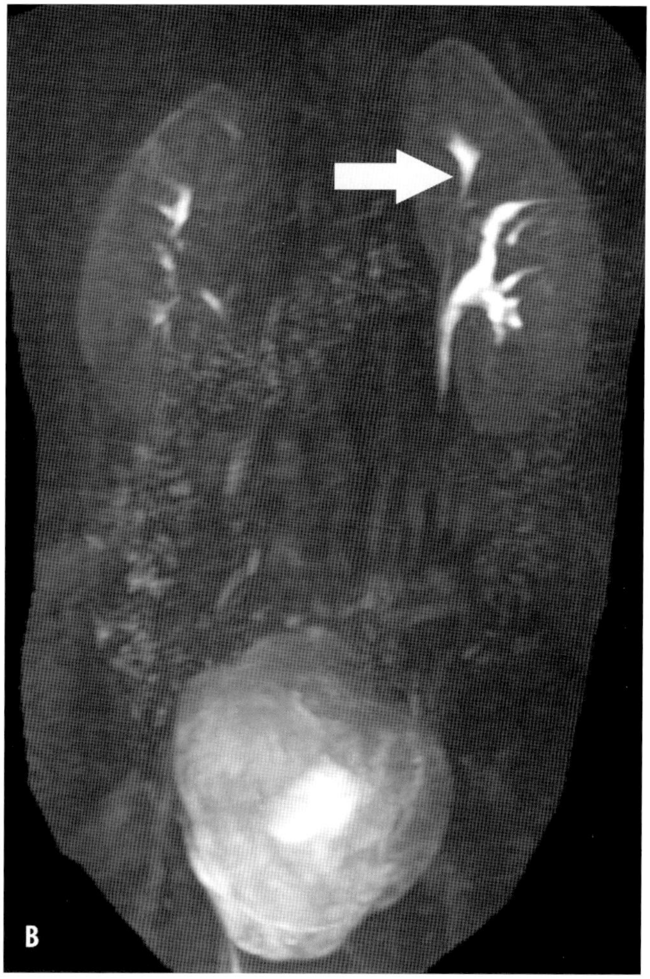

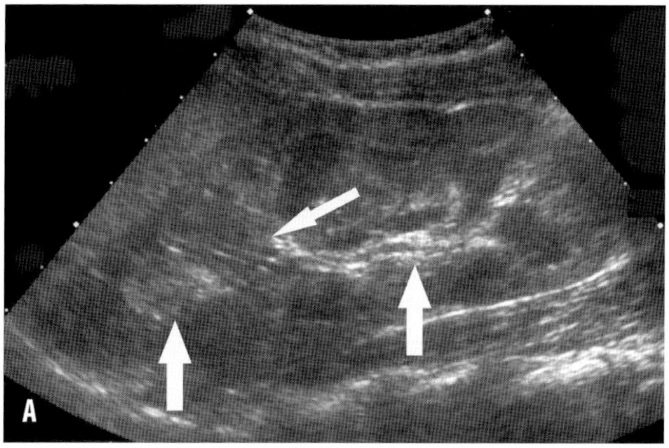

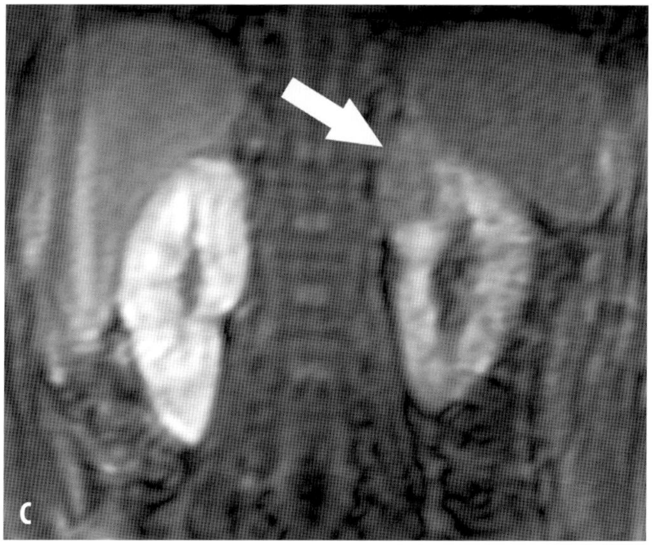

Figure 48.3 **A.** Left renal ultrasound image in a 5 year old girl with perineal wetness shows an enlarged left kidney. Subtle duplication of the renal collecting system with intervening renal parenchyma (thin arrow) dividing the renal sinus fat (thick arrows) into two compartments was not initially diagnosed. **B.** MR urogram shows a duplicated upper pole ureter (white arrow). **C.** Coronal post-gadolinium T1-weighted MR image shows mildly reduced enhancement in the upper pole moiety (arrow). Subsequent examination under anesthesia did not reveal an ectopic ureteral orifice. However, surgical exploration did confirm the presence of an ectopic ureter draining the upper pole moiety, which could be traced to the level of the bladder neck. The more inferior course of the ectopic ureter was not seen or dissected. A left ureteroureterostomy of the upper pole moiety ureter into the lower pole moiety ureter was performed. The patient recovered uneventfully and is now asymptomatic.

CASE 49 Retrocrural pseudotumor due to the cisterna chyli

Imaging description

The cisterna chyli is a variably sized sac at the commencement of the thoracic duct that receives lymph from the intestinal and lumbar lymphatic trunks. When present, the cistern chyli is located in the retrocrural space posterior to the aorta on the anterior aspect of the bodies of the upper lumbar vertebrae, usually on the right side. At cross-sectional imaging, the cisterna chyli is seen as a tubular or saccular fluid-filled retrocrural structure of variable length, diameter, and morphology [1–3] (Figures 49.1 and 49.2). The cisterna does not enhance on early and portal venous phase images, but enhancement or dependent layering of contrast can be seen on delayed phase images [4, 5] (Figure 49.3), presumably due to contrast that has leaked at a capillary level undergoing lymphatic resorption.

Importance

A large cisterna chyli may mimic retrocrural adenopathy at cross-sectional imaging [1].

Typical clinical scenario

Incidental visualization of the cisterna chyli has been reported in 1.7% of CT scans [6] and 15% of MRI scans [1].

Differential diagnosis

Fluid content helps to distinguish the cisterna chyli from solid retrocrural disease such as adenopathy. Lack of enhancement on non-delayed post-contrast images distinguishes the cisterna chyli from vascular structures such as the azygos or hemi-azygos vein or esophageal varices. Occasionally a cystic retroperitoneal mass may cause diagnostic confusion, but the presence of internal complexity or a masslike globular configuration should suggest a neoplastic etiology (Figure 49.4).

Teaching point

A fluid-filled tubular or saccular retrocrural structure is likely to be the cisterna chyli, and should not be mistaken for adenopathy or cystic tumors.

REFERENCES

1 Gollub MJ, Castellino RA. The cisterna chyli: a potential mimic of retrocrural lymphadenopathy on CT scans. *Radiology* 1996; **199**: 477–480.

2 Tamsel S, Ozbek SS, Sever A, Elmas N, Demirpolat G. Unusually large cisterna chyli: US and MRI findings. *Abdom Imaging* 2006; **31**: 719–721.

3 Pinto PS, Sirlin CB, Andrade-Barreto OA, et al. Cisterna chyli at routine abdominal MR imaging: a normal anatomic structure in the retrocrural space. *Radiographics* 2004; **24**: 809–817.

4 Lee KC, Cassar-Pullicino VN. Giant cisterna chyli: MRI depiction with gadolinium-DTPA enhancement. *Clin Radiol* 2000; **55**: 51–55.

5 Verma SK, Mitchell DG, Bergin D, et al. The cisterna chyli: enhancement on delayed phase MR images after intravenous administration of gadolinium chelate. *Radiology* 2007; **244**: 791–796.

6 Smith TR, Grigoropoulos J. The cisterna chyli: incidence and characteristics on CT. *Clin Imaging* 2002; **26**: 18–22.

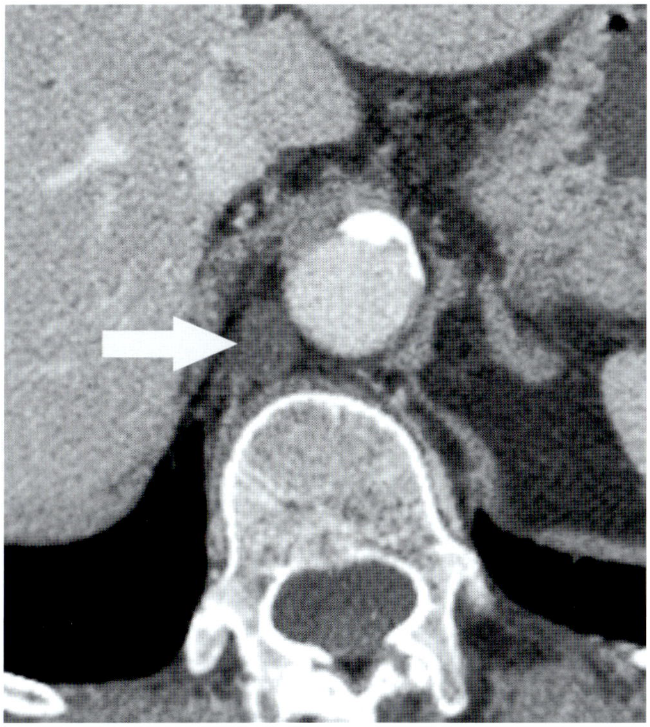

Figure 49.1 Axial contrast-enhanced CT image showing the typical appearance of the cisterna chyli as a fluid density saccular structure in the right retrocrural space.

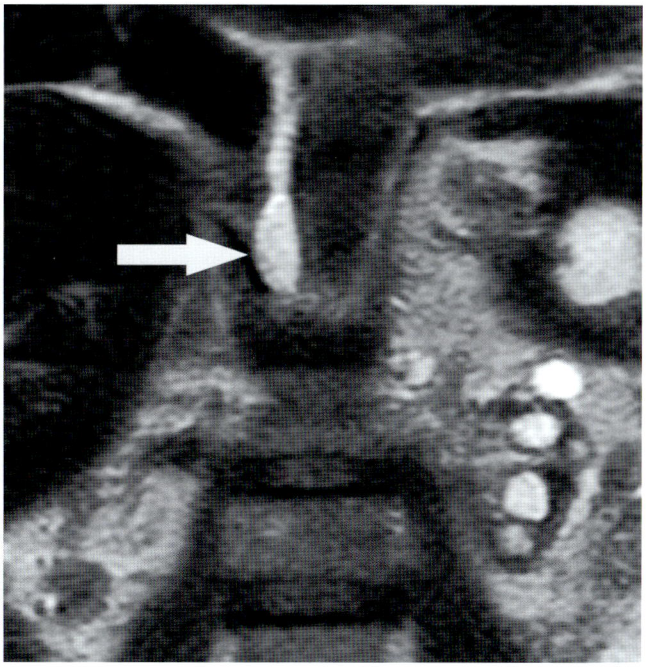

Figure 49.2 Coronal T2-weighted MR image demonstrates the tubular configuration and fluid signal intensity of the cisterna chyli (arrow). The thoracic dust is visible emanating from the superior aspect of the cisterna.

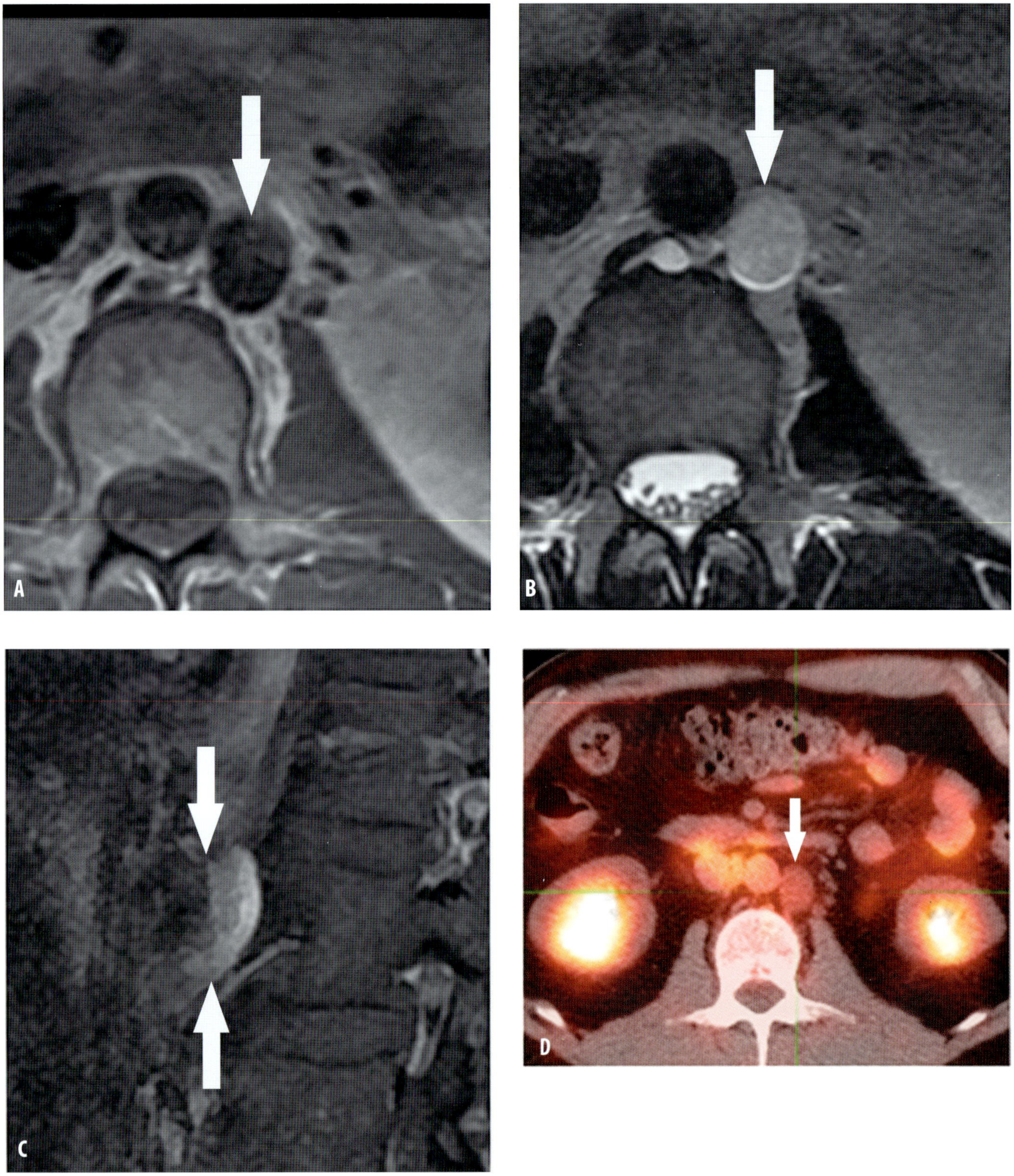

Figure 49.3 **A.** Axial T1-weighted MR image in a 57 year old woman undergoing lumbar spine MRI for low back pain. A structure (arrow) of low signal intensity is seen to the left of the aorta. **B.** Axial T2-weighted MR image shows the lesion (arrow) is of fluid signal intensity. **C.** Sagittal T1-weighted delayed post-gadolinium MR image shows a fluid-fluid level (between arrows) due to dependent layering of contrast (patient supine). The combination of findings is indicative of a cisterna chyli. **D.** Fused axial PET/CT image shows no increased FDG uptake in cisterna (arrow), as would be expected with such a benign entity. *Images for Figure 49.3 kindly provided by Dr Diego Ruiz, Palo Alto Medical Foundation.*

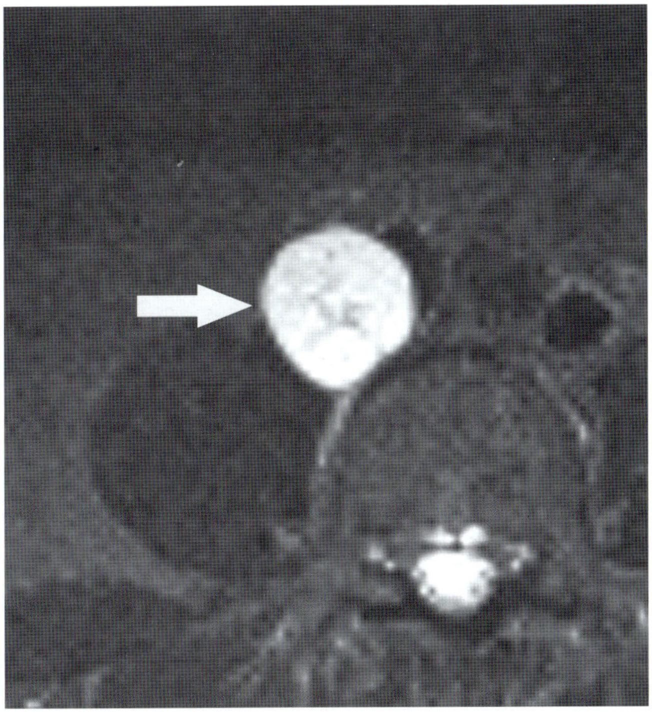

Figure 49.4 Axial T2-weighted and fat-saturated MR image shows a retroperitoneal lesion with hyperintense T2 signal intensity, somewhat suggestive of the cisterna chyli. However, the lesion demonstrated mass effect on the adjacent cava and had some internal complexity. A diagnosis of a predominantly cystic benign schwannoma was established after surgical resection.

CASE 50 Pseudothrombosis of the inferior vena cava

Imaging description

On early post-contrast CT or MRI studies of the abdomen, the inferior vena cava just above the renal veins often appears to contain a central ill-defined and poorly enhancing filling defect that tapers and disappears more superiorly. This pseudothrombosis is due to the laminar flow of enhanced blood from the renal veins streaming parallel to the column of unopacified blood returning from the lower body (Figure 50.1) [1, 2]. This pseudolesion disappears over time and is not seen on more delayed images, because the blood returning from the lower extremities through the inferior vena cava is then more opacified. Accordingly, this pseudolesion is commoner on spiral as compared to conventional CT scans [3].

Importance

Pseudothrombosis of the inferior vena cava may be mistaken for a true thrombus of the inferior vena cava, either tumor thrombus or bland thrombus, resulting in unnecessary follow-up investigations and patient anxiety.

Typical clinical scenario

Pseudothrombosis of the inferior vena cava is commonly seen on early post-contrast CT or MRI scans of the abdomen, particularly given the increasing use of spiral CT and multiphasic post-contrast imaging of the abdomen.

Differential diagnosis

Both tumor and bland thrombus can be seen in the inferior vena cava, but are typically better marginated and will not disappear on delayed post-contrast images. In addition, tumor thrombus will be contiguous with a primary tumor prone to venous invasion (such as renal cell carcinoma, adrenal cell carcinoma, or hepatocellular carcinoma) while bland thrombus will be contiguous with deep venous thrombus more inferiorly. The appearance of pseudothrombosis is usually characteristic, but occasionally problematic cases may require further evaluation with flow-sensitive MRI sequences (Figure 50.2). Other less well-recognized artifactual filling defects may also result from similar mixing of poorly and well enhanced blood, such as from an accessory hepatic vein flowing into an opacified inferior vena cava (Figure 50.3) [2], or from reflux of opacified blood from the heart into the periphery of the inferior vena cava in the setting of right heart disease or a high injection rate (Figure 50.4) [4]. Delayed images to show resolution of the filling defect are usually sufficient to confirm the artifactual nature of such pseudolesions. Very rarely, perihepatic fluid in the superior recess of the lesser sac may mimic an intracaval filling defect (in the same way as pericaval fat may give rise to the appearance of a pseudolipoma in the cava). Correlation with multiplanar reconstructed images can be helpful in recognizing this pitfall (Figure 50.5).

Teaching point

The appearance of a central ill-defined and poorly enhancing filling defect in the inferior vena cava just above the renal veins that tapers and disappears more superiorly on early post-contrast CT or MRI studies of the abdomen is typical of pseudothrombosis. Delayed images or flow-sensitive MRI sequences can be used to confirm this diagnosis in problematic cases.

REFERENCES

1 Vogelzang RL, Gore RM, Neiman HL, et al. Inferior vena cava CT pseudothrombus produced by rapid arm-vein contrast infusion. *Am J Roentgenol* 1985; **144**: 843–846.

2 Kaufman LB, Yeh BM, Breiman RS, et al. Inferior vena cava filling defects on CT and MRI. *Am J Roentgenol* 2005; **185**: 717–726.

3 McWilliams RG, Chalmers AG. Pseudothrombosis of the infra-renal inferior vena cava during helical CT. *Clin Radiol* 1995; **50**: 751–755.

4 Yeh BM, Kurzman P, Foster E, et al. Clinical relevance of retrograde inferior vena cava or hepatic vein opacification during contrast-enhanced CT. *Am J Roentgenol* 2004; **183**: 1227–1232.

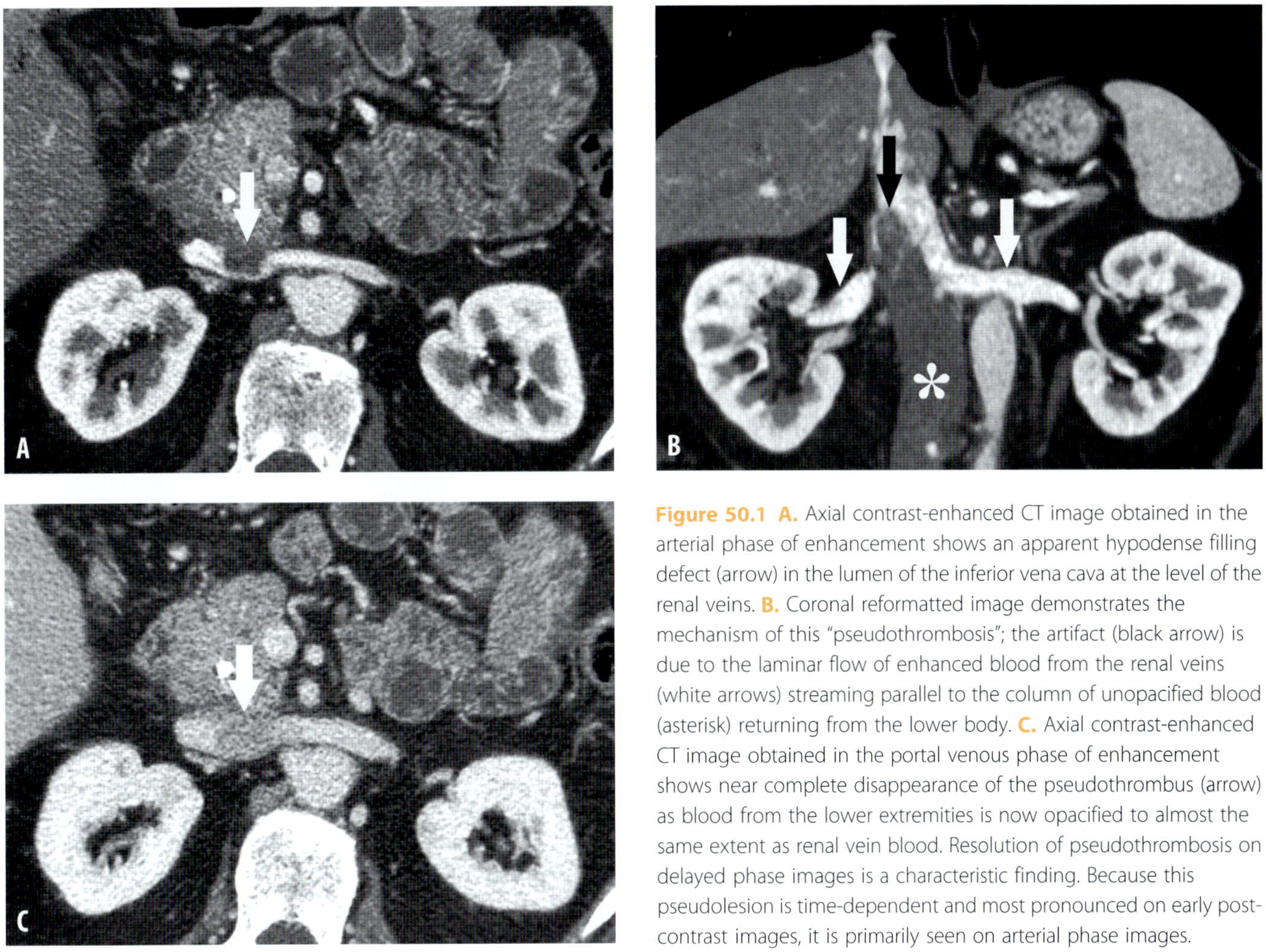

Figure 50.1 **A.** Axial contrast-enhanced CT image obtained in the arterial phase of enhancement shows an apparent hypodense filling defect (arrow) in the lumen of the inferior vena cava at the level of the renal veins. **B.** Coronal reformatted image demonstrates the mechanism of this "pseudothrombosis"; the artifact (black arrow) is due to the laminar flow of enhanced blood from the renal veins (white arrows) streaming parallel to the column of unopacified blood (asterisk) returning from the lower body. **C.** Axial contrast-enhanced CT image obtained in the portal venous phase of enhancement shows near complete disappearance of the pseudothrombus (arrow) as blood from the lower extremities is now opacified to almost the same extent as renal vein blood. Resolution of pseudothrombosis on delayed phase images is a characteristic finding. Because this pseudolesion is time-dependent and most pronounced on early post-contrast images, it is primarily seen on arterial phase images.

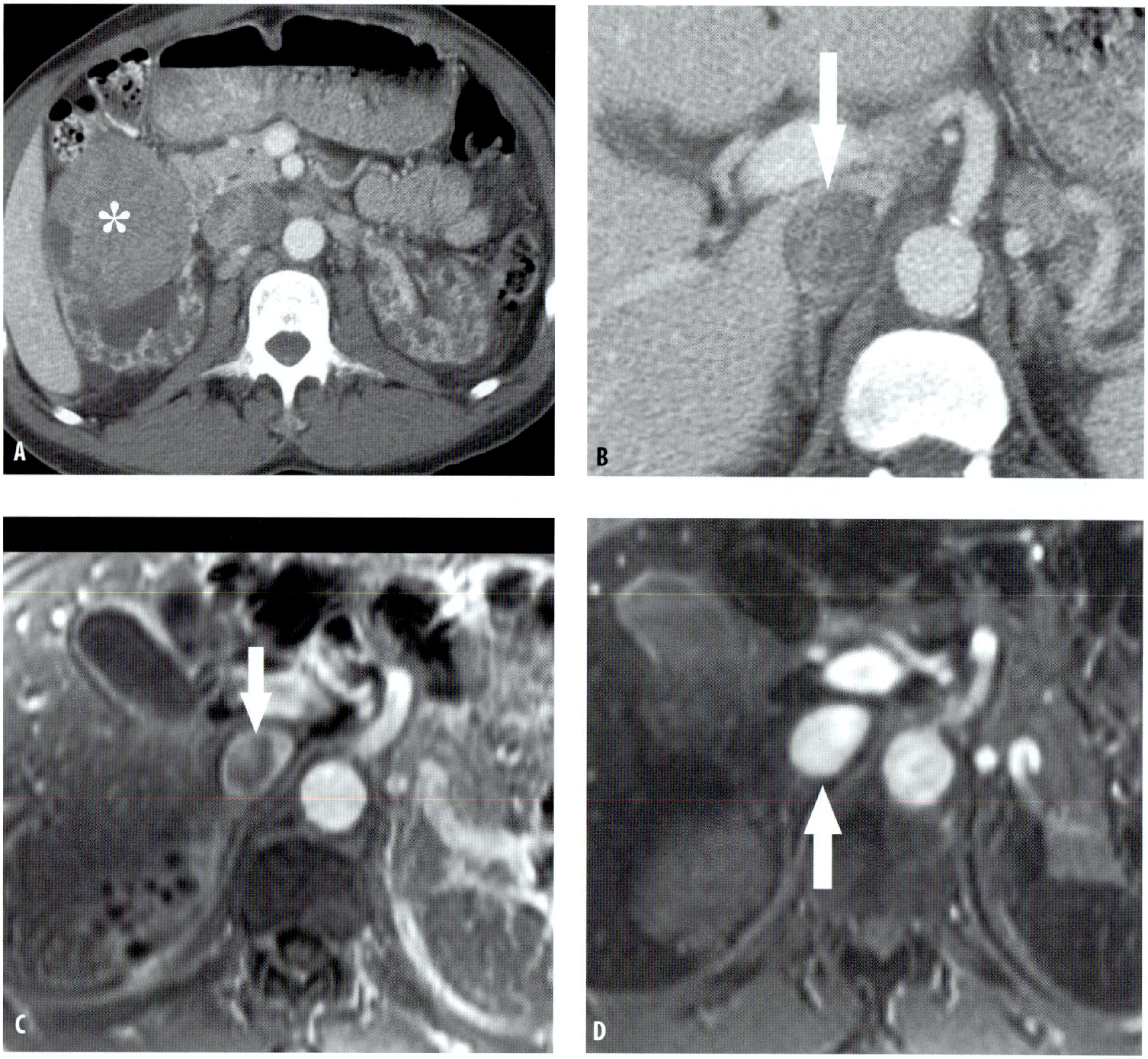

Figure 50.2 A. Axial contrast-enhanced CT image obtained in a 45 year-old man with a large renal cell carcinoma (asterisk) arising in the setting of acquired cystic kidney disease secondary to long term hemodialysis. **B.** Axial contrast-enhanced CT image obtained at a more superior level shows an apparent hypodense filling defect in the inferior vena cava, concerning for tumor thrombus in the setting of renal cell carcinoma. **C.** Axial spoiled gradient-echo T1-weighted post-gadolinium MR image shows a hypointense filling defect (arrow) in the inferior vena cava (note the study was obtained before the risk of nephrogenic systemic fibrosis related to gadolinium administration in renal failure was recognized). **D.** Axial flow-sensitive steady state gradient-echo T1-weighted post-gadolinium MR image shows normal flow in the inferior vena cava (arrow), indicating the filling defect seen on post-contrast CT and MR images was due to pseudothrombosis. In equivocal cases, flow-sensitive MRI can be used to distinguish pseudothrombus from true thrombus.

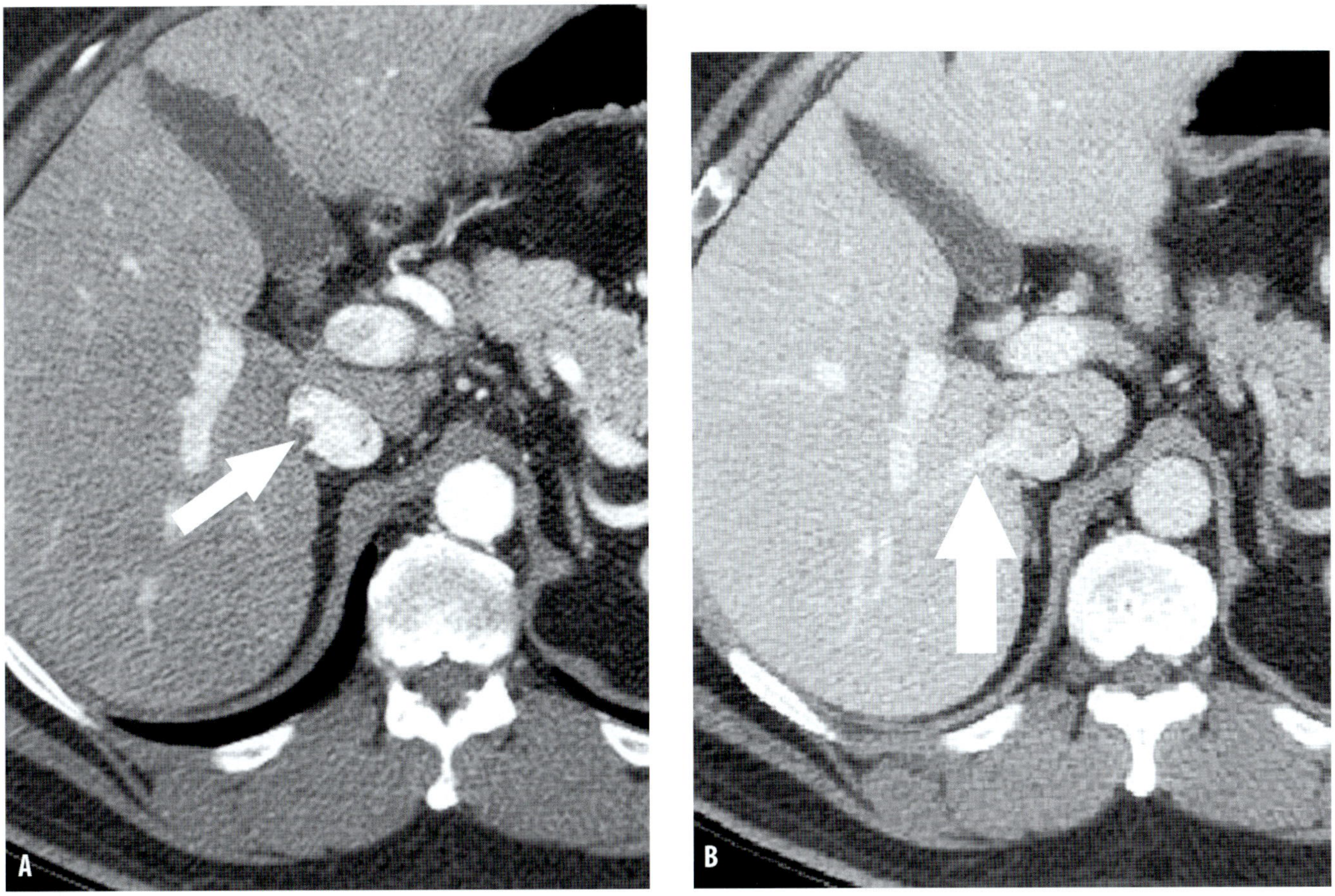

Figure 50.3 A. Axial contrast-enhanced CT image obtained in the early phase of enhancement in a 54 year old man with hepatitis C cirrhosis shows a filling defect (arrow) in the intrahepatic portion of the inferior vena cava. **B.** Axial contrast-enhanced CT image obtained in the portal venous phase of enhancement shows disappearance of the filling defect and an inferior accessory right hepatic vein draining to the inferior vena cava, confirming the pseudothrombus seen on early phase images was due to inflow of poorly opacified blood from the accessory vein.

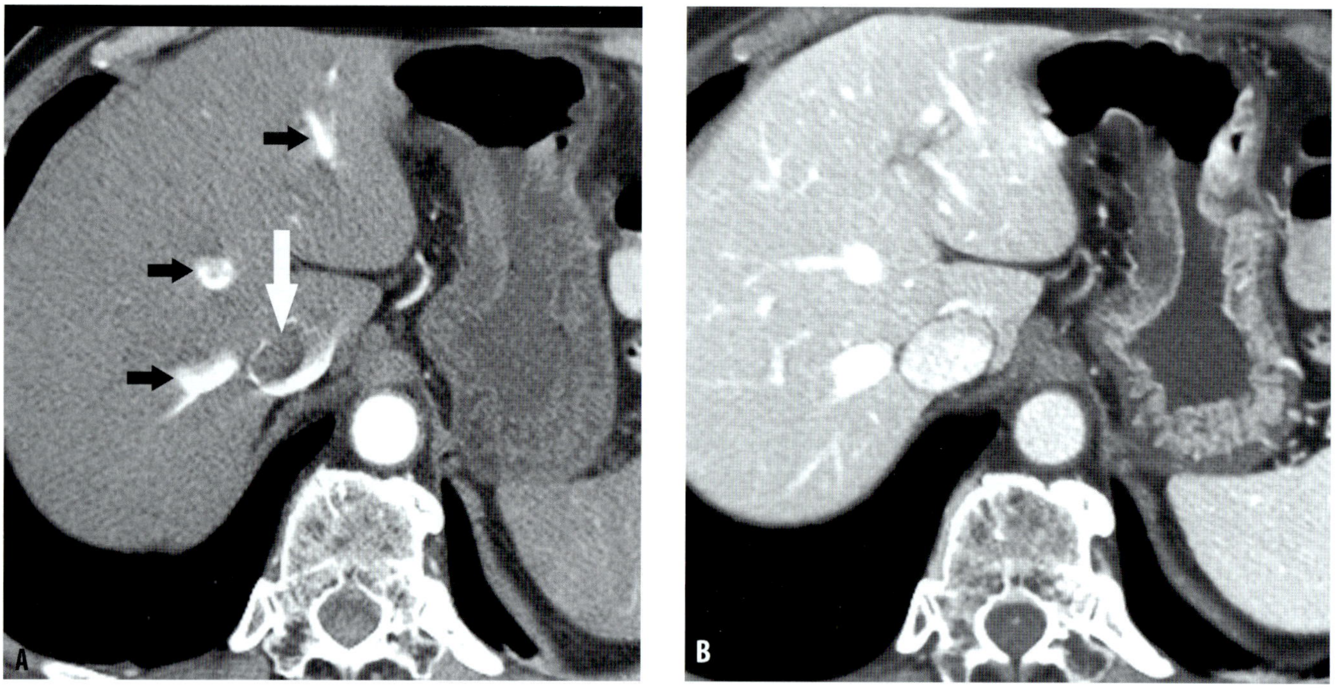

Figure 50.4 A. Axial contrast-enhanced CT image obtained in the early phase of enhancement for a multiphasic study of the liver performed at an injection rate of 5 cm^3 per second in a 66 year old woman with chronic hepatitis B. An apparent filling defect (white arrow) is seen in the inferior vena cava, due to reflux of contrast from the right atrium into the periphery of the cava. Note refluxed contrast is also seen in the hepatic veins (black arrows). **B.** Axial contrast-enhanced CT image obtained in the portal venous phase of enhancement shows disappearance of the pseudothrombus, confirming the artifactual nature of the finding.

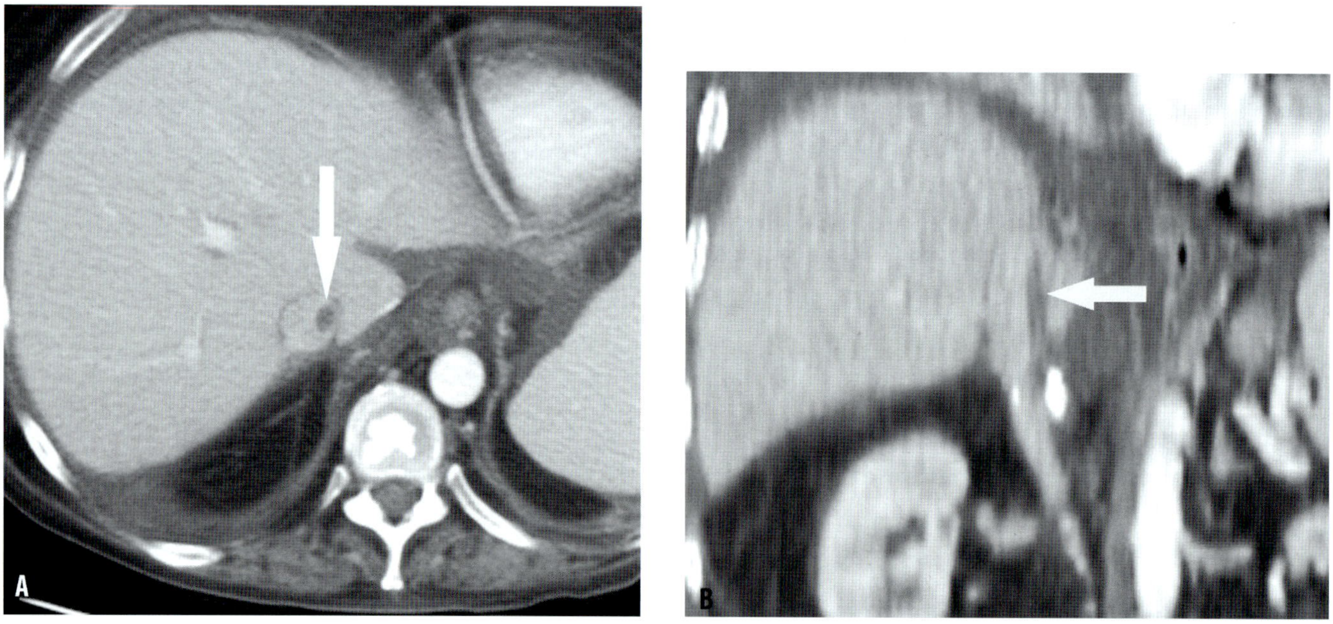

Figure 50.5 A. Axial contrast-enhanced CT image obtained in a 60 year old man 10 weeks after liver transplantation. An apparent filling defect (arrow) is seen in the inferior vena cava. **B.** Coronal reformatted image shows the filling defect (arrow) is actually due to pericaval fluid in the superior recess of the lesser sac.

CASE 51

Pseudoadenopathy due to venous anatomic variants

Imaging description

Several venous anatomic variants in the retroperitoneum may mimic adenopathy on CT or MRI [1–7], particularly if the veins are unenhanced or incompletely enhanced. Specifically, a duplicated or left-sided inferior vena cava may simulate para-aortic adenopathy (Figures 51.1 and 51.2). A prominent gonadal vein may mimic retroperitoneal adenopathy (Figure 51.3). A dilated left renal ascending lumbar communicant vein connecting the left renal vein to the lumbar or azygos system may mimic left para-aortic adenopathy (Figure 51.4). Finally, thrombosis of one of these retroperitoneal veins may simulate necrotic adenopathy (Figure 51.5) [8–10].

Importance

Misdiagnosis of retroperitoneal adenopathy may result in unnecessary surgery or treatment, particularly in patients with cancer [1–3].

Typical clinical scenario

Congenital anatomic variations of the inferior vena cava are relatively rare; the reported prevalence of a duplicated inferior vena cava is 0.2 to 3% and that of left-sided inferior vena cava is 0.2 to 0.5% [11, 12]. Dilated gonadal veins are common, and in one study dilated ovarian veins were found in 16 (47%) of 34 asymptomatic women [13]. At conventional left renal venography, a lumbar communicant vein was seen in 34 of 100 patients [6].

Differential diagnosis

The primary distinction is between venous anatomic variants and true retroperitoneal adenopathy. Venous variants are identified by their tubular nature and continuity with other vessels. Multiplanar reformatted images are often helpful in making these observations, and the increasing availability of spiral CT and three-dimensional post-processing has made misdiagnoses of these pseudotumors less common.

Teaching point

The possibility of venous anatomic variants should be considered in the differential diagnosis for apparent retroperitoneal adenopathy at CT or MRI. Such variants can be recognized because of their tubular nature and continuity with other vessels, and these observations are often facilitated by review of multiplanar reformatted images.

REFERENCES

1 Cohen SI, Hochsztein P, Cambio J, Sussett J. Duplicated inferior vena cava misinterpreted by computerized tomography as metastatic retroperitoneal testicular tumor. *J Urol* 1982; **128**: 389–391.

2 Klimberg I, Wajsman Z. Duplicated inferior vena cava simulating retroperitoneal lymphadenopathy in a patient with embryonal cell carcinoma of the testicle. *J Urol* 1986; **136**: 678–679.

3 Arisawa C, Kihara K, Fujii Y, et al. Possible misinterpretation on computed tomography of left inferior vena cava as retroperitoneal lymph node metastasis: a report of two cases. *Int J Urol* 1999; **6**: 215–218.

4 Einstein DM, Singer AA, Chilcote WA, Desai RK. Abdominal lymphadenopathy: spectrum of CT findings. *Radiographics* 1991; **11**: 457–472.

5 Meanock CI, Ward CS, Williams MP. The left ascending lumbar vein: a potential pitfall in CT diagnosis. *Clin Radiol* 1988; **39**: 565–566.

6 Lien HH, von Krogh J. Varicosity of the left renal ascending lumbar communicant vein: a pitfall in CT diagnosis. *Radiology* 1984; **152**: 484.

7 Yao Y, Okada Y, Yamato M, Ohtomo K. Communicating vein between the left renal vein and left ascending lumber vein: incidence and significance on abdominal CT. *Radiat Med* 2003; **21**: 252–257.

8 Silverman SG, Hillstrom MM, Doyle CJ, Tempany CM, Sica GT. Thrombophlebitic retroperitoneal collateral veins mimicking lymphadenopathy: MR and CT appearance. *Abdom Imaging* 1995; **20**: 474–476.

9 Evans JC, Earis J, Curtis J. Thrombosed double inferior vena cava mimicking paraaortic lymphadenopathy. *Br J Radiol* 2001; **74**: 192–194.

10 Kumar D, Kumar S, Lounsbury DE. Anomalous inferior vena cava with idiopathic thrombosis simulating a mass. *Comput Radiol* 1983; **7**: 223–227.

11 Phillips E. Embryology, normal anatomy, and anomalies. In: Ferris EJ, Hipona FA, Kahn PC, Phillips E, Shapiro JH, eds. *Venography of the inferior vena cava and its branches.* Baltimore, MD: Williams & Wilkins, 1969; 1–32.

12 Bass JE, Redwine MD, Kramer LA, Huynh PT, Harris JH Jr. Spectrum of congenital anomalies of the inferior vena cava: cross-sectional imaging findings. *Radiographics* 2000; **20**: 639–652.

13 Rozenblit AM, Ricci ZJ, Tuvia J, Amis ES Jr. Incompetent and dilated ovarian veins: a common CT finding in asymptomatic parous women. *Am J Roentgenol* 2001; **176**: 119–122.

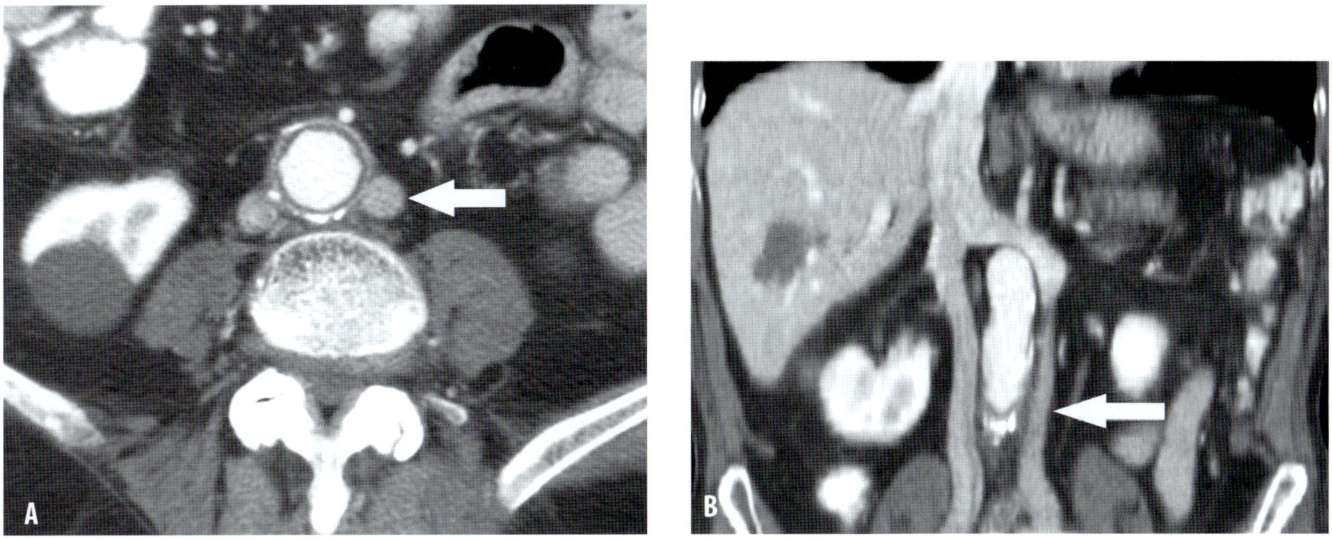

Figure 51.1 **A.** Axial contrast-enhanced CT image in a 72 year old man with metastatic renal cell carcinoma. An ovoid soft-tissue density (arrow) on the left side of the aorta resembles para-aortic adenopathy. **B.** Curved planar coronal reformatted CT image demonstrates the left para-aortic structure (arrow) is due to infrarenal duplication of the inferior vena cava. Note the duplicated portion of the cava courses superiorly before draining into the non-duplicated suprarenal inferior vena cava through the left renal vein.

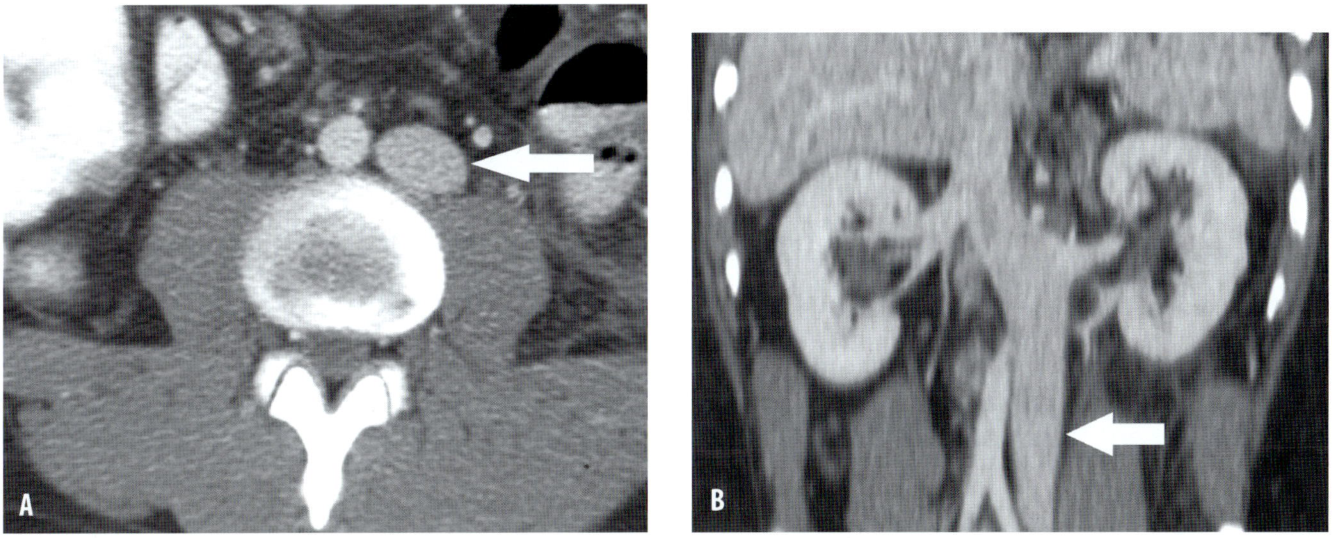

Figure 51.2 **A.** Axial contrast-enhanced CT image in a 28 year old man with Crohn's disease. An ovoid soft-tissue density (arrow) on the left side of the aorta resembles para-aortic adenopathy, but note that a normal right-sided inferior vena cava is not present. **B.** Curved planar coronal reformatted CT image demonstrates the left para-aortic structure (arrow) is due to an infrarenal left-sided inferior vena cava. Note the cava crosses the midline at the level of the kidneys and then drains superiorly in a normal right-sided location.

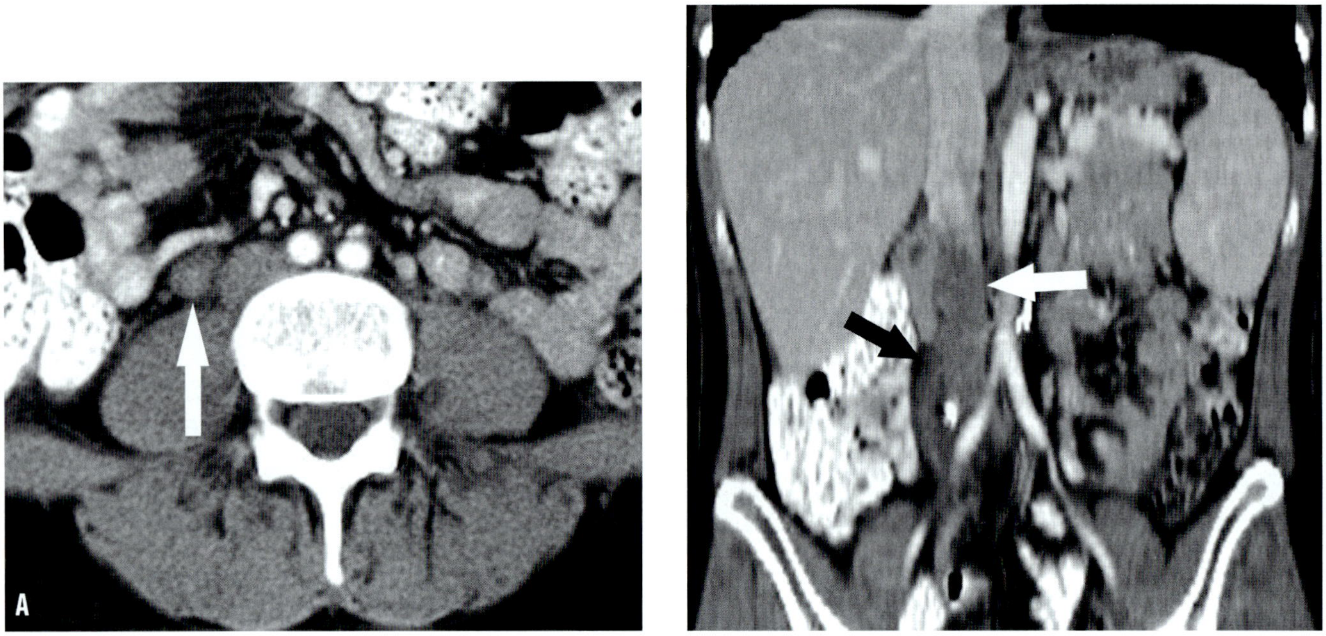

Figure 51.3 **A.** Axial contrast-enhanced CT image in a 37 year old woman with cervical cancer. An ovoid soft-tissue density (arrow) on the right side of the inferior vena cava resembles para-caval adenopathy. **B.** Curved planar coronal reformatted CT image demonstrates the right para-caval structure (black arrow) is due to a prominent right ovarian vein draining to the inferior vena cava (white arrow).

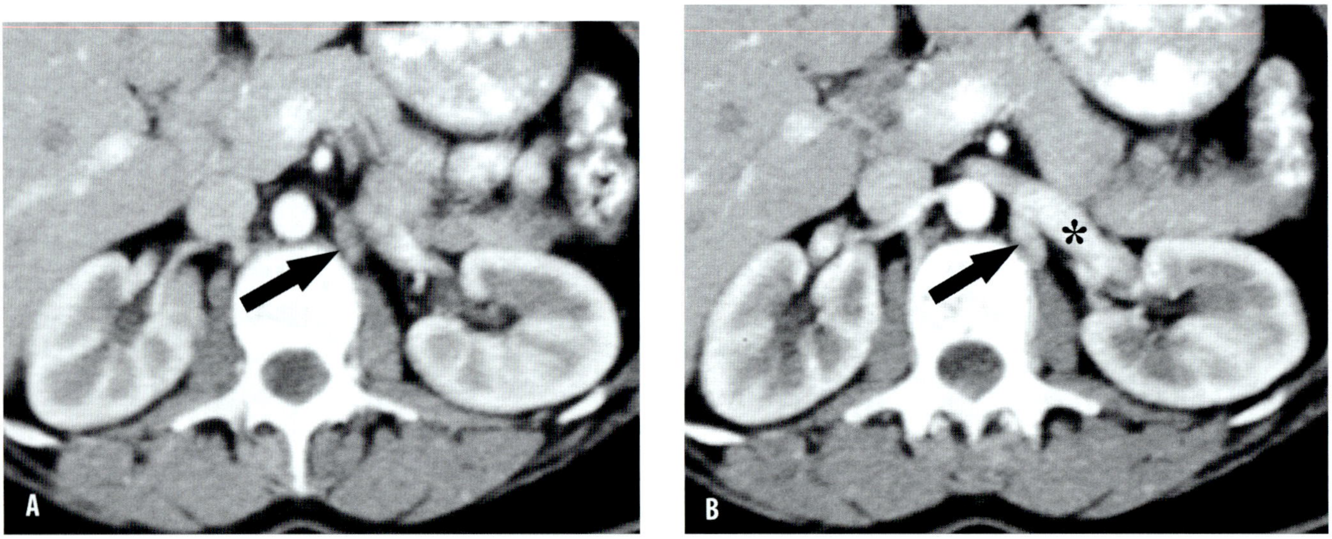

Figure 51.4 **A.** Axial contrast-enhanced CT image in a 26 year old man with testicular cancer. An ovoid soft-tissue density (arrow) on the left side of the aorta resembles para-aortic adenopathy. **B.** Axial contrast-enhanced CT image at a more superior level shows the apparent adenopathy is due to a prominent lumbar communicant vein (arrow) draining to the left renal vein (asterisk).

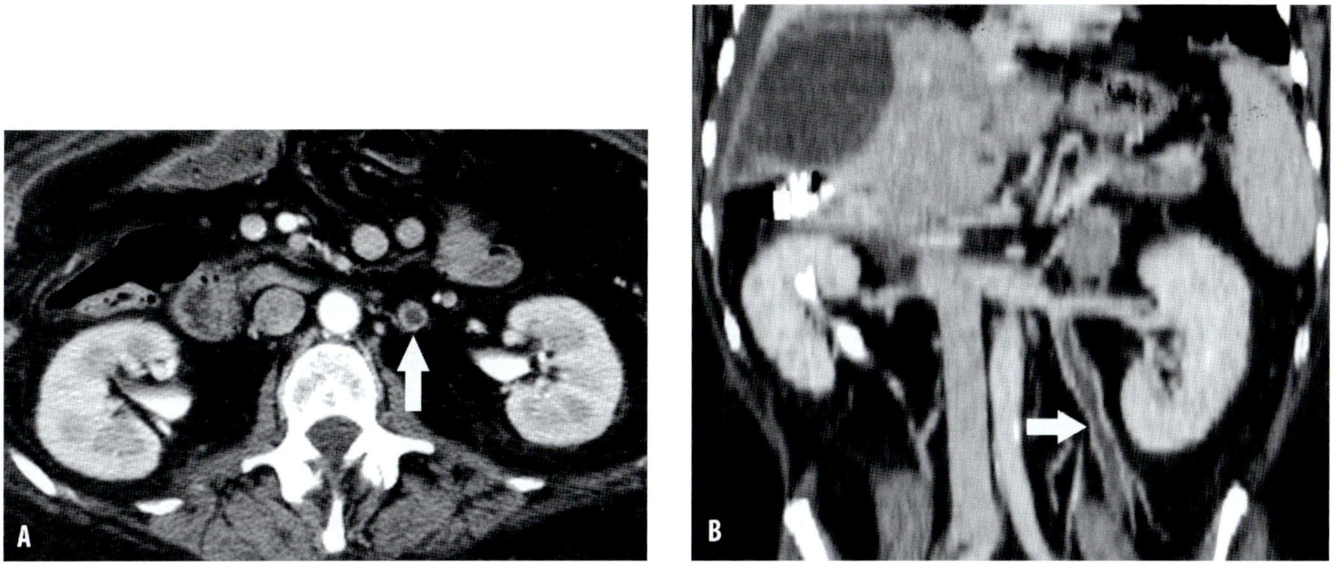

Figure 51.5 A. Axial contrast-enhanced CT image in a 56 year old woman after surgery for metastatic carcinoid tumor. An ovoid soft-tissue density (arrow) with a lower density center on the left side of the aorta resembles necrotic para-aortic adenopathy. **B.** Curved planar coronal reformatted CT image demonstrates the left para-aortic structure (arrow) is due to a thrombosed left ovarian vein. Note the ovarian vein drains superiorly to the left renal vein.

CASE

52 Pseudomass due to duodenal diverticulum

Imaging description

Duodenal diverticula are congenital anatomic variants and are found in up to 22% of the population at autopsy [1]. Fluid-filled or collapsed duodenal diverticula may mimic cystic or soft-tissue retroperitoneal or pancreatic masses at cross-sectional imaging (Figures 52.1–52.3) [2, 3]. Increased uptake of FDG within a duodenal diverticulum at PET imaging has also been reported [4].

Importance

Misdiagnosis of retroperitoneal adenopathy or a pancreatic mass may result in unnecessary surgery or treatment [2].

Typical clinical scenario

Duodenal diverticula are incidental findings that are likely to cause most diagnostic confusion when seen in patients with known malignancy (when they may suggest metastatic spread) or in the postoperative period (when they may resemble an abscess).

Differential diagnosis

Duodenal diverticula can usually be diagnosed by recognizing the characteristic location near the duodenum and by examination of all available studies that may allow identification of intradiverticular air [3].

Teaching point

The possibility of a duodenal diverticulum should be considered when an apparent cystic or soft-tissue mass or collection is seen adjacent to the duodenum. Comparison with prior studies or repeat examination after oral contrast may facilitate correct diagnosis.

REFERENCES

1 Lipsett P. Diverticular disease of the small bowel. In: Cameron JL, ed. *Current surgical therapy*, 6th edition. St. Louis, MO: Mosby, 1998; 148–150.

2 Hariri A, Siegelman SS, Hruban RH. Duodenal diverticulum mimicking a cystic pancreatic neoplasm. *Br J Radiol* 2005; **78**: 562–564.

3 Macari M, Lazarus D, Israel G, Megibow A. Duodenal diverticula mimicking cystic neoplasms of the pancreas: CT and MR imaging findings in seven patients. *Am J Roentgenol* 2003; **180**: 195–199.

4 Piesman M, Hwang I, Moses FM, Allen TW. Duodenal diverticulum presenting as a hypermetabolic mass on F-18 FDG PET/CT. *Clin Nucl Med* 2005; **30**: 747–748.

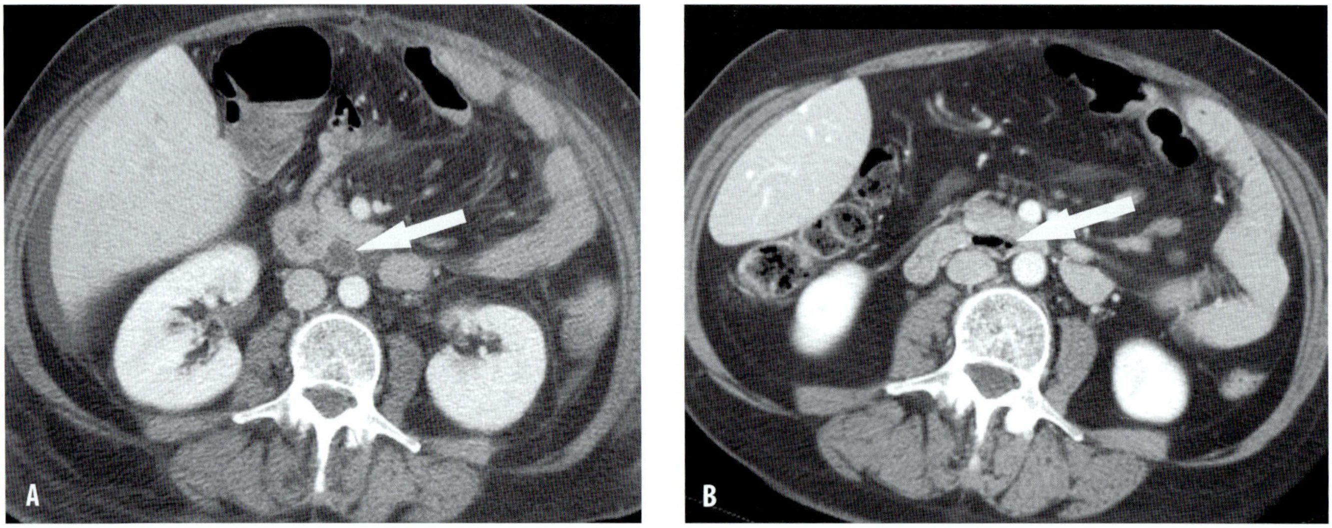

Figure 52.1 **A.** Axial contrast-enhanced CT image in a 58 year old woman with lymphoma shows an apparent cystic or necrotic mass anteriorly between the aorta and the inferior vena cava. Reasonable differential considerations would include necrotic lymphadenopathy or a cystic pancreatic mass. **B.** Axial contrast-enhanced CT image obtained one month earlier shows air and fluid within the lesion, consistent with a duodenal diverticulum.

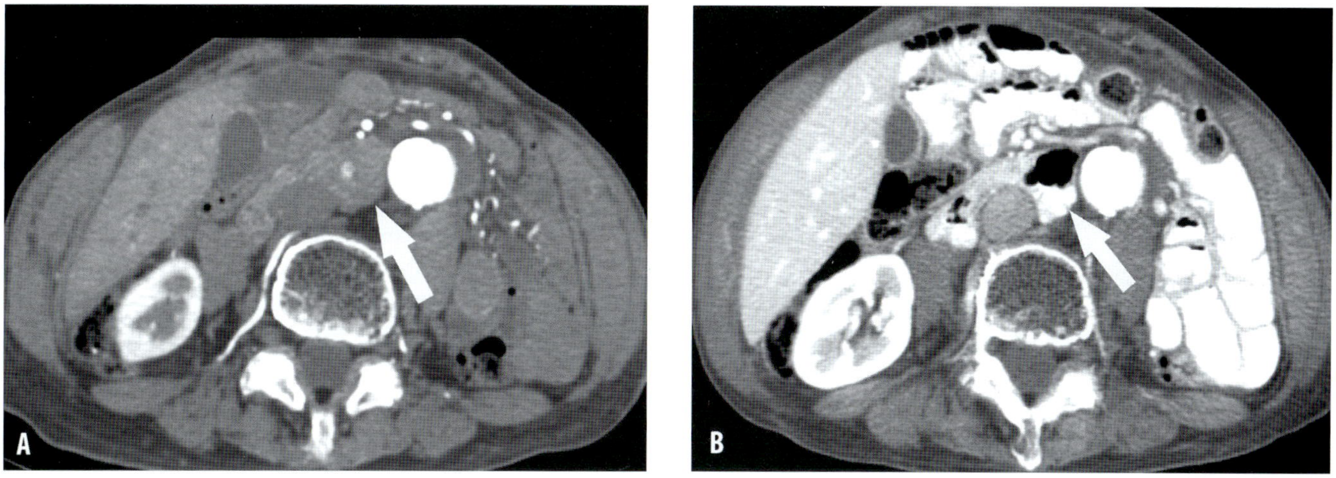

Figure 52.2 A. Axial contrast-enhanced arterial phase CT image performed prior to endovascular repair in an 86 year old man with an abdominal aortic aneurysm. A soft-tissue density between the aorta and the inferior vena cava was reported as a retroperitoneal mass. B. Axial contrast-enhanced portal venous CT image obtained after the administration of oral contrast for further evaluation of the apparent mass shows air and oral contrast within the lesion, confirming the diagnosis of a duodenal diverticulum.

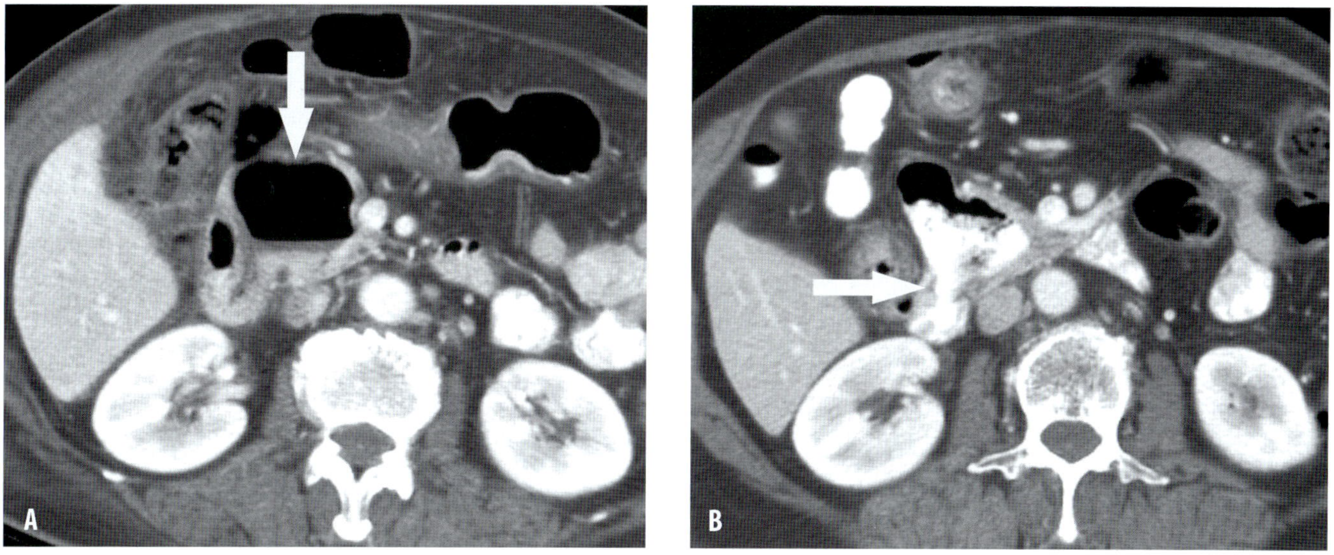

Figure 52.3 A. Axial contrast-enhanced arterial phase CT image in a 74 year old woman with fever and abdominal pain after cholecystectomy. An air and fluid-filled collection in the right upper quadrant could be interpreted as a postoperative abscess. B. Axial contrast-enhanced portal venous CT image obtained one year earlier shows the apparent collection is actually a large duodenal diverticulum, with oral contrast and air passing into the structure through a neck (arrow) connecting the duodenum and the diverticulum.

CASE 53 Segmental arterial mediolysis

Imaging description

Segmental arterial mediolysis is a rare idiopathic non-arteriosclerotic non-inflammatory disease first described in 1976 characterized by spontaneous degeneration in the medial layer of the visceral branches of the abdominal aorta, leading to varying combinations of intramural hemorrhage, peri-adventitial fibrin deposition, aneurysm formation, and dissection [1]. The diagnosis should be considered when CT arteriography shows isolated dissection, small saccular aneurysms, or a "string of beads" appearance in the visceral branches of the abdominal aorta (Figures 53.1–53.3) [2–4].

Importance

The frequency of segmental arterial mediolysis may be underestimated because angiography is not usually performed for gastrointestinal hemorrhage or abdominal pain. The findings may be subtle and overlooked at CT, although the diagnosis may become commoner with the increasing availability of high resolution multidetector CT.

Typical clinical scenario

Segmental arterial mediolysis typically presents with gastrointestinal hemorrhage or abdominal pain in middle-aged and elderly patients, although cerebral involvement has been reported in young adults [5]. The natural history of the disease is poorly understood. Ruptured aneurysms or segmental arterial thrombosis may be treated by reconstruction with a graft or patch, while stenoses may be managed with angioplasty.

Differential diagnosis

Segmental arterial mediolysis somewhat resembles fibromuscular dysplasia at imaging and some consider these to be related conditions [6, 7]. However, fibromuscular dysplasia is primarily a disease of young to middle-aged women and mainly affects the renal and carotid arteries. Associated clinical features, such as aphthous stomatitis in Behçet's syndrome, and laboratory findings, such as elevated inflammatory markers or autoantibodies, help in differentiating patients with systemic vasculidities such as polyarteritis nodosa, Takayasu's arteritis, Behçet's syndrome, and Henoch-Schönlein purpura from those with segmental arterial mediolysis [2]. The congenital vasculidities of neurofibromatosis type 1 and Ehlers-Danlos syndrome typically involve larger arteries and are associated with distinctive clinical manifestations. Mycotic aneurysms show a preference for bifurcations, while the lesions of segmental arterial mediolysis have a random distribution.

Teaching point

The visceral arteries should be reviewed with particular attention in patients undergoing CT for acute abdominal pain or gastrointestinal hemorrhage, and conventional catheter angiography should be suggested if there are findings suggestive of segmental arterial mediolysis.

REFERENCES

1 Slavin RE, Gonzalez-Vitale JC. Segmental mediolytic arteritis: a clinical pathologic study. *Lab Invest* 1976; **35**: 23–29.

2 Michael M, Widmer U, Wildermuth S, et al. Segmental arterial mediolysis: CTA findings at presentation and follow-up. *Am J Roentgenol* 2006; **187**: 1463–1469.

3 Heritz DM, Butany J, Johnston KW, Sniderman KW. Intraabdominal hemorrhage as a result of segmental mediolytic arteritis of an omental artery: case report. *J Vasc Surg* 1990; **12**: 561–565.

4 Slavin RE, Cafferty L, Cartwright J Jr. Segmental mediolytic arteritis: a clinicopathologic and ultrastructural study of two cases. *Am J Surg Pathol* 1989; **13**: 558–568.

5 Leu Hi. Cerebrovascular accidents resulting from segmental mediolytic arteriopathy of the cerebral arteries in young adults. *Cardiovasc Surg* 1994; **2**: 350–353.

6 Chan RJ, Goodman TA, Aretz TH, Lie JT. Segmental mediolytic arteriopathy of the splenic and hepatic arteries mimicking systemic necrotizing vasculitis. *Arthritis Rheum* 1998; **41**: 935–938.

7 Lie JT. Segmental mediolytic arteritis: not an arteritis but a variant of arterial fibromuscular dysplasia. *Arch Pathol Lab Med* 1992; **116**: 238–241.

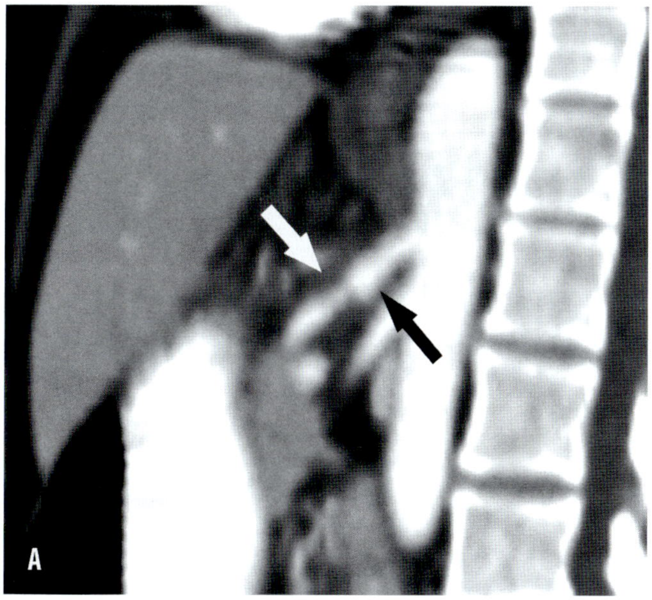

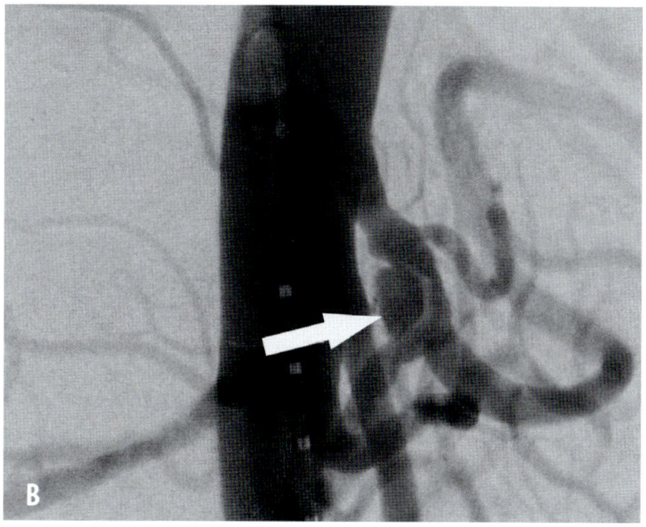

Figure 53.1 **A.** Sagittal contrast-enhanced arterial phase CT image in a 56 year old woman with sudden onset of severe abdominal pain shows the proximal celiac artery is mildly and focally dilated (black arrow). Just distal to the dilated segment, the artery is focally stenosed (white arrow) with adjacent soft-tissue cuffing. **B.** Conventional catheter aortogram performed two weeks later shows a focal dilatation (arrow) of the proximal celiac artery. At surgical reconstruction, the celiac artery was focally dilated due to a segmental dissection, and pathology confirmed the diagnosis of segmental arterial mediolysis.

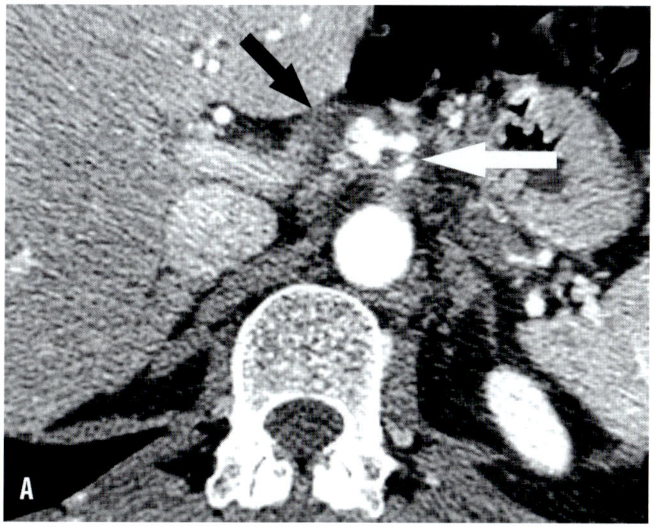

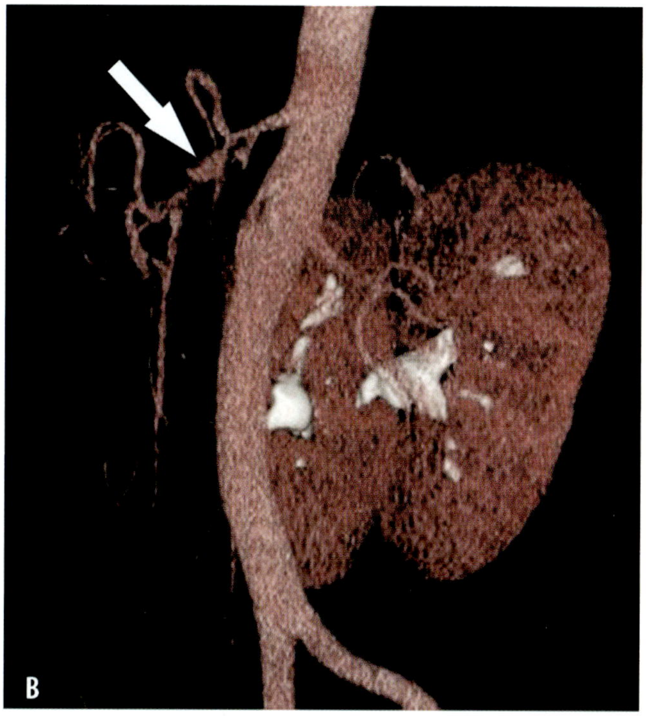

Figure 53.2 **A.** Axial contrast-enhanced arterial phase CT image in a 51 year old man with sudden onset of tearing epigastric abdominal pain. A dissection flap (white arrow) is seen in the proximal celiac artery, with surrounding soft-tissue density cuffing (black arrow). **B.** Three-dimensional CT reconstruction shows focal dilatation (arrow) of the celiac artery. At surgical reconstruction, the celiac artery was focally dilated due to a segmental dissection, and pathology confirmed the diagnosis of segmental arterial mediolysis.

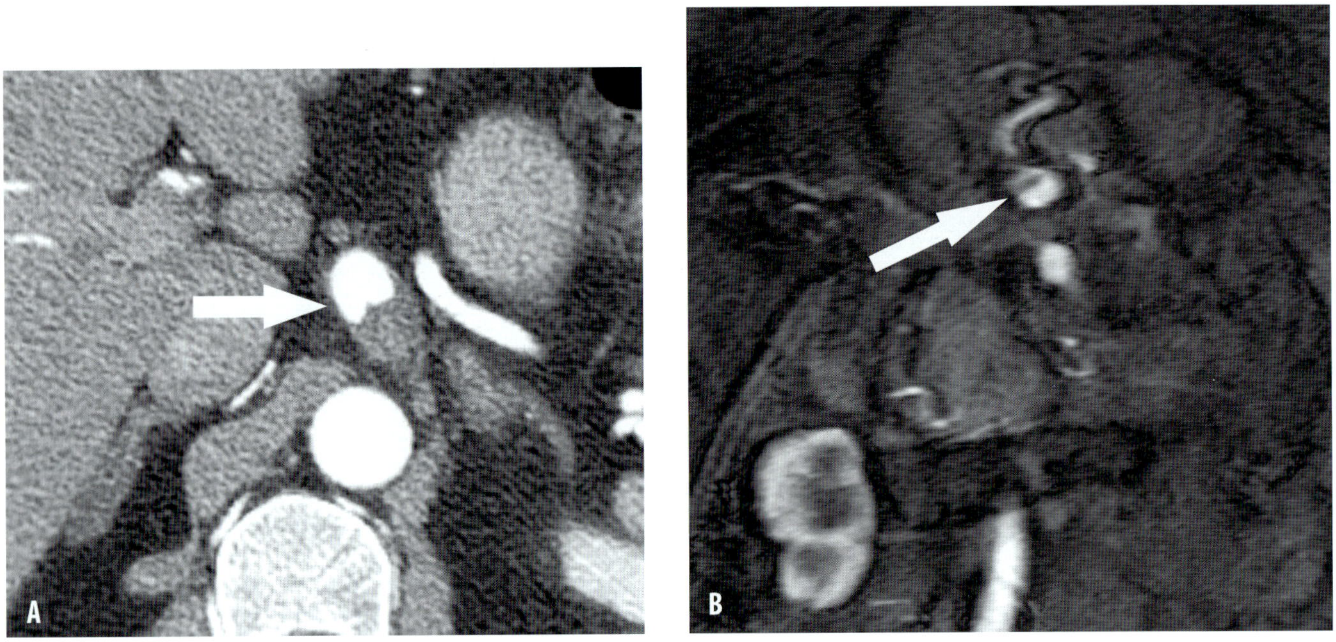

Figure 53.3 A. Axial contrast-enhanced arterial phase CT image in a 45 year old man with abdominal pain. The celiac artery (arrow) appears focally dilated and irregular with adjacent soft-tissue density cuffing. **B.** Coronal gadolinium-enhanced MR arteriogram image confirms the presence of a small dissection flap (arrow) in the celiac artery. A presumptive diagnosis of segmental arterial mediolysis was made after extensive workup excluded other possible etiologies. Symptoms resolved spontaneously and the patient did not undergo surgery. The arterial abnormality remained unchanged on imaging follow-up over four years.

CASE 54 Gastric antral wall thickening

Imaging description

The normal gastric antrum commonly measures over 5 mm and may measure as much as 12 mm in thickness at CT (Figures 54.1 and 54.2) [1].

Importance

Misinterpretation of antral wall thickening as inflammation or tumor may lead to unnecessary treatment or investigation.

Typical clinical scenario

Smooth thickening of the gastric antrum was seen in 152 of 153 (99%) consecutive patients without gastric disease undergoing CT [1]. The antral wall thickness exceeded 10 mm in seven patients (5%). Linear submucosal low attenuation (mural stratification) of the thickened portion of the gastric antrum was noted in 36 patients (24%), and in 14 of these cases the low density appeared to be of fat attenuation.

Differential diagnosis

Wall thickening of the gastric antrum is often misinterpreted as "antral gastritis". However, in a controlled study there was no association between antral wall thickness and *Helicobacter pylori* infection [2], and a prior study that suggested such an association lacked a control group, so the result may have been spurious [3].

Teaching point

Wall thickening of the gastric antrum is usually a normal finding and should not be misinterpreted as indicating inflammation or tumor.

REFERENCES

1 Pickhardt PJ, Asher DB. Wall thickening of the gastric antrum as a normal finding: multidetector CT with cadaveric comparison. *Am J Roentgenol* 2003; **181**: 973–979.

2 Kul S, Sert B, Sari A, et al. Effect of subclinical helicobacter pylori infection on gastric wall thickness: multislice CT evaluation. *Diagn Interv Radiol* 2008; **14**: 138–142.

3 Urban BA, Fishman EK, Hruban RH. Helicobacter pylori gastritis mimicking gastric carcinoma at CT evaluation. *Radiology* 1991; **179**: 689–691.

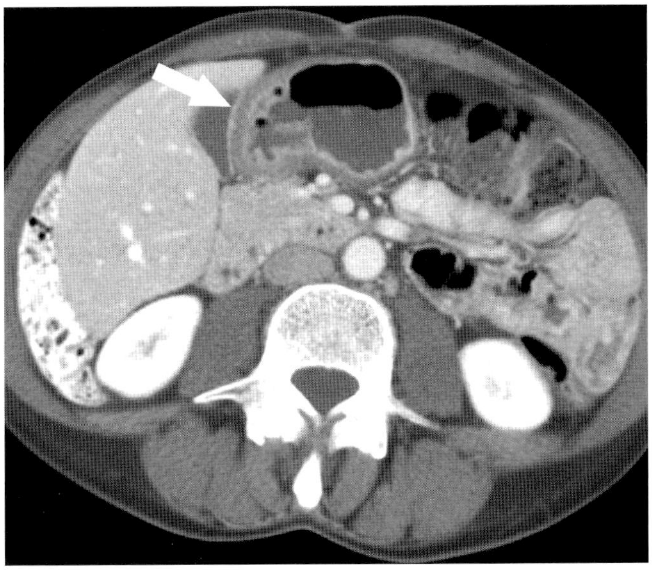

Figure 54.1 Axial contrast-enhanced CT image in a 59 year old woman undergoing surveillance CT 2 years after resection of a sigmoid adenocarcinoma. Uniform smooth thickening of the gastric antrum (arrow) is seen as an incidental finding.

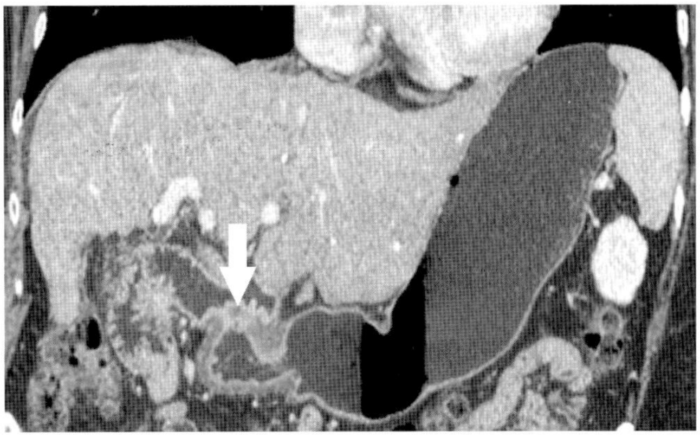

Figure 54.2 Curved planar coronal contrast-enhanced CT image in a 61 year old woman with cirrhosis due to hepatitis C shows uniform smooth thickening of the gastric antrum (arrow) as an incidental finding.

CASE 55 Pseudoabscess due to excluded stomach after gastric bypass

Imaging description

Roux-en-Y gastric bypass (in which a small gastric fundal pouch is created and connected to the rest of the bowel by a Roux loop of jejunum, and the majority of the stomach is excluded from the normal flow of food) is currently one of the commonest and most successful surgical treatments for obesity in the United States [1, 2]. The fluid-filled excluded stomach can mimic a rim-enhancing collection in the surgical bed on postoperative CT and may be mistaken for an abscess (Figure 55.1).

Importance

Misdiagnosis of the excluded stomach as an abscess could result in unnecessary workup, drainage, or even surgery.

Typical clinical scenario

This pitfall is most likely to result in misdiagnosis of abscess when a patient has a CT scan for fever or other symptoms in the early postoperative period after Roux-en-Y gastric bypass. In one study, the fundus of the excluded stomach mimicked a loculated fluid collection in 13 (18%) of 72 such patients [3].

Differential diagnosis

Roux-en-Y gastric bypass may be complicated by leak, abscess, or hematoma, so the correct identification of the excluded stomach is critical in the postoperative patient. The excluded stomach may be positively identified by the presence of gastric rugae and continuity with the duodenum (Figure 55.2).

Teaching point

The diagnosis of an abscess in the left upper quadrant after Roux-en-Y gastric bypass should be made with caution, since the excluded portion of the stomach can closely simulate a rim-enhancing fluid collection.

REFERENCES

1 Merkle EM, Hallowell PT, Crouse C, Nakamoto DA, Stellato TA. Roux-en-Y gastric bypass for clinically severe obesity: normal appearance and spectrum of complications at imaging. *Radiology* 2005; **234**: 674–683.

2 Carucci LR, Turner MA. Radiologic evaluation following Roux-en-Y gastric bypass surgery for morbid obesity. *Eur J Radiol* 2005; **53**: 353–365.

3 Yu J, Turner MA, Cho SR, et al. Normal anatomy and complications after gastric bypass surgery: helical CT findings. *Radiology* 2004; **231**: 753–760.

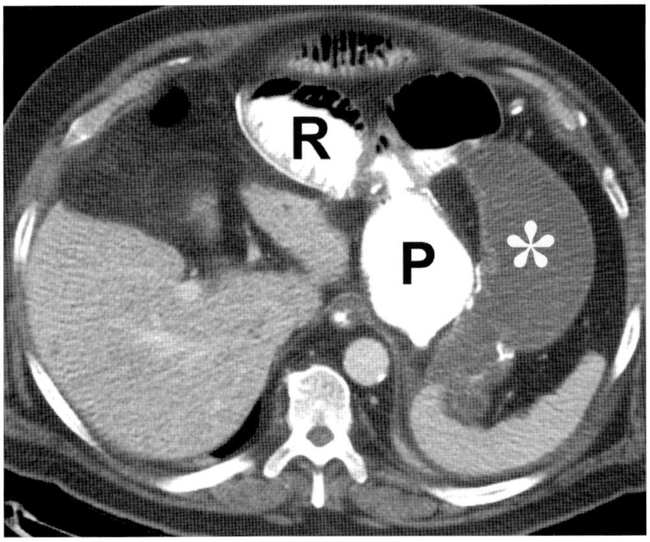

Figure 55.1 Axial contrast-enhanced CT image in a 45 year old woman with vomiting due to postoperative ileus after a Roux-en-Y gastric bypass. The fluid-filled excluded stomach (asterisk) could be mistaken for an abscess or collection. Note the presence of positive oral contrast in the gastric pouch (P) that empties into the jejunal Roux loop (R).

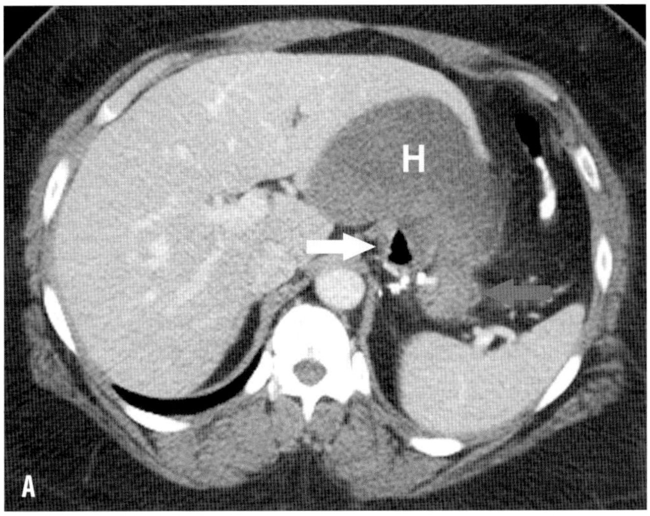

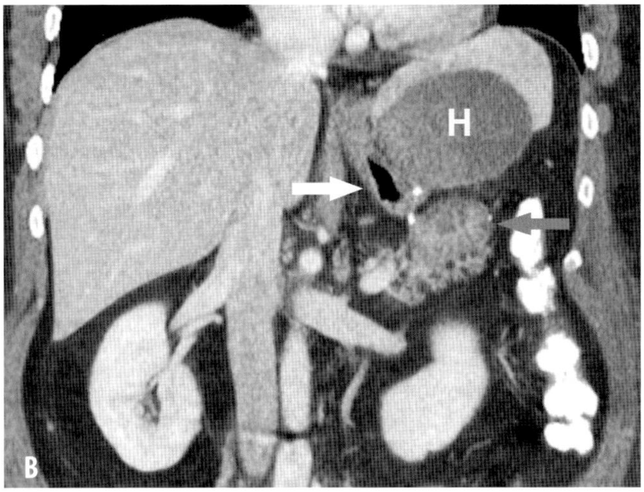

Figure 55.2 A. Axial contrast-enhanced CT image in a 38 year old woman with vomiting due to postoperative hematoma after Roux-en-Y gastric bypass. Note that the hematoma (H) is separate to both the gastric pouch (white arrow) and excluded stomach (gray arrow).
B. Coronal reformatted CT image confirms the hematoma (H) is separate to both the gastric pouch (white arrow) and excluded stomach (gray arrow). Note that gastric rugae can be identified within the collapsed excluded stomach.

CASE 56

Strangulated bowel obstruction

Imaging description

Bowel obstruction is the partial or complete blockage of the small or large intestine, while strangulated bowel obstruction is intestinal blockage accompanied by compromised blood flow. Non-strangulated or simple bowel obstruction is often successfully managed conservatively, while strangulated obstruction is a surgical emergency which can progress to infarction and gangrene in as little as six hours. Strangulation is usually associated with hernias (including internal hernias) or volvulus. In two large surgical series, strangulation occurred in 17 to 23% of small bowel obstructions [1, 2], but it should be remembered that many simple obstructions do not require surgery and so the true frequency of strangulation in "all-comers" with bowel obstruction is likely to be considerably lower. Irrespective of the presence or absence of strangulation, the cardinal imaging sign of obstruction is the finding of dilated bowel upstream to collapsed or non-dilated bowel. CT can usually detect the transition point between dilated and non-dilated bowel and suggest the likely etiology (such as hernia, mass, or intussusception – the lack of a visible cause suggests adhesions). The supplementary CT features that indicate strangulation are reduced bowel wall enhancement, mural thickening, mesenteric fluid or infiltration, congestion of small mesenteric veins, ascites, pneumatosis, and portal venous gas (Figures 56.1 and 56.2) [1–5]. The described supplementary CT signs of strangulated internal hernia are clustering of disproportionately dilated bowel segments, swirling or convergence of mesenteric vessels, and mesenteric vessel engorgement (Figure 56.3) [6, 7].

Importance

Failure to diagnose or suggest strangulation can be disastrous, since it may result in bowel infarction with its associated substantial morbidity and mortality. In my experience, a missed diagnosis of strangulation is the single commonest source of litigation in medical malpractice related to abdominal CT.

Typical clinical scenario

The possibility of strangulation should be considered for all patients with bowel obstruction on CT, since twisting or kinking of dilated loops can occur independent of any particular clinical scenario and no clinical indicators have been identified that reliably predict this life-threatening complication. That said, suggestive clinical features are pain disproportionate to physical findings, leukocytosis, and peritoneal signs [1]. Likewise, internal herniation can occur in patients with no predisposing factors, but patients with Roux-en-Y loops appear particularly at risk, because retrocolic or antecolic positioning of these loops creates anatomic windows through which other bowel loops can pass and become blocked. Such patients include those with Roux-en-Y loops for biliary drainage after liver transplantation and for passage of food after gastric bypass for obesity [8, 9].

Differential diagnosis

Closed loop obstruction is essentially synonymous with strangulated obstruction, but emphasizes the twisting or kinking effect on the bowel rather than the more important effect on the vasculature, and is therefore not the term preferred by most surgeons. Incarceration means that a hernia is irreducible (cannot be reversed by manipulation). As such, incarceration is a finding on physical examination and the term should not be used by radiologists. The published literature suggests CT findings are of variable but reasonable accuracy in the differentiation of strangulated from non-strangulated obstruction, and each publication tends to emphasize different signs [1–7]. Based on my clinical experience, the following general guidance is offered:

1. Ascites is not a feature of simple obstruction and anything more than a trace amount of free fluid should immediately raise concern for strangulation (but remembering that patients with cirrhosis, generalized edematous states, or other unrelated causes of ascites could develop obstruction, so the finding of obstruction and ascites is not specific for strangulated obstruction).
2. The combination of ascites, reduced mural enhancement, and mesenteric infiltration results in the bowel wall becoming difficult to discern (the "disappearing bowel wall" sign). Other bowel loops can be used as an internal control in evaluating the presence of reduced bowel wall enhancement.
3. Bowel wall thickening, pneumatosis, and portal venous air are findings of advanced ischemia and are not useful signs in the early identification of strangulation.
4. Congestion of small mesenteric veins, clustering of disproportionately dilated bowel segments, swirling or convergence of mesenteric vessels, and mesenteric vessel engorgement are of limited utility, and are rarely the primary clues to the diagnosis of strangulation.

Teaching point

The possibility of strangulation should be considered for all cases of bowel obstruction identified at CT. Suggestive findings are ascites, reduced mural enhancement, and mesenteric infiltration.

REFERENCES

1 Jancelewicz T, Vu LT, Shawo AE, et al. Predicting strangulated small bowel obstruction: an old problem revisited. *Journal of Gastrointest Surg* 2009; **13**: 93–99.

2 Zalcman M, Sy M, Donckier V, Closset J, Gansbeke DV. Helical CT signs in the diagnosis of intestinal ischemia in small-bowel obstruction. *Am J Roentgenol* 2000; **175**: 1601–1607.

3 Kim JH, Ha HK, Kim JK, et al. Usefulness of known computed tomography and clinical criteria for diagnosing strangulation in

small-bowel obstruction: analysis of true and false interpretation groups in computed tomography. *World J Surg* 2004; **28**: 63–68.

4 Frager D, Baer JW, Medwid SW, Rothpearl A, Bossart P. Detection of intestinal ischemia in patients with acute small-bowel obstruction due to adhesions or hernia: efficacy of CT. *Am J Roentgenol* 1996; **166**: 67–71.

5 Ha HK, Kim JS, Lee MS, et al. Differentiation of simple and strangulated small-bowel obstructions: usefulness of known CT criteria. *Radiology* 1997; **204**: 507–512.

6 Yen CH, Chen JD, Tui CM, et al. Internal hernia: computed tomography diagnosis and differentiation from adhesive small bowel obstruction. *J Chin Med Assoc* 2005; **68**: 21–28.

7 Gollub MJ, Yoon S, Smith LM, Moskowitz CS. Does the CT whirl sign really predict small bowel volvulus? Experience in an oncologic population. *J Comput Assist Tomogr* 2006; **30**: 25–32.

8 Blachar A, Federle MP. Bowel obstruction following liver transplantation: clinical and CT findings in 48 cases with emphasis on internal hernia. *Radiology* 2001; **218**: 384–388.

9 Lockhart ME, Tessler FN, Canon CL, et al. Internal hernia after gastric bypass: sensitivity and specificity of seven CT signs with surgical correlation and controls. *Am J Roentgenol* 2007; **188**: 745–750.

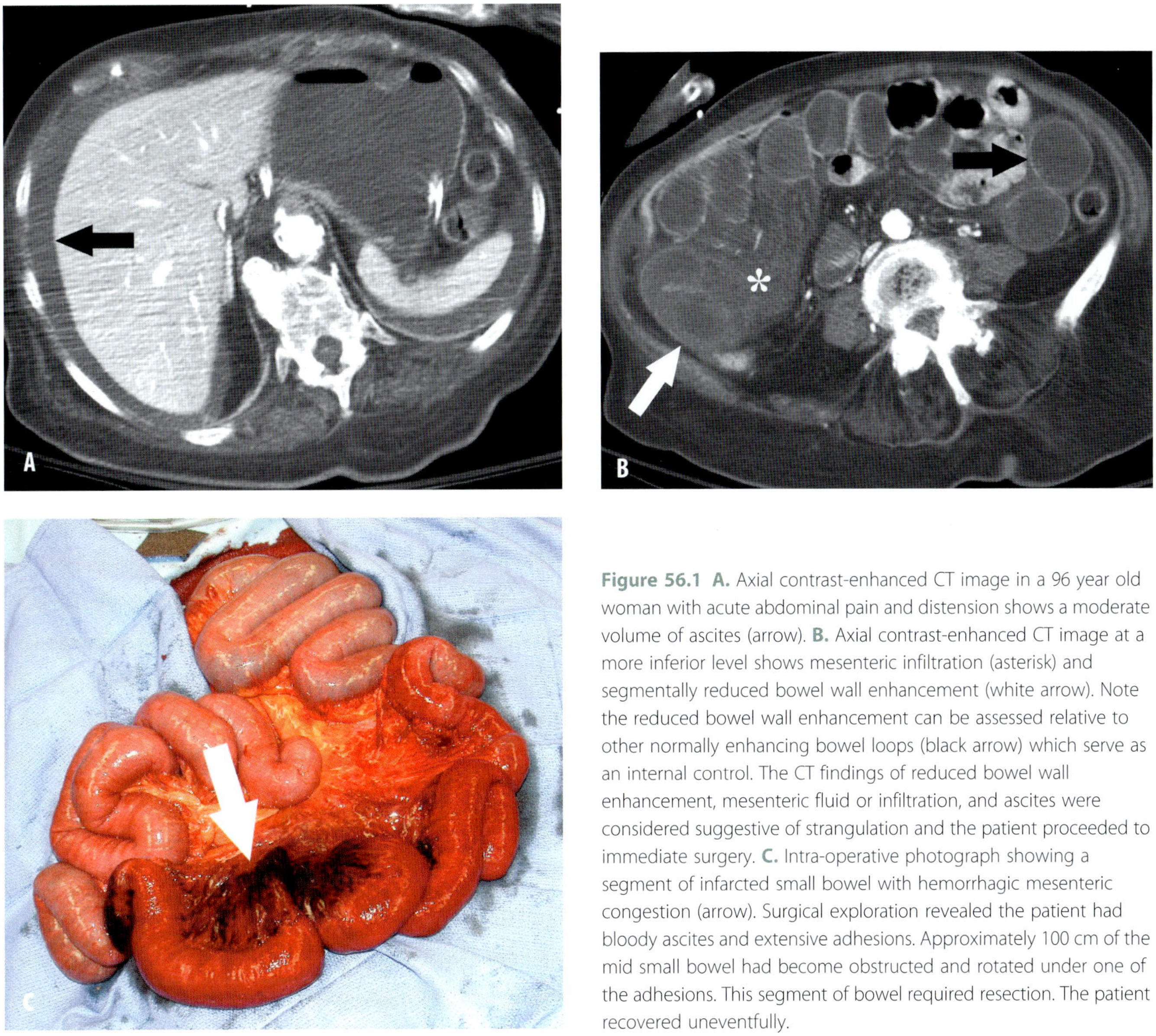

Figure 56.1 A. Axial contrast-enhanced CT image in a 96 year old woman with acute abdominal pain and distension shows a moderate volume of ascites (arrow). **B.** Axial contrast-enhanced CT image at a more inferior level shows mesenteric infiltration (asterisk) and segmentally reduced bowel wall enhancement (white arrow). Note the reduced bowel wall enhancement can be assessed relative to other normally enhancing bowel loops (black arrow) which serve as an internal control. The CT findings of reduced bowel wall enhancement, mesenteric fluid or infiltration, and ascites were considered suggestive of strangulation and the patient proceeded to immediate surgery. **C.** Intra-operative photograph showing a segment of infarcted small bowel with hemorrhagic mesenteric congestion (arrow). Surgical exploration revealed the patient had bloody ascites and extensive adhesions. Approximately 100 cm of the mid small bowel had become obstructed and rotated under one of the adhesions. This segment of bowel required resection. The patient recovered uneventfully.

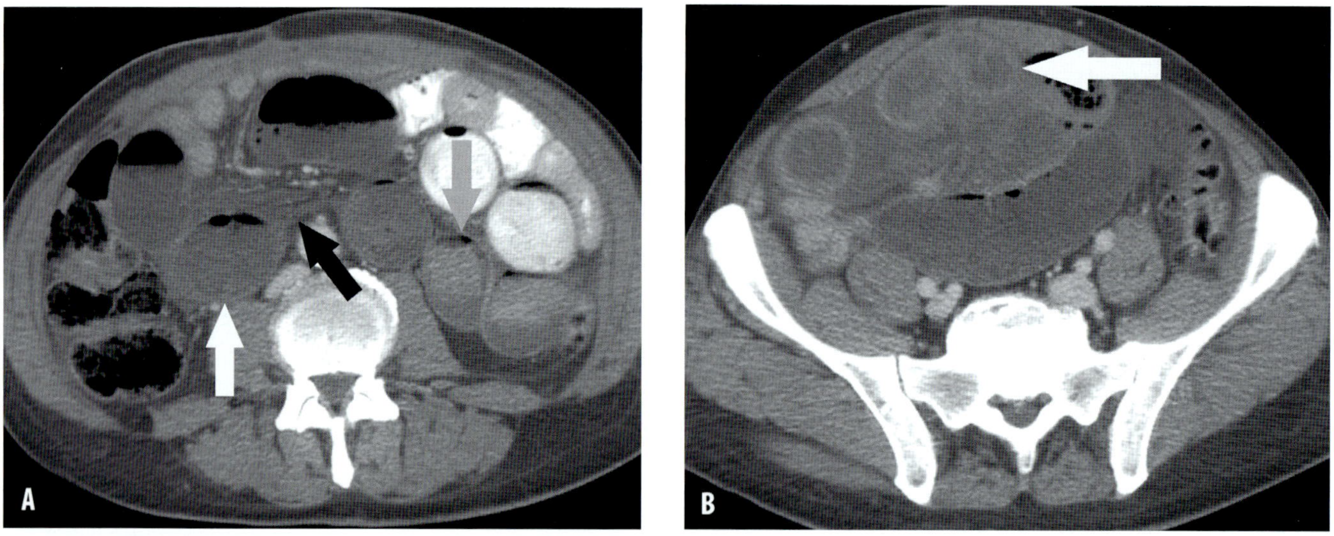

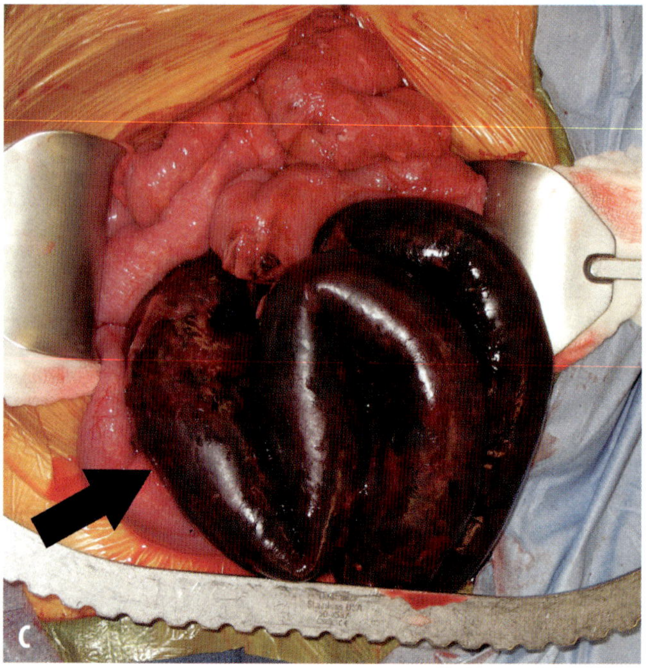

Figure 56.2 A. Axial contrast-enhanced CT image in a 64 year old man with a complicated surgical history and clinical features of bowel obstruction shows dilated small bowel loops with mesenteric infiltration (black arrow) and segmentally reduced bowel wall enhancement (white arrow) when compared to other normally enhancing bowel loops (gray arrow). **B.** Axial contrast-enhanced CT image at a more inferior level shows bowel wall thickening and reduced enhancement (arrow) and ascites. The CT findings of reduced bowel wall enhancement, mesenteric fluid or infiltration, and ascites were considered suggestive of strangulation and the patient proceeded to surgery. **C.** Intra-operative photograph showing a segment of infarcted small bowel (arrow). Surgical exploration revealed the patient had a 70 cm segment of mid-ileum which was infarcted with full-thickness necrosis due to a strangulated adhesive obstruction.

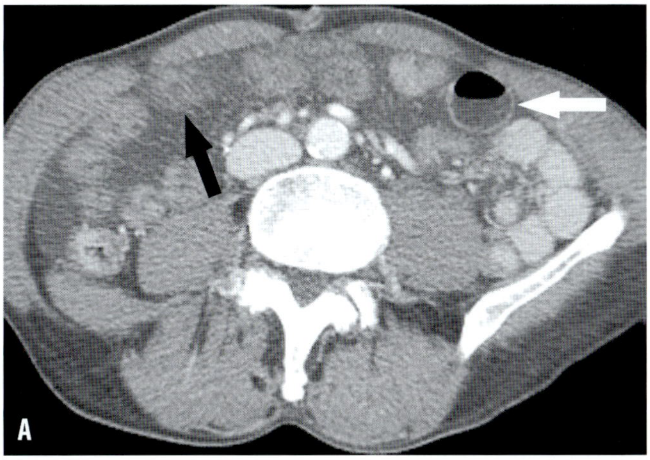

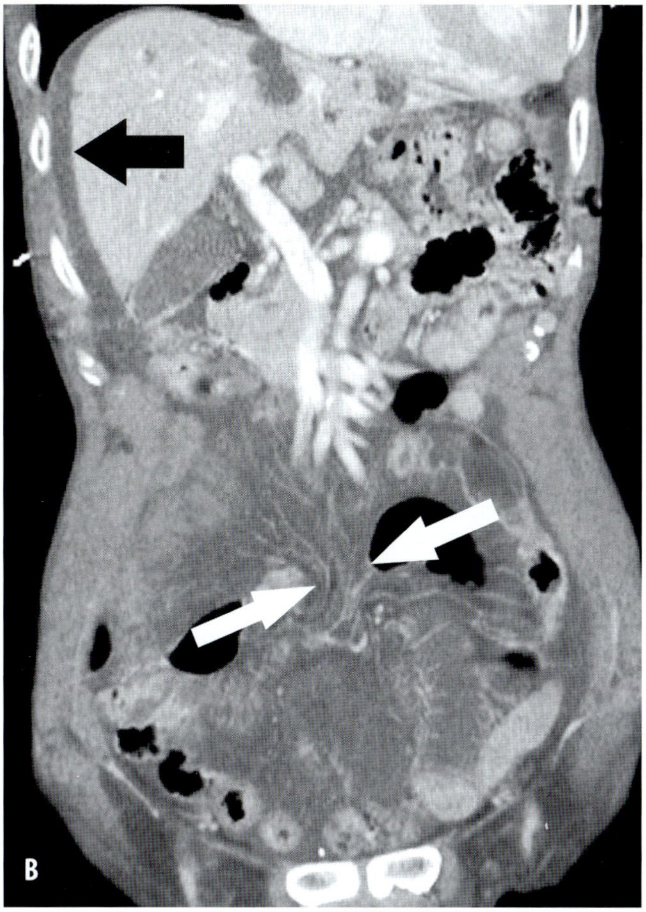

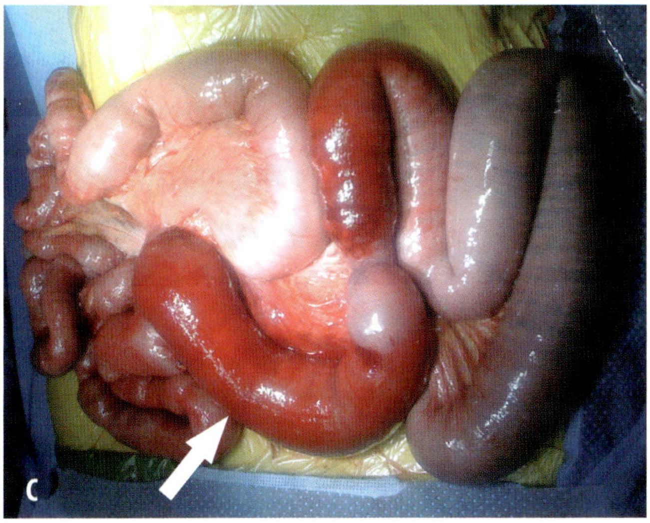

Figure 56.3 A. Axial contrast-enhanced CT image in an 83 year old man with acute severe lower abdominal pain shows dilated small bowel loops with a normally enhancing wall (white arrow). Small bowel loops in the right lower quadrant are surrounded by ascites and demonstrate reduced bowel wall enhancement such that the bowel is difficult to discern, the so-called "disappearing bowel wall" sign (black arrow). **B.** Coronal reformatted CT image again shows ascites (black arrow) and convergent mesenteric vessels (between white arrows) passing perpendicularly to a dilated bowel loop. The findings were considered concerning for a strangulated internal hernia. The patient proceeded to urgent surgery. **C.** Intra-operative photograph showing a segment of discolored ischemic small bowel (arrow). Surgical exploration revealed the patient had a strangulated internal hernia due to passage of a loop of jejunum through a mesenteric defect.

CASE 57 Transient ischemia of the bowel

Imaging description

Traditional CT signs of acute mesenteric ischemia such as visceral artery occlusion, pneumatosis, portomesenteric venous gas, and bowel wall thickening are generally derived from surgically proven series [1, 2]. This is scientifically rigorous but does not account for the fact that many patients with presumed thromboembolic mesenteric ischemia do not go to surgery, and are characterized by milder degrees of self-limiting segmental bowel dilatation, bowel wall thickening, mesenteric infiltration, and ascites (Figures 57.1–57.4) [3, 4]. The emerging concept that acute mesenteric ischemia covers a clinicoradiological spectrum varying from mild and self-limiting to severe and life-threatening [3] is analogous to the spectrum of neurological deficits due to cerebrovascular insufficiency varying from transient ischemic attack to full-blown stroke. The term "transient ischemia of the bowel" has reasonably been proposed to describe those with the milder forms of acute mesenteric ischemia.

Importance

It is likely that transient ischemia of the bowel is under-recognized. In one series, 8 of 30 patients with acute abdominal pain and atrial fibrillation had CT signs of end-organ ischemia or infarction [3]. Atrial fibrillation affects an estimated 2.3 million Americans [5], so the population at risk is substantial. Failure to recognize the diagnosis may result in a missed opportunity for reassessment of anticoagulation status.

Typical clinical scenario

Transient ischemia should be strongly considered as a differential possibility when a high-risk patient presents with acute abdominal pain. High-risk patients include those with atrial fibrillation or other cardiac conditions that predispose to thrombosis and embolism, particularly if they are not anticoagulated or are insufficiently anticoagulated.

Differential diagnosis

Other conditions that can produce segmental bowel wall thickening, dilatation, and mesenteric infiltration include angioedema, submucosal hemorrhage (particularly in patients who are anticoagulated and have a supratherapeutic international normalized ratio), Crohn's disease, and radiation enteritis. Like transient ischemia of the bowel, both angioedema and submucosal hemorrhage may be transient, while Crohn's disease and radiation enteritis are chronic.

Teaching point

The traditional radiological view of mesenteric ischemia needs to be broadened to accommodate the wider spectrum of CT appearances, which includes transient ischemia of the bowel characterized by segmental bowel wall thickening, dilatation, and mesenteric infiltration or ascites. Clinically, transient ischemia of the bowel is typically self-limiting and occurs in patients with atrial fibrillation.

REFERENCES

1 Wiesner W, Mortele KJ, Glickman JN, Ji H, Ros PR. Pneumatosis intestinalis and portomesenteric venous gas in intestinal ischemia: Correlation of CT findings with severity of ischemia and clinical outcome. *Am J Roentgenol* 2001; **177**: 1319–1323.

2 Aschoff AJ, Stuber G, Becker BW, et al. Evaluation of acute mesenteric ischemia: accuracy of biphasic mesenteric multi-detector CT angiography. *Abdom Imaging* 2009; **34**: 345–357.

3 Hunt SJ, Coakley FV, Webb EM, Westphalen AC, Yeh BM. CT of the acute abdomen in patients with atrial fibrillation. *J Comput Assist Tomogr* 2009; **33**: 280–285.

4 Barajas RF, Yeh BM, Webb EM, et al. CT of the non-traumatic acute abdomen in patients with atrial fibrillation. *Am J Roentgenol* 2009; **193**: 485–492.

5 Andrews M, Nelson BP. Atrial fibrillation. *Mt Sinai J Med* 2006; **73**: 482–492.

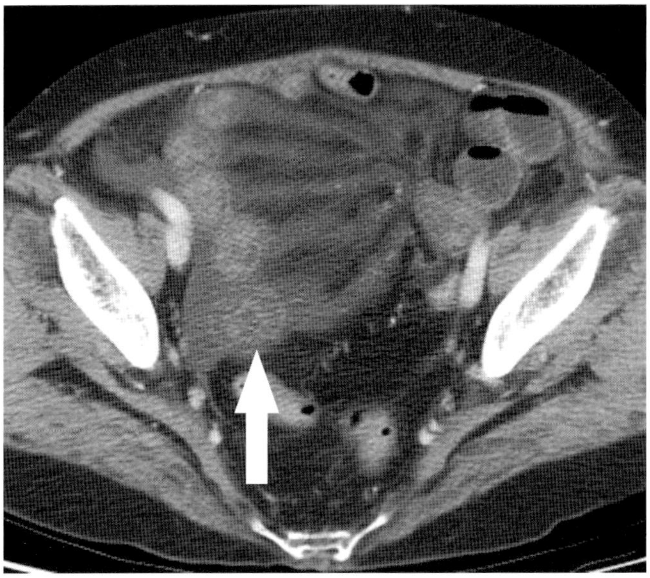

Figure 57.1 Axial contrast-enhanced CT image in an 80 year old man with atrial fibrillation and sudden onset of abdominal pain shows multiple dilated loops of mildly dilated distal small bowel in the pelvis, with associated mesenteric infiltration and free fluid. The affected small bowel wall is thickened with reduced enhancement (arrow). Patient was taking coumadin, with a sub-therapeutic international normalized ratio of 1.0. Symptoms resolved with supportive therapy, and the final presumptive clinical diagnosis was self-limiting embolic bowel ischemia.

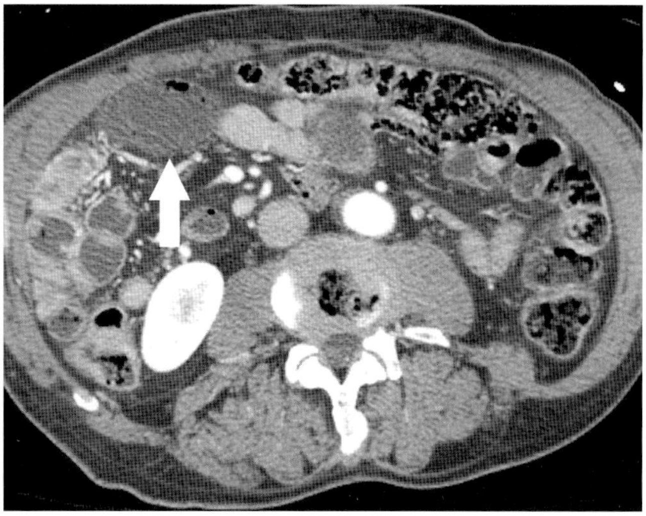

Figure 57.2 Axial contrast-enhanced CT image in a 79 year old man with atrial fibrillation and sudden onset of central abdominal pain shows a focally dilated loop of mildly small bowel in the right abdomen. The affected small bowel wall appears non-enhancing (arrow). The patient was taking coumadin, but with a subtherapeutic international normalized ratio of 1.3. Symptoms resolved with supportive therapy, and the final presumptive clinical diagnosis was self-limiting embolic bowel ischemia.

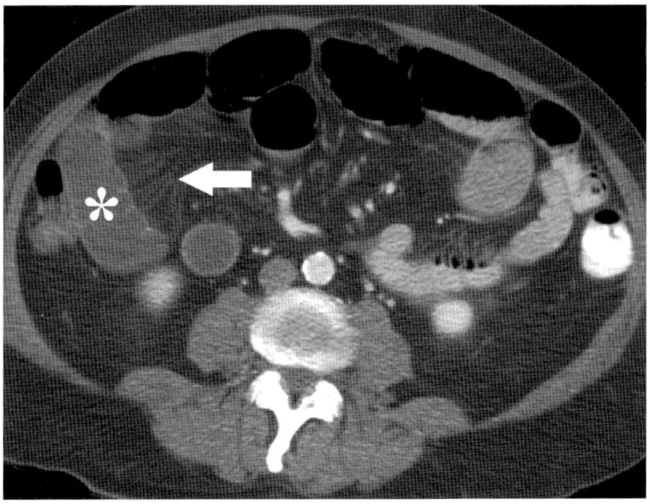

Figure 57.3 Axial contrast-enhanced CT image in a 67 year old man with atrial fibrillation and sudden onset of central abdominal pain shows focally dilated bowel loops in the right abdomen (asterisk) associated with subtle mesenteric infiltration (arrow). The patient was taking coumadin, with an international normalized ratio of 3.5. Symptoms resolved with supportive therapy, and the final presumptive clinical diagnosis was self-limiting embolic bowel ischemia.

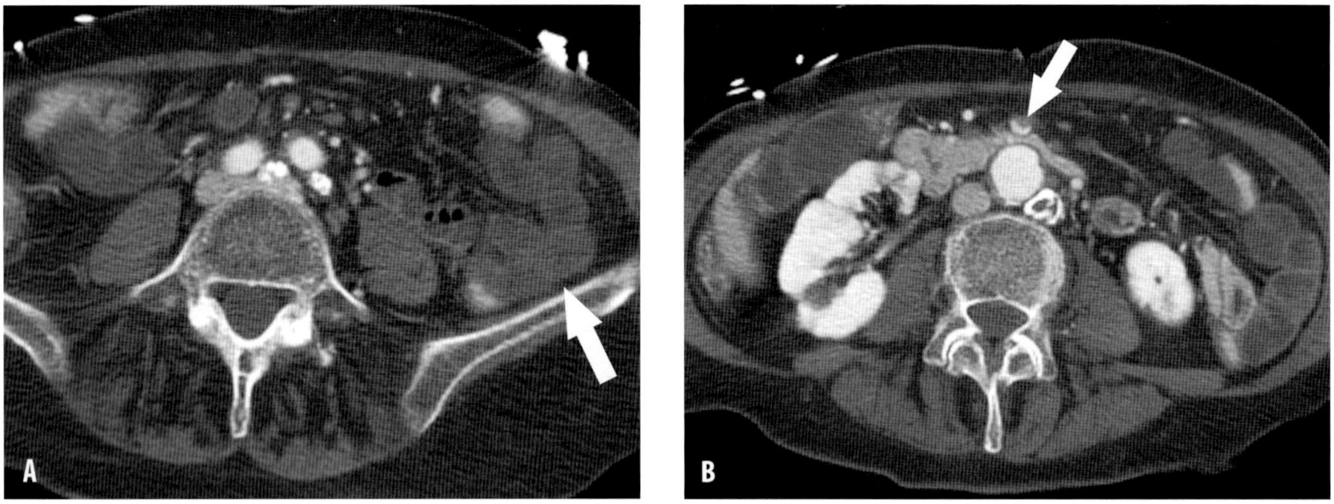

Figure 57.4 A. Axial contrast-enhanced CT image in a 75 year old woman with atrial fibrillation and sudden onset of central abdominal pain shows mildly dilated bowel loops in the left abdomen (arrow), with reduced wall enhancement (compare to the bowel wall in Figure 57.4B). The patient was taking coumadin, with an international normalized ratio of 2.3. **B.** Axial contrast-enhanced CT image at a more superior level shows a filling defect (arrow) in the superior mesenteric artery, presumably an embolus.

CASE 58 Angioedema of the bowel

Imaging description

Angioedema is a transient inflammatory reaction due to vascular leak that is related to urticaria, but affects the subcutaneous or submucosal tissues of the body while urticaria occurs in the skin only (although angioedema and urticaria may co-exist) [1]. Angioedema is likely due to the release of inflammatory mediators from mast cells and/or the activation of complement or kinin systems [2]. All parts of the body may be affected, including the eyes, lips, upper respiratory tract, and bowel. Angioedema of the bowel is often clinically non-specific and so awareness of the radiological findings is particularly important. At CT, angioedema of the bowel is characterized by transient wall thickening, mucosal hyperemia, mural stratification, prominent mesenteric vessels, and ascites (Figures 58.1 and 58.2) [3].

Importance

Angioedema is rare, and in one study accounted for only 1 to 4% of 6107 patients seen in a large emergency department with acute allergic reactions [4]. Hereditary angioedema has an estimated prevalence of 1 in 50,000 to 150,000 [3]. However, the condition is likely under-diagnosed and cases may go unrecognized for many years – in a series of 235 patients with C1-inhibitor deficiency, 34% had undergone abdominal surgery before the diagnosis was established [5, 6]. Knowledge of the CT findings may facilitate correct diagnosis, potentially averting unnecessary intervention and promoting appropriate workup and therapy (Figure 58.3).

Typical clinical scenario

Angioedema of the bowel typically presents with acute abdominal pain associated with distention, nausea, and vomiting followed by watery diarrhea [3]. Angioedema may be allergic, idiopathic, angiotensin converting enzyme inhibitor-induced, or may reflect hereditary or acquired deficiency or dysfunction of C1-esterase inhibitor [3]. Acute allergic angioedema is almost always associated with urticaria and occurs within one to two hours of exposure to an allergen (including drug-induced angioedema, with the exception of angiotensin converting enzyme inhibitor-induced angioedema). Multiple drugs have been associated with allergic angioedema, but the major culprits are aspirin and non-steroidal anti-inflammatory drugs [7]. Idiopathic angioedema resembles acute allergic angioedema, except that no known cause is found. Angiotensin converting enzyme inhibitor-induced angioedema appears to reflect a direct effect of the drug on the kinin system. Angioedema due to angiotensin converting enzyme inhibitors occurs with a frequency of 0.1 to 0.2% in those on these drugs and usually has an onset of less than 7 days after initiation of treatment, but can occur weeks to months later [8]. For unknown reasons, angiotensin converting enzyme inhibitor-induced angioedema appears to occur primarily in women [9]. Hereditary angioedema is an autosomal dominant condition in which abnormal activation of the complement and kinin systems results from either deficiency or dysfunction of C1-esterase inhibitor. Acquired C1-esterase inhibitor deficiency can occur as a paraneoplastic syndrome in malignancies such as lymphoma, multiple myeloma, and chronic lymphocytic leukemia or in association with autoimmune disease such as systemic lupus erythematosus.

Differential diagnosis

The differential considerations for the CT finding of segmental bowel wall thickening include ischemia, strangulated bowel obstruction, Henoch-Schönlein purpura (vasculitis), intramural bleeding (from trauma, anticoagulation, or bleeding diathesis such as hemophilia), infection, and inflammatory bowel disease [10]. In particular, strangulation and ischemia share many of the CT findings seen in angioedema, but mural stratification is rare in ischemia and strangulated bowel is usually moderately to markedly dilated. In patients with angioedema, the CT findings are entirely reversible and are seen only during an acute episode, which may help in the distinction from other conditions. The clinical setting may also allow for narrowing of the differential diagnosis.

Teaching point

Angioedema should be considered in the CT differential diagnosis for segmental bowel wall thickening, mucosal hyperemia, mural stratification, prominent mesenteric vessels, and ascites. Correlation with medication usage (particularly angiotensin converting enzyme inhibitors), history of malignancy or autoimmune disease, and C1-esterase levels may help.

REFERENCES

1 Kaplan AP, Greaves MW. Angioedema. *J Am Acad Dermatol* 2005; **53**: 373–388.

2 Frigas E, Nzeako UC. Angioedema. Pathogenesis, differential diagnosis, and treatment. *Clin Rev Allergy Immunol* 2002; **23**: 217–231.

3 De Backer AI, De Schepper AM, Vandevenne JE, et al. CT of angioedema of the small bowel. *Am J Roentgenol* 2001; **176**: 649–652.

4 Dantonio C, Galimberti M, Barbone B, et al. Suspected acute allergic reactions: analysis of admissions to the Emergency Department of the AOU Maggiore della Carità Hospital in Novara from 2003 to 2007. *Eur Ann Allergy Clin Immunol* 2008; **40**: 122–129.

5 Cicardi M, Conciato L, Agostoni A. Angioedema due to angiotensin-converting enzyme inhibition: an association frequently unrecognized. *Ann Ital Med Int* 1997; **12**: 8–10.

6 Agostoni A, Cicardi M. Hereditary and acquired C1-inhibitor deficiency: biological and clinical characteristics in 235 patients. *Medicine* 1992; **71**: 206–215.

7 Banerji A, Oren E, Hesterberg P, et al. Ten-year study of causes of moderate to severe angioedema seen by an inpatient allergy/immunology consult service. *Allergy Asthma Proc* 2008; **29**: 88–92.

8 Israili ZH, Hall WD. Cough and angioneurotic edema associated with angiotensin-converting enzyme inhibitor therapy: a review of the literature and pathophysiology. *Ann Intern Med* 1992; **117**: 234–242.

9 Chase MP, Fiarman GS, Scholz FJ, MacDermott RP. Angioedema of the small bowel due to an angiotensin-converting enzyme inhibitor. *J Clin Gastroenterol* 2000; **31**: 254–257.

10 Ciaccia D, Brazer SR, Baker ME. Acquired C1 esterase inhibitor deficiency causing intestinal angioedema: CT appearance. *Am J Roentgenol* 1993; **161**: 1215–1216.

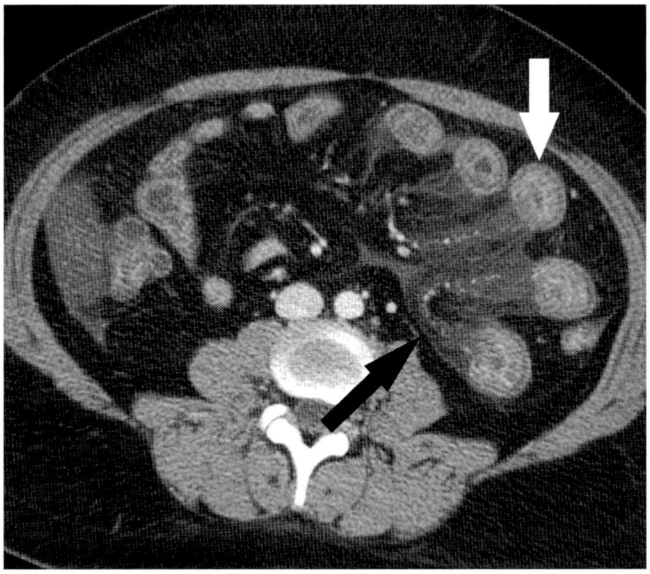

Figure 58.1 Axial contrast-enhanced CT image in a 30 year old woman with severe abdominal pain and a history of hypertension treated with lisinopril. Thick-walled bowel loops (white arrow) in the left abdomen are associated with mesenteric infiltration (black arrow). A repeat CT two days later (not shown) was entirely normal. A presumptive diagnosis of angiotensin converting enzyme inhibitor-induced angioedema was made and lisinopril was discontinued.

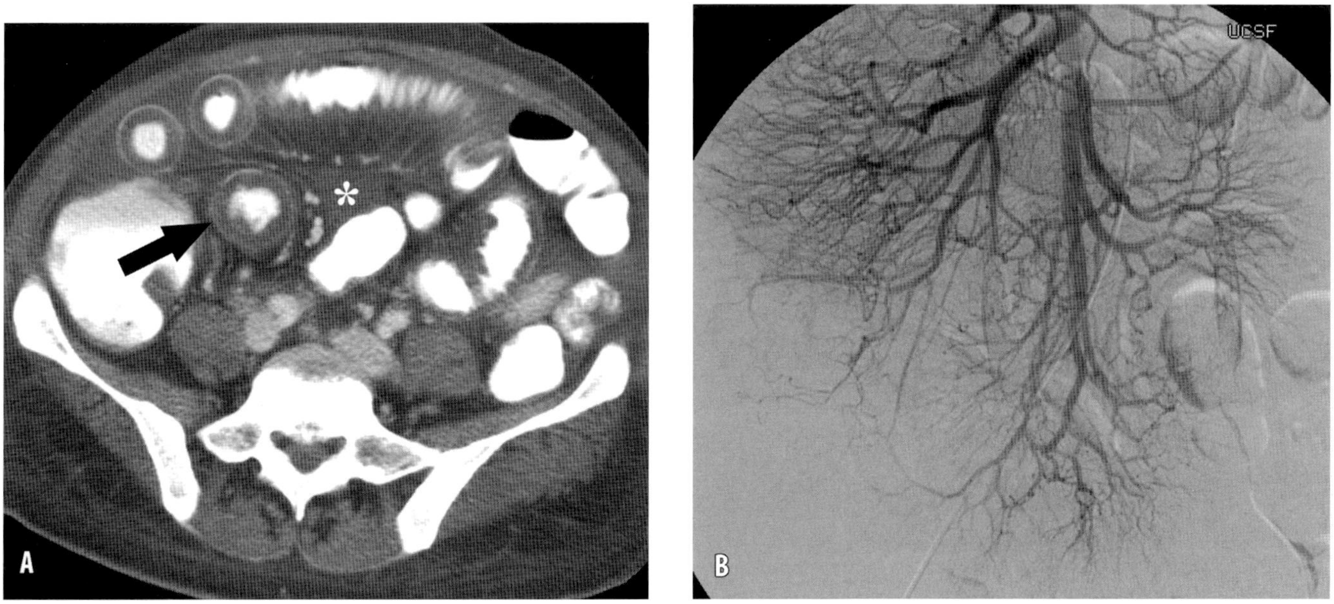

Figure 58.2 **A.** Axial contrast-enhanced CT image in a 22 year old woman with recurrent acute abdominal pain, vomiting, and diarrhea, and a history of systemic lupus erythematosus and hypertension treated by benazepril. Thick-walled bowel loops (arrow) are associated with ascites (asterisk) and mural stratification. The differential diagnosis included ischemia due to lupus vasculitis or angioedema. **B.** Conventional catheter angiography shows normal mesenteric vasculature. The constellation of findings was considered diagnostic of angioedema. Benazepril was discontinued.

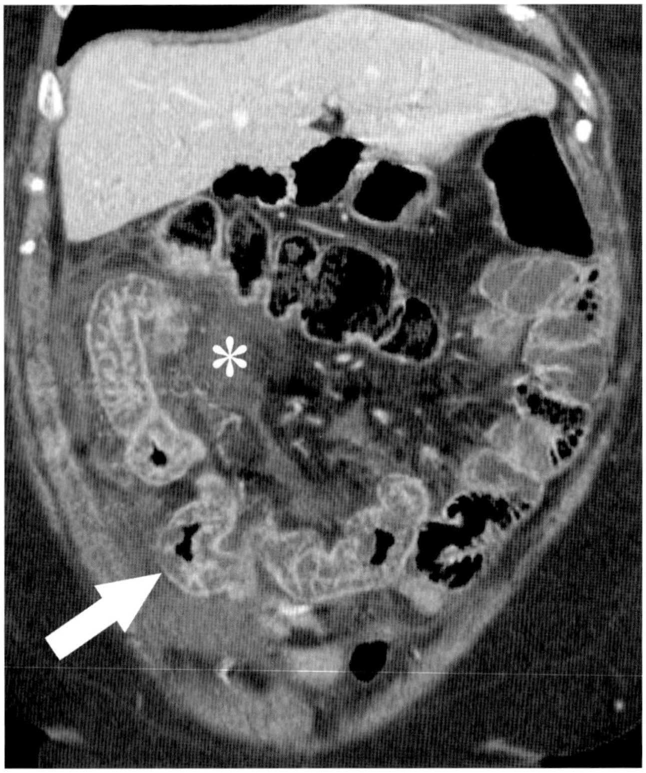

Figure 58.3 Coronal reformatted contrast-enhanced CT image in a 75 year old woman with acute abdominal pain and a history of rheumatoid arthritis. Thick-walled bowel loops with mural stratification (arrow) in the right abdomen are associated with ascites (asterisk). The findings were considered suggestive of angioedema. The patient proceeded to surgery because of clinical concern for ischemic bowel, and only mildly edematous bowel was found, with no surgical or pathological features of ischemia. The final diagnosis was of angioedema, based on the constellation of clinical, surgical, and CT findings.

CASE

59 Small bowel intramural hemorrhage

Imaging description

Intramural hemorrhage in the small bowel results in segmental circumferential bowel wall thickening. Thickening of the small bowel wall has been defined as a wall thickness greater than 4 mm [1], although in practice most radiologists use expert judgment to make this diagnosis. The wall thickening is usually homogeneous and may be of visibly high density. There may be associated bowel obstruction, adjacent fat stranding, or bloody ascites (Figures 59.1 and 59.2) [2–5].

Importance

Intramural hemorrhage in the small bowel is rare, but should be included in the differential of segmental small bowel wall thickening. It may be confused with ischemia, angioedema, or strangulated obstruction, and since management of these conditions is quite different, optimal radiological evaluation is critical.

Typical clinical scenario

Supratherapeutic anticoagulation with coumadin is the single commonest reason for spontaneous or non-traumatic intramural hemorrhage in the bowel, accounting for 8 cases in one series of 13 such patients [4]. In the same study, small bowel obstruction was present in 11 (85%) patients. A single hematoma was present in 85% of patients, and multiple hematomas were present in 15%. The jejunum was the most common site of hematoma, followed by the ileum and duodenum. Most patients with intestinal intramural hemorrhage can be treated conservatively with a good outcome.

Differential diagnosis

The primary differential consideration for intramural hematoma is segmental ischemia, since both conditions commonly occur in patients who are anticoagulated because of a predisposition to arterial thromboemboli (e.g., patients with atrial fibrillation). Such patients are prone to both thromboembolic and hemorrhagic complications [6]. Angioedema and strangulated obstruction are also in the radiological differential, but typically occur in patients without risk factors for hemorrhage or thromboembolism, and strangulated obstruction is typically associated with more marked bowel dilatation. Although some of the CT features overlap, a short segment involvement with wall thickening of 1 cm or greater is typical of intramural hemorrhage; a long segment involvement with wall thickening of less than 1 cm is typical of ischemia (Figure 59.3).

Teaching point

Segmental bowel wall thickening in a patient who is anticoagulated is likely due to ischemia or intramural hemorrhage; short segment involvement with wall thickening in excess of 1 cm or associated with a supratherapeutic international normalized ratio favors hemorrhage.

REFERENCES

1 Desai RK, Tagliabue JR, Wegryn SA, Einstein DM. CT evaluation of wall thickening in the alimentary tract. *Radiographics* 1991; **11**: 771–783.

2 Balthazar EJ, Hulnick D, Megibow AJ, Opulencia JF. Computed tomography of intramural intestinal hemorrhage and bowel ischemia. *J Comput Assist Tomogr* 1987; **11**: 67–72.

3 Macari M, Chandarana H, Balthazar E, Babb J. Intestinal ischemia versus intramural hemorrhage: CT evaluation. *Am J Roentgenol* 2003; **180**: 177–184.

4 Abbas MA, Collins JM, Olden KW. Spontaneous intramural small-bowel hematoma: imaging findings and outcome. *Am J Roentgenol* 2002; **179**: 1389–1394.

5 Lane MJ, Katz DS, Mindelzun RE, Jeffrey RB Jr. Spontaneous intramural small bowel haemorrhage: importance of non-contrast CT. *Clin Radiol* 1997; **52**: 378–380.

6 Hunt SJ, Coakley FV, Webb EM, et al. Computed tomography of the acute abdomen in patients with atrial fibrillation. *J Comput Assist Tomogr* 2009; **33**: 280–285.

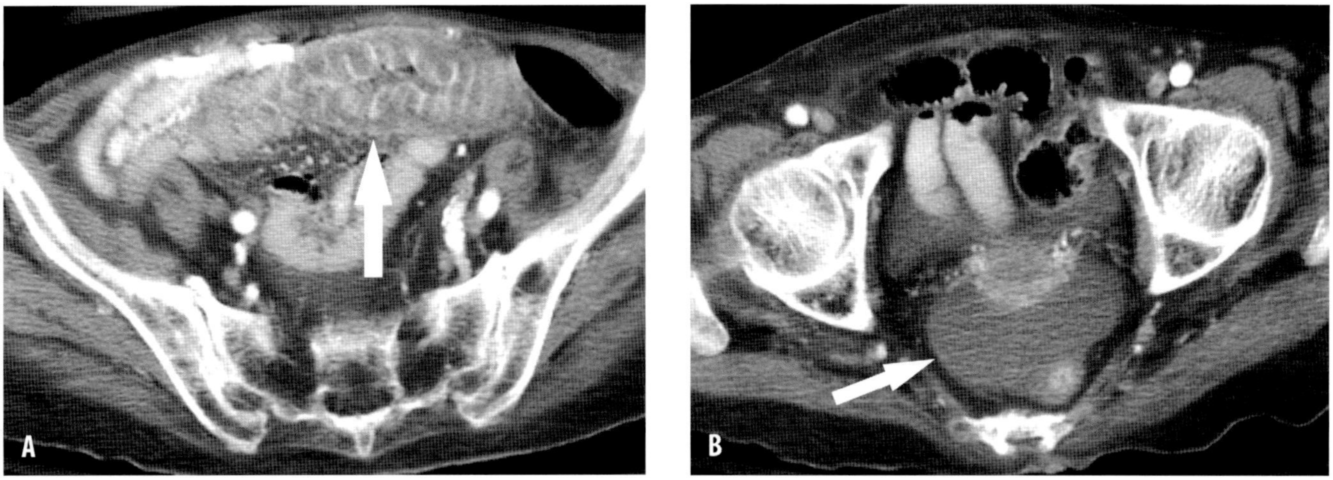

Figure 59.1 **A.** Axial contrast-enhanced CT image in an 88 year old woman with acute abdominal pain and an international normalized ratio of 20 secondary to poorly controlled anticoagulation with coumadin. Circumferential wall thickening (arrow) due to presumptive intramural hemorrhage is seen in a segment of small bowel. **B.** Axial contrast-enhanced CT image at a more inferior level shows a small volume of mildly dense ascites (arrow), consistent with hemoperitoneum. The patient recovered with conservative management.

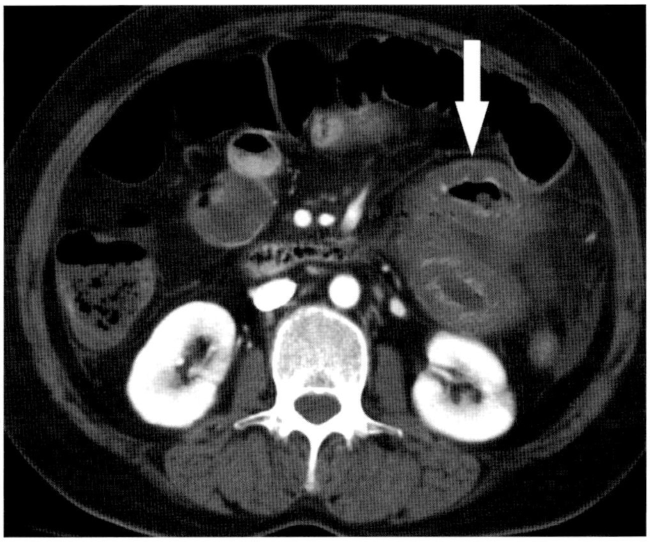

Figure 59.2 Axial contrast-enhanced CT image in an 88 year old woman with acute abdominal pain and tenderness. The patient was receiving coumadin because of prior aortic and mitral valve replacements, and on admission had an uncontrolled international normalized ratio of over 10. Segmental small bowel wall thickening (arrow) is consistent with intramural hemorrhage. The patient recovered with conservative management.

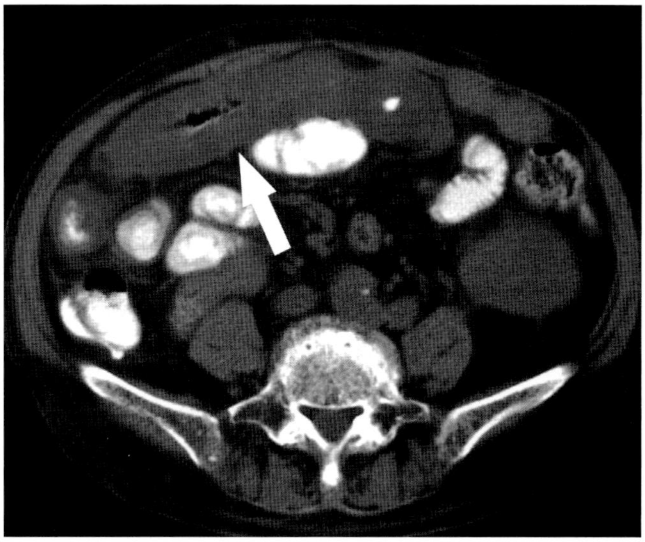

Figure 59.3 Axial non-enhanced CT image in a 75 year old man with acute abdominal pain and a history of anticoagulation with coumadin for atrial fibrillation. A small bowel segment demonstrates marked bowel wall thickening (arrow). In this setting, both segmental ischemia and intramural hemorrhage are radiological considerations, but the degree of thickening favors intramural hemorrhage. The latter was also favored clinically, because the international normalized ratio was elevated at 5.5 (a soft-tissue density in the left lower quadrant is a renal transplant).

CASE 60 Pseudopneumatosis

Imaging description

Gas trapped against the mucosal surface of the bowel by viscous bowel content or small gas bubbles clinging to the mucosal surface can generate a CT appearance that resembles small bubbles of gas within the bowel wall itself (Figures 60.1 and 60.2). The term pseudopneumatosis has been applied to this mimic of true pneumatosis, and the phenomenon appears to occur most frequently in the cecum and ascending colon [1–4].

Importance

A false positive diagnosis of pneumatosis is likely to lead to unnecessary testing and anxiety, because true pneumatosis indicates ischemic or non-ischemic bowel injury (some breakdown in mucosal integrity is required to allow the introduction or development of intramural gas) [3].

Typical clinical scenario

Pseudopneumatosis is relatively common, but usually only becomes of clinical concern when it is detected in a patient with acute abdominal pain and is mistaken for ischemic pneumatosis.

Differential diagnosis

The distinction of pseudopneumatosis from true pneumatosis can be challenging, but only pneumatosis results in mucosal dissection, which is usually circumferential. The identification of dissected mucosa, visible as a curvilinear density outlined by air on the mural side and by air or other bowel content on the luminal side, allows for a confident diagnosis of true pneumatosis (Figure 60.3). Conversely, absence of mucosal dissection generally indicates pseudopneumatosis. Gas within vessels is also diagnostic of pneumatosis, and so the mesenteric and portal veins should be closely examined for intravenous air (Figure 60.4) [3]. Rescanning the patient in a decubitus position sometimes facilitates the distinction of pseudopneumatosis from true pneumatosis [5].

Teaching point

Pseudopneumatosis is due to gas bubbles abutting the bowel that mimic true intramural gas, but can usually be distinguished from true pneumatosis by the absence of mucosal dissection in pseudopneumatosis.

REFERENCES

1 Simon AM, Birnbaum BA, Jacobs JE. Isolated infarction of the cecum: CT findings in two patients. *Radiology* 2000; **214**: 513–516.

2 Taourel P, Garibaldi F, Arrigoni J, et al. Cecal pneumatosis in patients with obstructive colon cancer: correlation of CT findings with bowel viability. *Am J Roentgenol* 2004; **183**: 1667–1671.

3 Wittenberg J, Harisinghani MG, Jhaveri K, Varghese J, Mueller PR. Algorithmic approach to CT diagnosis of the abnormal bowel wall. *Radiographics* 2002; **2**: 1093–1107.

4 Macari M, Balthazar EJ. CT of bowel wall thickening: significance and pitfalls of interpretation. *Am J Roentgenol* 2001; **176**: 1105–1116.

5 Balthazar EJ. CT of the gastrointestinal tract: principles and interpretation. *Am J Roentgenol* 1991; **156**: 23–32.

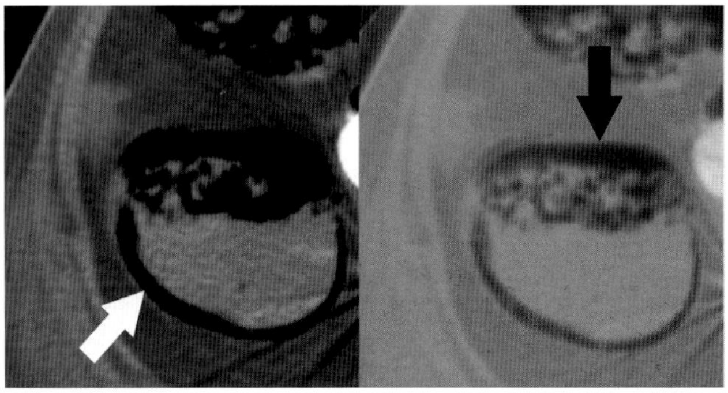

Figure 60.1 Axial non-enhanced CT image (shown as a photomontage at both soft-tissue and colon windows) in a 43 year old man with impaired bowel motility secondary to cystic fibrosis. Gas trapped against the mucosal surface of the bowel mimics pneumatosis (white arrow). Note that mucosal dissection cannot be identified and that the abnormality does not extend circumferentially to involve the non-dependent wall of the bowel (black arrow), as would be expected with true pneumatosis.

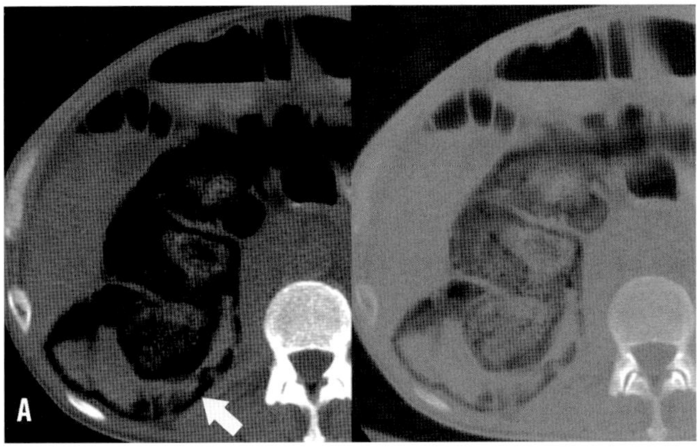

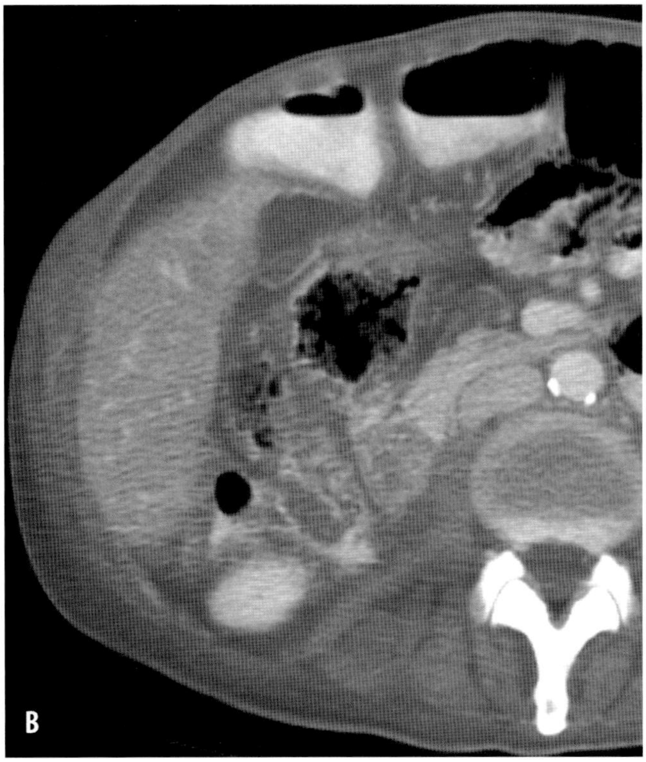

Figure 60.2 **A.** Axial non-enhanced CT image (shown as a photomontage at both soft-tissue and colon windows) in a 56 year old man with impaired bowel motility secondary to pneumonia and narcotic use. Gas trapped against the mucosal surface of the bowel by viscous bowel content creates in the distended right colon an appearance that resembles pneumatosis (arrow). **B.** Axial contrast-enhanced CT image three days later shows the right colon is no longer distended and the pseudopneumatosis has resolved.

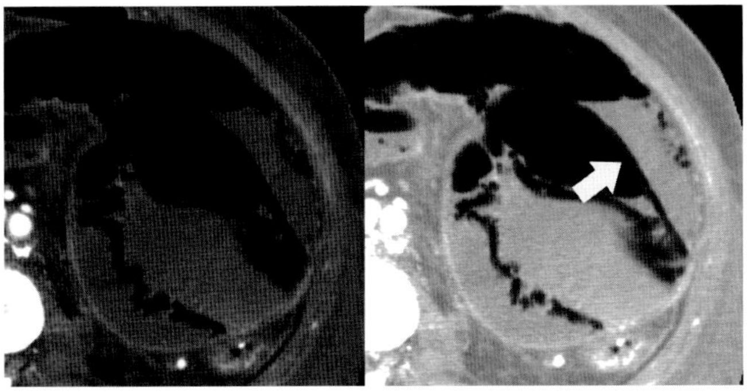

Figure 60.3 Axial non-enhanced CT image (shown as a photomontage at both soft-tissue and colon windows) in a 78 year old woman with ischemic pneumatosis. Circumferential mucosal dissection is visible, with dissected mucosa (arrow) visible on the non-dependent surface of the bowel.

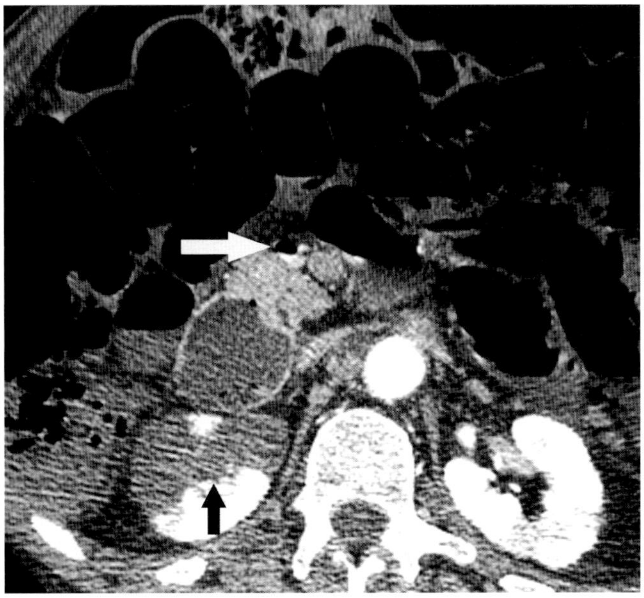

Figure 60.4 Axial contrast-enhanced CT image in an 82 year old man with atrial fibrillation and acute embolic mesenteric ischemia. Air in the superior mesenteric vein (white arrow) confirms the diagnosis of mesenteric ischemia. Venous gas can help distinguish true pneumatosis from pseudopneumatosis. A co-existent embolic infarct (black arrow) is seen in the right kidney.

CASE

61 Meckel's diverticulitis

Imaging description

Meckel's diverticulum, named after the German anatomist Johann Friedrich Meckel who first described it in 1809, is a blind sac arising from the antimesenteric border of distal ileum approximately 40–100 cm upstream from the ileocecal junction that occurs as a congenital anomaly in about 2% of the population [1, 2]. The diverticulum is a developmental remnant due to incomplete regression of the vitelline (or omphalomesenteric) duct that connects the midgut to the yolk sac during early fetal life. The diverticulum can become inflamed and present as an acute abdomen. Historically, a preoperative diagnosis of complicated Meckel's diverticulum has only been made in 6 to 10% of cases [3, 4], but with the emergence of CT the diagnosis of Meckel's diverticulitis can be made with high accuracy. At CT, an inflamed Meckel's diverticulum appears as a blind-ending pouch of variable size with mural thickening and surrounding mesenteric inflammation (Figures 61.1 and 61.2) [5]. That is, an inflamed Meckel's diverticulum resembles an inflamed appendix, except that it is attached to the distal ileum rather than the cecum.

Importance

The total lifetime complication rate of a Meckel's diverticulum is approximately 4% [6], and common symptomatic complications include inflammation, bleeding, obstruction, and inversion with intussusception. Inflammation may result from obstruction with infection (like acute appendicitis), torsion, or peptic inflammation secondary to ectopic gastric mucosa within the diverticulum [3].

Typical clinical scenario

Meckel's diverticulum occurs with equal frequency in both sexes, but symptoms from complications are more common in male patients. Meckel's diverticulitis typically presents as an acute abdomen of uncertain etiology, and in one series, appendicitis was the most common preoperative diagnosis [4].

Differential diagnosis

While Meckel's diverticulitis is a clinical mimic of acute appendicitis, the CT distinction of these two conditions should be straightforward because in Meckel's diverticulitis the inflamed structure attaches to the distal ileum and not the cecum and a normal appendix can usually be identified separately. Conceivably, Meckel's diverticulitis located in the midline anteriorly might suggest an infected urachal cyst, but urachal remnants are generally closely related to the bladder dome rather than the distal ileum. Occasionally other congenital or acquired diverticula of the small bowel may become inflamed, and such small bowel diverticulitis could mimic Meckel's diverticulitis at CT [7, 8]. Finally, inflammatory changes and fluid related to the distal ileum might also suggest a localized contained small bowel perforation or complicated Crohn's disease [5].

Teaching point

A tubular enhancing structure arising from the distal ileum with adjacent fatty infiltration seen at CT in a patient with acute abdominal pain is suggestive of Meckel's diverticulitis.

REFERENCES

1 Rossi P, Gourtsoyiannis N, Bezzi M, et al. Meckel's diverticulum: Imaging diagnosis. *Am J Roentgenol* 1996; **166**: 567–573.

2 Elsayes KM, Menias CO, Harvin HJ, Francis IR. Imaging manifestations of Meckel's diverticulum. *Am J Roentgenol* 2007; **189**: 81–88.

3 Yamaguchi M, Takeuchi S, Awazu S. Meckel's diverticulum: investigation of 600 patients in Japanese literature. *Am J Surg* 1978; **136**: 247–249.

4 Weinstein EC, Cain JC, ReMine WH. Meckel's diverticulum: 55 years of clinical and surgical experience. *JAMA* 1962; **182**: 131–133.

5 Bennett GL, Birnbaum BA, Balthazar EJ. CT of Meckel's diverticulitis in 11 patients. *Am J Roentgenol* 2004; **182**: 625–629.

6 Leijonmarck CE, Bonman-Sandelin K, Frisell J, Räf L. Meckel's diverticulum in the adult. *Br J Surg* 1986; **73**: 146–149.

7 Greenstein S, Jones B, Fishman EK, Cameron JL, Siegelman SS. Small-bowel diverticulitis: CT findings. *Am J Roentgenol* 1986; **147**: 271–274.

8 Graña L, Pedraja I, Mendez R, Rodríguez R. Jejuno-ileal diverticulitis with localized perforation: CT and US findings. *Eur J Radiol* 2009; **71**: 318–323.

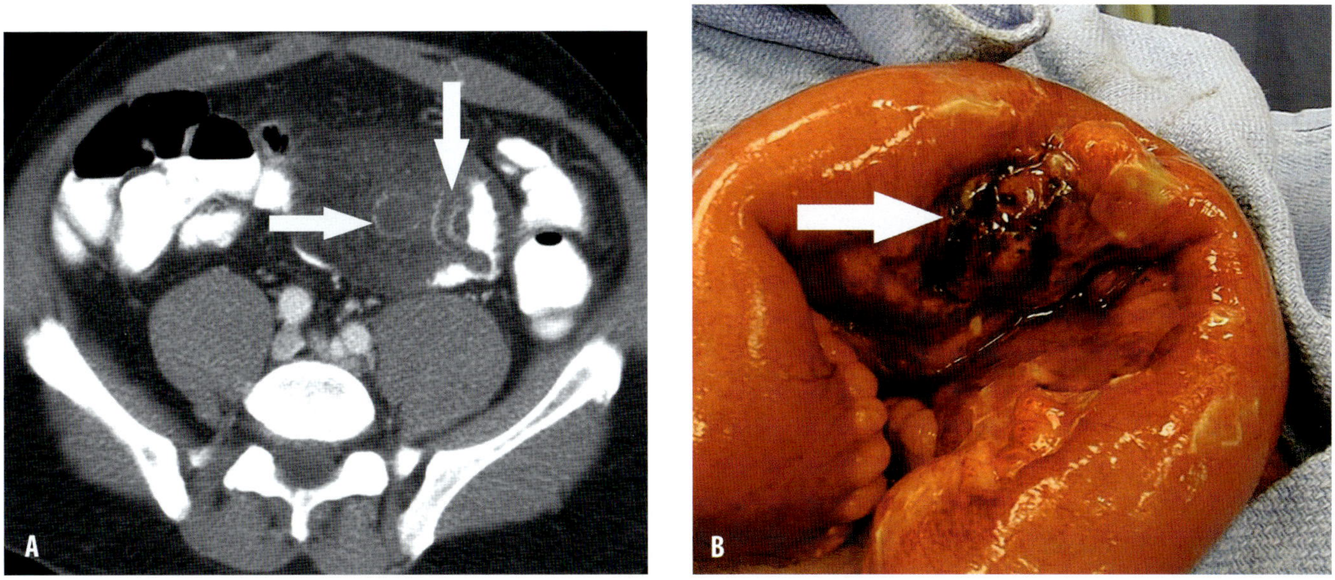

Figure 61.1 A. Axial contrast-enhanced CT image in a 32 year old otherwise healthy woman with a 2-day history of acute abdominal pain shows a tubular structure (vertical arrow) arising from the distal ileum and surrounded by extensive fat stranding. An incomplete ring of enhancement (horizontal arrows) lies adjacent to but apparently disconnected from the tubular structure. A diagnosis of perforated Meckel's diverticulitis was suggested. **B.** Intra-operative photograph confirming a gangrenous, perforated, and inflamed Meckel's diverticulum (arrow).

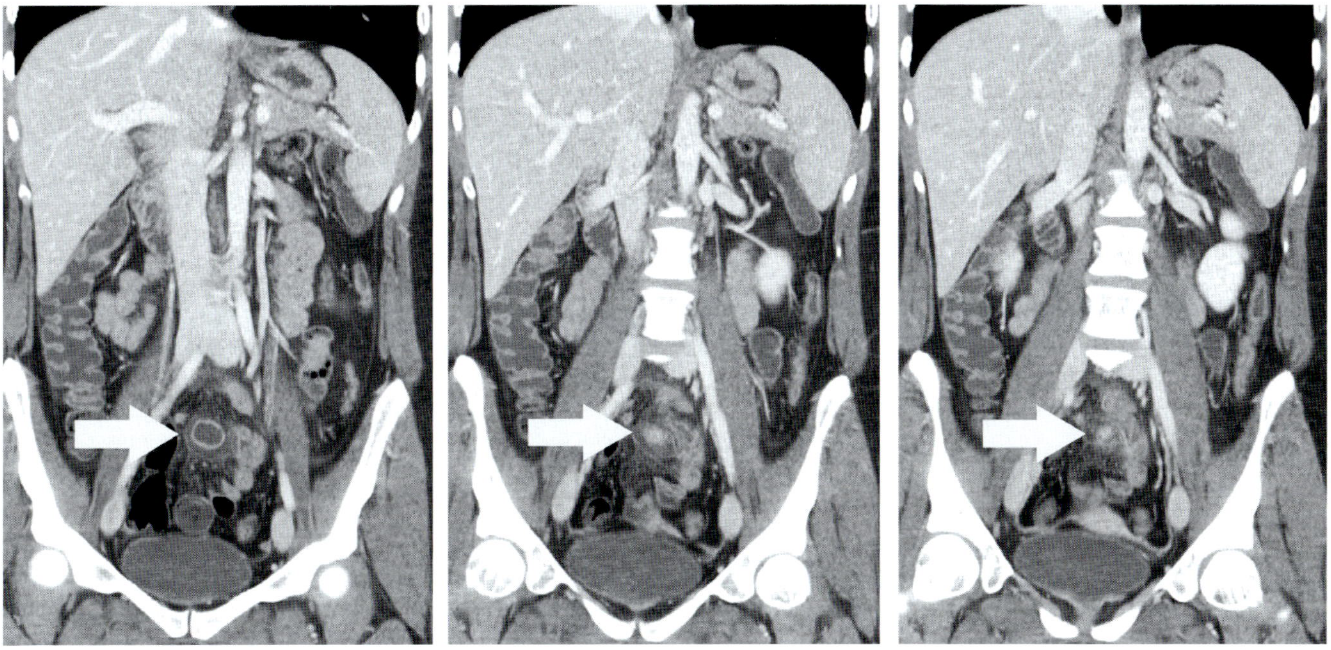

Figure 61.2 Photomontage of three coronal reformatted contrast-enhanced CT images arranged from anterior to posterior in a 26 year old otherwise healthy woman with acute abdominal pain shows a rim-enhancing fluid-filled structure (arrow) arising from the distal small bowel. Surgery confirmed a diagnosis of acute Meckel's diverticulitis.

CASE 62

Small bowel intussusception

Imaging description

Intussusception is the telescoping or prolapse of one portion of the bowel into the adjacent downstream segment. Small bowel intussusception appears at CT as a targetlike or sausage-shaped intraluminal soft-tissue mass with fat attenuation due to invaginated mesentery; the so-called bowel-within-bowel appearance (Figures 62.1–62.4) [1–4]. Intussusception is the commonest cause of acute intestinal obstruction in young children, and pediatric intussusception can usually be successfully managed by pressure reduction using barium, air, or saline enema [5]. Conversely, small bowel intussusception in adults is sometimes detected unexpectedly at CT and may result in considerable uncertainty as to appropriate management.

Importance

Adult small bowel intussusception has traditionally been regarded as a surgical condition, because up to 90% of cases are said to be associated with a lead point tumor or other abnormality [6–12]. These older reports related to the frequency of lead point pathology are generally based on surgical series where patients presented with obstructive symptoms and the diagnosis of intussusception was made intra-operatively. These results are not applicable to adult intussusception identified on CT, since CT detects many subclinical cases. More recent studies confirm many CT-detected small bowel intussusceptions are transient. The reported rates of small bowel intussusception that is self-limiting in the CT era range from 79% (23 of 29) to 84% (31 of 37) to 96% (143 of 149) [13–15].

Typical clinical scenario

Adult small bowel intussusception is occasionally seen unexpectedly at CT, with a reported frequency in large unselected series of patients undergoing abdominal CT ranging from 1 in 1865 to 1 in 2557 [14, 15].

Differential diagnosis

The fact that many cases of CT-detected adult small bowel intussusception are self-limiting highlights the need for a reliable method of distinguishing surgical cases with lead point pathology from those that lack lead point pathology and can be safely managed conservatively. Direct visualization of a lead point mass is helpful, but many lead point masses cannot be seen due to the complex appearance of invaginated edematous bowel loops [16]. In one study, a lead point mass was seen preoperatively in only two of six surgically confirmed neoplastic intussusceptions [14]. Several studies have investigated other clinical or imaging features that distinguish surgical from self-limiting cases of adult small bowel intussusception, and consistently show that short intussusceptions (less than 3.5 cm in length) are likely to be self-limiting, while longer intussusceptions may be self-limiting or surgical [9, 14, 15]. The good news is that a reported 54% (20 of 37) of small bowel intussusceptions are short, and based on published data can safely be described as presumably self-limiting [14]. The bad news is that the management of longer intussusceptions that do not go to surgery is not established, and the role of modalities such as interval CT, small bowel follow-through examination, enteroclysis, and capsule endoscopy is undefined [16].

Teaching point

Adult small bowel intussusception is seen in approximately 1 in 2000 unselected abdominal CT examinations. Intussusception length is the main factor in distinguishing the majority of CT-detected small bowel intussusceptions that are self-limiting from the minority that require surgery; an intussusception less than 3.5 cm in length is likely to be self-limiting. An intussusception more than 3.5 cm in length may be self-limiting or surgical in nature, and management should be tailored to the individual clinical situation.

REFERENCES

1 Curcio CM, Feinstein RS, Humphrey RL, et al. Computed tomography of entero-enteric intussusception. *J Comput Tomogr* 1982; **6**: 969–974.

2 Gourtsoyiannis NC, Papakonstantinou O, Bays D, Malamas M. Adult enteric intussusception: additional observations on enteroclysis. *Abdom Imaging* 1994; **19**: 11–17.

3 Lorigan JG, DuBrow RA. The computed tomographic appearances and clinical significance of intussusception in adults with malignant neoplasms. *Br J Radiol* 1990; **63**: 257–262.

4 Merine DS, Fishman EK, Jones B. Enteroenteric intussusception: CT findings in nine patients. *Am J Roentgenol* 1987; **148**: 1129–1132.

5 Huppertz HI, Soriano-Gabarró M, Grimprel E, et al. Intussusception among young children in Europe. *Pediatr Infect Dis J* 2006; **25**: S22–S29.

6 Agha FP. Intussusception in adults. *Am J Roentgenol* 1986; **146**: 527–531.

7 Azar T, Berger DL. Adult intussusception. *Ann Surg* 1997; **226**: 134–138.

8 Begos DG, Sandor A, Modlin IM. The diagnosis and management of adult intussusception. *Am J Surg* 1997; **173**: 88–94.

9 Desai N, Wayne MG, Taub PJ, et al. Intussusception in adults. *Mt Sinai J Med* 1999; **66**: 336–340.

10 Khurrum BM, Hussain S, Wise M, et al. Controversy in the treatment of adult long ileocolic intussusception: case report. *Am Surg* 2000; **66**: 742–743.

11 Nagorney DM, Sarr MG, McIlrath DC. Surgical management of intussusception in the adult. *Ann Surg* 1981; **193**: 230–236.

12 Weillbaecher D, Bolin JA, Hearn D, Ogden W. Intussusception in adults: review of 160 cases. *Am J Surg* 1971; **121**: 531–535.

13 Warshauer DM, Lee JK. Adult intussusception detected at CT or MR imaging: clinical-imaging correlation. *Radiology* 1999; **212**: 853–860.

14 Lvoff N, Breiman RS, Coakley FV, Lu Y, Warren RS. Distinguishing features of self-limiting adult small-bowel intussusception identified at CT. *Radiology* 2003; **227**: 68–72.

15 Rea JD, Lockhart ME, Yarbrough DE, et al. Approach to management of intussusception in adults: a new paradigm in the computed tomography era. *Am Surg* 2007; **73**: 1098–1105.

16 Kim YH, Blake MA, Harisinghani MG, et al. Adult intestinal intussusception: CT appearances and identification of a causative lead point. *Radiographics* 2006; **26**: 733–744.

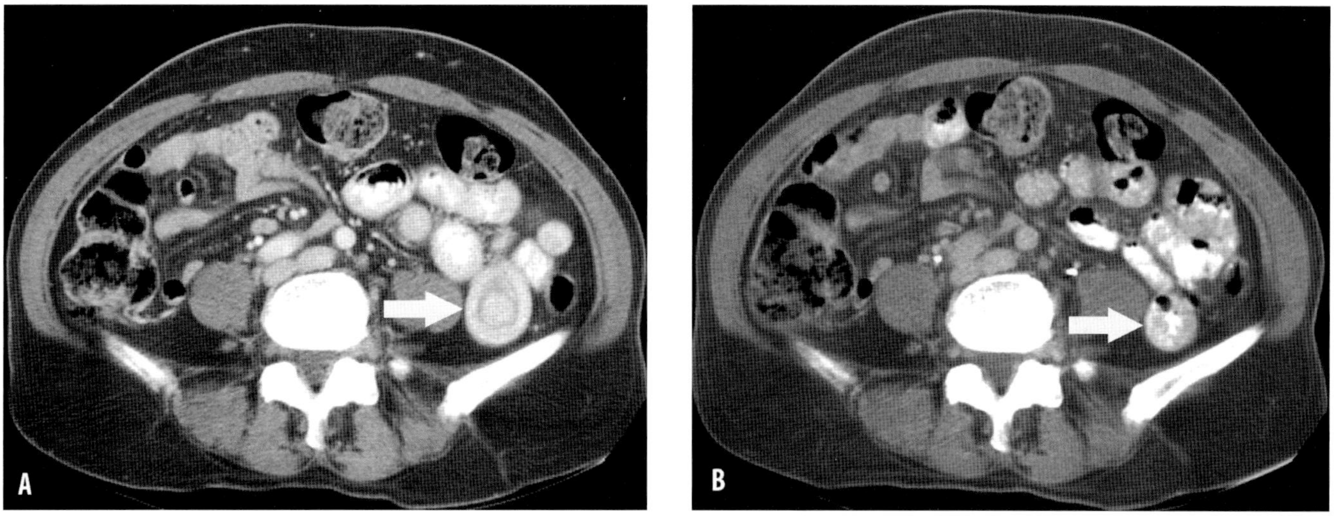

Figure 62.1 **A.** Axial contrast-enhanced CT image in a 57 year old woman with vague abdominal pain shows a typical small bowel intussusception (arrow), with a "bowel-within-bowel" appearance. **B.** Axial contrast-enhanced CT image obtained a few minutes later during the same study as Figure 62.1A shows the intussusception has spontaneously resolved, consistent with a transient and self-limiting small bowel intussusception.

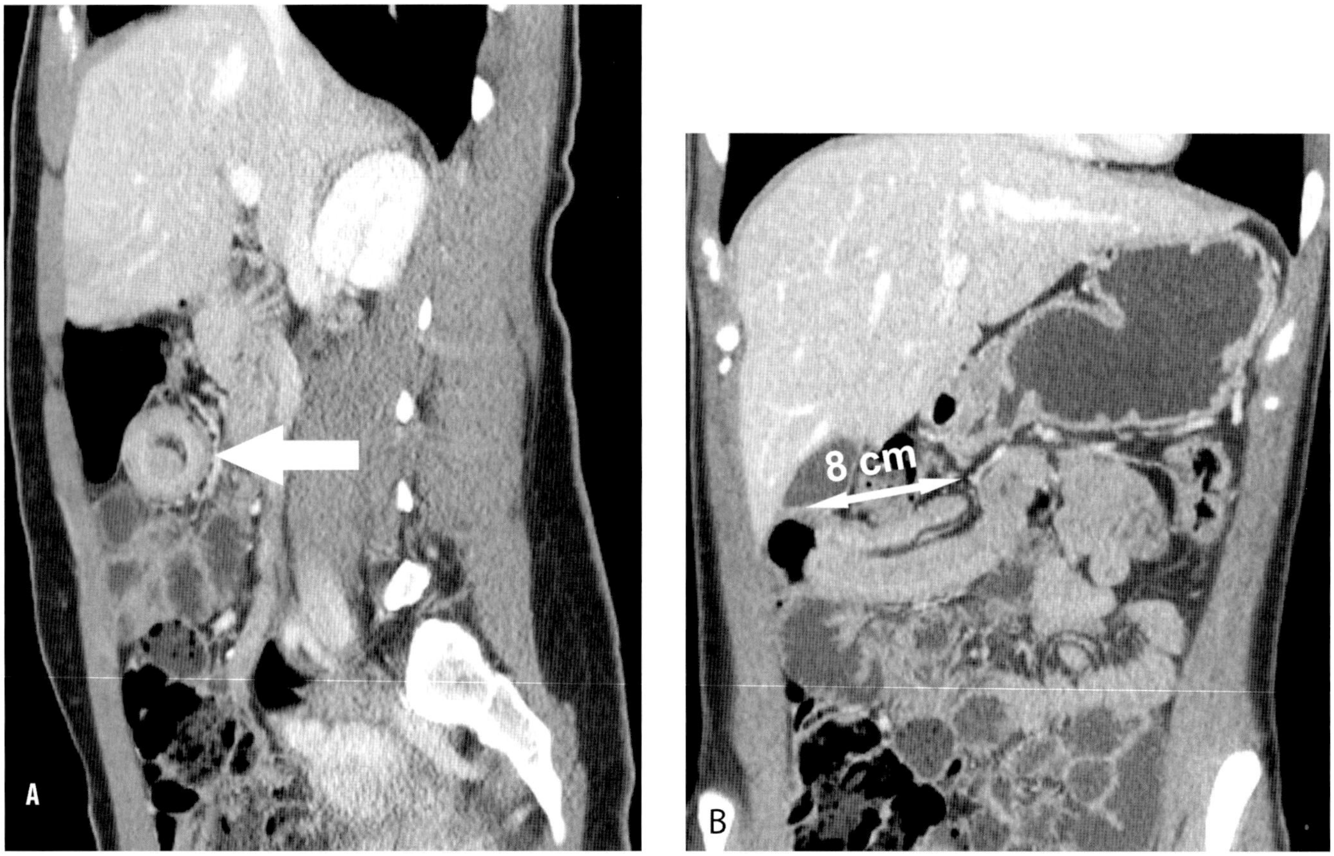

Figure 62.2 A. Sagittal reformatted contrast-enhanced CT image in a 26 year old woman with severe episodic epigastric pain shows the typical "bowel-within-bowel" appearance of an intussusception (arrow). Note the finding of intraluminal fat due to mesenteric invagination. **B.** Coronal reformatted contrast-enhanced CT image shows the intussusception is approximately 8 cm in length. At diagnostic laparoscopy, the small bowel was unremarkable, with no visible or palpable tumor or other abnormality. The intussusception was presumed to have reduced spontaneously. Subsequent endoscopy showed multiple duodenal ulcers.

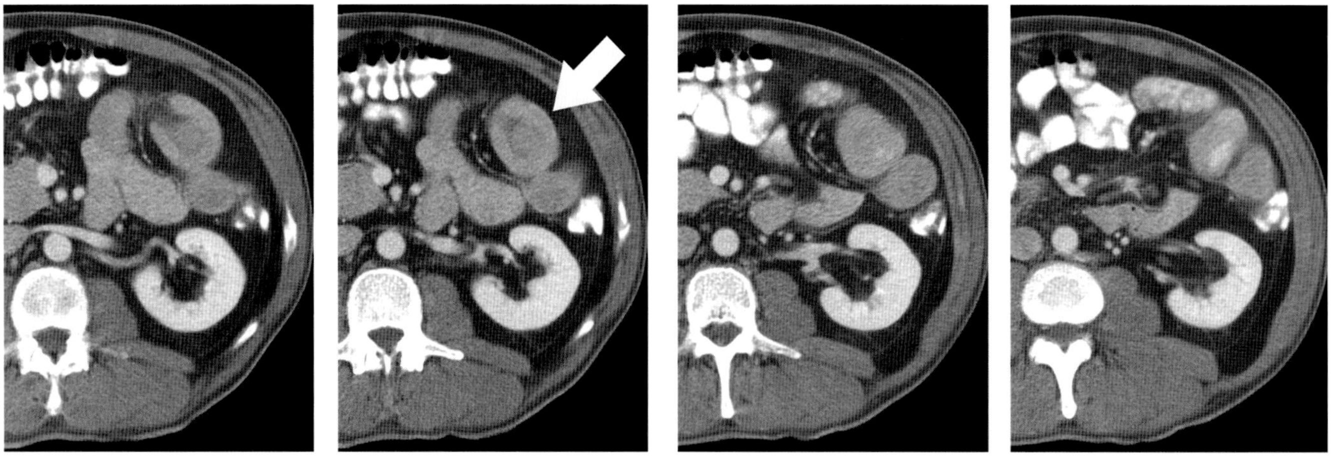

Figure 62.3 Photomontage of four successive axial contrast-enhanced CT images in a 52 year old man with chronic gastrointestinal hemorrhage and human immunodeficiency virus infection shows a relatively short intussusception (arrow) that was measured as 4 cm in length. No lead point mass is evident. The patient had undergone resection of a right maxillary sinus plasmacytoma two years before. Surgical exploration demonstrated an intussuscepting 3 cm lead point mass that was found to be a plasmacytoma on pathology examination. Lead point pathology may be present, but be invisible by imaging, even in relatively inconsequential-appearing intussusceptions.

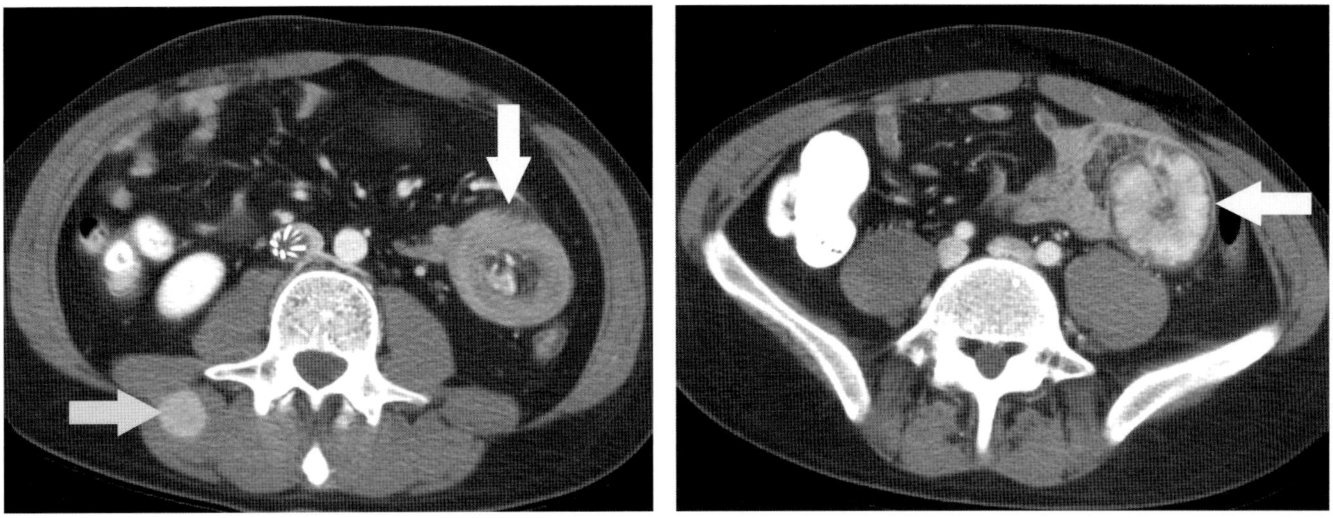

Figure 62.4 Photomontage of two axial contrast-enhanced CT images in a 30 year old man with widely metastatic osteosarcoma shows an intussusception (vertical white arrow) secondary to a partially calcified lead point mass (horizontal white arrow). Note the soft-tissue metastasis (gray arrow). A 6 cm intussuscepting metastasis was subsequently resected with palliative intent.

CASE 63 Pseudoappendicitis

Imaging description

The primary CT signs of acute appendicitis are appendiceal dilatation (the upper limit of appendiceal dilatation has been variably reported, but is generally regarded as between 6 and 10 mm) with or without periappendiceal fat stranding or fluid [1]. Occasionally, such CT findings are due to other causes, including normal variation in appendiceal diameter, appendiceal dilatation associated with malignancy, hematoma, spontaneously resolving acute appendicitis, Crohn's appendicitis, ischemic appendicitis, granulomatous appendicitis, lymphoid hyperplasia, mucocele, appendiceal tumor, and right-sided colon cancer (Figures 63.1–63.9) [2–10]. In such settings, the term pseudoappendicitis can be appropriately applied to the imaging findings.

Importance

A false positive diagnosis of acute appendicitis may lead to unnecessary appendectomy.

Typical clinical scenario

Pseudoappendicitis is rare. In patients with abdominal pain, spontaneously resolving acute appendicitis, Crohn's appendicitis, ischemic appendicitis, and granulomatous appendicitis should be considered as potential additional causes of appendiceal dilatation with or without periappendiceal fat stranding or fluid. In patients without abdominal pain, considerations include lymphoid hyperplasia, mucocele, appendiceal tumor, and right-sided colon cancer.

Differential diagnosis

There are several forms of pseudoappendicitis:

- Appendiceal diameter is variable, and a diameter over 6 mm has been reported to occur in up to 42% of normal appendixes, either with visible or indiscernible content [2]. Such variation may account for a contemporary false positive rate of 3% for the diagnosis of acute appendicitis by CT [11]. It is also possible that a dilated appendix on CT in patients in whom a diagnosis of acute appendicitis is rejected after surgical consultation is still due to spontaneously resolving acute appendicitis, because up to 38% of these patients ultimately require appendectomy [5].
- Asymptomatic thickening of the appendix may also be seen in patients with extra-intestinal cancers, although the histopathological basis of this association is unknown [3].
- Hematoma has been reported as a cause of appendiceal thickening in hemophilia, and could presumably occur in other bleeding diatheses [4].
- Inflammatory conditions other than typical acute appendicitis may cause abnormalities at CT, including Crohn's appendicitis, ischemic appendicitis, granulomatous appendicitis, and lymphoid hyperplasia. Isolated granulomatous inflammation of the appendix is extremely rare. Etiology is unknown, as is the potential relationship to or distinction from isolated Crohn's disease of the appendix [12]. The relative frequency of these other inflammatory conditions was indicated in a study of 106 patients undergoing appendectomy for presumed appendicitis [9]. Ten patients (9%) had chronic appendiceal inflammatory conditions, including three with lymphoid hyperplasia and two with granulomatous appendicitis. In another series of almost 1400 appendectomy specimens, 25 patients (1.8%) were identified who presented with appendicitis as the initial manifestation of Crohn's disease [13]. Review of the preoperative radiologic findings suggested visualization of a markedly thickened appendiceal wall with a patent or irregularly narrowed lumen supporting the diagnosis of Crohn's appendicitis over suppurative appendicitis.
- Mucocele of the appendix is a descriptive term that implies a dilated appendiceal lumen caused by abnormal accumulation of mucus. The prevalence of mucoceles in appendectomy specimens is only 0.2 to 0.3% [14]. Appendiceal mucoceles may be idiopathic, due to benign neoplastic epithelium similar to that seen in villous adenomas and adenomatous polyps of the colon, or due to malignant neoplastic epithelium similar to that seen in adenocarcinomas of the colon [15].
- Right-sided colon cancer is associated with both appendicitis and CT changes that mimic appendicitis, presumably reflecting some degree of obstruction with "back pressure" changes in the appendix [10]. A reported 0.85% (16/1873) of patients with acute appendicitis harbor a colorectal cancer, and routine postoperative optical colonoscopy has been recommended for patients over 40 years with acute appendicitis [16].

Teaching point

Appendiceal dilatation to a diameter of over 6 to 10 mm with or without periappendiceal fat stranding or fluid is not always indicative of acute appendicitis; clinical correlation and inspection of the right colon are important steps when considering the various forms of pseudoappendicitis.

REFERENCES

1 Birnbaum BA, Wilson SR. Appendicitis at the Millenium. *Radiology* 2000; **215**: 337–348.

2 Tamburrini S, Brunetti A, Brown M, Sirlin CB, Casola G. CT appearance of the normal appendix in adults. *Eur J Radiol* 2005; **15**: 2096–2103.

3 Webb EM, Joe BN, Coakley FV, et al. Appendiceal wall thickening at CT in asymptomatic patients with extraintestinal malignancy may mimic appendicitis. *Clin Imaging* 2009; **33**: 200–203.

4 McCoy HE 3rd, Kitchens CS. Small bowel hematoma in a hemophiliac as a cause of pseudoappendicitis: diagnosis by CT imaging. *Am J Hematol* 1991; **38**: 138–139.

5 Stengel JW, Webb EW, Poder L, et al. Clinical outcome in patients with an apparent false positive diagnosis of acute appendicitis at CT who do not undergo surgery. Radiological Society of North America Scientific Assembly and Annual Meeting Program 2008; 519.
6 Cobben LP, de Van Otterloo AM, Puylaert JB. Spontaneously resolving appendicitis: frequency and natural history in 60 patients. *Radiology* 2000; **215**: 349–352.
7 Moyana TN. Necrotizing arteritis of the vermiform appendix. A clinicopathologic study of 12 cases. *Arch Pathol Lab Med* 1988; **112**: 738–741.
8 Bak M, Andersen JC. Crohn's disease limited to the vermiform appendix. *Acta Chir Scand* 1987; **153**: 441–446.
9 Checkoff JL, Wechsler RJ, Nazarian LN. Chronic inflammatory appendiceal conditions that mimic acute appendicitis on helical CT. *Am J Roentgenol* 2002; **179**: 731–734.
10 Watchorn RE, Poder L, Wang ZJ, et al. CT findings mimicking appendicitis as a manifestation of colorectal cancer. *Clin Imaging* 2009; **33**: 430–432.
11 Raman SS, Osuagwu FC, Kadell B, et al. Effect of CT on false positive diagnosis of appendicitis and perforation. *N Engl J Med* 2008; **358**: 972–973.
12 Dudley TH Jr, Dean PJ. Idiopathic granulomatous appendicitis, or Crohn's disease of the appendix revisited. *Hum Pathol* 1993; **24**: 595–601.
13 Agha FP, Ghahremani GG, Panella JS, Kaufman MW. Appendicitis as the initial manifestation of Crohn's disease: radiologic features and prognosis. *Am J Roentgenol* 1987; **149**: 515–518.
14 Dachman A, Lichtenstein J, Friedman A. Mucocele of the appendix and pseudomyxoma peritonei. *Am J Roentgenol* 1985; **144**: 923–929.
15 Madwed D, Mindelzun R, Jeffrey RB Jr. Mucocele of the appendix: imaging findings. *Am J Roentgenol* 1992; **159**: 69–72.
16 Lai HW, Loong CC, Tai LC, Wu CW, Lui WY. Incidence and odds ratio of appendicitis as first manifestation of colon cancer: a retrospective analysis of 1873 patients. *J Gastroenterol Hepatol* 2006; **21**: 1693–1696.

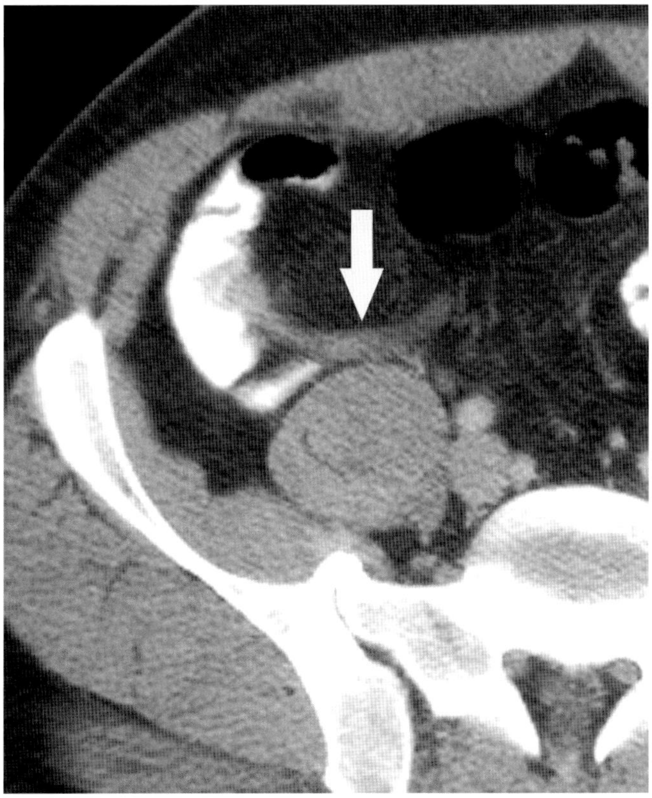

Figure 63.1 Axial non-enhanced CT image in a 38 year old asymptomatic potential renal donor. The appendix (arrow) measures 10 mm in diameter. In the absence of symptoms, this finding presumably reflects normal variation in appendiceal diameter.

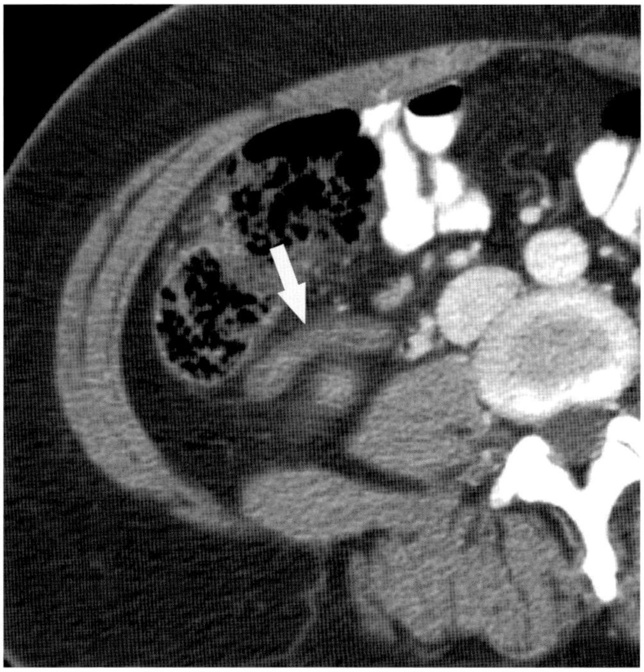

Figure 63.2 Axial contrast-enhanced CT image in a 45 year old woman with newly diagnosed breast cancer. The appendix (arrow) measures 12 mm in diameter. Appendiceal dilatation in the absence of acute appendicitis has been reported in association with malignancy. The histopathological basis of this association is unknown.

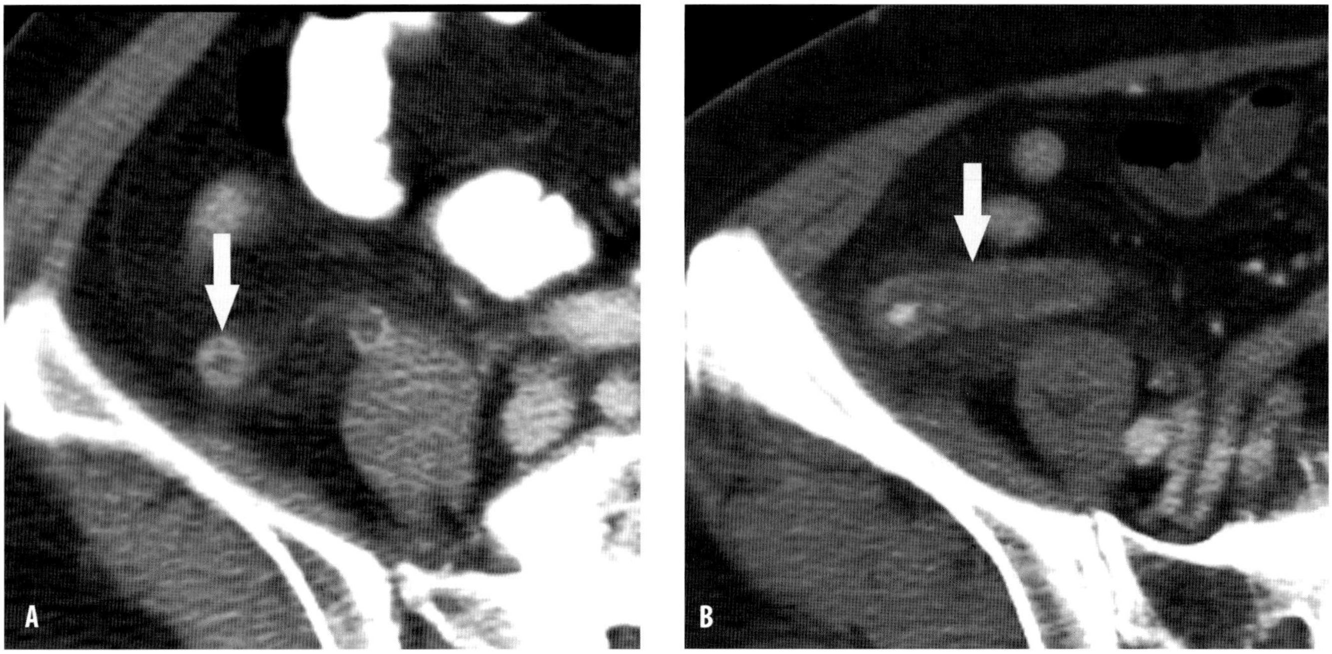

Figure 63.3 **A.** Axial contrast-enhanced CT image in a 68 year old woman with acute abdominal pain that was ascribed to "gastroenteritis" after surgical consultation. The appendix (arrow) measures 12 mm in diameter, and there is associated periappendiceal stranding. **B.** Axial contrast-enhanced CT image performed 16 months later because of recurrent abdominal pain. The appendix (arrow) is dilated to 15 mm with more marked periappendiceal stranding. On this occasion, appendectomy was performed and confirmed acute appendicitis. The sequence of events strongly suggests that the initial presentation reflected an episode of self-limiting acute appendicitis. Some "false positive" diagnoses of acute appendicitis based on surgical evaluation as the reference standard may actually be true positives!

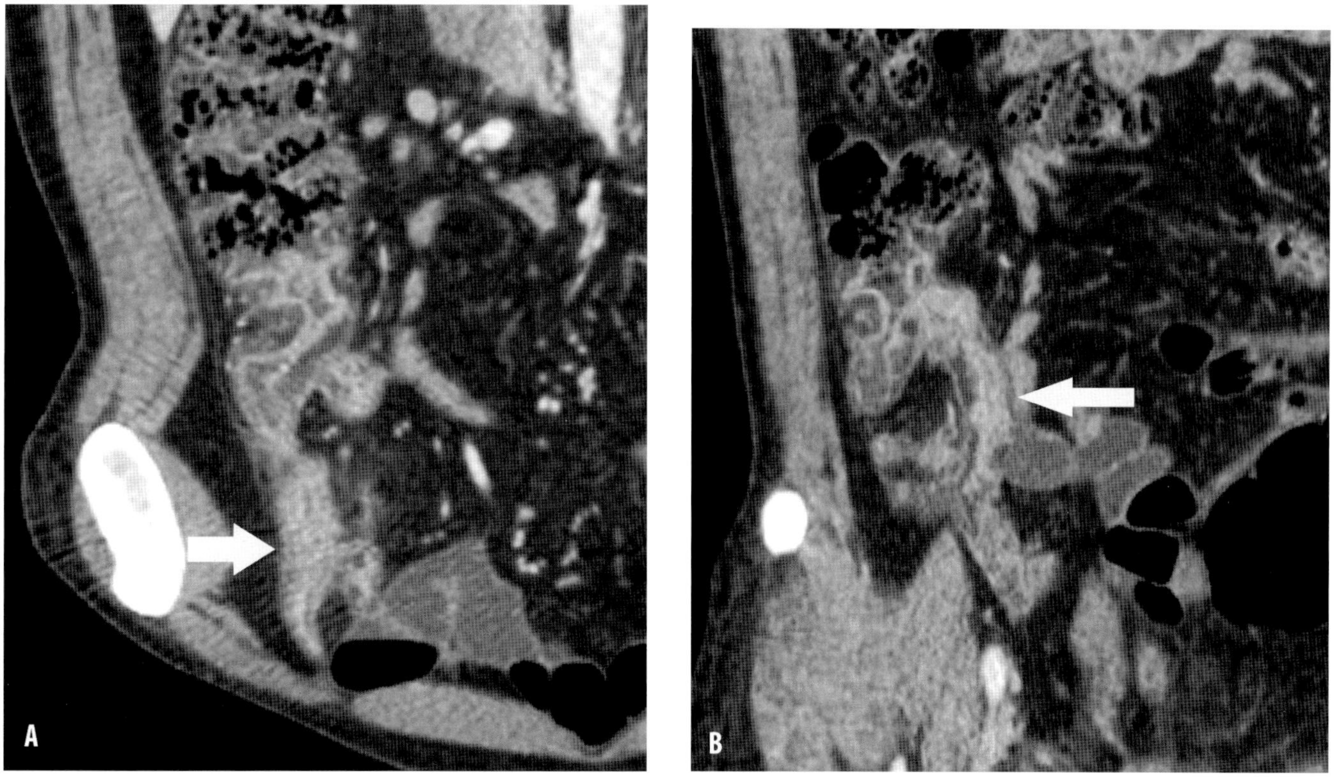

Figure 63.4 **A.** Curved planar coronal reformatted CT image in a 29 year old man with known ileocecal Crohn's disease and right lower quadrant pain. The appendix (arrow) is thickened to a diameter of 14 mm, with subtle surrounding periappendiceal infiltration. **B.** Curved planar coronal reformatted CT image in an adjacent plane shows wall thickening of the terminal ileum (arrow) due to Crohn's disease. The findings were considered consistent with Crohn's appendicitis, and the patient was managed medically with symptomatic resolution.

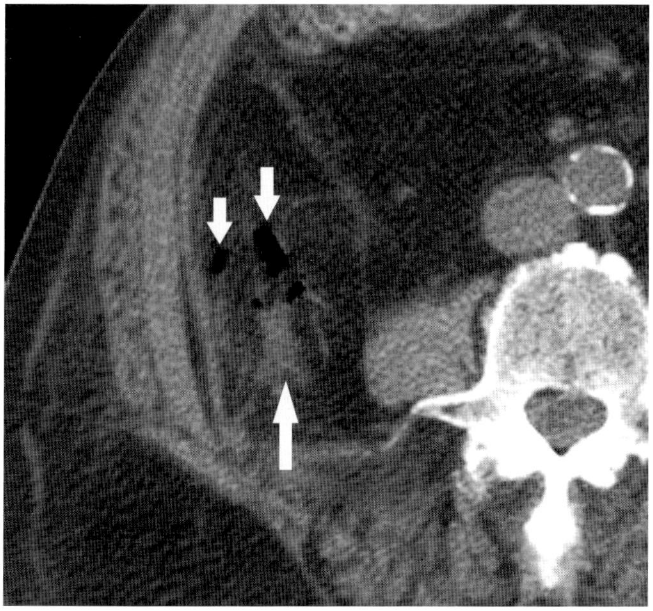

Figure 63.5 Axial non-enhanced CT image in an 87 year old man with atrial fibrillation and severe right lower quadrant pain. The appendix (upward pointing arrow) is thickened to a diameter of 13 mm and there are bubbles of periappendiceal free air (downward pointing arrows). Appendectomy demonstrated periappendicitis and perforation of the appendiceal tip, without the classical changes of appendicitis. The patient later sustained embolic events elsewhere, confirming that the appendiceal changes were likely ischemic in origin.

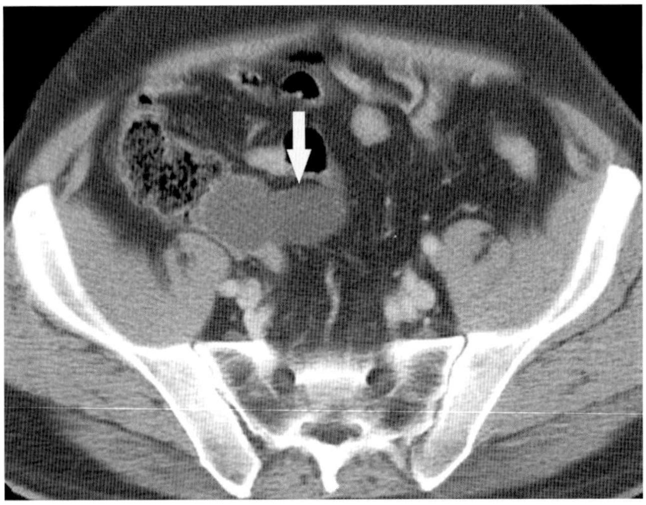

Figure 63.6 Axial contrast-enhanced CT image in a 66 year old man with mild right lower quadrant pain. The appendix (arrow) is dilated to a diameter of 26 mm, without periappendiceal stranding. Appendectomy demonstrated mucinous distension of the appendix due to focal adenomatous change in the epithelium of the appendix. Mucocele is a descriptive term used to refer to such mucinous distension of the appendix.

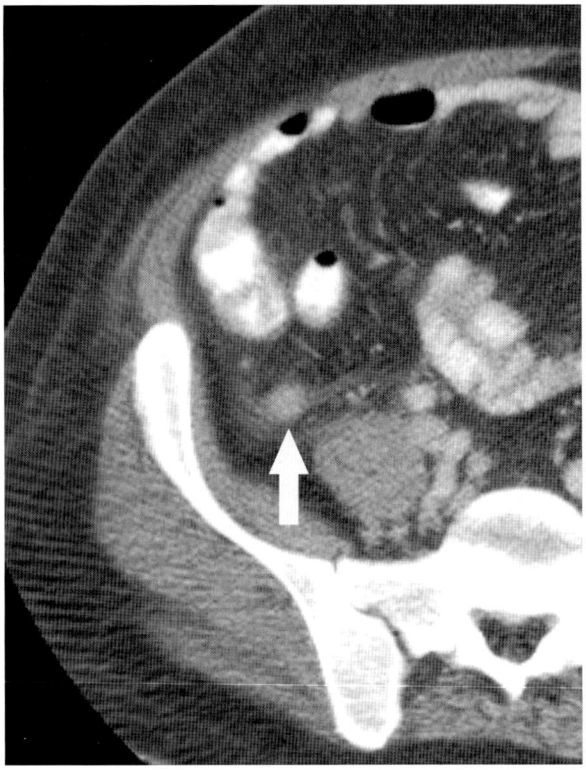

Figure 63.7 Axial contrast-enhanced CT image in a 35 year old man with lower abdominal pain and a history of human immunodeficiency virus infection. The appendix (arrow) is thickened to a diameter of 13 mm with periappendiceal fat stranding. Appendectomy was performed for presumed acute appendicitis. Histopathological review demonstrated Kaposi sarcoma of the appendix.

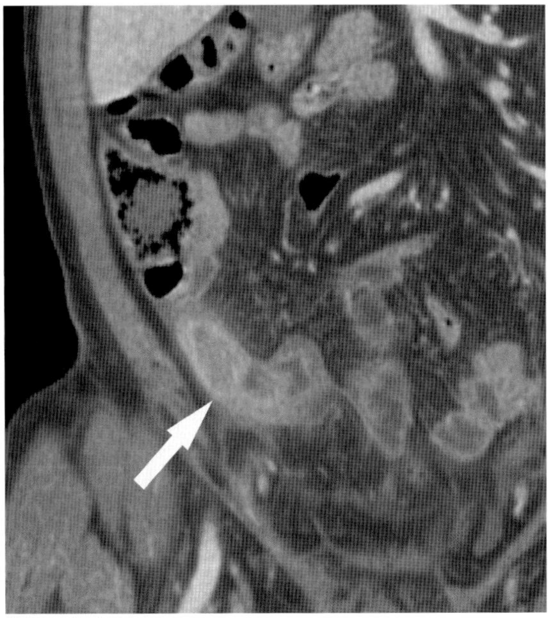

Figure 63.8 Coronal reformatted contrast-enhanced CT image in a 55 year old man with acute right lower quadrant abdominal pain and tenderness. The appendix (arrow) is thickened to a diameter of 25 mm with periappendiceal fat stranding. Histopathological review demonstrated ruptured appendiceal carcinoma.

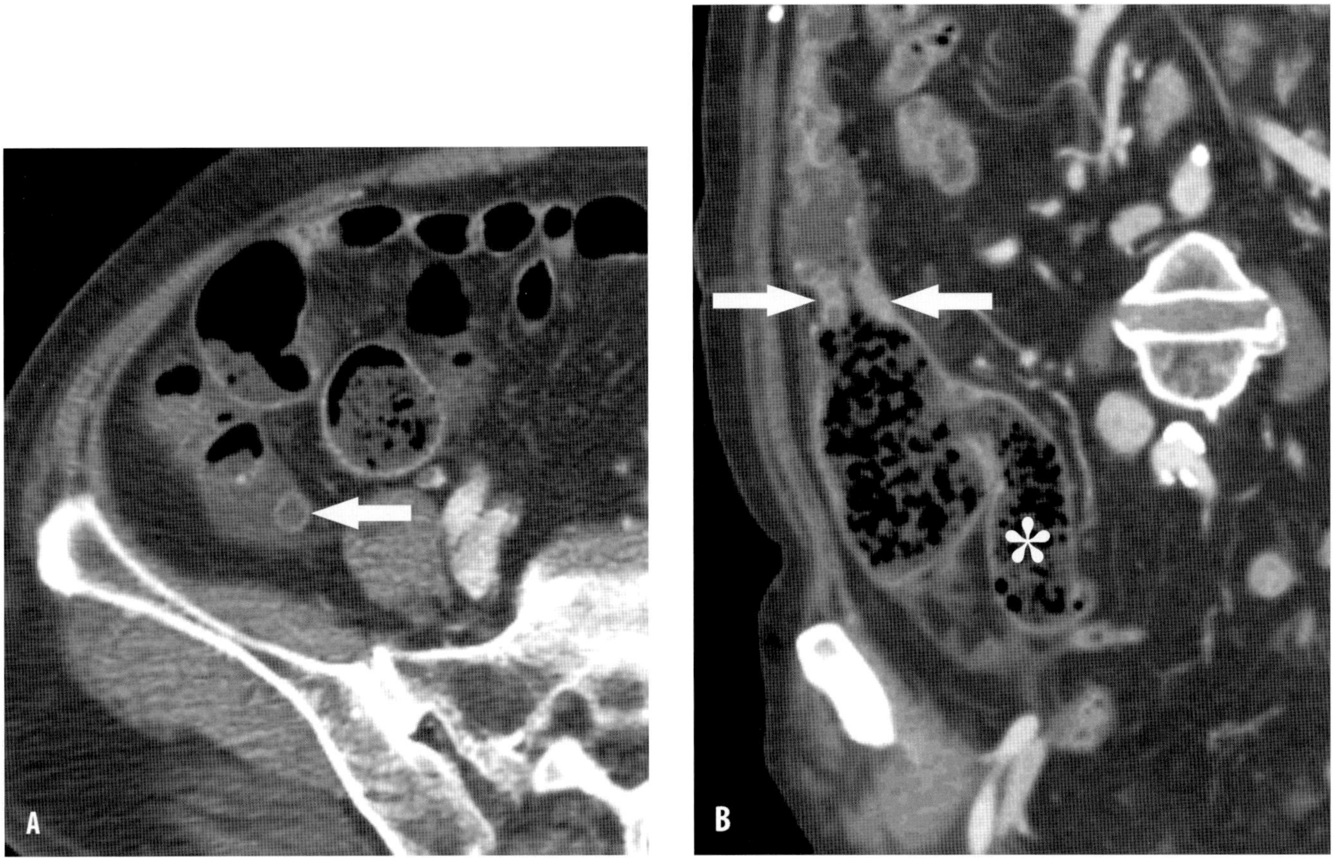

Figure 63.9 **A.** Axial contrast-enhanced CT section of the lower abdomen in a 78 year old woman with right sided abdominal pain. The appendix (arrow) is fluid-filled and distended to a diameter of 10 mm, with surrounding periappendiceal fluid. **B.** Coronal curved planar reformatted contrast-enhanced CT section shows the terminal ileum (asterisk) is mildly dilated and demonstrates the small bowel feces sign. A subtle short segment of concentric wall thickening (between arrows) is seen in the ascending colon, suggestive of primary colonic carcinoma (subsequently confirmed on endoscopy and biopsy).

CASE 64

Portal hypertensive colonic wall thickening

Imaging description

Colonic wall thickening (often defined as a colonic wall thickness greater than 6 mm, although most radiologists use expert judgment to make this diagnosis in practice) is a common finding in patients with cirrhosis and portal hypertension, and appears to reflect the congestive effect of elevated portal venous pressure (Figures 64.1 and 64.2) [1, 2]. While any or all of the large bowel may be affected, the right colon is frequently the only or dominant site of involvement [1]. It has been postulated that such preferential involvement of the right colon reflects drainage of the left colon through the inferior mesenteric vein to the splenic vein, with relative decompression through short gastric, lienorenal, or gastroesophageal venous collaterals [1]. The condition is variable and, for example, can occur in non-cirrhotic portal hypertension or affect the left colon disproportionately (Figure 64.3).

Importance

Colonic wall thickening in cirrhosis should not be mistaken for colitis or colon cancer, since this may provoke unnecessary investigation and patient anxiety. Portal hypertensive colonic wall thickening is correlated with the degree of cirrhosis and portal hypertension [3].

Typical clinical scenario

The reported frequency of colonic wall thickening in cirrhotic patients varies between 33% (21 of 63) and 37% (21 of 57) [1, 2]. As such, it is common to see colonic wall thickening as an incidental finding in cirrhotic patients.

Differential diagnosis

The term portal hypertensive colopathy is sometimes used to refer to the CT finding of colonic wall thickening in cirrhosis, however, this presupposes an association that is plausible but unproven. Strictly, portal hypertensive colopathy refers to endoscopic findings of colitis-like abnormalities (edema, erythema, granularity, friability, or spontaneous bleeding) or vascular lesions (cherry red spots, telangiectasias, or angiodysplasia-like lesions) that are commonly seen in the large bowel mucosa of patients with cirrhosis and portal hypertension [4]. It has yet to be shown that patients with portal hypertensive colopathy at endoscopy are the same patients that have colonic wall thickening at CT. Of course, it is possible for a patient with cirrhosis to have an ischemic, infectious, or inflammatory colitis (Figure 64.4), but this seems relatively rare. In a series of 57 patients with cirrhosis, 21 were found to have colonic wall thickening and only three were due to a true colitis (infection in two and ischemia in one).

Teaching point

Colonic wall thickening in patients with cirrhosis is usually due to the congestive effect of portal hypertension on the colonic wall. The absence of diarrhea helps distinguish portal hypertensive colonic wall thickening from a true colitis.

REFERENCES

1 Guingrich JA, Kuhlman JE. Colonic wall thickening in patients with cirrhosis: CT findings and clinical implications. *Am J Roentgenol* 1999; **172**: 919–924.

2 Quílez C, Palazón JM, Arenas J, et al. Colonic wall thickening in patients with cirrhosis and portal hypertension. *Rev Esp Enferm Dig* 2003; **95**: 269–272.

3 Baba Y, Hokotate H, Inoue H, Nakajo M. Correlations between colonic wall thickening in patients with virally induced cirrhosis on CT and clinical status. *J Comput Assist Tomogr* 2001; **25**: 786–791.

4 Bini EJ, Lascarides CE, Micale PL, Weinshel EH. Mucosal abnormalities of the colon in patients with portal hypertension: an endoscopic study. *Gastrointest Endosc* 2000; **52**: 511–516.

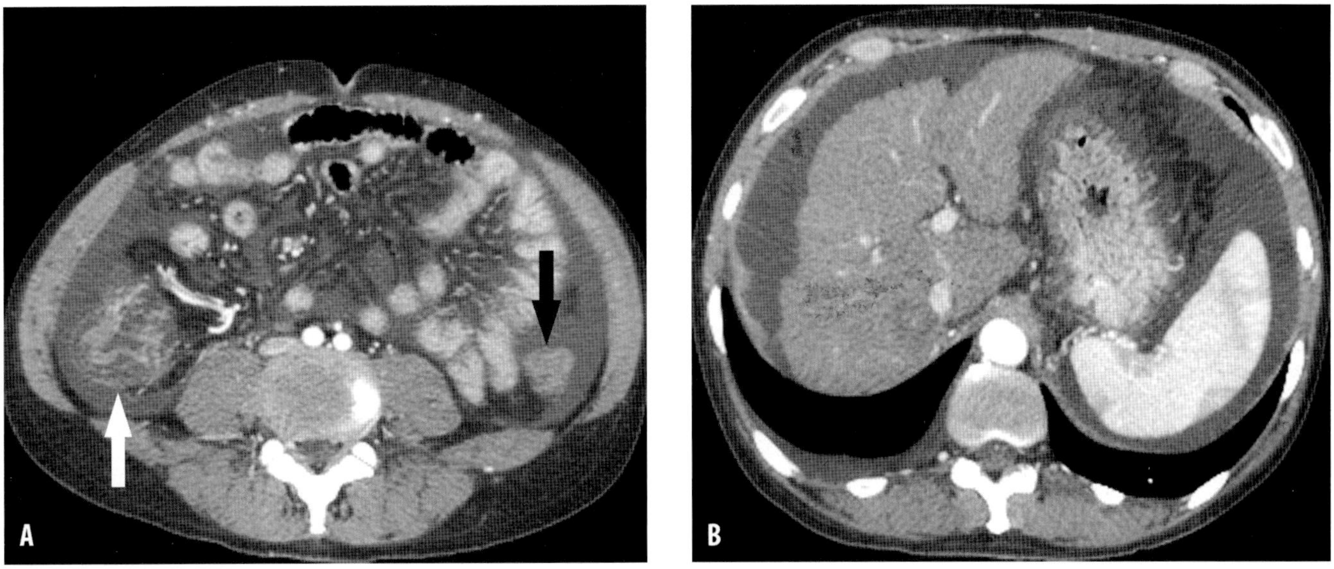

Figure 64.1 A. Axial contrast-enhanced CT image in a 49 year old man being considered for hepatic transplantation. The right colonic wall (white arrow) is diffusely thickened, while the left colon appears relatively normal (black arrow). B. Axial contrast-enhanced CT image at a more superior level demonstrates cirrhosis and ascites. The patient had no gastrointestinal symptoms or diarrhea. The findings are those of portal hypertensive colonic wall thickening.

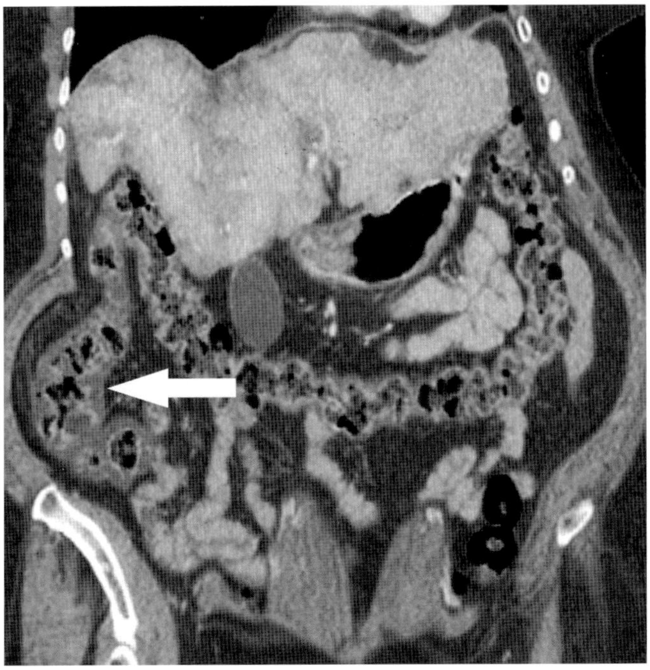

Figure 64.2 Coronal curved planar reformatted CT image in a 61 year old woman with cirrhosis due to alcohol and hepatitis C. The wall of the right colon (arrow) is diffusely thickened. The patient had no gastrointestinal symptoms or diarrhea.

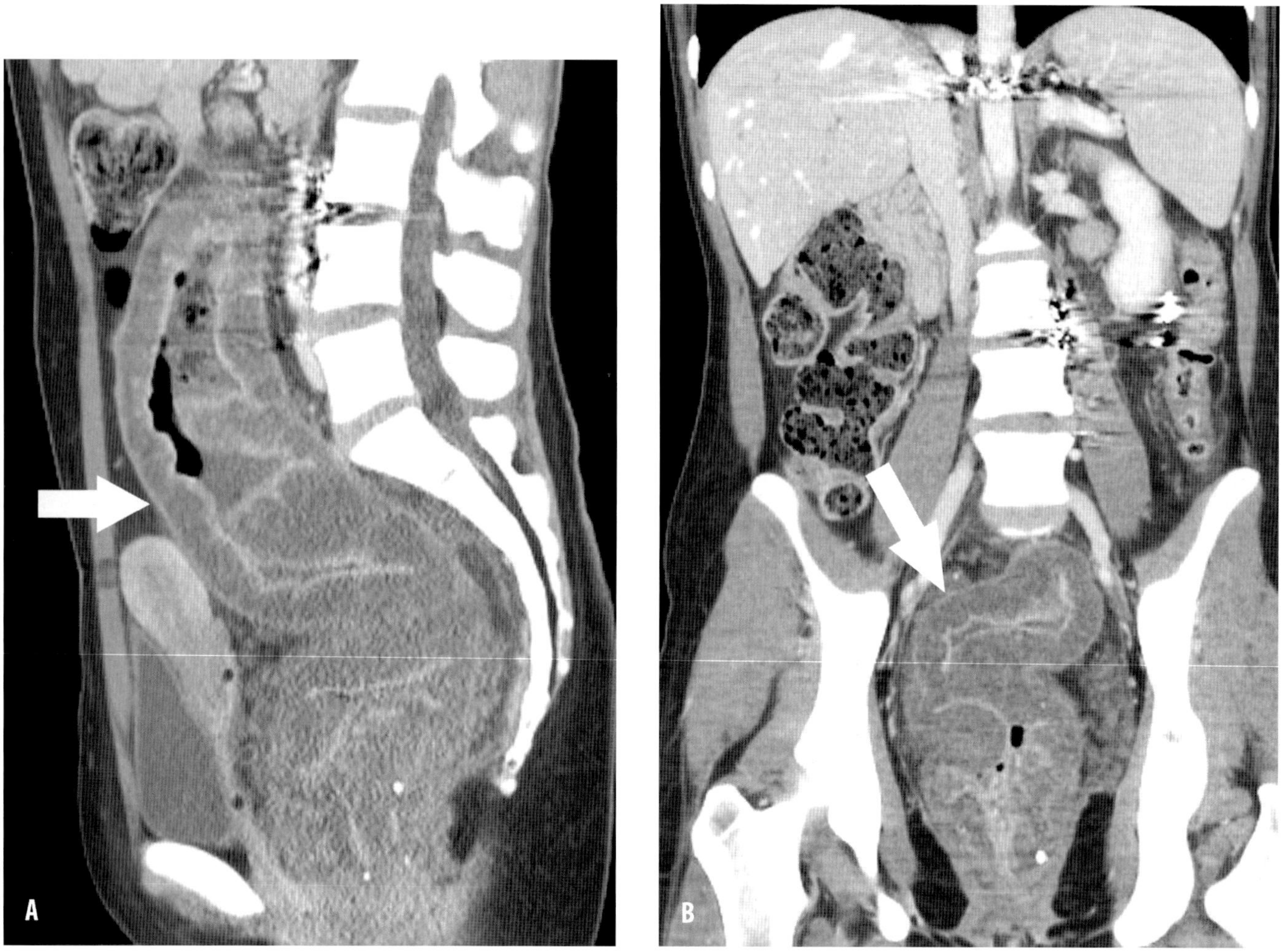

Figure 64.3 A. Sagittal reformatted CT image in a 22 year old woman with non-cirrhotic portal hypertension from portal venous thrombosis secondary to neonatal umbilical venous catheterization and line sepsis. Marked thickening of the rectum and sigmoid (arrow) is seen. Recurrent lower gastrointestinal hemorrhage did not respond to mesocaval shunt creation, transjugular intrahepatic portosystemic shunt placement, or inferior mesenteric vein embolization (venography confirmed hepatofugal flow prior to embolization). Surgical resection of the rectum and sigmoid was ultimately required, and confirmed portal hypertensive colopathy. B. Coronal reformatted CT image again shows the marked thickening of the rectum and sigmoid (arrow). Note the liver is not cirrhotic and the right colon is unaffected.

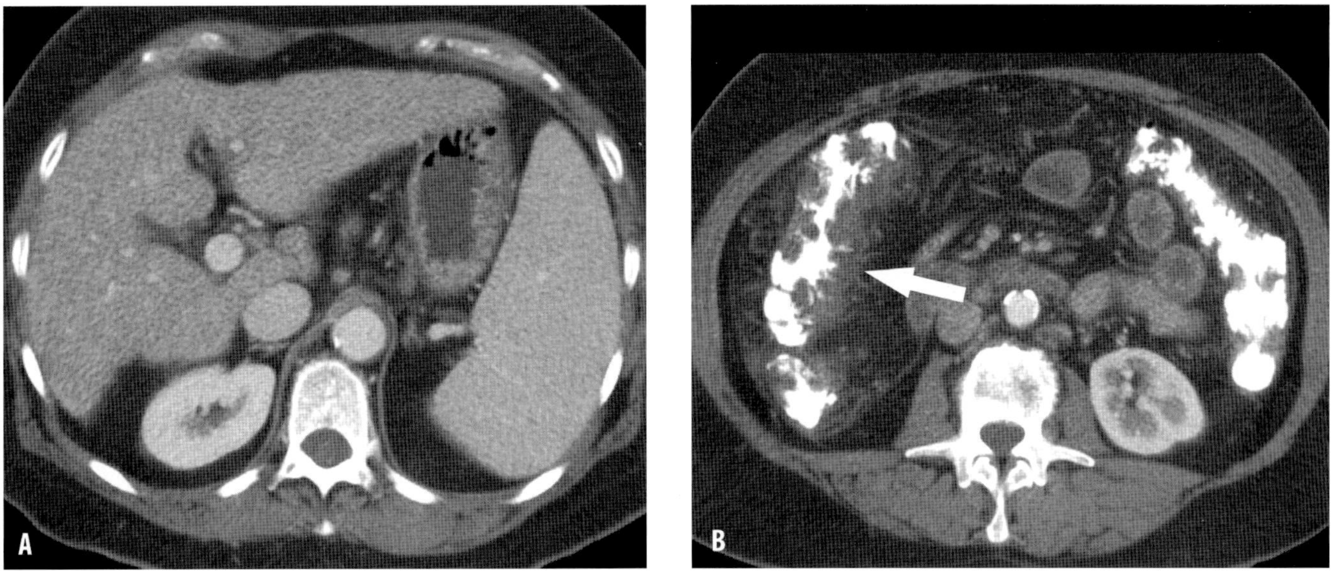

Figure 64.4 **A.** Axial contrast-enhanced CT image in a 58 year old woman being considered for hepatic transplantation, showing cirrhosis and splenomegaly. **B.** Axial contrast-enhanced CT image at a more inferior level shows marked colonic wall thickening (arrow), particularly in the right colon. The findings were initially considered indicative of portal hypertensive colonic wall thickening, but shortly after the scan the patient developed diarrhea that responded to oral metronidazole. The final diagnosis was of infectious colitis.

CASE 65 Pseudotumor due to undistended bowel

Imaging description

Undistended or contracted bowel may appear as focally narrowed segments with circumferential wall thickening and so mimic stricture or tumor (Figures 65.1 and 65.2).

Importance

A false positive diagnosis of bowel pathology may lead to unnecessary testing and anxiety.

Typical clinical scenario

Pseudotumor due to undistended bowel is most commonly seen in the colon or rectum as an incidental finding at CT performed for unrelated reasons.

Differential diagnosis

Physiological bowel wall thickening due to under-distension or contraction is a well-recognized finding at fluoroscopy, but is usually easily recognized on such a dynamic investigation because it is transient. This pitfall can be much more problematic on a static study such as CT. While the upper limit of colonic wall thickness is usually considered to be 3–5 mm [1, 2], a thickness of up to 8 mm has been reported in contracted normal colonic segments [3]. As such, wall thickening in a non-distended segment of bowel should be interpreted cautiously. The absence of additional signs of pathology such as altered wall density or enhancement or perimural fat stranding may point towards a physiological etiology. Comparison with prior studies may also help. Occasionally, administration of rectal or oral contrast (Figures 65.3 and 65.4), intravenous injection of spasmolytic agent, or follow-up endoscopy may be required to confidently confirm the absence of pathology.

Teaching point

The diagnosis of tumor or stricture in an undistended bowel segment should be made with caution, particularly in the large bowel where normal contractions can closely simulate tumor.

REFERENCES

1 Fisher JK. Normal colonic wall thickness on CT. *Radiology* 1982; **145**: 415–418.

2 Desai RK, Tagliabue JR, Wegryn SA, et al. Computed tomographic evaluation of wall thickening in the alimentary tract. *Radiographics* 1991; **11**: 771–783.

3 Wiesner W, Mortelé KJ, Ji H, Ros PR. Normal colonic wall thickness at CT and its relation to colonic distension. *J Comput Assist Tomogr* 2002; **26**: 102–106.

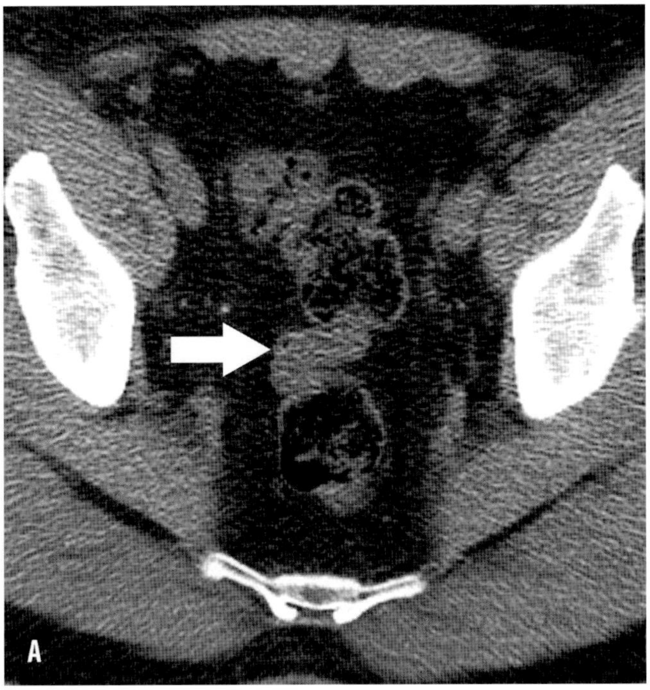

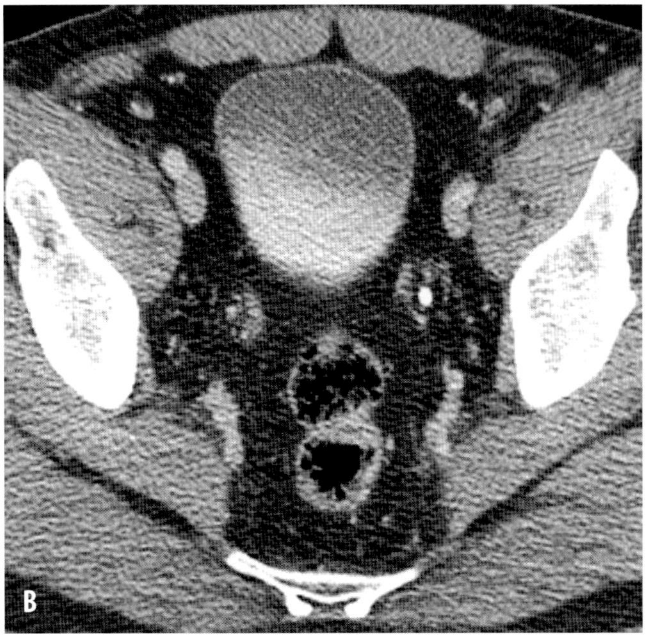

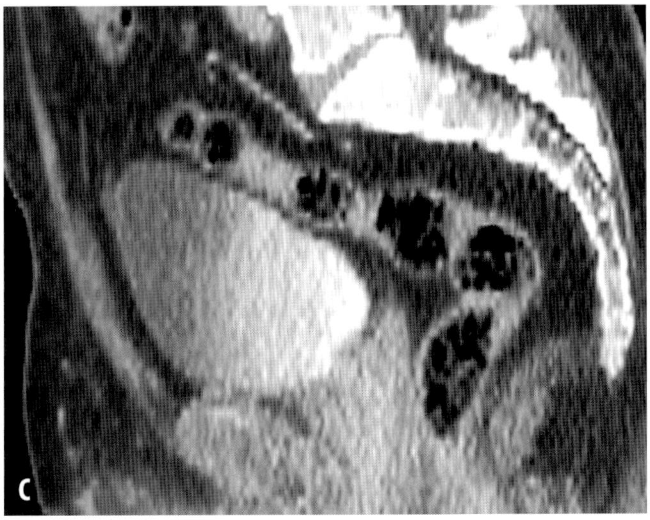

Figure 65.1 **A.** Axial non-enhanced CT image in a 50 year old man undergoing multiphase CT because of hematuria. A focal narrowed segment (arrow) of rectum might be mistaken for a tumor. **B.** Axial delayed phase CT image performed 10 minutes later as part of the CT urogram shows the abnormality has resolved, indicating it was due to rectal contraction. **C.** Sagittal reformatted delayed phase CT image confirms the absence of any rectal abnormality.

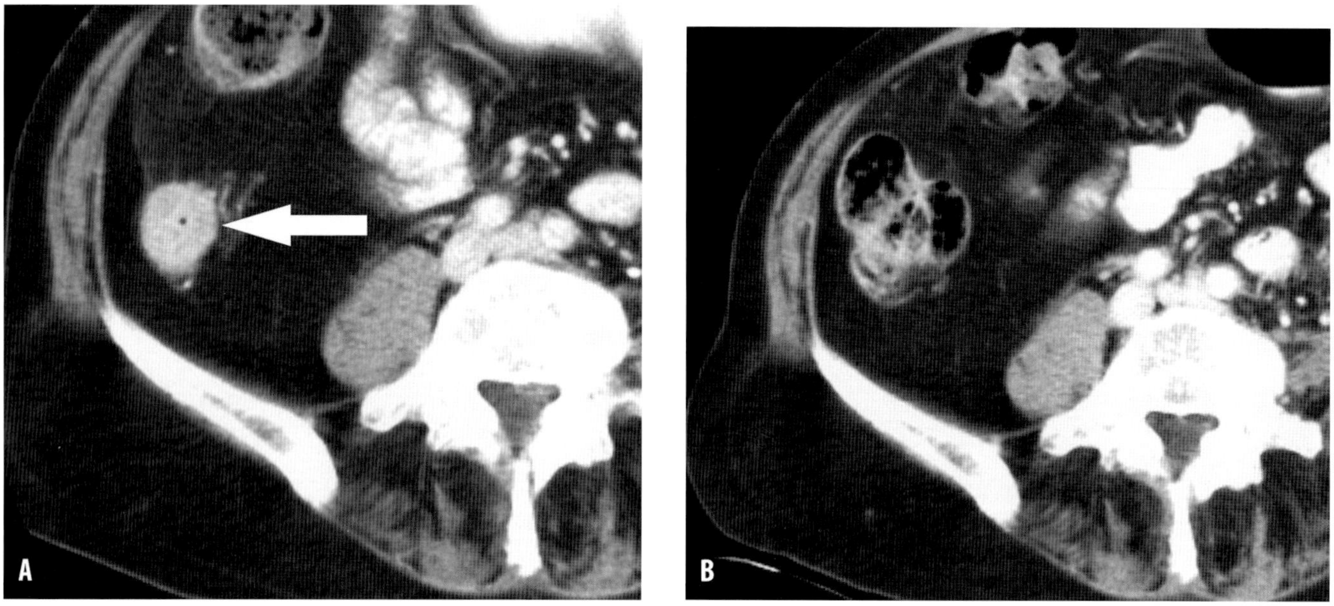

Figure 65.2 **A.** Axial contrast-enhanced CT image in an 88 year old man undergoing CT surveillance for lymphoma. **B.** Axial contrast-enhanced CT image performed six months later (and with no intervening treatment) shows resolution of the cecal abnormality, indicating it was due to colonic contraction or underdistension.

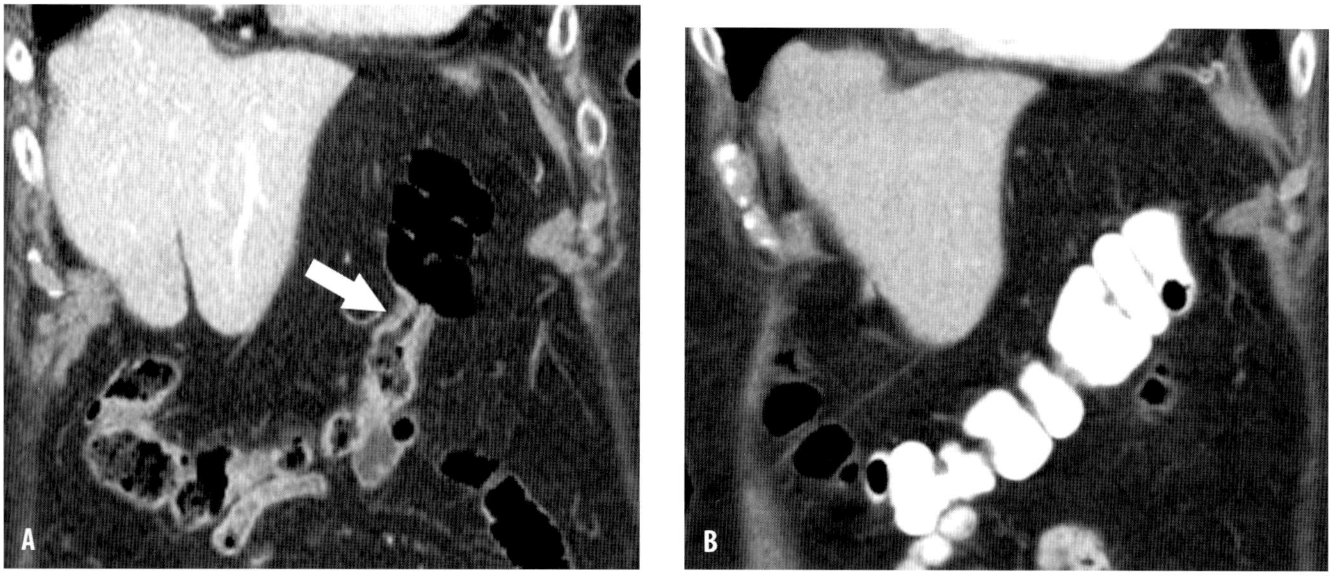

Figure 65.3 **A.** Coronal contrast-enhanced CT image in an 87 year old woman being staged for newly diagnosed gastric cancer. A focal segment (arrow) of narrowing and wall thickening is visible in the transverse colon, and resembles an "apple-core" stricture of primary colorectal cancer. **B.** Coronal CT image after the administration of rectal contrast shows resolution of the abnormality, indicating it was due to colonic contraction.

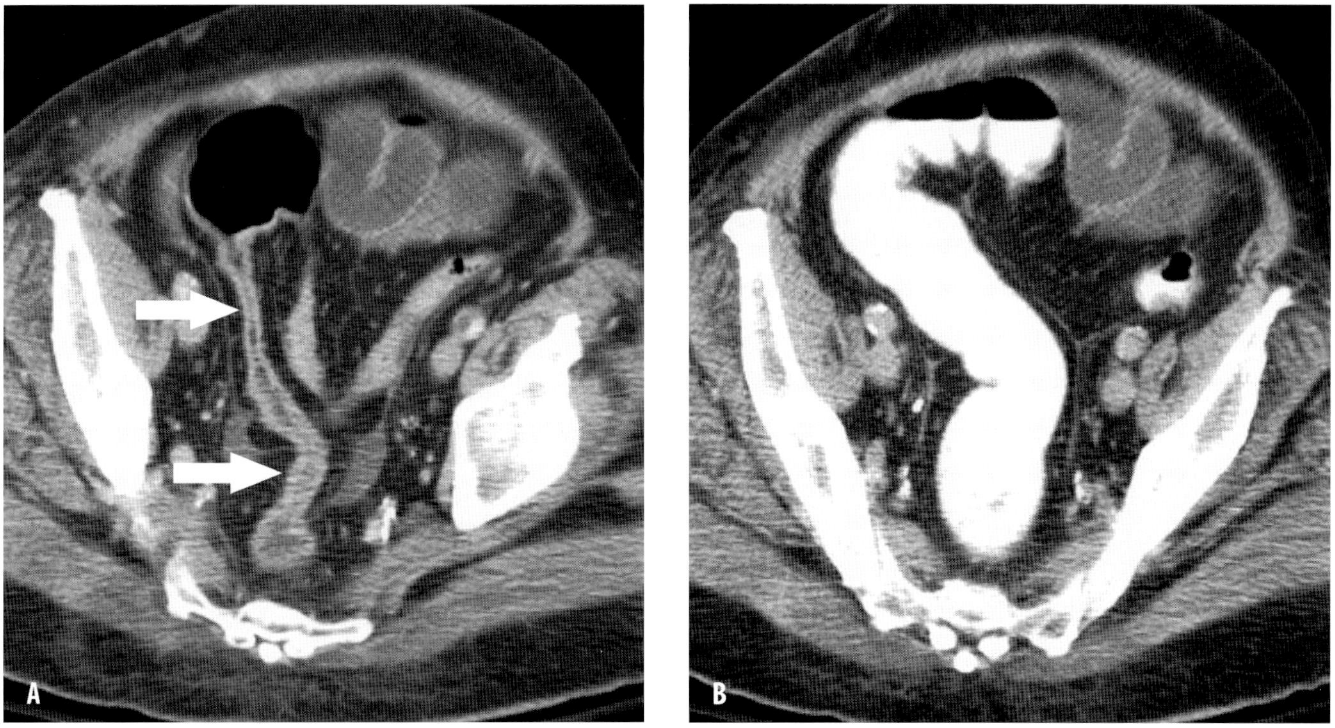

Figure 65.4 A. Axial contrast-enhanced CT image in a 62 year old man with acute abdominal pain shows a long segment (arrows) of sigmoid colon appears narrowed and thick-walled. B. Axial CT image after the administration of rectal contrast shows resolution of the abnormality, indicating it was due to colonic underdistension. Apparent stricturing of the sigmoid colon due to underdistension is particularly common when the sigmoid is long and redundant, as in this patient.

CASE

66 Gastrointestinal pseudolesions due to oral contrast mixing artifact

Imaging description

Incomplete mixing of oral contrast and bowel content may create a spurious appearance of an intraluminal mass or bowel wall thickening at CT (Figures 66.1–66.4).

Importance

A false positive diagnosis of bowel pathology may lead to unnecessary testing and anxiety.

Typical clinical scenario

While the potential for incomplete mixing of positive oral content with gastrointestinal content to create pseudolesions has been described in the stomach [1], I have encountered this pitfall most commonly in the right colon, likely due to the entry of liquid positive oral contrast from the terminal ileum into the semi-viscous content in the large bowel.

Differential diagnosis

Artifactual masses or wall thickening due to oral contrast mixing artifact can usually be recognized by the presence of air bubbles within the apparently thickened wall (Figures 66.3 and 66.4), or by failure of the abnormalities to extend onto the non-dependent portion of the bowel wall (Figure 66.2). Occasionally, follow-up imaging (Figure 66.1) or other testing may be required to confirm the absence of pathology.

Teaching point

An apparent mass or wall thickening in the gastrointestinal tract of a patient who has received positive oral contrast, especially if seen in the cecum or right colon, should be examined carefully for the possibility of a pseudolesion due to incomplete mixing of oral contrast and bowel content.

REFERENCE

1 Horton KM, Fishman EK. Current role of CT in imaging of the stomach. *Radiographics* 2003; **23**: 75–87.

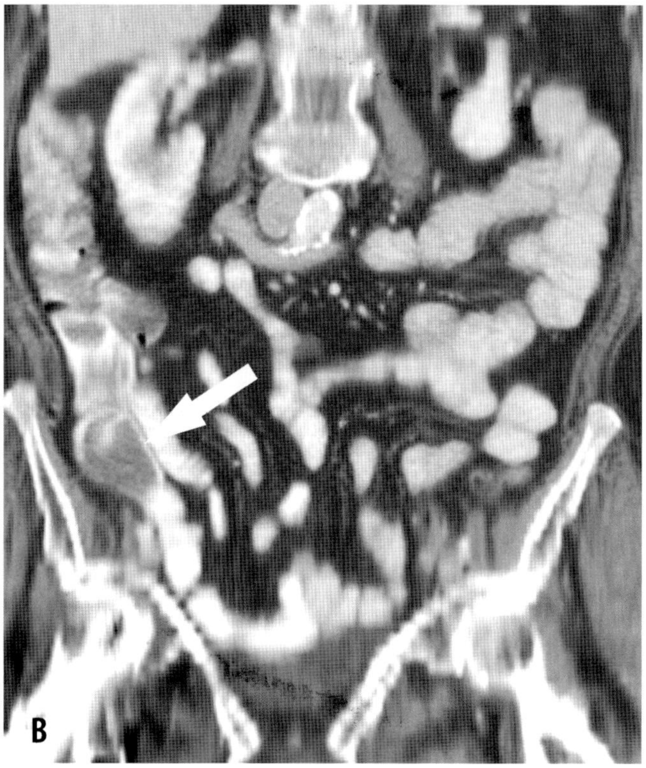

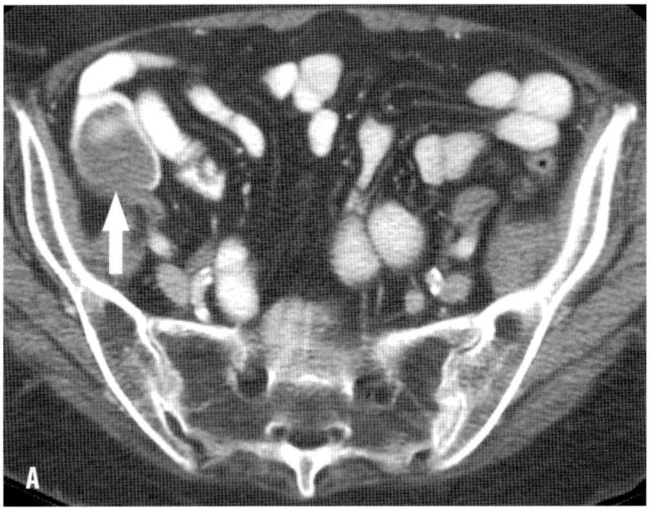

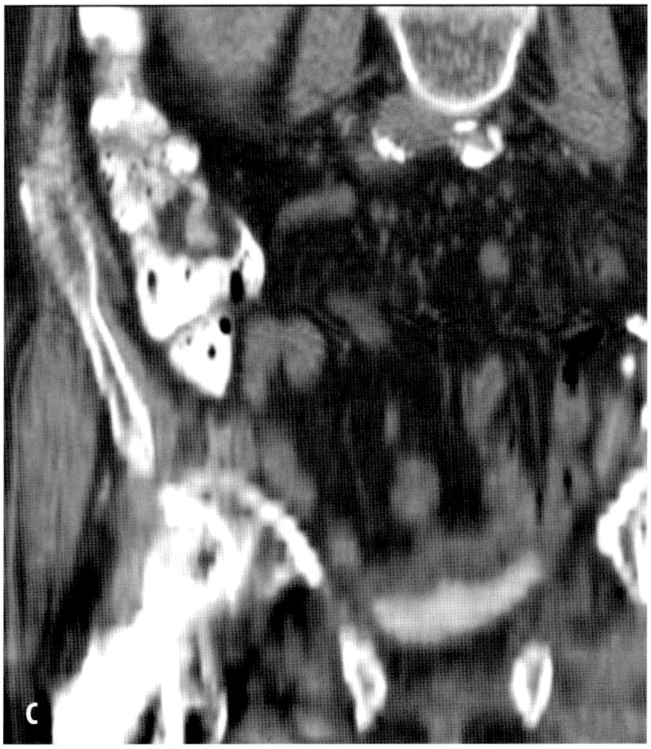

Figure 66.1 A. Axial contrast-enhanced CT image in a 79 year old woman being staged for newly diagnosed non-small cell lung cancer. An apparent mass (arrow) is seen in the cecum. B. Curved planar coronal contrast-enhanced CT image again shows an apparent cecal mass (arrow). C. Curved planar coronal CT image performed eight hours later shows the apparent mass has resolved, with contrast now uniformly distributed through the cecum, confirming the initial finding was a pseudomass due to incomplete mixing.

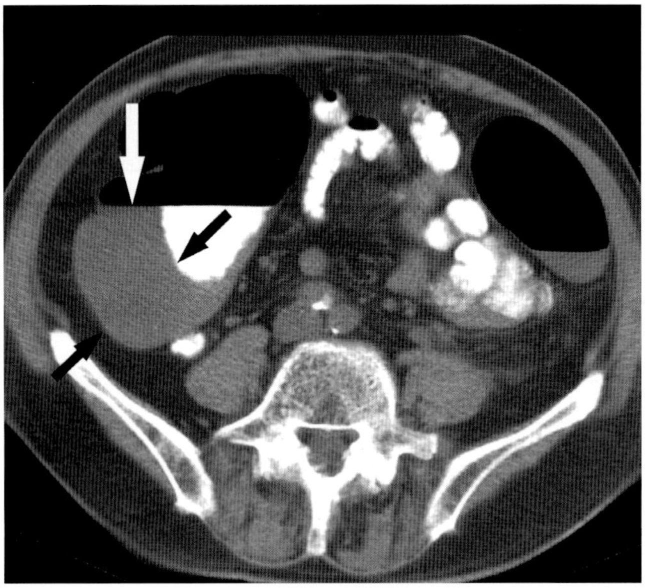

Figure 66.2 Axial non-enhanced CT image in a 75 year old man with postoperative colonic pseudo-obstruction after radical cystectomy for bladder cancer. Apparent wall thickening (between black arrows) stops (white arrow) at the air-fluid level in the dilated colon and does not extend circumferentially onto the non-dependent portion of the colonic wall, confirming it is artifactual due to incomplete mixing of oral contrast and colonic content.

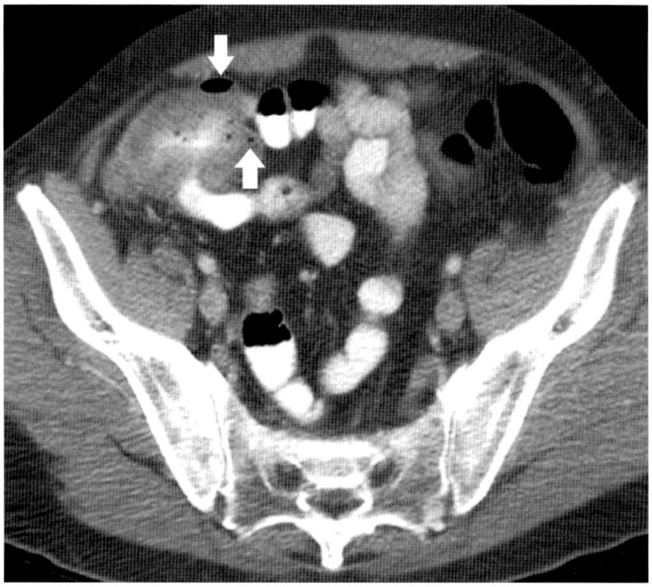

Figure 66.3 Axial contrast-enhanced CT image in a 59 year old man with fever while being treated for acute myeloblastic leukemia. Positive oral contrast confined to the central lumen of the right colon gives an appearance of circumferential wall thickening, but small air bubbles (arrows) are seen within the apparently thickened wall, confirming the finding is artifactual and due to incomplete mixing of oral contrast and bowel content.

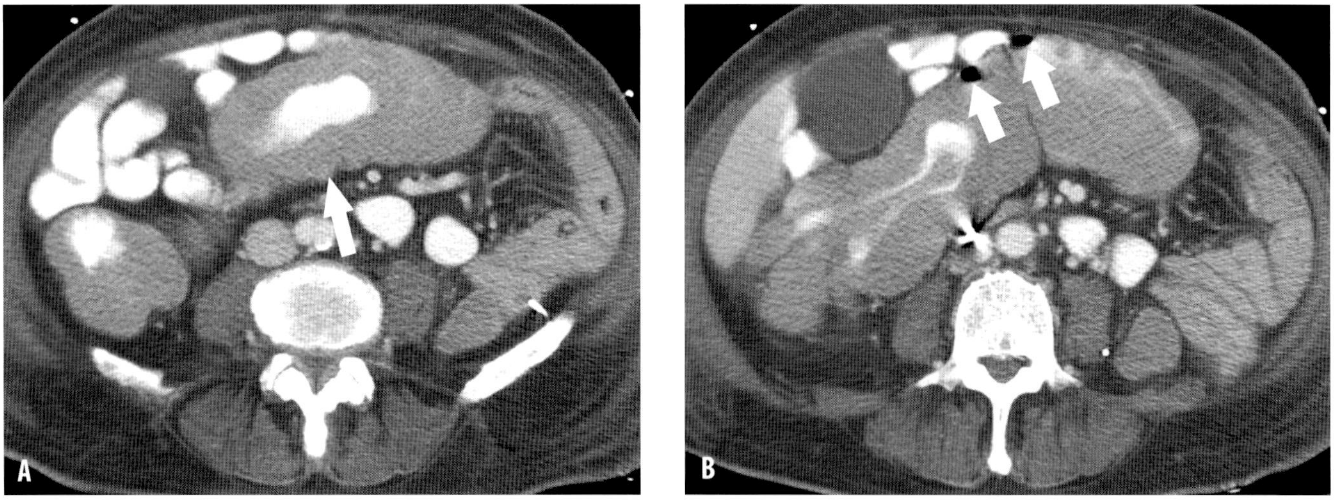

Figure 66.4 **A.** Axial non-enhanced CT image in a 73 year old man with fever after left nephrectomy for renal cell carcinoma. The colonic wall (arrow) appears diffusely and circumferentially thickened. **B.** Axial CT image at a more superior level shows air bubbles (arrows) in the apparently thickened wall, confirming the finding is artifactual and due to incomplete mixing of oral contrast and bowel content.

CASE

67 Perforated colon cancer mimicking diverticulitis

Imaging description

The combination of segmental colonic wall thickening with adjacent pericolonic fat stranding usually indicates acute diverticulitis. Perforated diverticulitis may result in extraluminal air or fluid as additional findings. These signs are highly accurate [1], but occasionally similar findings can be seen in colorectal cancer (Figures 67.1 and 67.2), where the wall thickening is due to the primary tumor and the pericolonic findings of stranding with or without free air or fluid presumably reflect some degree of contained or sealed perforation.

Importance

An incorrect diagnosis of diverticulitis in the setting of a perforated colon cancer may result in a missed chance for diagnosis and potential cure of the malignancy, and create a substantial medicolegal liability. Accordingly, the American Society of Colon and Rectal Surgeons recommends that after resolution of an initial episode of acute diverticulitis, the colon should be adequately evaluated by colonoscopy or contrast enema to exclude other diagnoses, primarily cancer, ischemia, or inflammatory bowel disease [2]. While the recommendation is for colonoscopy or contrast enema, I prefer to recommend follow-up endoscopy since this seems more likely to definitively confirm or refute the diagnosis of colorectal cancer.

Typical clinical scenario

The frequency with which colon cancer mimics diverticulitis at CT is not well established, but in one study of 64 patients with clinically suspected diverticulitis, CT findings suggested diverticulitis in 37 cases [3]. Two (5.4%) of these 37 apparent cases at CT were ultimately found to be due to colorectal carcinoma. Conversely, a reported 1.9 to 3.3% of colorectal cancers present with perforation [4, 5].

Differential diagnosis

A number of studies have investigated whether the presence or absence of pericolonic lymphadenopathy or perfusion imaging of the thickened colonic wall can help distinguish diverticulitis from colorectal cancer [6–8]. While the presence of adjacent adenopathy favors malignancy, none of these discriminating features is sufficiently accurate to preclude the need for follow-up endoscopy. In the study which showed 2 of 37 cases of suspected diverticulitis at CT had colorectal cancer [3], there were 5 other false positives in which the final diagnoses were ischemic colitis (n=2), indeterminate colitis (n=2), and non-specific abdominal pain (n=1), indicating that the described CT findings of diverticulitis are not completely specific and that the differential diagnosis is not confined to colorectal carcinoma.

Teaching point

Perforated colon cancer can resemble diverticulitis at CT, and a CT diagnosis of diverticulitis should always be accompanied by a recommendation to perform follow-up endoscopy (or contrast enema) after resolution of acute symptoms to exclude the possibility of a colon cancer.

REFERENCES

1 Werner A, Diehl SJ, Farag-Soliman M, Düber C. Multi-slice spiral CT in routine diagnosis of suspected acute left-sided colonic diverticulitis: a prospective study of 120 patients. *Eur Radiol* 2003; **13**: 2596–2603.

2 Rafferty J, Shellito P, Hyman NH, Buie WD. Standards Committee of American Society of Colon and Rectal Surgeons. Practice parameters for sigmoid diverticulitis. *Dis Colon Rectum* 2006; **49**: 939–944.

3 Pradel JA, Adell JF, Taourel P, et al. Acute colonic diverticulitis: prospective comparative evaluation with US and CT. *Radiology* 1997; **205**: 503–512.

4 Chen HS, Sheen-Chen SM. Obstruction and perforation in colorectal adenocarcinoma: an analysis of prognosis and current trends. *Surgery* 2000; **127**: 370–376.

5 Mandava N, Kumar S, Pizzi WF, Aprile IJ. Perforated colorectal carcinomas. *Am J Surg* 1996; **172**: 236–238.

6 Chintapalli KN, Chopra S, Ghiatas AA, et al. Diverticulitis versus colon cancer: differentiation with helical CT findings. *Radiology* 1999; **210**: 429–435.

7 Chintapalli KN, Esola CC, Chopra S, Ghiatas AA, Dodd GD 3rd. Pericolic mesenteric lymph nodes: an aid in distinguishing diverticulitis from cancer of the colon. *Am J Roentgenol* 1997; **169**: 1253–1255.

8 Goh V, Halligan S, Taylor SA, et al. Differentiation between diverticulitis and colorectal cancer: quantitative CT perfusion measurements versus morphologic criteria–initial experience. *Radiology* 2007; **242**: 456–462.

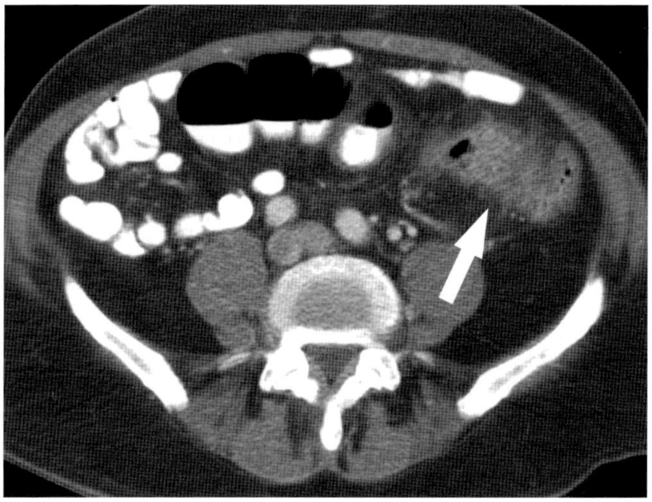

Figure 67.1 Axial contrast-enhanced CT image in a 71 year old man with left lower quadrant pain and a reportedly normal but incomplete colonoscopy 4 months earlier. A short segment (arrow) of the sigmoid colon demonstrates circumferential wall thickening with mild adjacent fat stranding. The CT appearances resemble acute diverticulitis, but colorectal carcinoma was found at repeat endoscopy (which was suggested to exclude underlying malignancy).

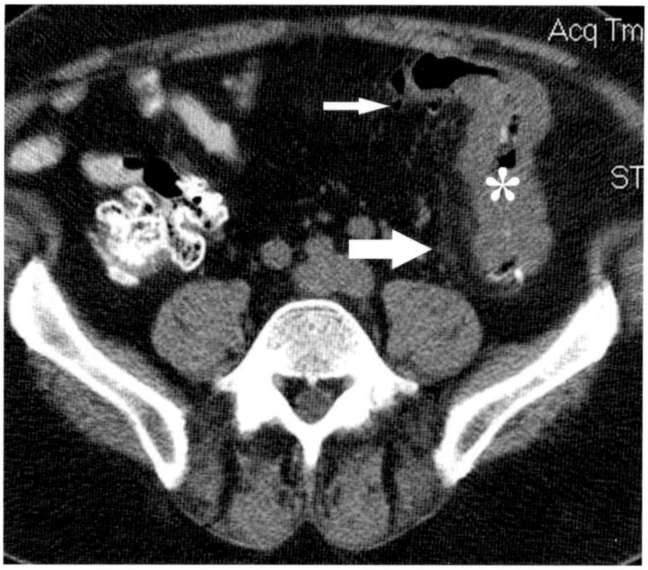

Figure 67.2 Axial non-enhanced CT image requested by the emergency department to exclude urinary stones in a 60 year old man with left flank pain. A short segment (asterisk) of the sigmoid colon demonstrates marked circumferential wall thickening. Diverticula (thin white arrow) are seen in the adjacent colon and pericolonic fat stranding (thick white arrow) is also visible. The findings were interpreted as diverticulitis, with no recommendation for follow-up. Six months later, the patient was diagnosed with an adenocarcinoma of the sigmoid colon.

CASE 68 Pseudoabscess due to absorbable hemostatic sponge

Imaging description

Commercially available bioabsorbable hemostatic sponges are sometimes intentionally left in the operative bed at surgery in order to reduce bleeding. These hemostatic sponges may be made of absorbable gelatin (Gelfoam®), cellulose (Surgicel®, one form of which is marketed as Surgicel Nu-Knit®), or collagen (Helistat® and Ultrafoam®). At postoperative CT or US, these sponges may appear as complex air and fluid-containing collections that resemble abscesses (Figures 68.1–68.5) [1, 2].

Importance

Misidentification of a retained hemostatic sponge as an abscess may result in unnecessary attempted aspiration, drainage procedures, or even reoperation.

Typical clinical scenario

This pitfall is typically seen when a surgical patient is scanned for evaluation of postoperative fever or other complications. The use of these sponges appears to depend heavily on local practice and surgical preference, but in general they are used to maintain hemostasis when postoperative oozing is anticipated, such as after splenectomy, transplantation, vascular procedures, or gynecological or urological surgery. According to the manufacturers, these sponges become liquefied within a week or less and are then gradually absorbed in two to six weeks. Absorbable cellulose is reportedly absorbed within two weeks while absorbable gelatin may persist for four to six weeks [3].

Differential diagnosis

CT findings that may help differentiate a hemostatic sponge from an abscess include lack of an air-fluid level, lack of an enhancing wall, linear arrangement of tightly packed gas bubbles, and fixed position of gas bubbles on serial studies [1, 4]. Hemostatic sponges can become infected, and in one series this was seen in 5 of 18 patients (22%) [4]. Absorbable cellulose is distinguished by being of low T2 signal intensity on MRI [5]. To date, the MRI findings for the other types of hemostatic sponge have not been described.

Teaching point

The possibility of an absorbable sponge should be considered when an apparent abscess is seen on postoperative imaging, especially when there is no distinct air-fluid level or rim-enhancing wall. Correlation with the operative record or direct discussion with the surgeon may also be helpful.

REFERENCES

1 Young ST, Paulson EK, McCann RL, Baker ME. Appearance of oxidized cellulose (Surgicel) on postoperative CT scans: similarity to postoperative abscess. *Am J Roentgenol* 1993; **160**: 275–277.

2 Melamed JW, Paulson EK, Kliewer MA. Sonographic appearance of oxidized cellulose (Surgicel): pitfall in the diagnosis of postoperative abscess. *J Ultrasound Med* 1995; **14**: 27–30.

3 Arnold AC, Sodickson A. Postoperative Surgicel mimicking abscesses following cholecystectomy and liver biopsy. *Emerg Radiol* 2008; **15**: 183–185.

4 Sandrasegaran K, Lall C, Rajesh A, Maglinte DT. Distinguishing gelatin bioabsorbable sponge and postoperative abdominal abscess on CT. *Am J Roentgenol* 2005; **184**: 475–480.

5 Oto A, Remer EM, O'Malley CM, Tkach JA, Gill IS. MR characteristics of oxidized cellulose (Surgicel). *Am J Roentgenol* 1999; **172**: 1481–1484.

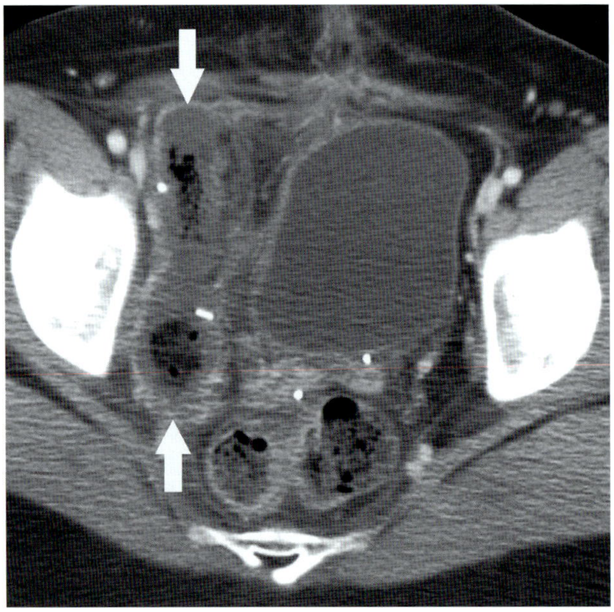

Figure 68.1 Axial contrast-enhanced CT image in a 48 year old woman with fever 8 days after radical hysterectomy and lymphadenectomy for cervical cancer shows an air and fluid-containing structure (between arrows) due to absorbable cellulose (Surgicel®) abutting the right pelvic sidewall. The appearance mimics an abscess.

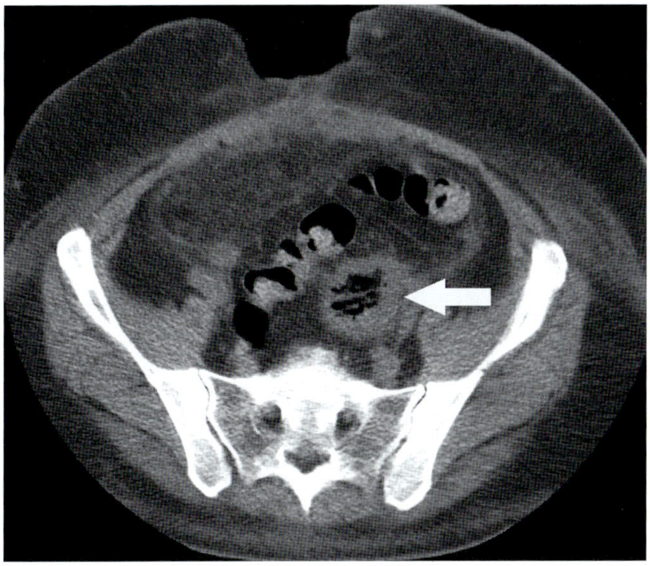

Figure 68.2 Axial contrast-enhanced CT image in a 56 year old woman with fever 5 days after bilateral salpingoophorectomy and hysterectomy for ovarian cancer shows an air-containing structure (arrow) due to absorbable cellulose (Surgicel®) in the left pelvis. The appearance mimics an abscess.

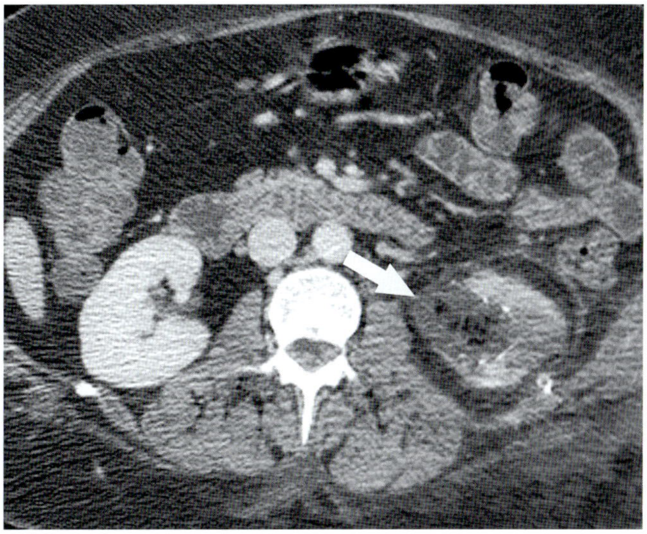

Figure 68.3 Axial contrast-enhanced CT image in a 36 year old woman with a superficial wound infection 15 days after an open partial nephrectomy for an incidentally discovered renal cell carcinoma shows an air and fluid-containing collection (arrow) due to absorbable gelatin (Gelfoam®) in the surgical defect. The appearance mimics an abscess.

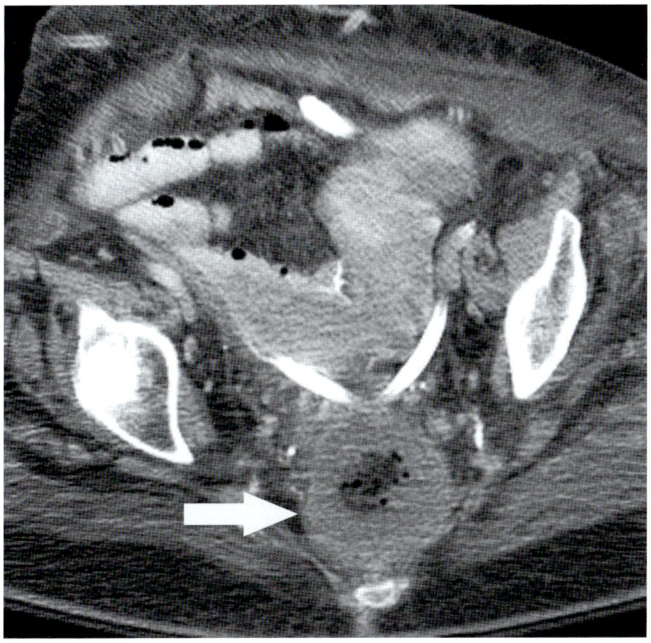

Figure 68.4 Axial contrast-enhanced CT image in an 83 year old woman with fever 7 days after resection of a large pelvic carcinosarcoma shows an air and fluid-containing collection (arrow) due to absorbable gelatin (Gelfoam®) used intra-operatively to control hemorrhage in the cul-de-sac. The appearance mimics an abscess.

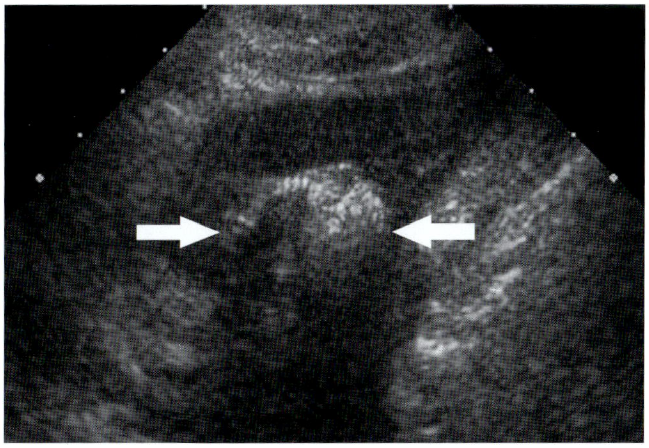

Figure 68.5 Ultrasound image through the left hepatic lobe in a 47 year old woman with elevated liver enzymes 5 days after hepatic transplantation. Surgery was complicated by bleeding from the left hepatic lobe which required electrocautery and placement of an absorbable hemostatic sponge. The sponge is visible as a brightly echogenic lesion (between arrows) with acoustic shadowing abutting the posterior aspect of the left hepatic lobe.

CASE 69 Pseudoperforation due to enhancing ascites

Imaging description

Delayed contrast-enhanced CT in patients with ascites frequently shows increased density of the ascitic fluid when compared to the initial post-contrast images (Figure 69.1), likely due to permeation of intravenous contrast into the peritoneal cavity. Enhancing ascites could be misinterpreted as indicating gastrointestinal perforation with leakage of positive oral contrast into the peritoneal cavity, leakage of excreted contrast from the urinary tract into the peritoneal cavity, or active bleeding with extravasation of intravenous contrast from the vascular system [1, 2].

Importance

Enhancing ascites may result in a misdiagnosis of gastrointestinal perforation, potentially resulting in unnecessary surgery.

Typical clinical scenario

Enhancing ascites is likely to be most problematic when patients with ascites develop abdominal pain. In a study of 50 patients with ascites and delayed contrast-enhanced CT, 27 (54%) showed an interval increase in density of ascites (mean increase of 25 Hounsfield Units) on delayed images (obtained after a mean interval of 33 minutes). Enhancement of ascitic fluid was unrelated to history of malignancy, renal function, or serum albumin level. The authors speculated that passage of intravenous contrast into the peritoneal cavity might reflect altered membrane permeability. This seems plausible because the peritoneum is a dynamic organ and substances may equilibrate between the vascular compartment and the peritoneal space (this is the basis for peritoneal dialysis).

Differential diagnosis

Enhancing ascites is a diffuse uniform process, and lacks the focality that is seen with leakage of positive oral contrast from the gastrointestinal tract, leakage of excreted contrast from the urinary tract, or active bleeding with extravasation of intravenous contrast from the vascular system into the peritoneal cavity.

Teaching point

Enhancing ascites should be the primary consideration when delayed CT images show a diffuse uniform increase in density of ascitic fluid.

REFERENCES

1 Cooper C, Silverman PM, Davros WJ, Zeman RK. Delayed contrast enhancement of ascitic fluid on CT: frequency and significance. *Am J Roentgenol* 1993; **161**: 787–790.

2 Minutoli A, Volta S, Gaeta M. Delayed enhancement of ascites following high-dose contrast CT for liver metastases. *J Comput Assist Tomogr* 1989; **13**: 916–917.

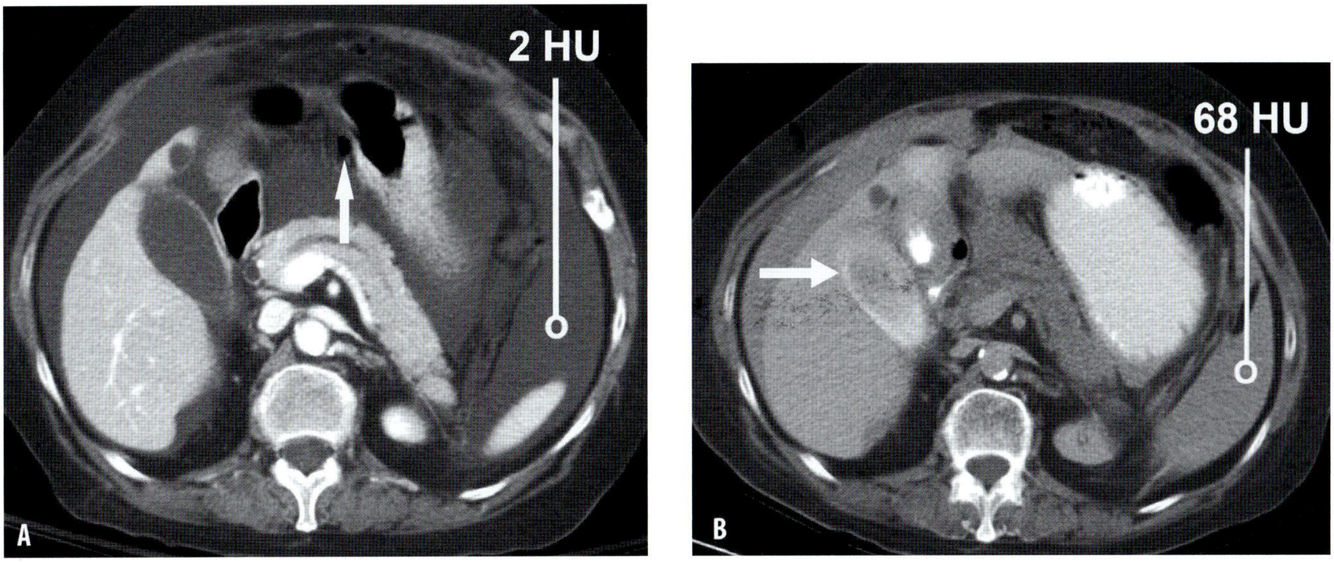

Figure 69.1 **A.** Axial oral and intravenous contrast-enhanced CT image in a 75 year old woman with severe acute abdominal pain and a history of peritoneal dialysis for end-stage renal disease shows free fluid of unremarkable CT density (2 Hounsfield Units) and a focus of free air (arrow). These findings were considered non-specific in the setting of peritoneal dialysis. **B.** CT image obtained six hours later because of ongoing clinical concern for possible gastrointestinal perforation shows a diffuse and uniform increase in density (to 68 Hounsfield Units) of the ascitic fluid. Because of high clinical suspicion and the possibility that the increase in density could reflect leakage of oral contrast into the peritoneal cavity, the patient was taken to surgery. No gastrointestinal perforation or active extravasation into the peritoneal cavity was found.

CASE 70 Pseudomyxoma peritonei

Imaging description

Pseudomyxoma peritonei (colloquially but aptly called "jelly-belly") refers to the progressive accumulation of copious, thick, mucinous or gelatinous material in the peritoneal cavity, and as such is a morphological description rather than a specific pathological diagnosis [1]. The mucin is produced by mucinous tumor cells in the peritoneal cavity, believed to arise from rupture of an intra-abdominal mucinous tumor, usually of the appendix or ovary [2]. Imaging findings that suggest pseudomyxoma peritonei include ascitic fluid that contains non-mobile echoes or septations at ultrasound or is slightly above simple fluid density at CT, or that produces masslike changes on the intraperitoneal contents at cross-sectional imaging, particularly scalloping of the hepatic and splenic margins (Figures 70.1–70.3)[2–5]. Occasionally, faint, curvilinear, or amorphous calcifications may be visible [2].

Importance

While pseudomyxoma peritonei is a rare condition, it is important to recognize and distinguish it from simple ascites and standard peritoneal carcinomatosis. While a primary site in the appendix or ovary is the presumed etiology in most cases, a primary tumor is usually not separately identifiable by the time of diagnosis [6]. Sometimes both the appendix and ovaries are involved, and some believe that the ovarian tumors are then metastases from the appendix, whereas others do not [7, 8]. Independent of these considerations, the appendix and ovaries should be closely inspected in all newly diagnosed cases.

Typical clinical scenario

The incidence of pseudomyxoma peritonei is one per million population per year [9], with a mean age at diagnosis of 49 years (range 23 to 83) [10]. Common presenting symptoms are progressive painful abdominal distension and weight loss.

Differential diagnosis

Microscopically, the mucinous tumor cells in pseudomyxoma peritonei may be benign, borderline, or malignant in appearance. The benign to low-grade malignant forms constitute "classical" pseudomyxoma peritonei and some authorities use the term disseminated peritoneal adenomucinosis for this form of the disease [11–13]. Some controversy exists as to whether the term pseudomyoxoma should be used when the mucinous cells are frankly malignant and derived from an intra-abdominal mucinous adenocarcinoma (known pathologically as peritoneal mucinous carcinomatosis), but the imaging findings overlap and so using the term pseudomyxoma peritonei seems reasonable from a radiological perspective. While disseminated peritoneal adenomucinosis and mucinous carcinomatosis may be indistinguishable at gross inspection, the distinction is important because the former has a five-year survival of 84% compared to less than 10% for the latter [13]. Given the confusion with respect to terminology, it is worth knowing that a myxoma is a benign tumor of connective tissue containing jellylike material (and most occur in the heart). A pseudomyxoma is a gelatinous mass resembling a myxoma but composed of mucus.

Teaching point

Scalloping of the visceral surfaces of the intraperitoneal organs, particularly of the liver or spleen, is an important diagnostic finding that helps to differentiate pseudomyxoma peritonei from simple ascites.

REFERENCES

1 Levy AD, Shaw JC, Sobin LH. Secondary tumors and tumorlike lesions of the peritoneal cavity: imaging features with pathologic correlation. *Radiographics* 2009; **29**: 347–373.

2 Smeenk RM, Bruin SC, van Velthuysen ML, Verwaal VJ. Pseudomyxoma peritonei. *Curr Probl Surg* 2008; **45**: 527–575.

3 Hanbidge AE, Lynch D, Wilson SR. US of the peritoneum. *Radiographics* 2003; **23**: 663–684.

4 Yeh HC, Shafir MK, Slater G, et al. Ultrasonography and computed tomography in pseudomyxoma peritonei. *Radiology* 1984; **153**: 507–510.

5 Seshul MB, Coulam CM. Pseudomyxoma peritonei: computed tomography and sonography. *Am J Roentgenol* 1981; **136**: 803–806.

6 Sugarbaker PH, Ronnett BM, Archer A, et al. Pseudomyxoma peritonei syndrome. *Adv Surg* 1996; **30**: 233–280.

7 Seidman JD, Elsayed AM, Sobin LH, Tavassoli FA. Association of mucinous tumors of the ovary and appendix: a clinicopathologic study of 25 cases. *Am J Surg Pathol* 1993; **17**: 22–34.

8 Young RH, Gilks CB, Scully RE. Mucinous tumors of the appendix associated with mucinous tumors of the ovary and pseudomyxoma peritonei: a clinicopathological analysis of 22 cases supporting an origin in the appendix. *Am J Surg Pathol* 1991; **15**: 415–429.

9 Moran BJ, Cecil TD. The etiology, clinical presentation, and management of pseudomyxoma peritonei. *Surg Oncol Clin N Am* 2003; **12**: 585–603.

10 Fenoglio-Preiser CM. *Gastrointestinal pathology: an atlas and text.* Philadelphia, PA: Wolters Kluwer/Lippincott Williams & Wilkins, 2008.

11 Bradley RF, Stewart JH, Russell GB, Levine EA, Geisinger KR. Pseudomyxoma peritonei of appendiceal origin: a clinicopathologic analysis of 101 patients uniformly treated at a single institution, with literature review. *Am J Surg Pathol* 2006; **30**: 551–559.

12 Carr NJ, Arends MJ, Deans GT, Sobin LH. Adenocarcinoma of the appendix. In: Aaltonen LA, Hamilton SR, eds. *World Health Organization classification of tumours: pathology and genetics of tumours of the digestive system.* Lyon, France: IARC, 2000; 95–102.

13 Ronnett BM, Zahn CM, Kurman RJ, et al. Disseminated peritoneal adenomucinosis and peritoneal mucinous carcinomatosis: a clinicopathologic analysis of 109 cases with emphasis on distinguishing pathologic features, site of origin, prognosis, and relationship to "pseudomyxoma peritonei." *Am J Surg Pathol* 1995; **19**: 1390–1408.

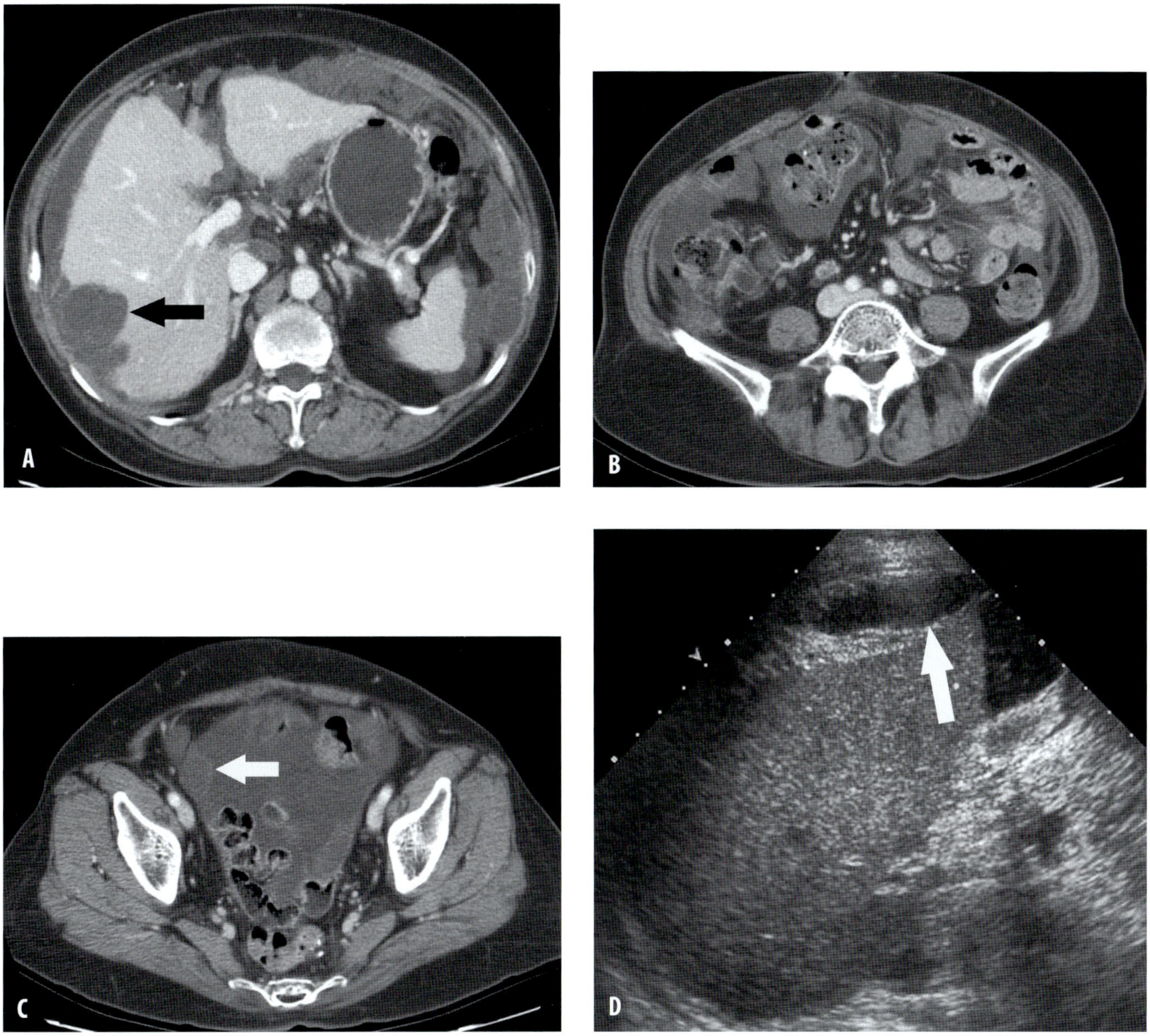

Figure 70.1 **A.** Axial contrast-enhanced CT image in a 68 year old woman with disseminated peritoneal mucinous adenomucinosis arising from an appendiceal mucinous adenoma shows the typical appearance of pseudomyxoma peritonei, with mucinous ascites indenting the liver (arrow). **B.** Axial contrast-enhanced CT image at a more inferior level shows low-density lobulated mucinous ascites filling the peritoneal cavity. **C.** Axial contrast-enhanced CT image through the pelvis shows ascites containing several low-density nodules (arrow). **D.** Sagittal ultrasound image through the right hepatic lobe shows hypoechoic ascitic fluid surrounding and indenting (arrow) the liver.

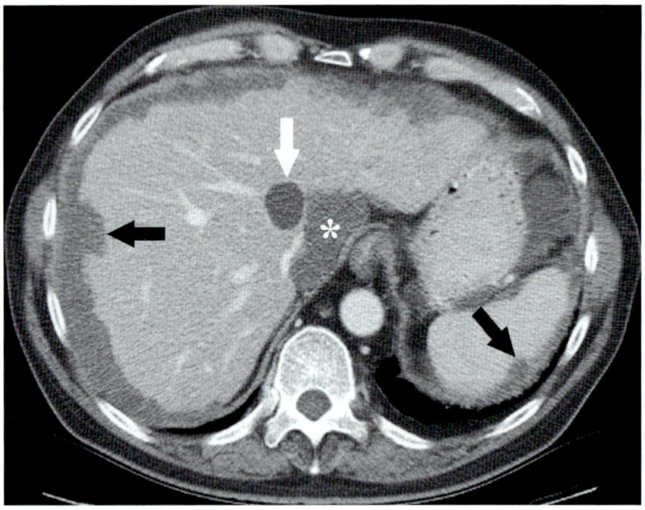

Figure 70.2 Axial contrast-enhanced CT image in a 56 year old man with metastatic mucinous adenocarcinoma of the appendix shows the typical appearance of pseudomyxoma peritonei, with low-density gelatinous material filling the peritoneal cavity and scalloping the liver and spleen (black arrows). Note that the intraperitoneal fluid is of low density, but not of simple fluid density. This is best appreciated by comparing the gelatinous fluid in the superior recess of the lesser sac (asterisk) with simple fluid in a hepatic cyst (white arrow).

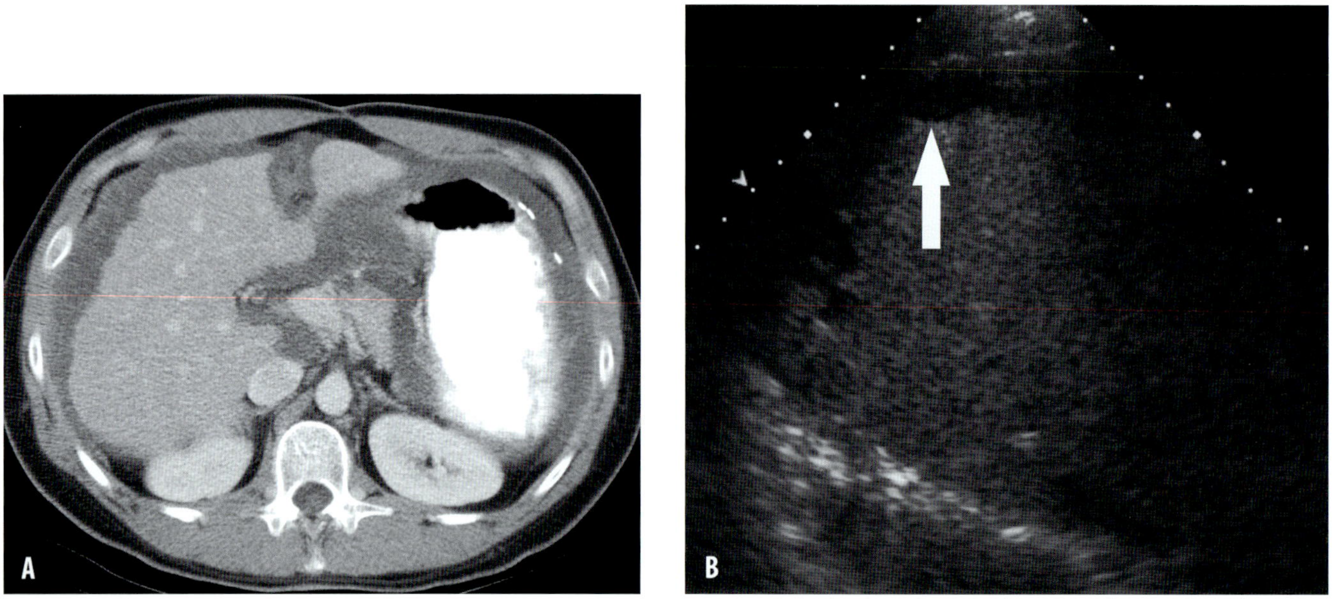

Figure 70.3 **A.** Axial contrast-enhanced CT image in a 33 year old man with metastatic mucinous low-grade adenocarcinoma of the appendix shows the typical appearance of pseudomyxoma peritonei, with obvious scalloping of the liver. The spleen had been removed as part of a prior debulking surgery. **B.** Transverse ultrasound image through the right hepatic lobe shows hypoechoic ascitic fluid indenting the liver surface (arrow).

CASE **71** Gossypiboma

Imaging description

A gossypiboma is a retained surgical sponge or swab, derived from gossypium (Latin; cotton) and boma (Swahili; place of concealment) [1]. Sponges are manufactured with various radiopaque markers. The two commonest markers look like a short piece of crumpled radiopaque scotch-tape (Figure 71.1) or a small tangle of wire (Figure 71.2). While these markers are made to be easily seen, in reality they may be obscured by extraneous paraphernalia and poor image quality due to the challenges of x-raying a patient with portable equipment in the operative or early postoperative setting (Figure 71.3). Retained sponges or swabs may also go unrecognized because they are rare and do not "belong" to any of the usual radiology subspecialties, so that many radiologists have limited training in the recognition of gossypibomas. At CT, retained sponges appear as soft-tissue density masses (Figure 71.4), and may show a whorled texture or a "spongiform" pattern with contained gas bubbles [2]. The latter may persist for many months, even when sterile [3]. At MRI, a retained sponge is typically seen as a soft-tissue density mass with a thick well-defined capsule and a whorled internal configuration on T2-weighted imaging (Figure 71.5) [4].

Importance

Radiologically, gossypibomas may go unrecognized or be confused with postoperative collections or tumors. Retained surgical foreign bodies are of major administrative importance because they are regarded as a "never event" by the Center for Medicare and Medicaid Services. The associated medical and liability costs have been estimated at over $200,000 per incident [5].

Typical clinical scenario

Retained sponges may be asymptomatic and be first recognized at postoperative imaging performed for unrelated reasons or result in symptoms secondary to granulomatous response, abscess development, intestinal obstruction, or fistula formation. Retention of surgical sponges or swabs in the abdomen or pelvis occurs with a frequency of 1 in 1000 to 5000 operations [6, 7]. A history of an abnormal surgical sponge or swab count cannot be relied upon to indicate the diagnosis. In one series, the sponge count was reported as correct in 22 (76%) of 29 patients with retained sponges in the abdomen [7].

Differential diagnosis

A variety of calcified or radiopaque structures may simulate a sponge at radiography (Figures 71.6 and 71.7). Correlation with prior or other imaging, surgical records, and the known appearance of retained sponges or swabs may be required to make the distinction. Unfortunately, not all sponges have a radiopaque marker. In one series, 3 of 29 retained radiopaque sponges lacked a visible radiopaque marker [7]. Very rarely, a surgical towel is retained, and these also lack markers. Sponges without visible radiopaque markers may be identified by radiographic visualization of mottled lucencies due to air trapping [8], or by cross-sectional imaging. Not all retained sponges are accidental, and the possibility of bioabsorbable hemostatic sponges should be considered for an apparent abscess or collection seen in the early postoperative period.

Teaching point

Awareness of the typical radiological appearances is critical to the diagnosis of retained surgical sponges or swabs. Accidentally retained sponges are often clinically unsuspected and may be first recognized by imaging. A high index of suspicion is required, because a history of an incorrect sponge count is frequently lacking and because a radiopaque marker is not always visible. CT or MR imaging may be helpful in problematic cases.

REFERENCES

1 O'Connor AR, Coakley FV, Meng MV, Eberhardt SC. Imaging of retained surgical sponges in the abdomen and pelvis. *Am J Roentgenol* 2003; **180**: 481–489.

2 Kalovidouris A, Kehagias D, Moulopoulos L, et al. Abdominal retained surgical sponges: CT appearances. *Eur Radiol* 1999; **9**: 1407–1410.

3 Kopka L, Fischer U, Gross AJ, et al. CT of retained surgical sponges (textilomas): pitfalls in detection and evaluation. *J Comput Assist Tomogr* 1996; **20**: 919–923.

4 Furukawa H, Hara T, Taniguchi T. Two cases of retained foreign bodies after cholecystectomy: diagnosis by sonography, CT, angiography and MRI. *Jpn J Surg* 1991; **21**: 566–570.

5 Regenbogen SE, Greenberg CC, Resch SC, et al. Prevention of retained surgical sponges: a decision-analytic model predicting relative cost-effectiveness. *Surgery* 2009; **145**: 527–535.

6 Lauwers PR, Van Hee RH. Intra-peritoneal gossypibomas: the need to count sponges. *World J Surg* 2000; **24**: 521–527.

7 Kaiser CW, Friedman S, Spurling KP, Slowick T, Kaiser HA. The retained surgical sponge. *Ann Surg* 1996; **224**: 79–84.

8 Barriga P, Garcia C. Ultrasonography in the detection of intra-abdominal retained surgical sponges. *J Ultrasound Med* 1984; **3**: 173–176.

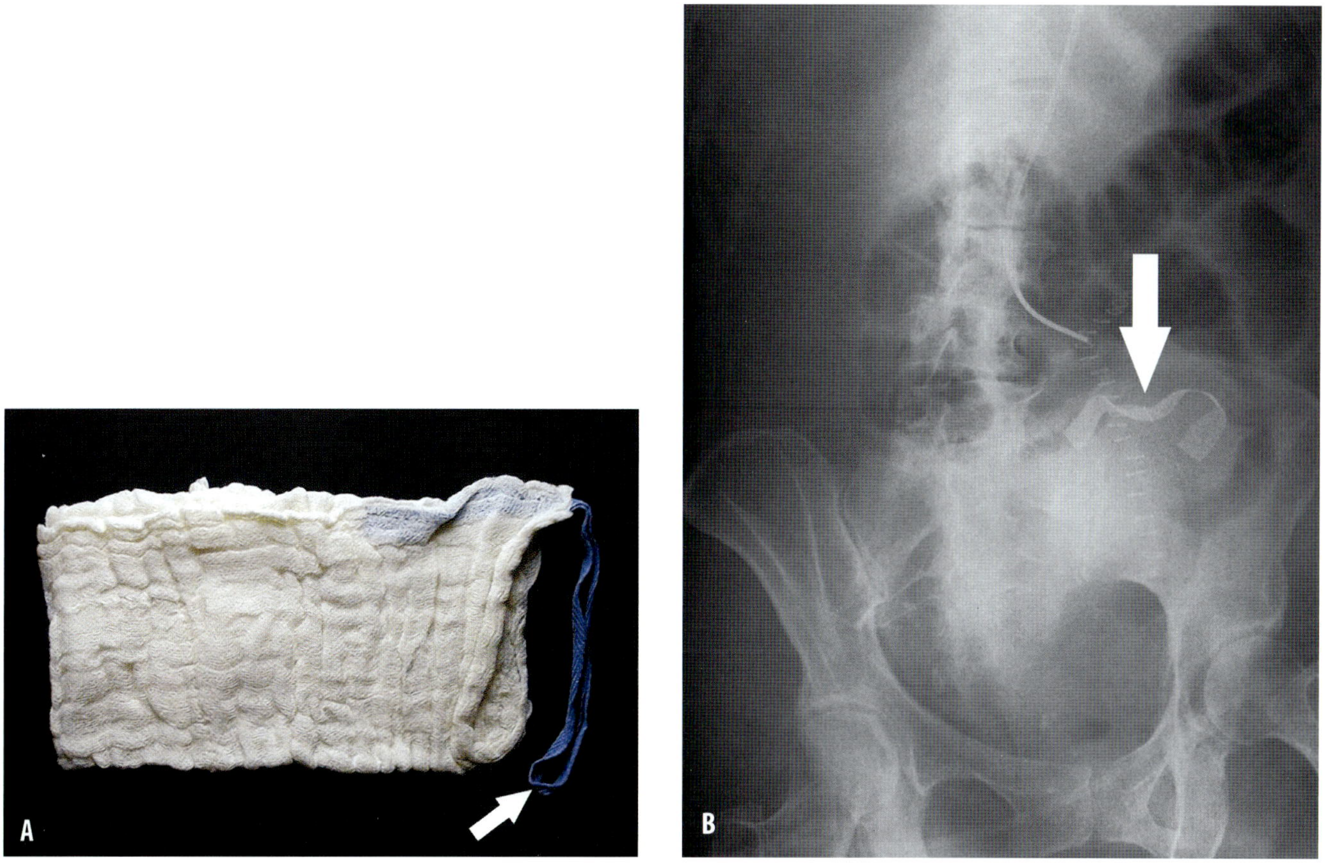

Figure 71.1 A. Photograph of typical laparotomy sponge (commonly called a "lap" by surgical personnel) shows an attached blue strip of material (arrow) that is radiopaque due to impregnated barium. **B.** Abdominal radiograph performed to confirm feeding tube placement in a 65 year old woman 9 days after uneventful laparotomy with lysis of adhesions for small bowel obstruction shows a radiopaque marker projected over the left iliac blade. Repeat surgery with removal of a retained laparotomy sponge was performed the next day.

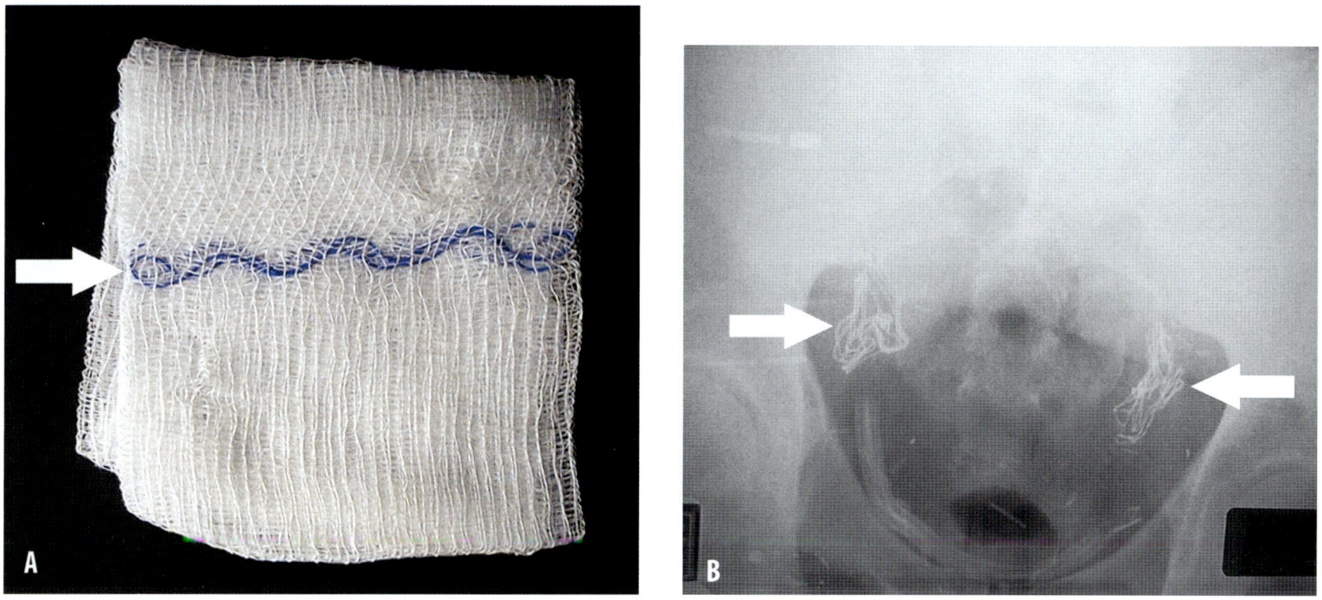

Figure 71.2 A. Photograph of another typical laparotomy sponge (commonly called a "RayTec"® or "4×4" by surgical personnel, after the manufacturer and size in inches, respectively) shows an interwoven and barium-impregnated blue marker (arrow). **B.** Abdominal radiograph performed because of an incorrect count during a radical retropubic prostatectomy shows two retained RayTec® sponges (arrows).

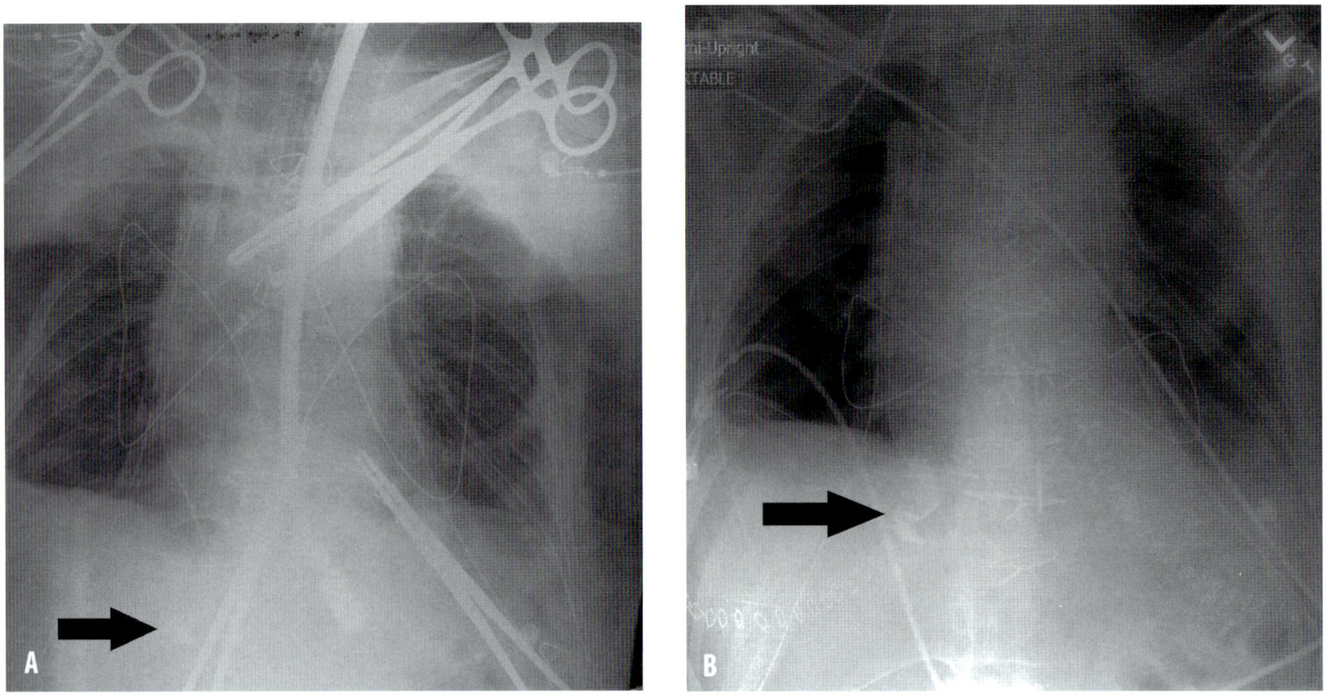

Figure 71.3 A. Intra-operative chest radiograph in a 25 year old man performed because of an incorrect count after bilateral lung transplantation shows a barely visible retained sponge (arrow) projected over the upper abdomen. The sponge is obscured by underpenetration and surrounding surgical paraphernalia, and was not recognized. **B.** On a postoperative chest radiograph, the sponge (arrow) is a little easier to see, and was reported and removed.

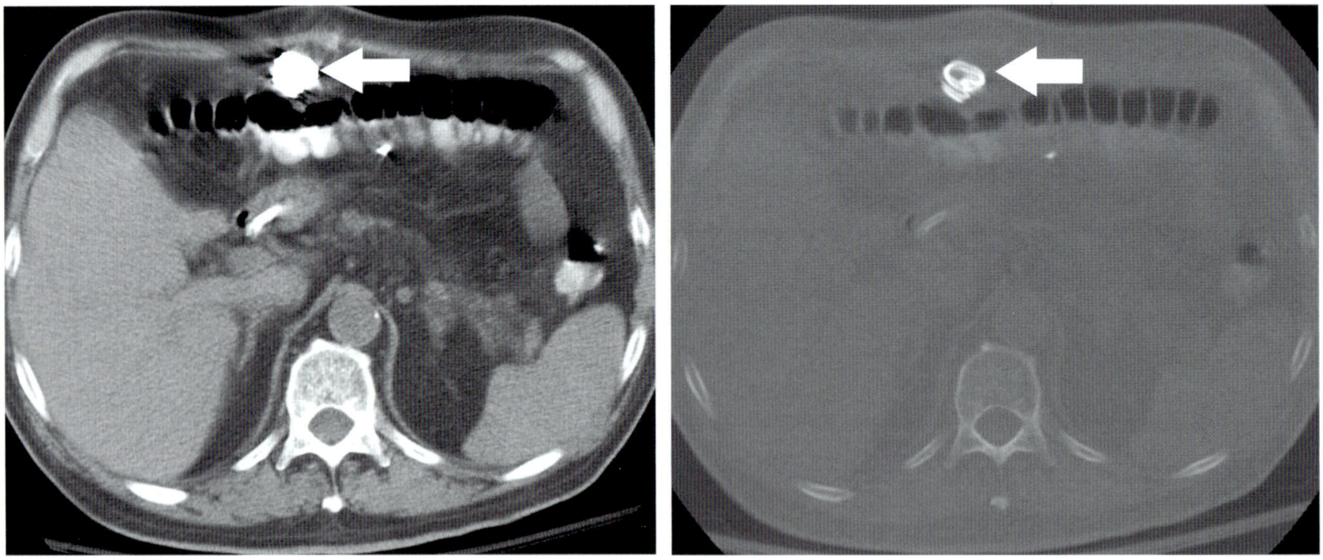

Figure 71.4 Photomontage of axial non-enhanced CT image displayed at soft-tissue and bone windows in a 69 year old man 4 weeks after aortofemoral bypass shows a retained sponge (arrow) anterior to the transverse colon.

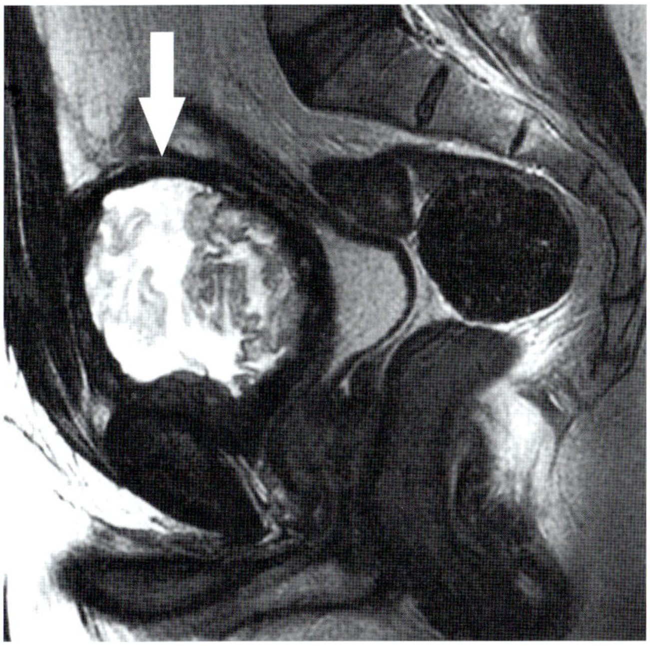

Figure 71.5 Sagittal fast spin-echo T2-weighted MR image in a 56 year old man complaining of urinary frequency 5 months after radical retropubic prostatectomy shows the whorled configuration of a retained sponge (arrow). The sponge was surgically removed.

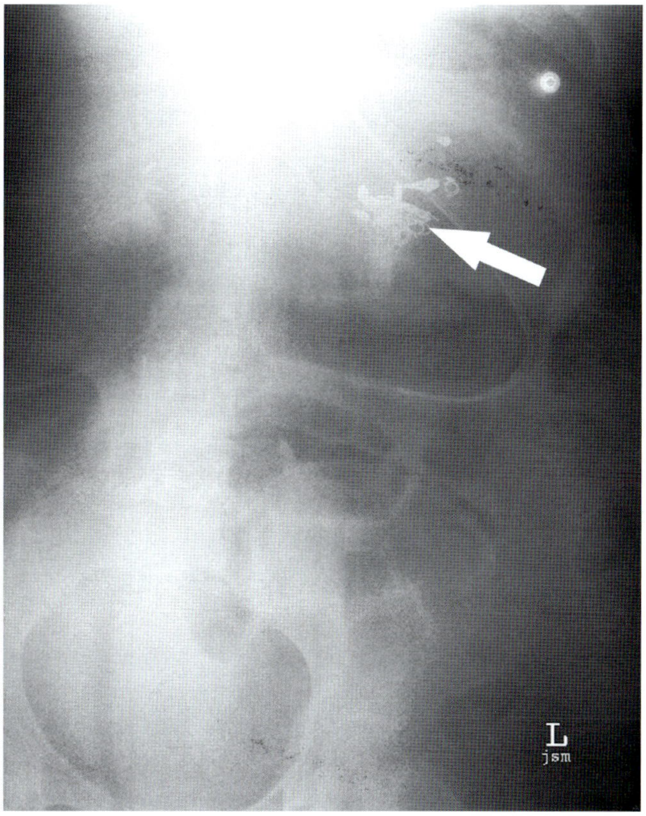

Figure 71.6 Abdominal radiograph in a 63 year old woman with primary biliary cirrhosis and a previously embolized splenic artery aneurysm shows a cluster of embolization coils (arrow) in the left upper quadrant that resembles a retained sponge marker.

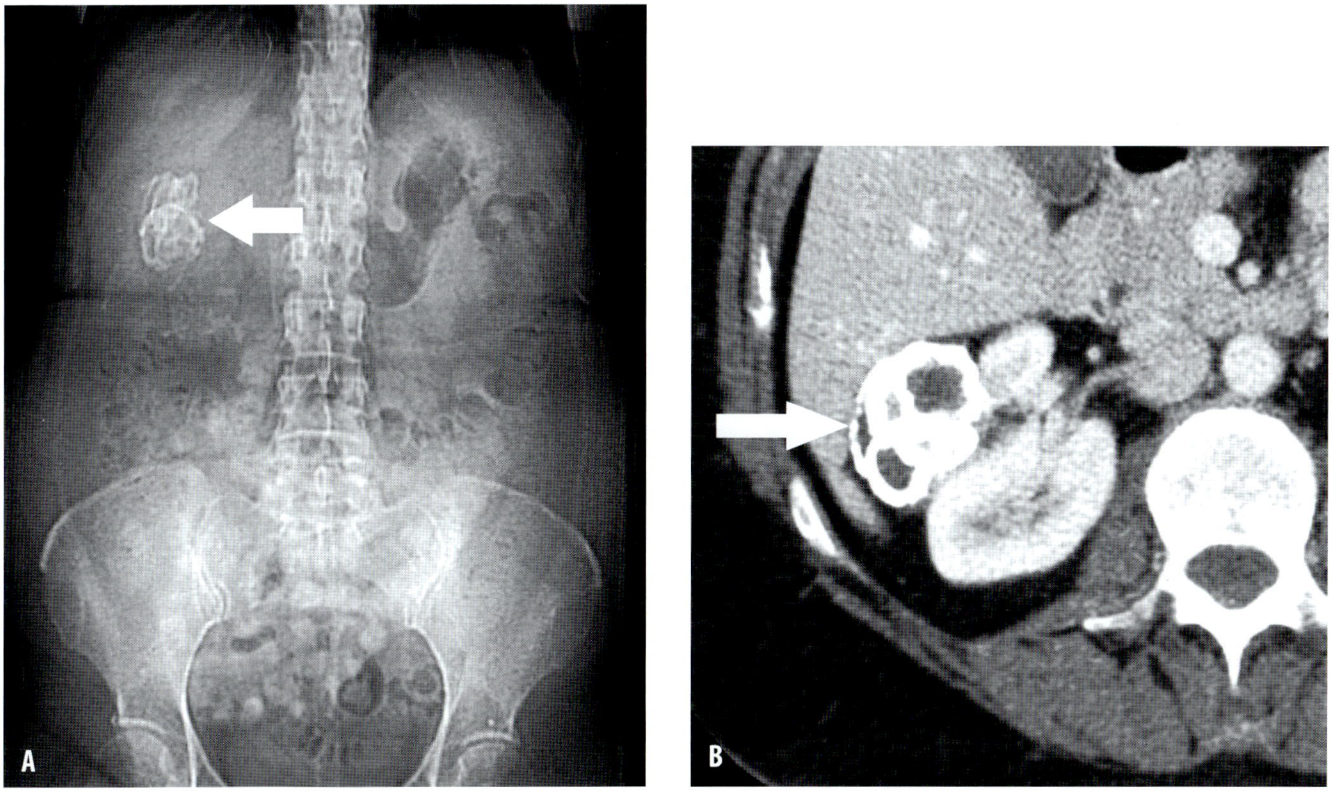

Figure 71.7 **A.** Scout view from an abdominal CT performed in a 55 year old woman for evaluation of non-specific abdominal pain shows a radiodense structure (arrow) in the right flank that somewhat resembles a sponge marker. **B.** Axial contrast-enhanced CT image shows the structure is a densely calcified right renal mass (arrow). A 4-cm calcified renal cell carcinoma was removed at surgery.

CASE 72

Corpus luteum cyst

Imaging description

Corpus luteum cysts are normal physiologic ovarian structures formed after ovulation by the dominant follicle when the partially collapsed follicular wall becomes vascularized and thickened (these changes in the follicular wall are known as luteinization) [1, 2]. Corpus luteum cysts function as endocrine organs, because the luteinized cells secrete estrogens and progesterones in the second half of the menstrual cycle. Corpus luteum cysts are typically seen at ultrasound as 1–3 cm ovarian cysts with echogenic crenulated walls and variable internal low-level echogenicity that may demonstrate dependent layering [3–6]. Doppler insonation demonstrates a characteristic "ring of fire" due to the vascular nature of the cyst wall (Figure 72.1). CT or MRI shows an ovarian cyst with a thick, crenulated, and briskly enhancing wall (Figures 72.2 and 72.3) [7]. Corpus luteum cysts are metabolically active and so may show increased uptake on PET (Figure 72.4). A small amount of physiological free fluid is a common associated finding. Rarely, a large volume of hemoperitoneum may be seen, which presumably reflects rupture with bleeding from the vascular wall of the cyst (Figure 72.5). Such hemorrhagic corpus luteum cysts may occur spontaneously or in patients with bleeding diatheses [8, 9]. Corpus luteum cysts resolve spontaneously at the end of the menstrual cycle. In the event of pregnancy, corpus luteum cysts persist as normal physiological structures up to approximately 14 weeks' gestation (Figure 72.6).

Importance

The commonest error is for the CT findings of an enhancing thick-walled corpus luteum cyst to be misinterpreted as an ovarian tumor or abscess, potentially leading to unnecessary workup and treatment.

Typical clinical scenario

Corpus luteum cysts are not pathologic but can be associated with periovulatory pain that precipitates radiological evaluation. Given the increasing use of CT for the evaluation of acute abdominal or pelvic pain, it is common to see corpus luteum cysts on CT scans requested in women of reproductive age referred from the emergency department. The reported predominance of corpus luteum cysts on the right side [7] likely reflects selection bias, since patients with right-sided abdominal pain are more likely to undergo imaging because of suspected appendicitis.

Differential diagnosis

The imaging features of corpus luteum cysts are relatively characteristic, but occasionally a similar appearance of a briskly rim-enhancing adnexal structure may be the result of ectopic pregnancy, tubo-ovarian or pelvic abscess, or hypervascular neoplasm such as Sertoli-Leydig cell tumor (Figures 72.7–72.9). These differential considerations are rare and usually clinically distinctive. While a woman of reproductive age with acute abdominal pain should always have a pregnancy test performed by the clinical team, it is still good radiological practice to consider the possibility of ectopic pregnancy for any hypervascular adnexal mass seen at imaging in this population.

Teaching point

A cyst with a thick crenulated vascular wall in the ovary of a woman of reproductive age is likely to be a corpus luteum cyst, and may be associated with acute pain.

REFERENCES

1 Young RH, Scully RE. The ovary. In: Sternberg SS, ed. *Diagnostic surgical pathology.* New York, NY: Raven Press, 1994; 2197–2202.

2 Russel P, Farnsworth A. Dysfunctional cysts. In: Russel P, Farnsworth A, eds. *Surgical pathology of the ovaries.* New York: Churchill Livingstone, 1997; 115–130.

3 Ritchie WGM. Sonographic evaluation of normal and induced ovulation. *Radiology* 1986; **161**: 1–10.

4 Frates MC, Doubilet PM, Durfee SM, et al. Sonographic and doppler characteristics of the corpus luteum: can they predict pregnancy outcome? *J Ultrasound Med* 2001; **20**: 821–827.

5 Frates MC, Visweswaran A, Laing FC. Comparison of tubal ring and corpus luteum echogenicities. *J Ultrasound Med* 2001; **20**: 27–31.

6 Durfee SM, Frates MC. Sonographic spectrum of the corpus luteum in early pregnancy: gray-scale, color, and pulsed doppler appearance. *J Clin Ultrasound* 1999; **27**: 55–59.

7 Borders RJ, Breiman RS, Yeh BM, Qayyum A, Coakley FV. Computed tomography of corpus luteal cysts. *J Comput Assist Tomogr* 2004; **28**: 340–342.

8 Wilbur AC, Goldstein LD, Prywitch BA. Hemorrhagic ovarian cysts in patients on anticoagulation therapy: CT findings. *J Comput Assist Tomogr,* 1993; **17**: 623–625.

9 Hertzberg BS, Kliewer MA, Paulson EK. Ovarian cyst rupture causing hemoperitoneum: imaging features and the potential for misdiagnosis. *Abdom Imaging* 1999; **24**: 304–308.

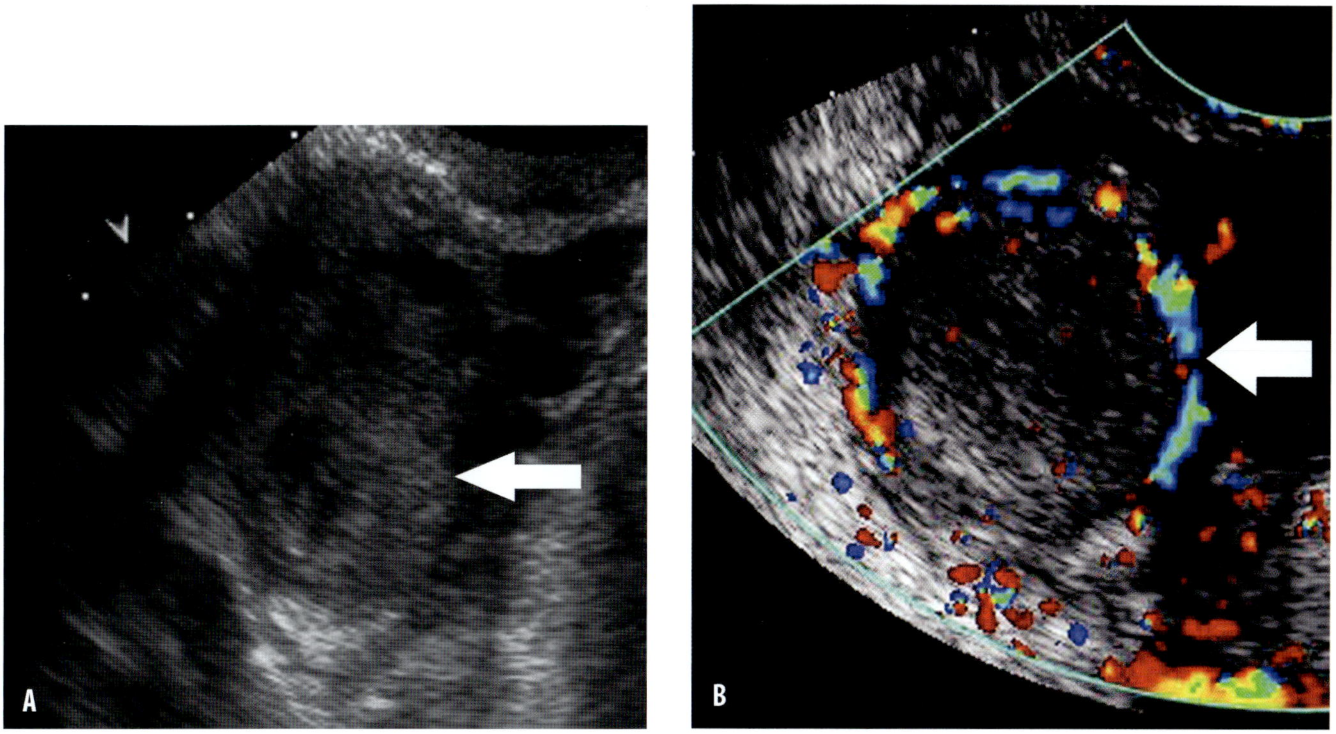

Figure 72.1 A. Endovaginal ultrasound image of the right ovary in a 42 year old woman with right lower quadrant pain shows a 2 cm hypoechoic crenulated thick-walled corpus luteum cyst (arrow) in the right ovary. **B.** Doppler ultrasound image confirms the characteristic appearance of prominent peripheral blood flow (arrow), the so-called "ring of fire".

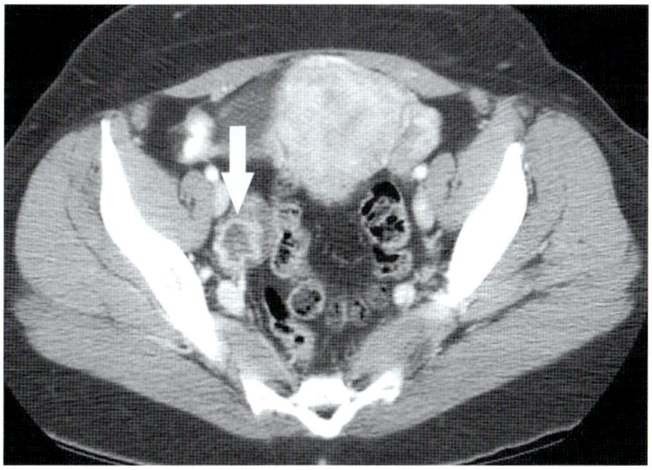

Figure 72.2 Axial contrast-enhanced CT image in a 35 year old woman with right lower quadrant pain shows the characteristic thick, crenulated, and hyperdense wall of a corpus luteum cyst (arrow) in the right ovary.

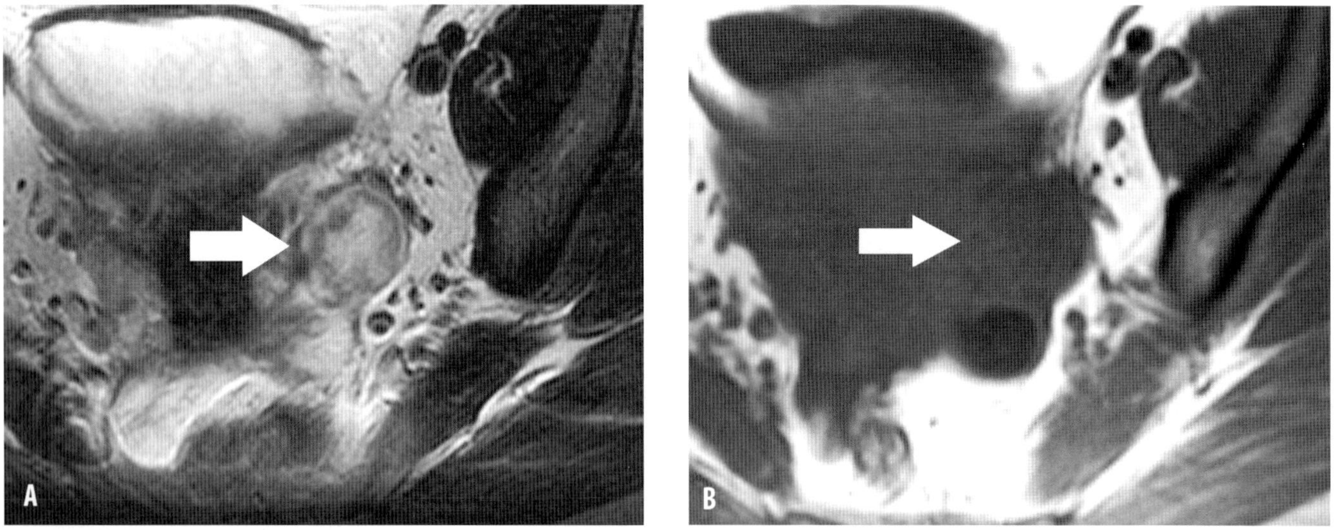

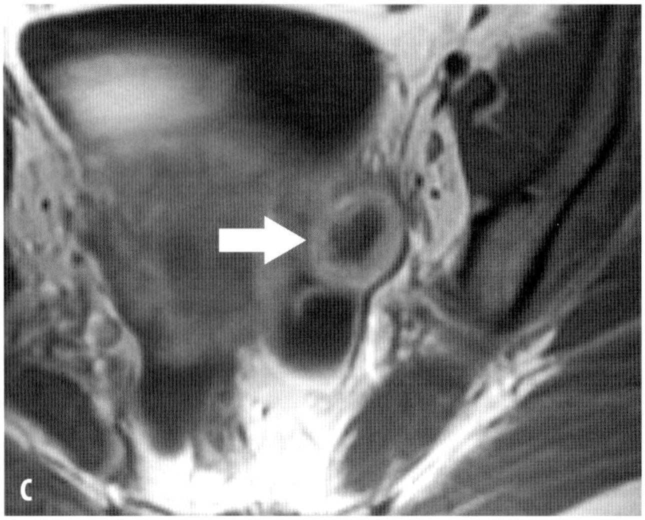

Figure 72.3 A. Axial T2-weighted MR image in a 38 year old woman with pelvic inflammatory disease shows a thick-walled corpus luteum cyst (arrow) in the left ovary. **B.** Axial T1-weighted MR image shows the cyst (arrow) to be isointense to the uterus. **C.** Axial T1-weighted MR image after intravenous gadolinium shows the typical brisk enhancement of the cyst wall (arrow). The cyst had resolved on an ultrasound performed six weeks later.

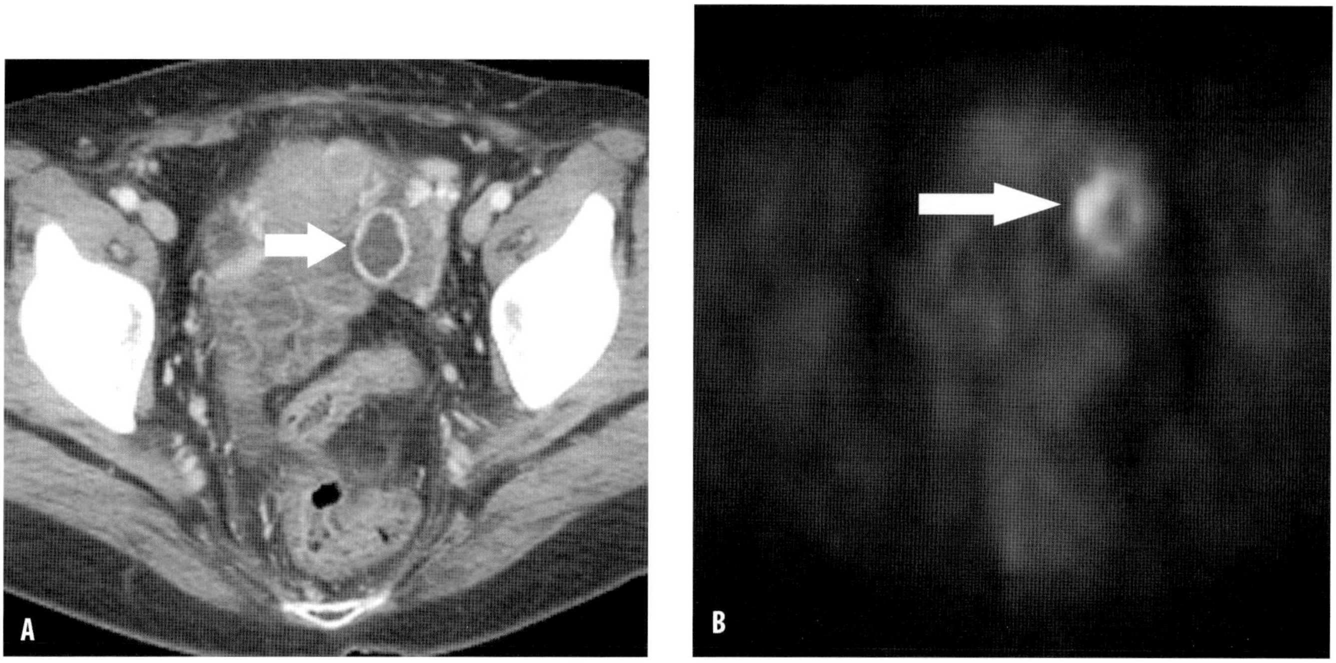

Figure 72.4 A. Axial contrast-enhanced CT image performed as part of a whole-body PET/CT in a 38 year old woman with weakness and hyperthyroidism and a history of cardiac transplantation. A corpus luteum cyst (arrow) is seen in the left ovary. **B.** Corresponding axial PET image shows a ring of increased FDG uptake in the cyst (arrow). Corpus luteum cysts are endocrine organs that are metabolically active and may show focal increased uptake on FDG PET.

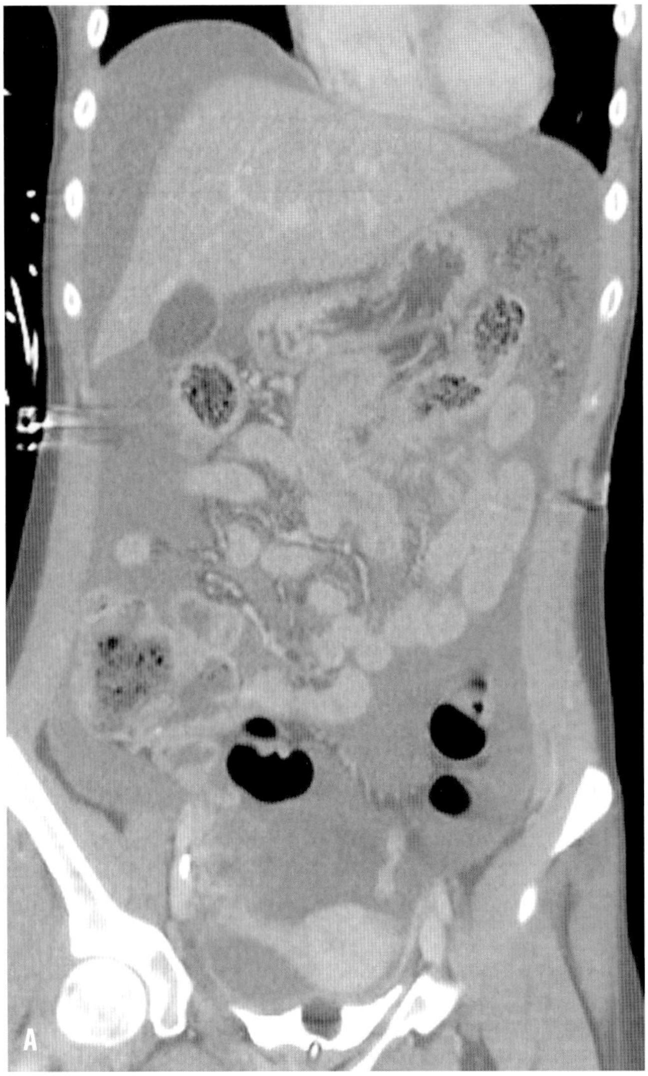

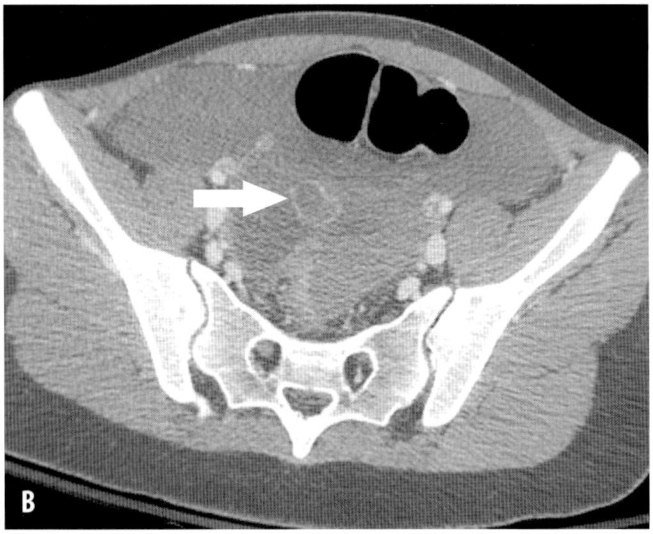

Figure 72.5 **A.** Reformatted coronal contrast-enhanced CT image in a 22 year old woman with acute abdominal pain and an elevated international normalized ratio of 13 (on coumadin for systemic lupus erythematosus). A large volume of high-density ascites is visible throughout the abdomen and pelvis. **B.** Axial contrast-enhanced CT image shows a corpus luteum cyst (arrow) in the right ovary. Because of hemodynamic instability, the patient required an exploratory laparotomy and right oophorectomy, with a confirmed histopathological diagnosis of ruptured corpus luteum cyst.

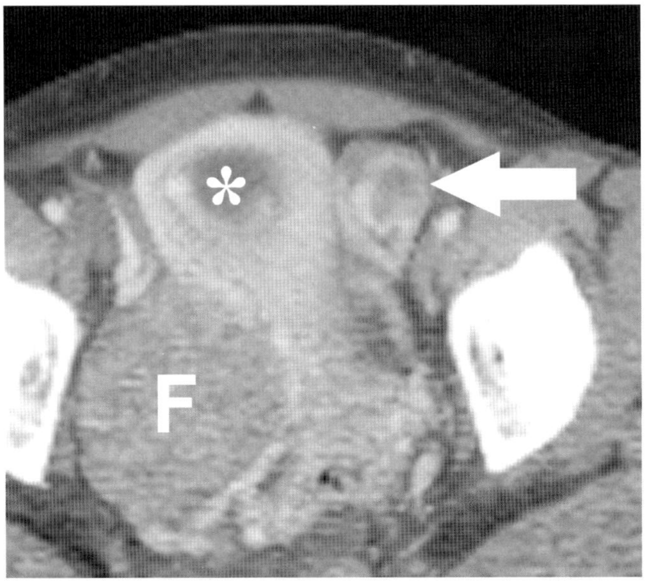

Figure 72.6 Axial contrast-enhanced CT image performed after failed termination of pregnancy (cervical os could not be cannulated) in a 36 year old woman at 8 weeks' gestation shows a corpus luteum cyst (arrow) in the left ovary; a normal physiological finding in early pregnancy. A large exophytic fibroid (F) is incidentally seen compressing the cervix. An intrauterine gestation (asterisk) is present. Pregnancy was subsequently terminated by open surgery, with resection of the fibroid.

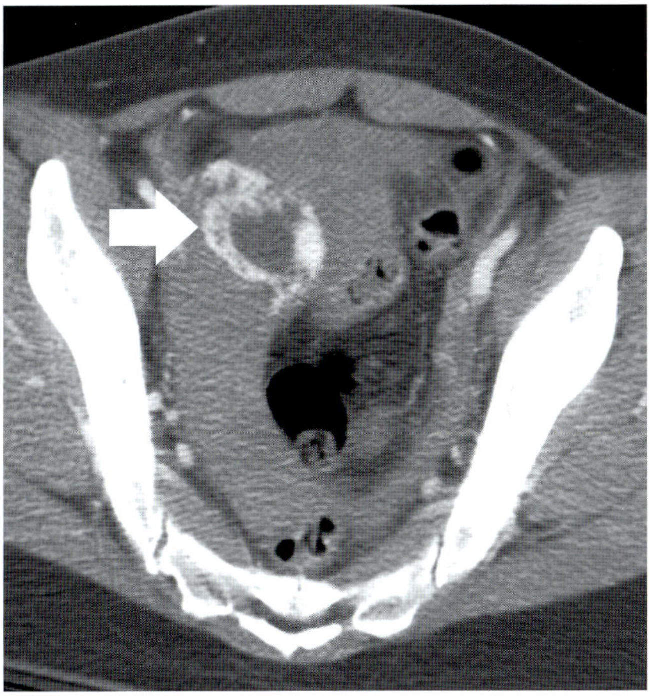

Figure 72.7 Axial contrast-enhanced CT image in a 40 year old woman with acute abdominal pain six weeks after a total hysterectomy for persistent high-grade cervical dysplasia shows high-density free fluid in the pelvis associated with a briskly rim-enhancing structure (arrow) in the right adnexa. The appearance resembles a ruptured corpus luteum cyst, but the serum beta hCG was 4300 IU/mL and a right salpingoophorectomy confirmed the diagnosis of a ruptured ectopic pregnancy.

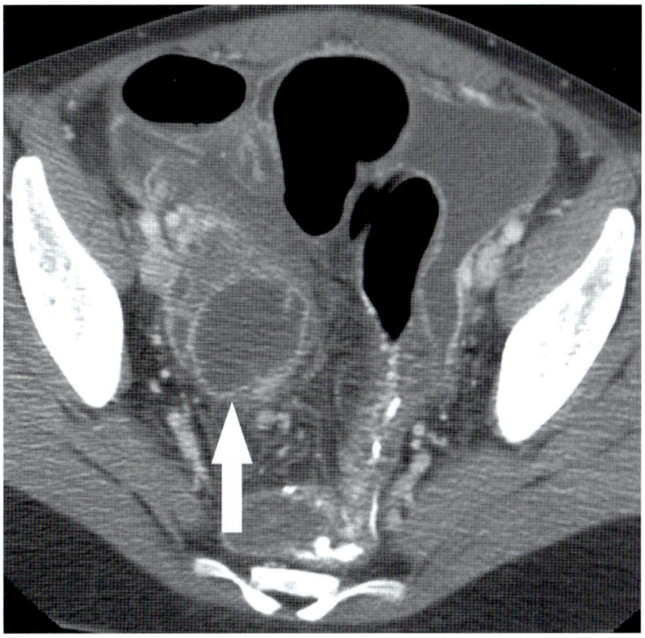

Figure 72.8 Axial contrast-enhanced CT image in a 32 year old woman with persistent pelvic pain and fever one year after proctocolectomy with ileal J pouch formation for Crohn's colitis shows a rim-enhancing fluid-filled structure (arrow) in the right pelvis that could be mistaken for a corpus luteum cyst on imaging alone. Surgery demonstrated a pelvic abscess (unrelated to the right adnexal) secondary to a small contained leak from the J pouch.

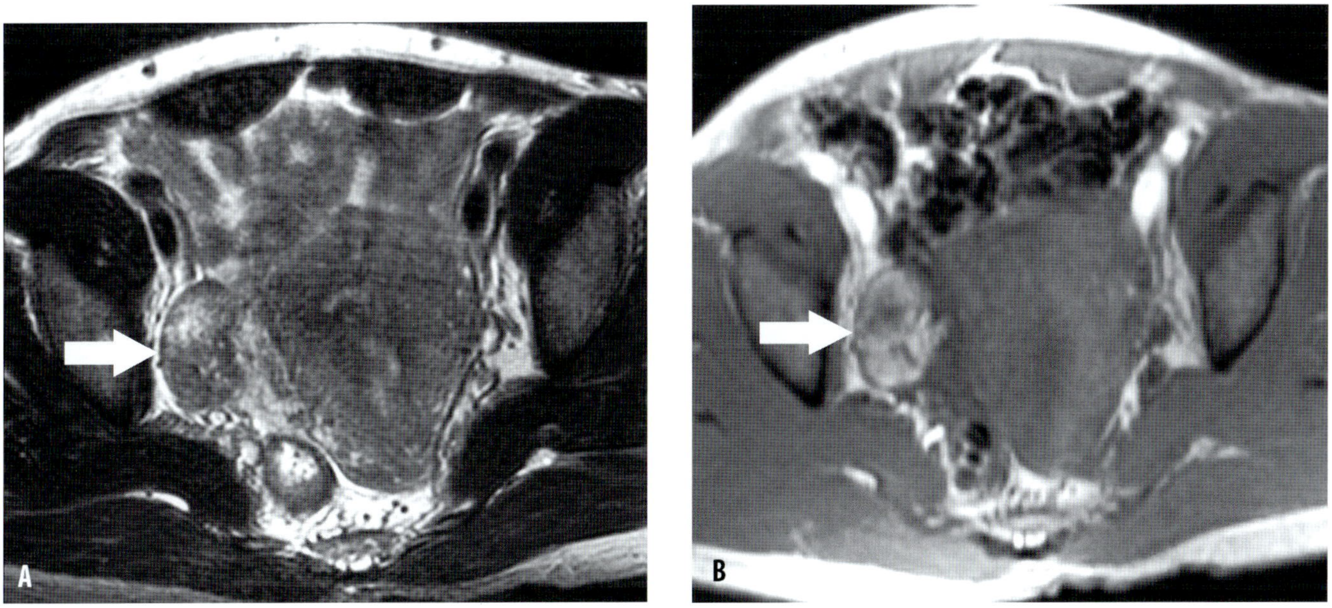

Figure 72.9 A. Axial T2-weighted MR image in a 55 year old woman with virilization and a highly elevated serum testosterone level shows a centrally hyperintense lesion (arrow) in the right ovary. **B.** Axial T1-weighted MR image after intravenous gadolinium shows brisk peripheral enhancement (arrow). The lesion resembles a corpus luteum cyst, but the patient is postmenopausal and this diagnosis would not explain virilization. Surgical resection confirmed a Sertoli-Leydig cell tumor.

CASE 73 Peritoneal inclusion cyst

Imaging description

Peritoneal inclusion cysts are collections of ovulatory fluid that are trapped by peritoneal adhesions around the ovary. This pathophysiology leads naturally to the three features that suggest the diagnosis of a peritoneal inclusion cyst:

- The patient is of reproductive age (i.e., ovulating).
- The patient usually has known risk factors for adhesions, such as prior pelvic surgery, endometriosis, or pelvic inflammatory disease.
- A fluid-filled structure that surrounds or abuts an ovary is seen in the pelvis at cross-sectional imaging. The fluid typically conforms to the outline of the pelvic peritoneal cavity (Figure 73.1) [1, 2]. Ultrasound may demonstrate a spiderweb pattern of fine internal septations (Figure 73.2) [3].

Importance

The cystic and solid appearance of a peritoneal inclusion cyst in combination with the surrounded ovary may result in consideration of adnexal malignancy, resulting in unnecessary surgery and patient anxiety. Correct identification of a peritoneal inclusion cyst should result in conservative therapy rather than salpingoophorectomy.

Typical clinical scenario

Peritoneal inclusion cysts are typically detected when pelvic imaging is performed in a woman of reproductive age with a history of prior pelvic surgery (Figure 73.3), endometriosis (Figure 73.4), or pelvic inflammatory disease.

Differential diagnosis

In a large surgical series of 213 patients with 13 proven peritoneal inclusion cysts, 8 cases were correctly identified by preoperative ultrasound. An additional 8 cases were incorrectly identified as peritoneal inclusion cysts on preoperative ultrasound, with final pathological diagnoses of hemorrhagic cyst, hydrosalpinx, mucinous cystadenoma, dermoid cyst, serous cystadenoma, benign Brenner tumor, simple cyst, and paraovarian cyst. While this suggests some limitations in the diagnosis of peritoneal inclusion cyst by imaging, it is reassuring that all the false positive diagnoses were benign.

Teaching point

A fluid-filled mass that conforms to the shape of the pelvis and surrounds an ovary in a woman of reproductive age with a history of pelvic surgery, endometriosis, or pelvic inflammatory disease is likely to be a peritoneal inclusion cyst.

REFERENCES

1 Jain KA. Imaging of peritoneal inclusion cysts. *Am J Roentgenol* 2000; **174**: 1559–1563.

2 Kim JS, Lee HJ, Woo SK, Lee TS. Peritoneal inclusion cysts and their relationship to the ovaries: evaluation with sonography. *Radiology* 1997; **204**: 481–484.

3 Guerriero S, Ajossa S, Mais V, et al. Role of transvaginal sonography in the diagnosis of peritoneal inclusion cysts. *J Ultrasound Med* 2004; **23**: 1193–1200.

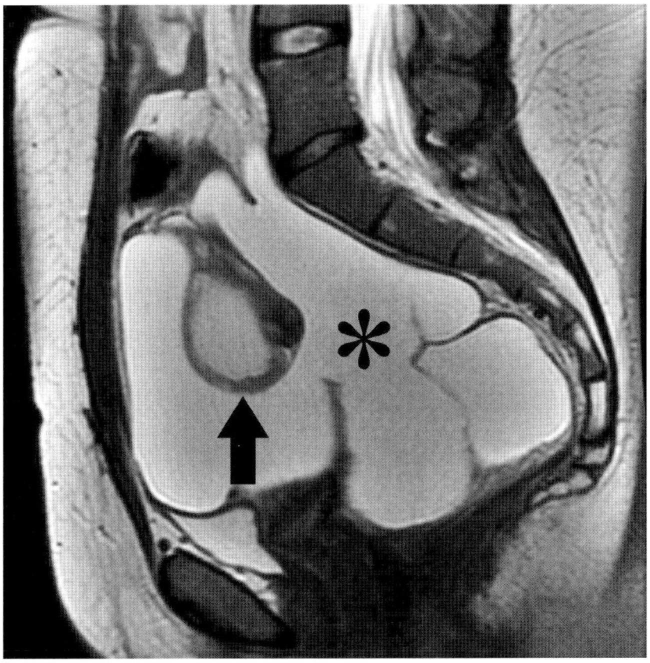

Figure 73.1 Sagittal T2-weighted MR image through the left pelvis in a 28 year old woman with a left adnexal mass detected at ultrasound. The patient underwent total colectomy for ulcerative colitis 11 years earlier. The fluid-filled left pelvic mass (asterisk) is seen to conform to the outline of the pelvic cavity, and to surround the left ovary (arrow). The appearances are typical of a peritoneal inclusion cyst.

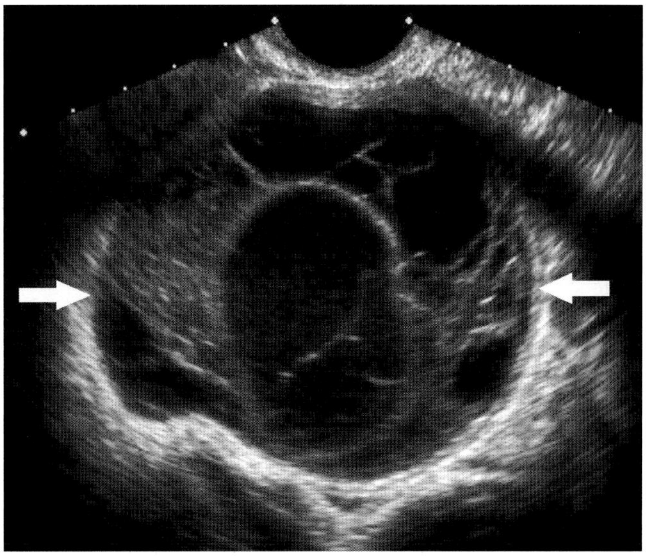

Figure 73.2 Axial transabdominal ultrasound in a 41 year old woman complaining of pelvic pressure 4 months after radical hysterectomy and pelvic node dissection for cervical cancer shows the typical spiderweb appearance of a peritoneal inclusion cyst (between arrows). The lesion subsequently resolved spontaneously.

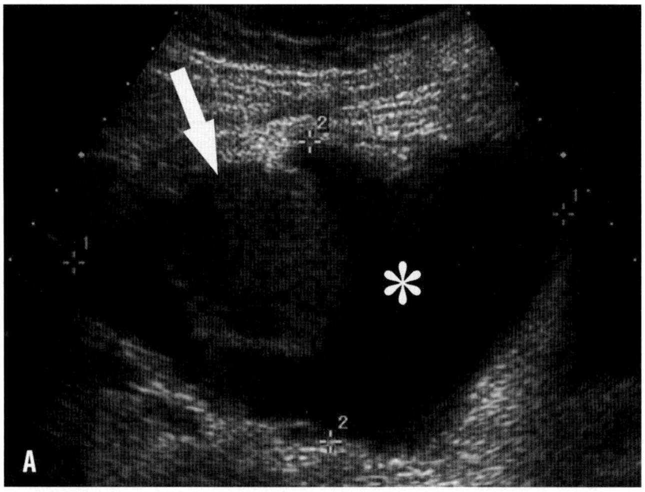

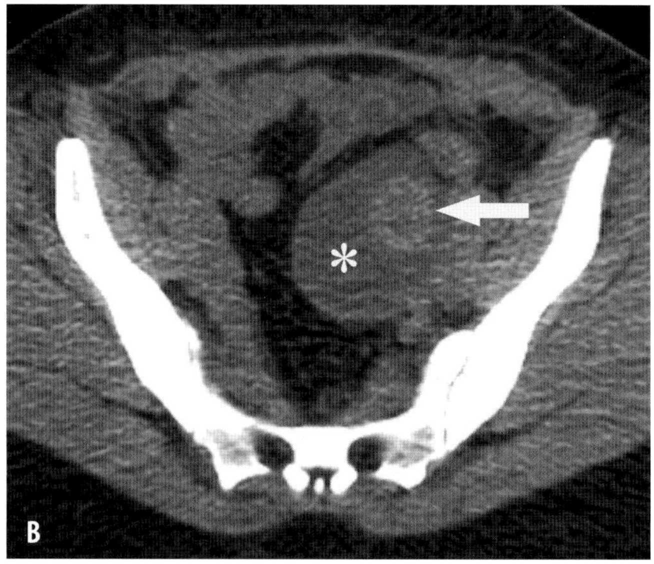

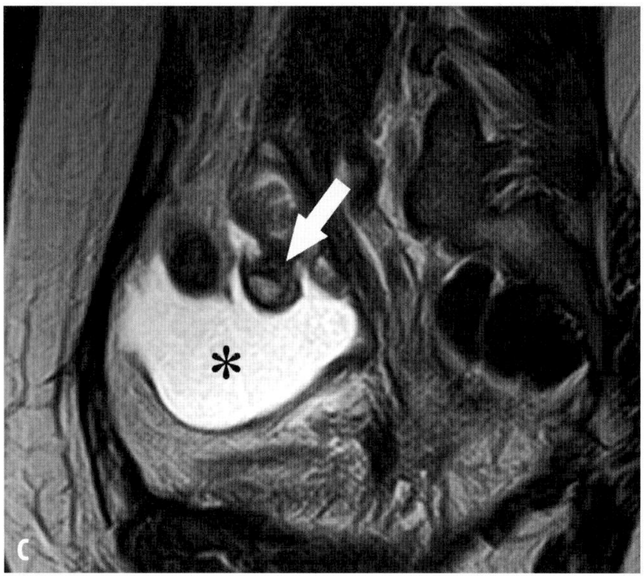

Figure 73.3 **A.** Sagittal transabdominal ultrasound image through the left pelvis in a 45 year old woman with a history of radical hysterectomy for cervical cancer shows a large fluid-filled structure (asterisk) surrounding an ovoid mass (arrow). **B.** Axial non-enhanced CT image of the pelvis also demonstrates a left-sided fluid-filled structure (asterisk) surrounding an ovoid mass (arrow). **C.** Sagittal T2-weighted MR image through the left pelvis shows the fluid-filled left pelvic mass (asterisk) conforms to the outline of the pelvic cavity, and that the ovoid mass (arrow) contains a small cyst, likely a follicle. The findings are those of a peritoneal inclusion cyst around the left ovary.

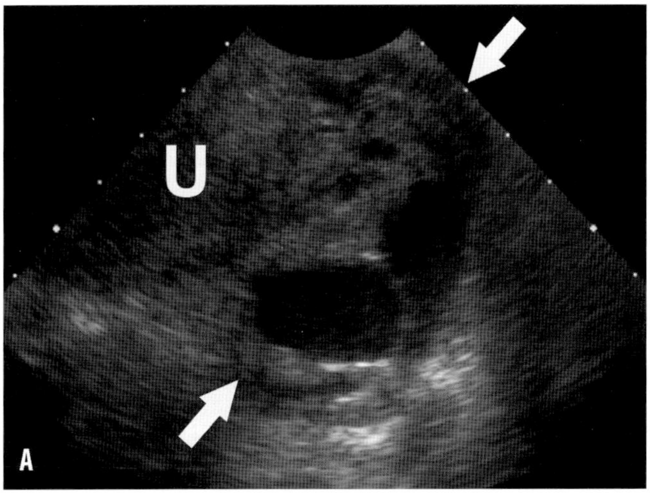

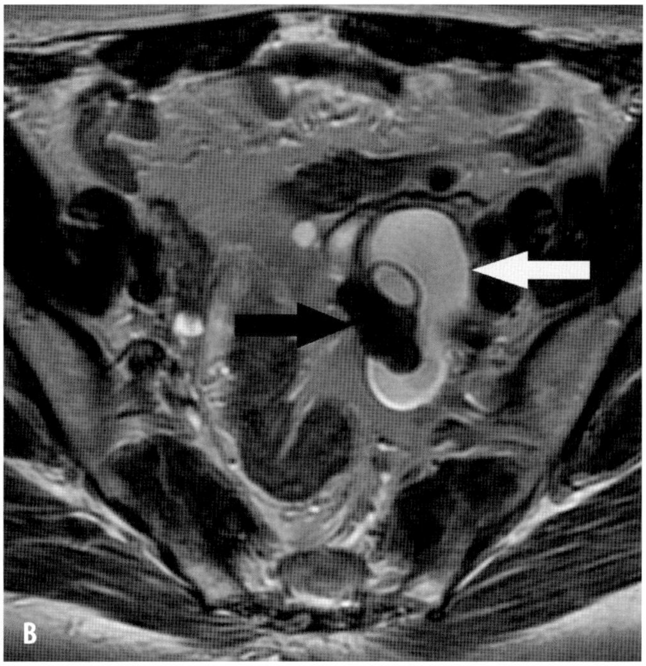

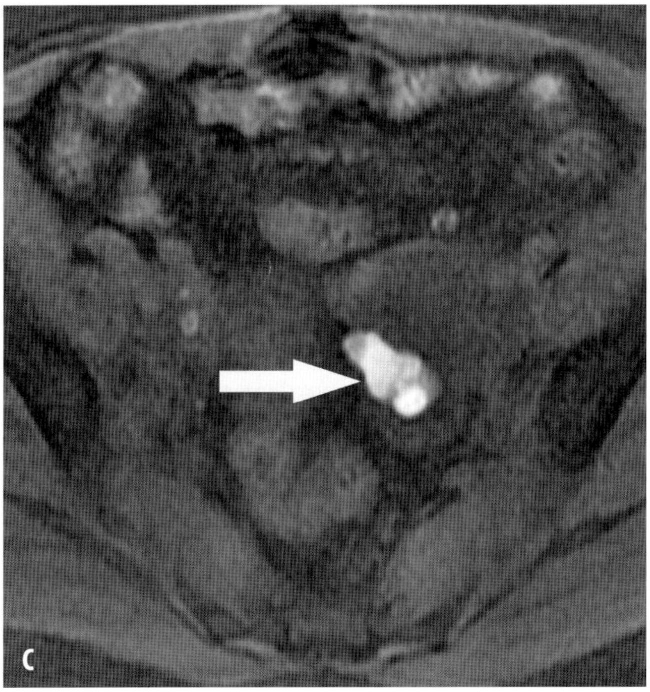

Figure 73.4 A. Axial transvaginal ultrasound image in a 48 year old woman with a history of endometriosis shows a complex left adnexal mass (arrows) adjacent to the uterus (U). **B.** Axial T2-weighted MR image through the left pelvis shows the adnexal mass consists of fluid (white arrow) surrounding the left ovary (black arrow). Note much of the left ovary is of markedly reduced T2 signal intensity. **C.** Axial T1-weighted MR image with fat saturation shows that the parts of the ovary that were of low T2 signal intensity are of high T1 signal intensity (arrow), consistent with blood products. The overall findings suggest the diagnosis of a peritoneal inclusion cyst associated with endometriosis.

CASE

74 Adnexal pseudotumor due to exophytic uterine fibroid

Imaging description

At ultrasound, it is usually straightforward to determine that a mass adjacent to the uterus is an exophytic fibroid rather than an adnexal mass. However, sometimes this distinction can be difficult or impossible and then an exophytic uterine fibroid can mimic an adnexal mass (Figures 74.1–74.4) [1–3]. It is useful to remember the aphorism that "the commonest cause of an apparent solid adnexal mass at ultrasound is an exophytic fibroid." Exophytic fibroids, like other fibroids, can undergo cystic degeneration, and can then mimic a cystic adnexal mass (Figure 74.2) [4–6].

Importance

The recognition that a mass adjacent to the uterus is an exophytic fibroid can be of major clinical importance, since it eliminates adnexal malignancy as a differential consideration. Occasionally an exophytic fibroid can torse and be a cause of pain. The finding of an exophytic fibroid with little or no enhancement on CT or MRI and with or without some degree of ascites in a patient with acute pelvic pain is suggestive of torsion (Figure 74.5) [7, 8], remembering that a poorly enhancing mass adjacent to the uterus can be an ovarian fibroma (Figure 74.6). In the latter case, the history of acute pain will typically be absent.

Typical clinical scenario

Like other fibroids, exophytic fibroids are primarily seen in women of older reproductive age.

Differential diagnosis

Several imaging findings are helpful in the accurate characterization of an exophytic uterine fibroid:

- Identification of the ipsilateral ovary as a separate structure is probably the easiest and simplest way of establishing that a mass adjacent to the uterus is likely to be an exophytic fibroid.
- If the ipsilateral ovary is not demonstrable or further evaluation is required, MRI is often very helpful by demonstrating the mass to be of low T2 signal intensity (characteristic of uncomplicated fibroids) [1].
- Even if the mass is truly of adnexal origin, intralesional low T2 signal is still reassuring because it strongly suggests the diagnosis of ovarian fibroma or primary leiomyoma of the adnexa (and mistaking one benign tumor for another benign tumor is not usually of great clinical consequence) [1].
- Ovarian fibromas demonstrate slower and lesser enhancement on dynamic-enhanced MRI, which may be a useful distinguishing feature [9].
- In addition to low T2 signal intensity, MRI may show connection between the mass and the uterus, a beak or claw-shaped interface between the mass and the uterus, or bridging vessels crossing between the two – all signs that indicate a uterine origin (Figures 74.1 to 74.3). The "bridging vascular sign" refers to the detection of blood vessels running across the interface between the uterus and the adjacent mass at Doppler ultrasound or MRI (Figure 74.1). The presence of such vessels indicates a uterine origin for a mass adjacent to the uterus with reportedly high positive predictive value [10, 11].

Teaching point

The possibility of an exophytic fibroid should be considered for any pelvic mass adjacent to the uterus, particularly if solid and with a separately visible ipsilateral ovary. MRI can be helpful in confirming this diagnosis.

REFERENCES

1 Weinreb JC, Barkoff ND, Megibow A, Demopoulos R. The value of MR imaging in distinguishing leiomyomas from other solid pelvic masses when sonography is indeterminate. *Am J Roentgenol* 1990; **154**: 295–299.

2 Baltarowich OH, Kurtz AB, Pennell AC, et al. Pitfalls in the sonographic diagnosis of uterine fibroids. *Am J Roentgenol* 1988; **151**: 725–728.

3 Landrum LM, Rutledge TL, Osunkoya AO, Mannel R. Uterine leiomyoma with associated pseudo-Meigs syndrome mimicking ovarian carcinoma. *J Okla State Med Assoc* 2008; **101**: 38–39.

4 Low SC, Chong CL. A case of cystic leiomyoma mimicking an ovarian malignancy. *Ann Acad Med Singapore* 2004; **33**: 371–374.

5 Yarwood RL, Arroyo E. Cystic degeneration of a uterine leiomyoma masquerading as a postmenopausal ovarian cyst. A case report. *J Reprod Med* 1999; **44**: 649–652.

6 Dancz CE, Macdonald HR. Massive cystic degeneration of a pedunculated leiomyoma. *Fertil Steril* 2008; **90**: 1180–1181.

7 Roy C, Bierry G, El Ghali S, Buy X, Rossini A. Acute torsion of uterine leiomyoma: CT features. *Abdom Imaging* 2005; **30**: 120–123.

8 Marcotte-Bloch C, Novellas S, Buratti MS, et al. Torsion of a uterine leiomyoma: MRI features. *Clin Imaging* 2007; **31**: 360–362.

9 Thomassin-Naggara I, Daraï E, Nassar-Slaba J, et al. Value of dynamic enhanced magnetic resonance imaging for distinguishing between ovarian fibroma and subserous uterine leiomyoma. *J Comput Assist Tomogr* 2007; **31**: 236–242.

10 Madan R. The bridging vascular sign. *Radiology* 2006; **238**: 371–372.

11 Kim JC, Kim SS, Park JY. "Bridging vascular sign" in the MR diagnosis of exophytic uterine leiomyoma. *J Comput Assist Tomogr* 2000; **24**: 57–60.

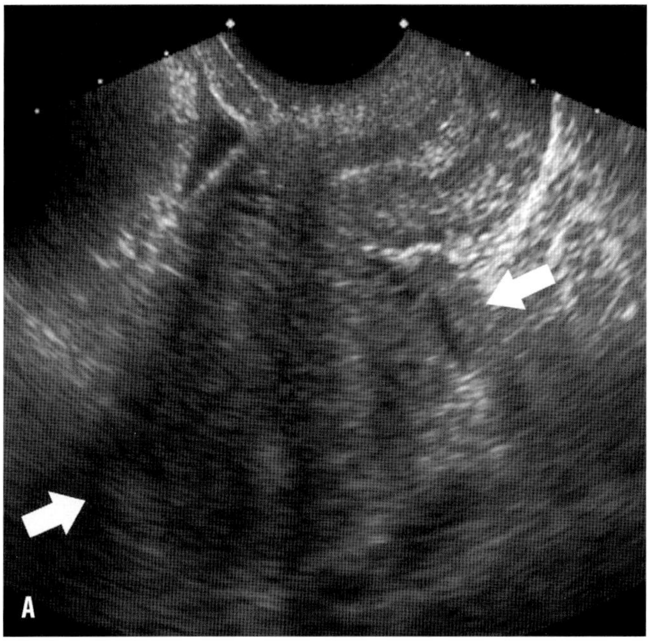

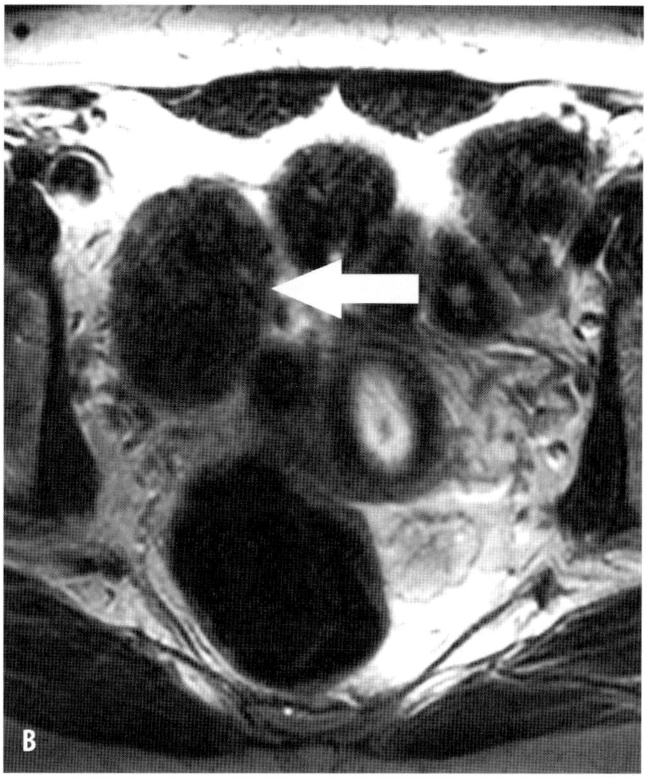

Figure 74.1 A. Endovaginal ultrasound image in a 34 year old woman with left lower quadrant pain shows an echogenic solid mass (between arrows) adjacent to the uterus. The possibility of an adnexal mass was raised. **B.** Axial T2-weighted MR image shows the mass (arrow) is of low T2 signal intensity, suggesting either an exophytic leiomyoma or an ovarian fibroma as the leading diagnostic possibilities. **C.** Axial T2-weighted MR image at a more superior level shows the follicle-containing right ovary (arrow) is separate to the mass, favoring an exophytic fibroid. **D.** Axial post-gadolinium T1-weighted MR image shows the mass (arrow) is strongly enhancing, also favoring an exophytic fibroid over an ovarian fibroma (the latter are typically minimally enhancing). **E.** Axial post-gadolinium T1-weighted MR image in the early phase of enhancement shows a blood vessel (arrow) connecting the uterus and the parauterine mass. The constellation of findings and the presence of a bridging vessel were considered diagnostic of an exophytic uterine fibroid.

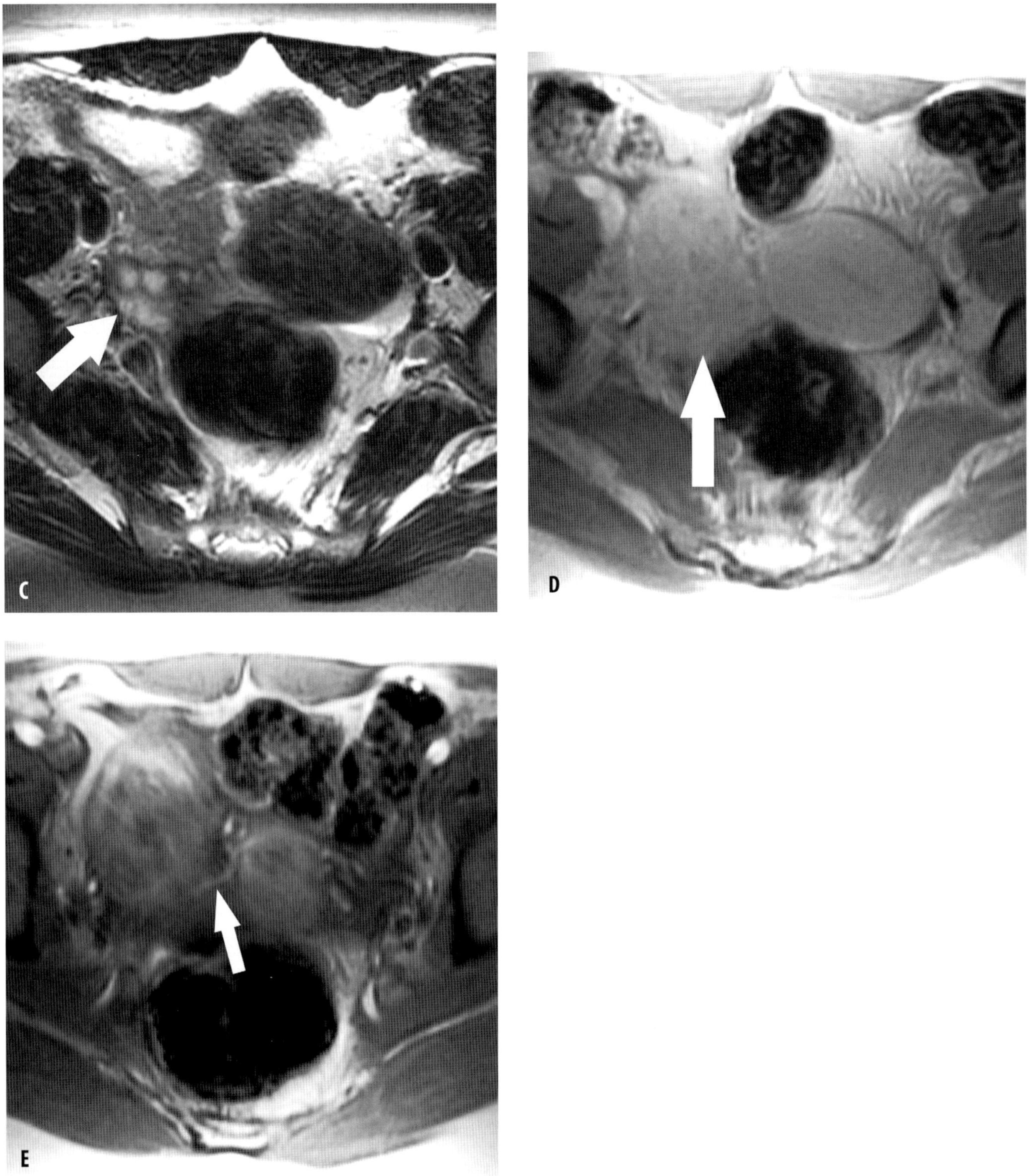

Figure 74.1 (cont.)

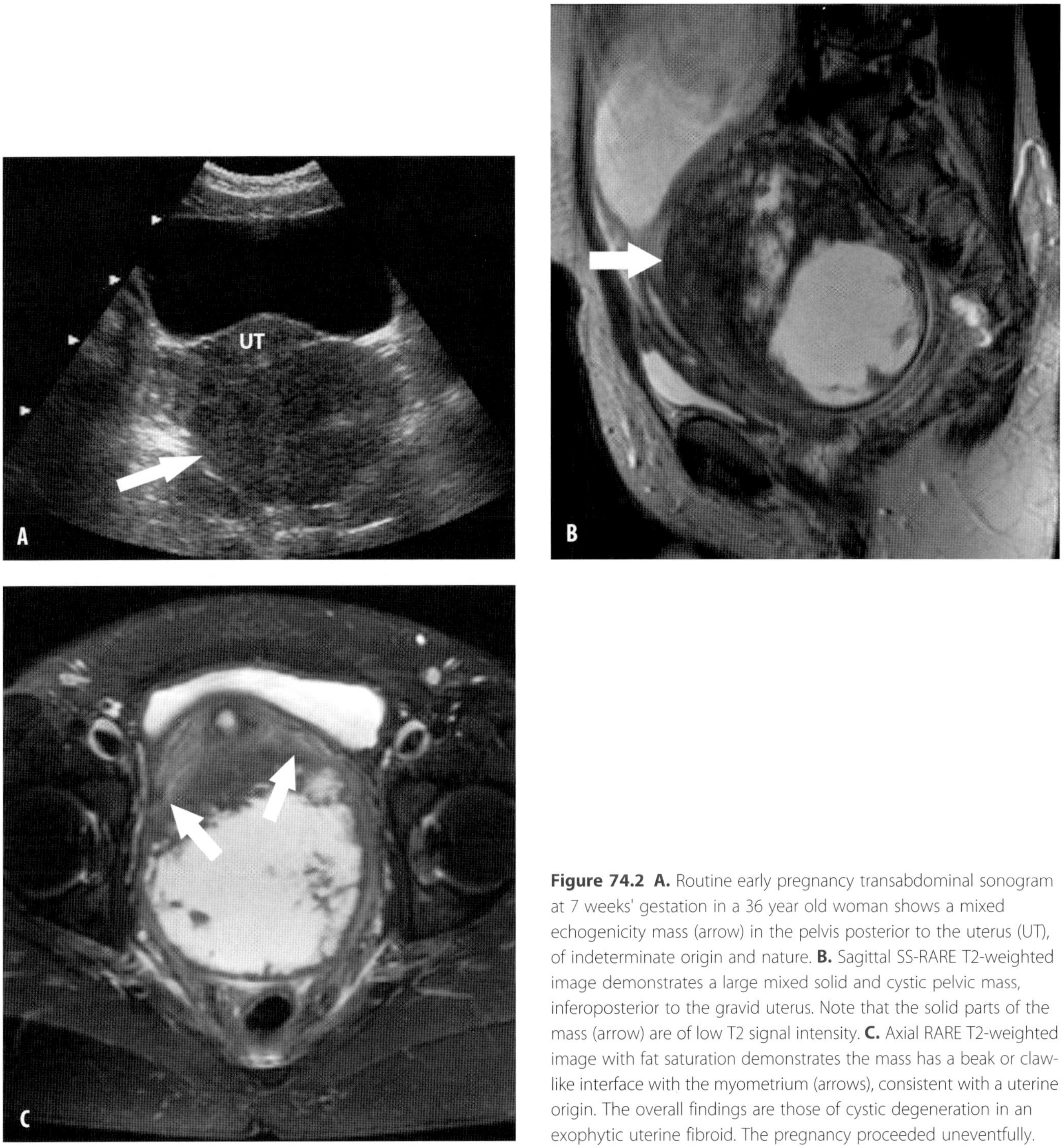

Figure 74.2 A. Routine early pregnancy transabdominal sonogram at 7 weeks' gestation in a 36 year old woman shows a mixed echogenicity mass (arrow) in the pelvis posterior to the uterus (UT), of indeterminate origin and nature. **B.** Sagittal SS-RARE T2-weighted image demonstrates a large mixed solid and cystic pelvic mass, inferoposterior to the gravid uterus. Note that the solid parts of the mass (arrow) are of low T2 signal intensity. **C.** Axial RARE T2-weighted image with fat saturation demonstrates the mass has a beak or claw-like interface with the myometrium (arrows), consistent with a uterine origin. The overall findings are those of cystic degeneration in an exophytic uterine fibroid. The pregnancy proceeded uneventfully.

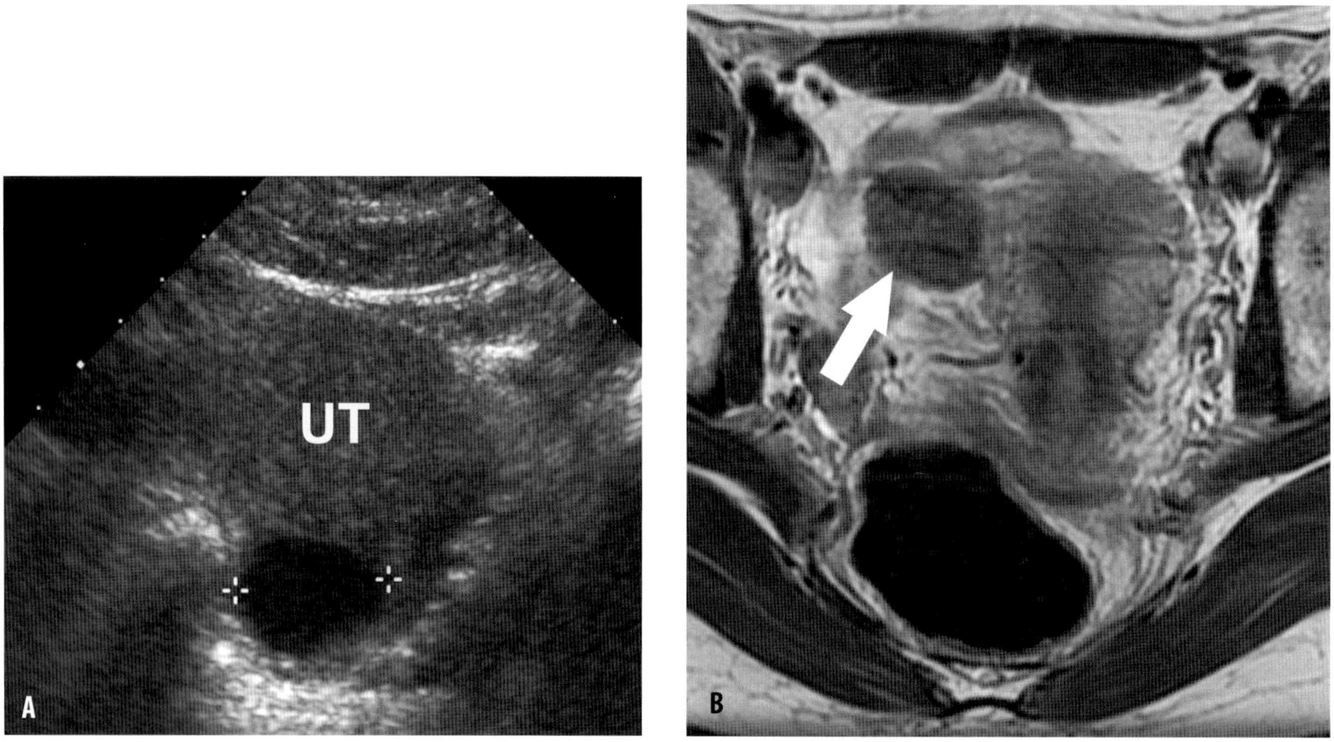

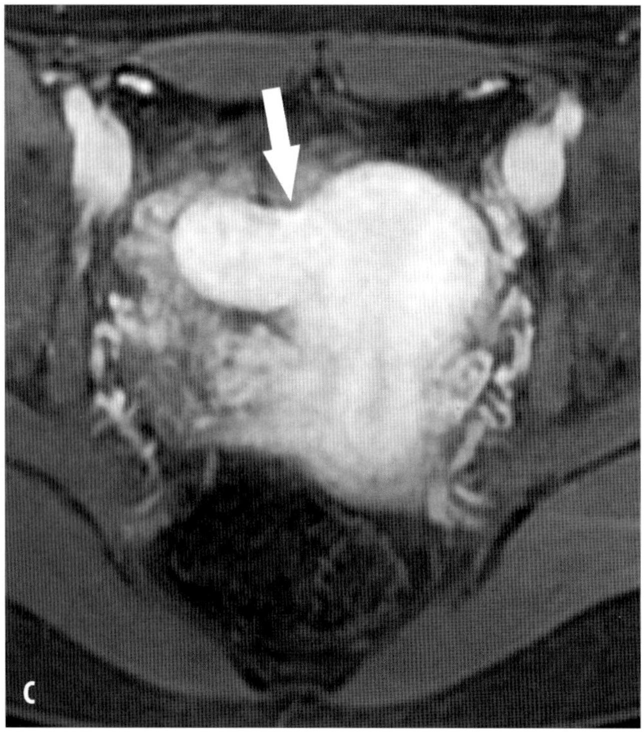

Figure 74.3 A. Endovaginal ultrasound image in a 38 year old woman with irregular periods shows an echogenic solid mass (between calipers) adjacent to the uterus (UT). The possibility of an adnexal mass was raised. **B.** Axial T2-weighted MR image shows the mass (arrow) is of low T2 signal intensity, suggesting either an exophytic leiomyoma or an ovarian fibroma. **C.** Axial post-gadolinium T1-weighted MR image shows the mass is strongly enhancing and connects (arrow) to the uterus, establishing an exophytic fibroid as by far the most likely diagnosis.

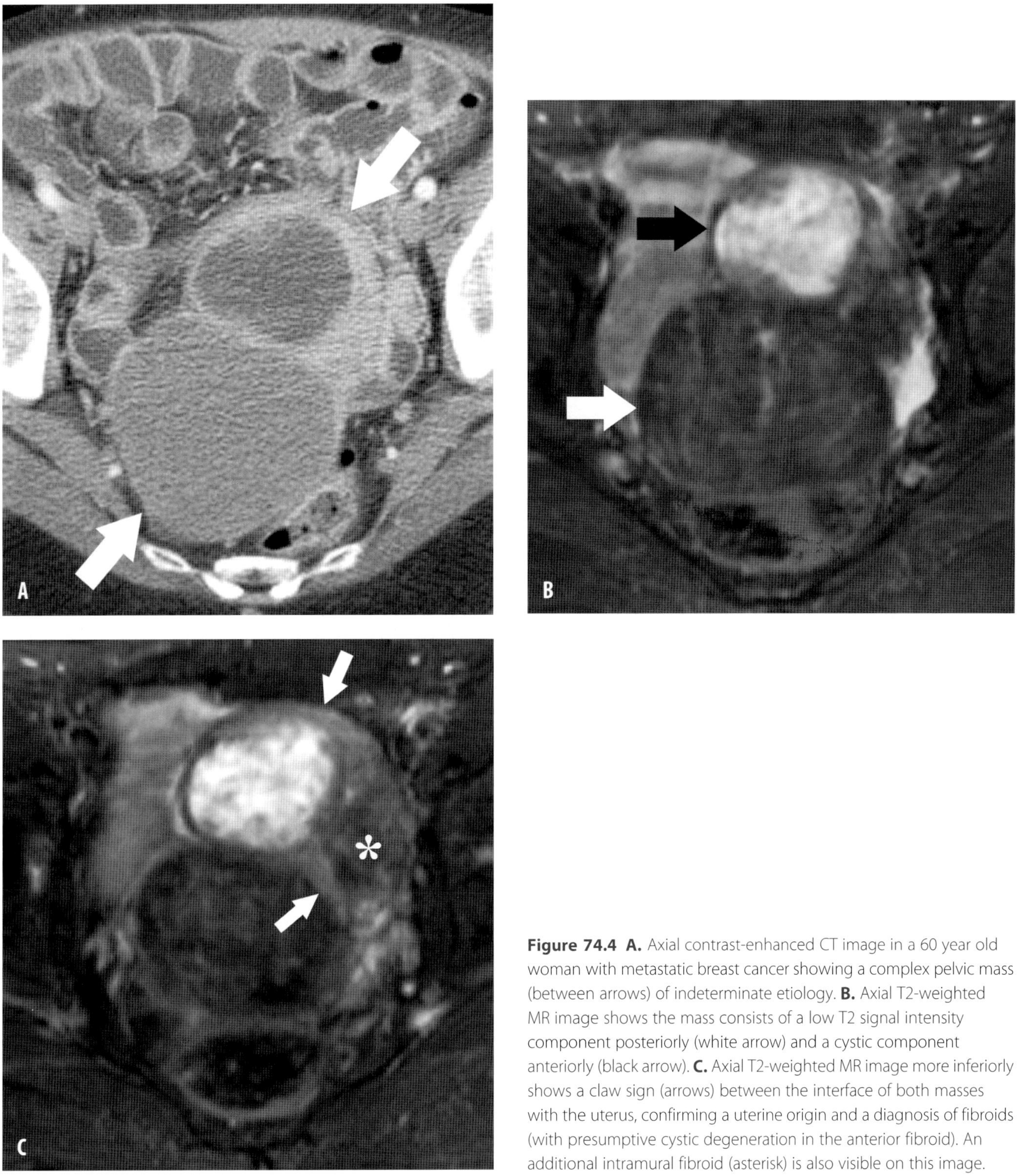

Figure 74.4 A. Axial contrast-enhanced CT image in a 60 year old woman with metastatic breast cancer showing a complex pelvic mass (between arrows) of indeterminate etiology. **B.** Axial T2-weighted MR image shows the mass consists of a low T2 signal intensity component posteriorly (white arrow) and a cystic component anteriorly (black arrow). **C.** Axial T2-weighted MR image more inferiorly shows a claw sign (arrows) between the interface of both masses with the uterus, confirming a uterine origin and a diagnosis of fibroids (with presumptive cystic degeneration in the anterior fibroid). An additional intramural fibroid (asterisk) is also visible on this image.

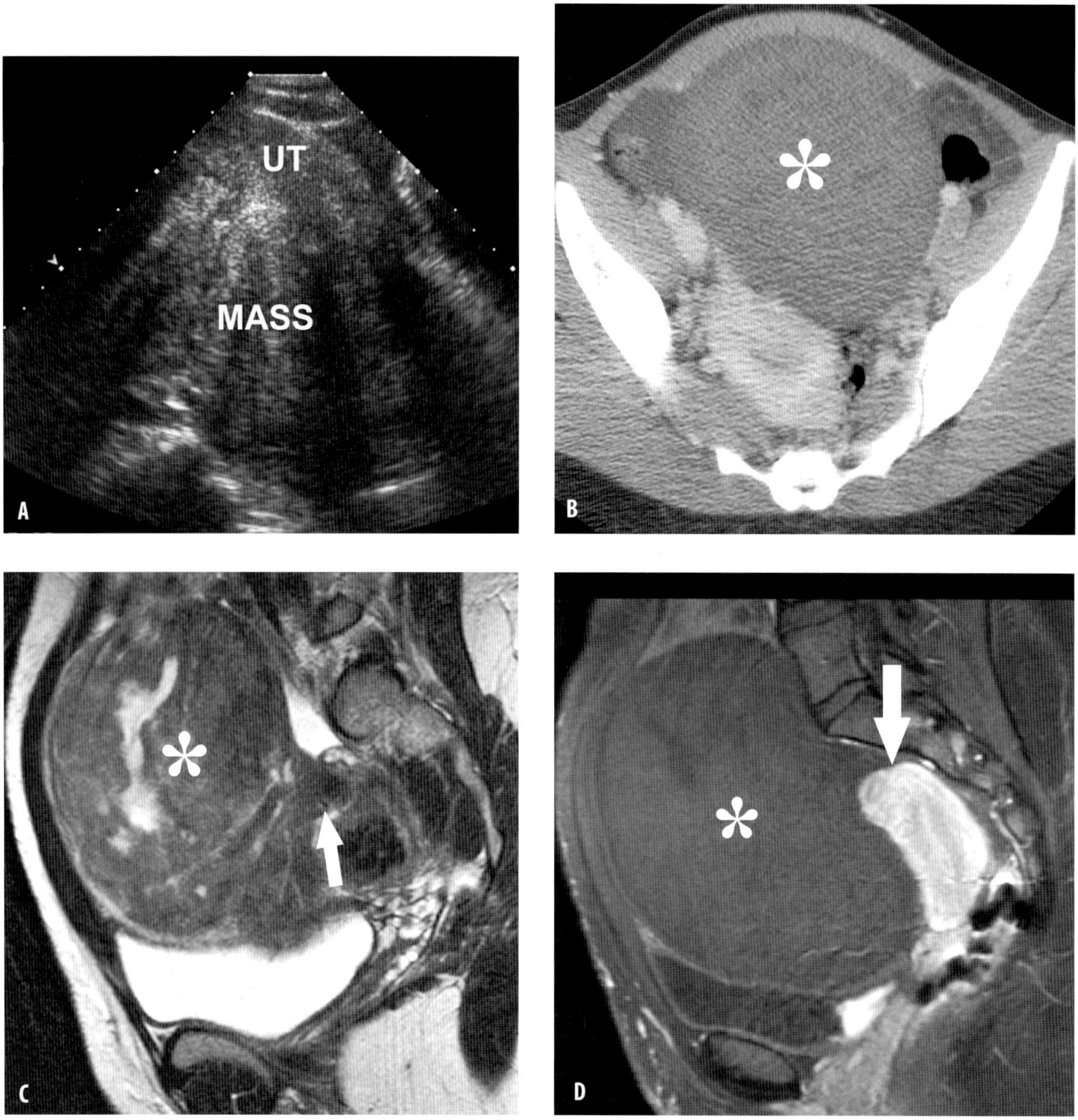

Figure 74.5 A. Endovaginal ultrasound image in a 29 year old woman with sudden onset of severe pelvic pain and a negative pregnancy test. An ill-defined solid echogenic mass (MASS) is visible adjacent to the uterus (UT). **B.** Axial contrast-enhanced CT image shows the mass (asterisk) appears hypodense and non-enhancing. **C.** Sagittal T2-weighted MR image shows the mass (asterisk) is of generally low T2 signal intensity and has a narrow stalk-like connection (arrow) to the uterus. These features strongly suggest the diagnosis of an exophytic fibroid. **D.** Sagittal post-gadolinium T1-weighted MR image shows the mass (asterisk) has no perceptible enhancement, despite the normal brisk physiological enhancement seen in the uterus (arrow). The lack of enhancement at CT and MRI in conjunction with the history of sudden-onset pain suggested torsion of a pedunculated fibroid, and this diagnosis was confirmed at surgery.

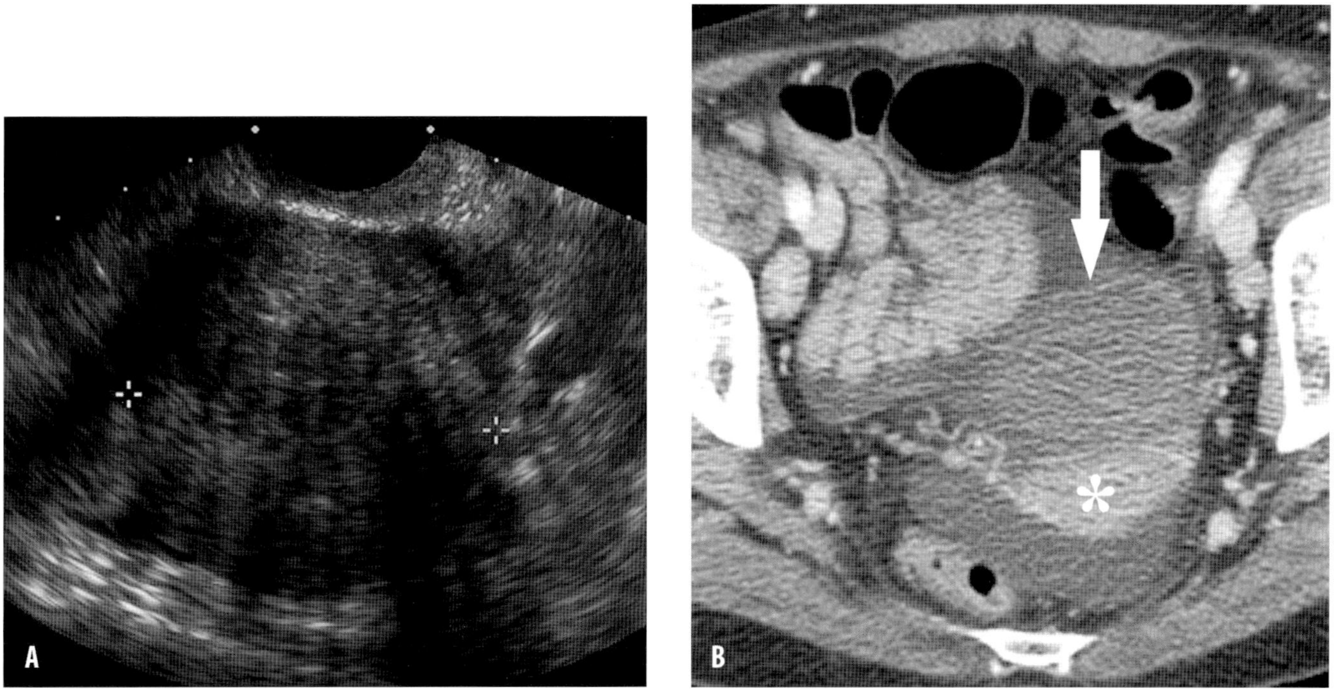

Figure 74.6 A. Endovaginal ultrasound image in a 74 year old woman with a two-year history of intermittent upper abdominal pain shows a solid mass (between calipers) adjacent to the uterus. **B.** Axial contrast-enhanced CT image showing a solid poorly enhancing mass (arrow) adjacent to the uterus (asterisk) with a small amount of free fluid. In isolation, the CT findings might suggest a torsed exophytic fibroid, but the patient did not have acute pelvic pain. Resection revealed an ovarian fibroma.

CASE 75 Malignant transformation of endometrioma

Imaging description

Malignant degeneration of an endometrioma (i.e., endometriosis involving the ovary) should be suspected when enhancing nodules are seen in an ovarian cyst with features otherwise suggestive of an endometrioma at MRI (Figures 75.1 and 75.2) [1, 2]. Features that suggest an endometrioma at MRI are increased T1 signal intensity and variable degrees of reduced T2 signal intensity ("shading") [3]. These signal changes reflect the presence of blood products.

Importance

Failure to recognize features suggestive of malignant degeneration in an endometrioma may result in a missed opportunity for cure.

Typical clinical scenario

Malignant transformation occurs in approximately 0.6 to 1% of women with ovarian endometriosis [2]. Clear cell adenocarcinoma is the most common endometriosis-associated ovarian cancer, followed by endometroid cancer. Only a few studies have described the clinical features of endometriosis-related malignancies, and the clinical presentation and appropriate preoperative imaging of these patients are not well established [4]. In one study of two patients [1], CA-125 was normal in each, suggesting tumor markers may be of limited utility and highlighting the importance of imaging in making the diagnosis of superimposed malignancy.

Differential diagnosis

The role of PET in detecting malignant transformation of endometriosis has not yet been described, although increased uptake of FDG by uncomplicated endometriomas can occur and suggests PET uptake may be of limited discriminative value in this setting [5]. Occasionally, a blood-filled ovarian cystadenocarcinoma may mimic ovarian cancer arising in an endometrioma (Figure 75.3).

Teaching point

An apparent endometrioma at MRI should be carefully scrutinized for solid enhancing elements that may indicate complicating malignancy.

REFERENCES

1 Wu TT, Coakley FV, Qayyum A, et al. Magnetic resonance imaging of ovarian cancer arising in endometriomas. *J Comput Assist Tomogr* 2004; **28**: 836–868.

2 Tanaka YO, Yoshizako T, Nishida M, et al. Ovarian carcinoma in patients with endometriosis: MR imaging findings. *Am J Roentgenol* 2000; **175**: 1423–1430.

3 Glastonbury CM. The shading sign. *Radiology* 2002; **224**: 199–201.

4 Leiserowitz GS, Gumbs JL, Oi R, et al. Endometriosis-related malignancies. *Int J Gynecol Cancer* 2003; **13**: 466–471.

5 Rieber A, Nüssle K, Stöhr I, et al. Preoperative diagnosis of ovarian tumors with MR imaging: comparison with transvaginal sonography, positron emission tomography, and histologic findings. *Am J Roentgenol* 2001; **177**: 123–129.

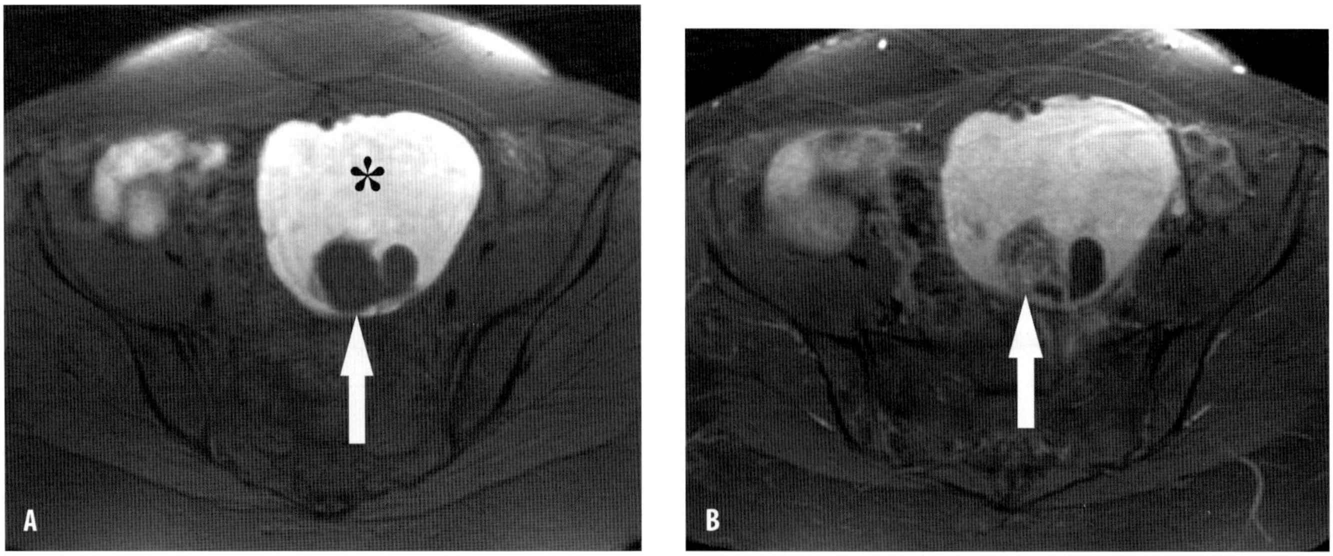

Figure 75.1 A. Axial T1-weighted MR image in a 39 year old woman shows a 9 cm cystic mass (asterisk) with high signal intensity. A 4 cm lobulated hypointense mural nodule (arrow) is seen in the posterior part of the mass. **B.** Axial fat-suppressed post-gadolinium T1-weighted MRI section shows partial enhancement of the mural nodule (arrow). Surgical resection demonstrated an endometrioma with intracystic nodules of stage IC serous carcinoma.

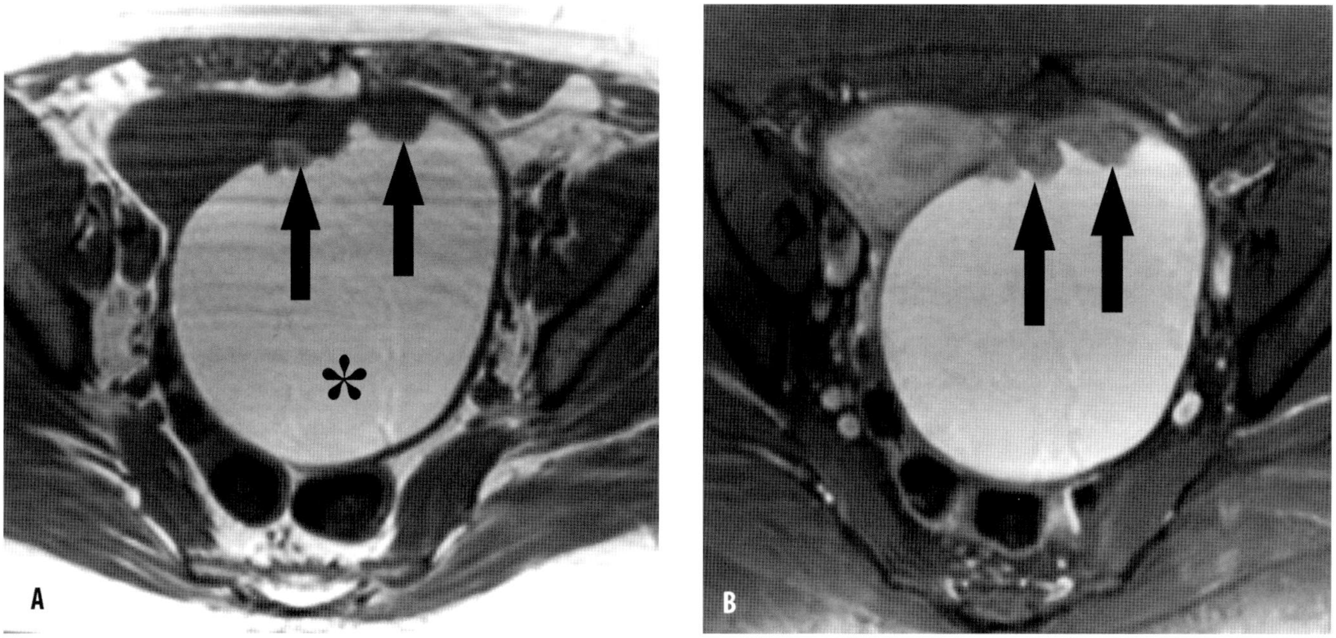

Figure 75.2 A. Axial T1-weighted MR image in a 56 year old woman shows a 14 cm hyperintense cystic mass (asterisk) with hypointense mural nodules (arrows) along the anterior aspect of the mass. **B.** Axial fat-suppressed post-gadolinium T1-weighted MRI section shows enhancement in the mural nodules (arrows). Surgical resection demonstrated an endometrioma with intracystic nodules of stage I clear cell carcinoma.

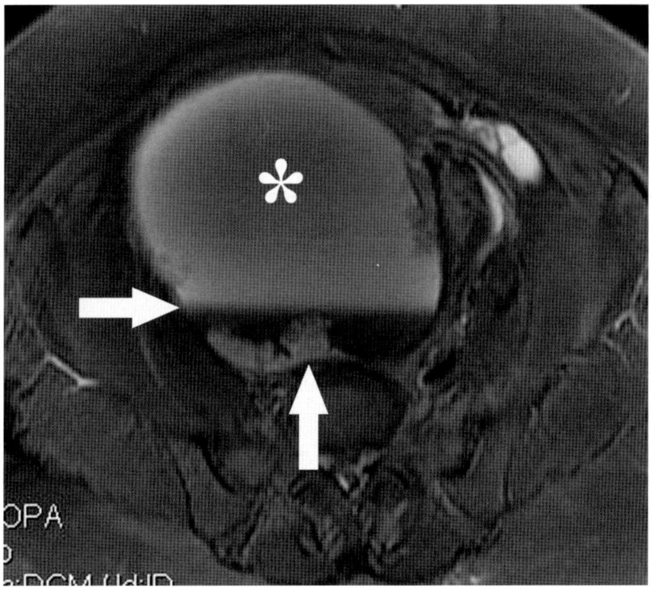

Figure 75.3 Axial fat-suppressed post-gadolinium T1-weighted MR image in a 57 year old woman with a palpable lower abdominal mass and a CA 125 of 1291 U/mL. After surgery, the mass was found to be a serous cystadenocarcinoma containing hemorrhagic watery fluid and friable papillary hemorrhagic excrescences. Occasionally, a blood-filled ovarian cystadenocarcinoma might be a potential mimic of ovarian cancer arising in endometrioma (Figure 75.3).

CASE 76 Ovarian transposition

Imaging description

Ovarian transposition (or oophoropexy) is the surgical relocation of one or both ovaries into a fixed anatomic position, and is usually performed to shield the ovaries from radiation therapy to the pelvis and hence preserve gonadal function in premenopausal women or to prevent recurrent ovarian torsion [1–3]. Typically the ovaries are sutured superolaterally in the paracolic gutters, up to the level of the lowest ribs [4]. At cross-sectional imaging, the displaced ovaries typically appear as mixed solid and cystic ovoid masses in the iliac fossae (Figure 76.1) and may be mistaken for primary or secondary tumors [4]. Specifically, the displaced ovaries may be misinterpreted as peritoneal implants in patients who have a history of pelvic radiation for malignancy [5]. The transposed ovaries are commonly marked with metallic surgical clips [2, 6], which may facilitate correct identification. Other helpful signs are the presence of multiple small intralesional cysts (presumably follicles) and the identification of a vascular pedicle in continuity with the gonadal vessels [7]. The frequency of benign cysts and peritoneal inclusion cysts is increased in transposed ovaries (Figures 76.2 and 76.3) [4, 7].

Importance

Misdiagnosis of ovarian transposition as primary or secondary ovarian tumor may result in unnecessary surgery or treatment, particularly when recurrent malignancy is suggested in patients with a history of irradiated pelvic cancer.

Typical clinical scenario

Transposed ovaries are typically seen when surveillance imaging is performed in women of reproductive age with a history of radiation for pelvic malignancies such as cervical cancer, rectal cancer, or lymphoma.

Differential diagnosis

Differential considerations for a paracolic gutter lesion detected at cross-sectional imaging include a peritoneal implant, colonic mass, abscess, hematoma, lymphocele, or cystic neoplasm. On the right side, possibilities would also include appendiceal tumor or mucocele. Imaging clues to the correct diagnosis of ovarian transposition include bilaterality, surgical markers, vascular pedicle in continuity with the gonadal vessels, and intralesional follicles. After hysterectomy, the position of the ovaries is somewhat variable and they may be up to 7 cm from the vaginal apex [8], so the diagnosis of ovarian transposition should not be considered for ovaries that only appear mildly displaced from their normal location (Figure 76.4).

Teaching point

Solid or cystic paracolic gutter masses in a woman who has received radiation to the pelvis are likely to represent transposed ovaries. The diagnosis can be confirmed by correlation with surgical history and inspection for bilaterality, surgical markers, vascular pedicle in continuity with the gonadal vessels, and intralesional follicles.

REFERENCES

1 Morice P, Juncker L, Rey A, et al. Ovarian transposition for patients with cervical carcinoma treated by radiosurgical combination. *Fertil Steril* 2000; **74**: 743–748.

2 Bisharah M, Tulandi T. Laparoscopic preservation of ovarian function: an underused procedure. *Am J Obstet Gynecol* 2003; **188**: 367–370.

3 Abeş M, Sarihan H. Oophoropexy in children with ovarian torsion. *Eur J Pediatr Surg* 2004; **14**: 168–171.

4 Sella T, Mironov S, Hricak H. Imaging of transposed ovaries in patients with cervical carcinoma. *Am J Roentgenol* 2005; **184**: 1602–1610.

5 Reed DH, Dixon AK, Williams MV. Ovarian conservation at hysterectomy: a potential diagnostic pitfall. *Clin Radiol* 1989; **40**: 274–276.

6 Anderson B, LaPolla J, Turner D, Chapman G, Buller R. Ovarian transposition in cervical cancer. *Gynecol Oncol* 1993; **49**: 206–214.

7 Bashist B, Friedman WN, Killackey MA. Surgical transposition of the ovary: radiologic appearance. *Radiology* 1989; **173**: 857–860.

8 Parulekar SG. Ultrasound evaluation of the posthysterectomy pelvis. *J Clin Ultrasound* 1982; **10**: 265–269.

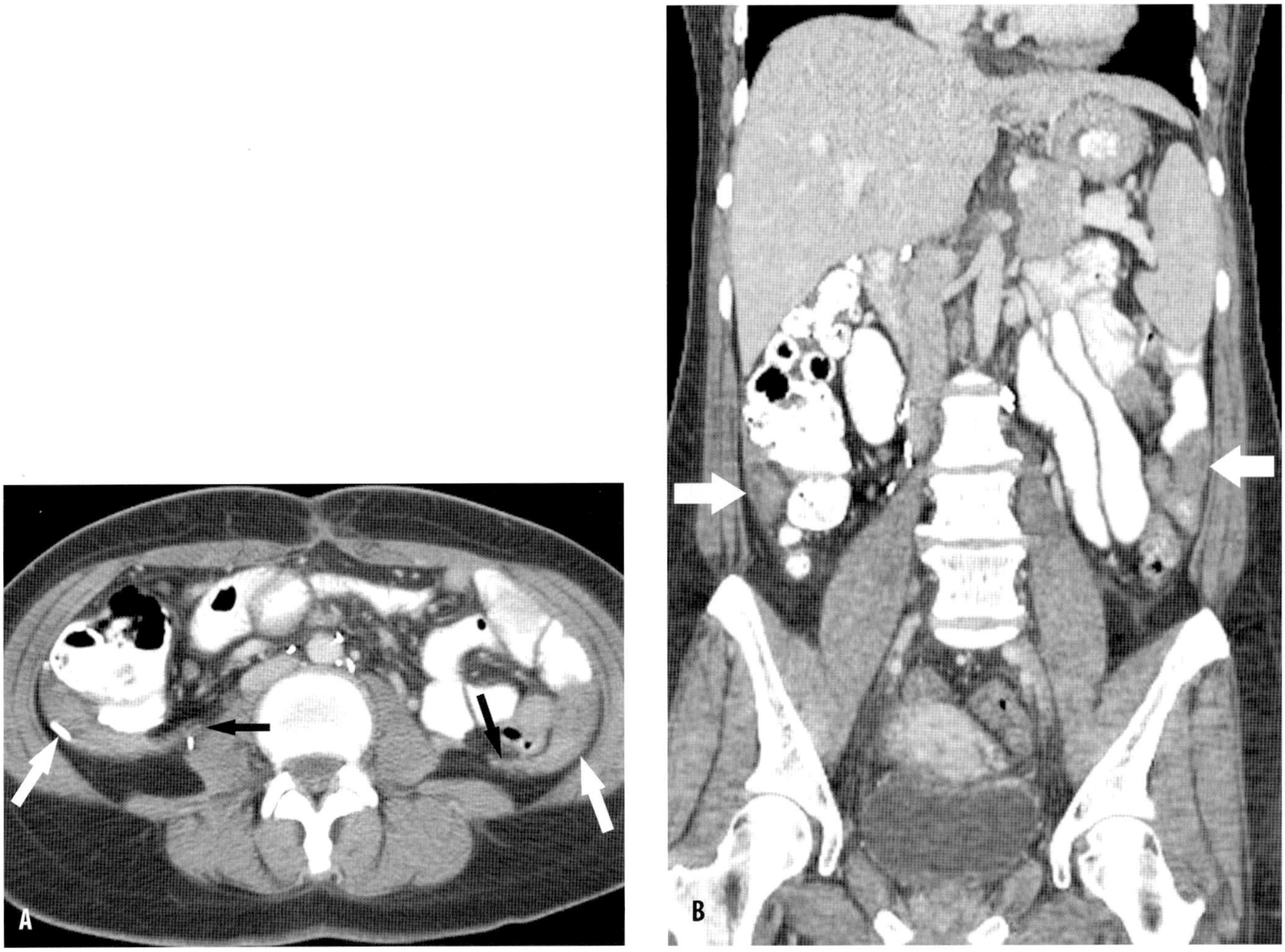

Figure 76.1 A. Axial contrast-enhanced CT image in a 38 year old woman one year after ovarian transposition and external beam radiotherapy for cervical cancer shows the transposed ovaries as ovoid soft-tissue masses (white arrows) in the paracolic gutters. Note a surgical marker adjacent to the right ovary, surgical clips related to prior lymph node dissection in the retroperitoneum, and the presence of vascular pedicles (black arrows) at the medial aspect of the masses. **B.** Coronal contrast-enhanced CT image shows the lateral location of the transposed ovaries (arrows). Note the presence of smaller cyst-like structures in the right ovary, presumably follicles.

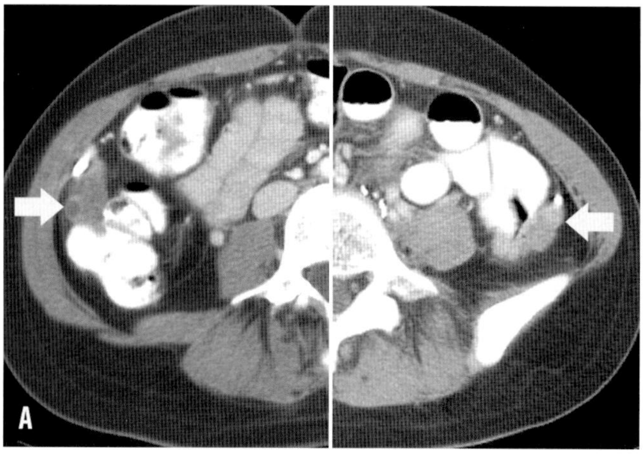

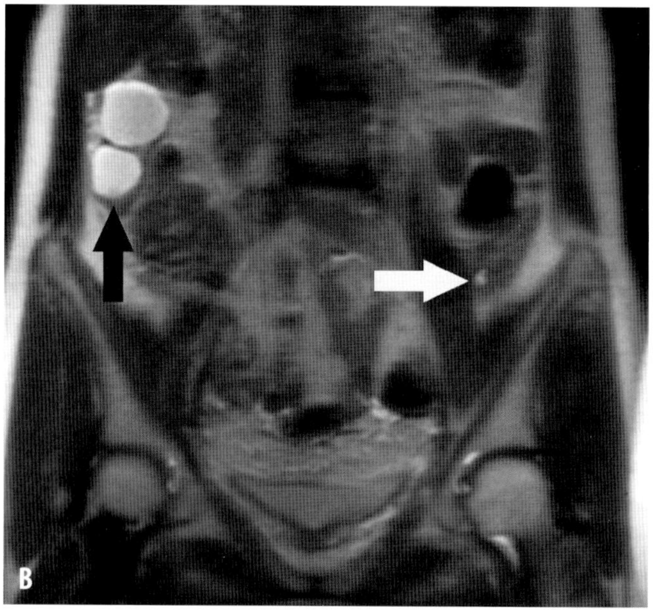

Figure 76.2 **A.** Photomontage of two axial contrast-enhanced CT images in a 35 year old woman two years after radical hysterectomy with pelvic lymphadenectomy, and bilateral ovarian transposition. The transposed ovaries are seen as soft-tissue density masses (arrows) in the paracolic gutters. Note the adjacent surgical markers and the presence of small follicles in the right ovary. **B.** Coronal T2-weighted MR image obtained six months later shows two adjacent benign-appearing cysts in the right ovary (black arrow) and a small follicle in the left ovary (white arrow).

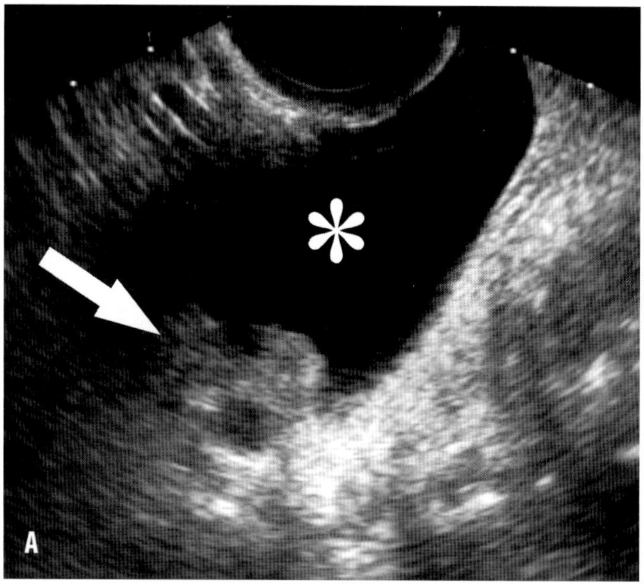

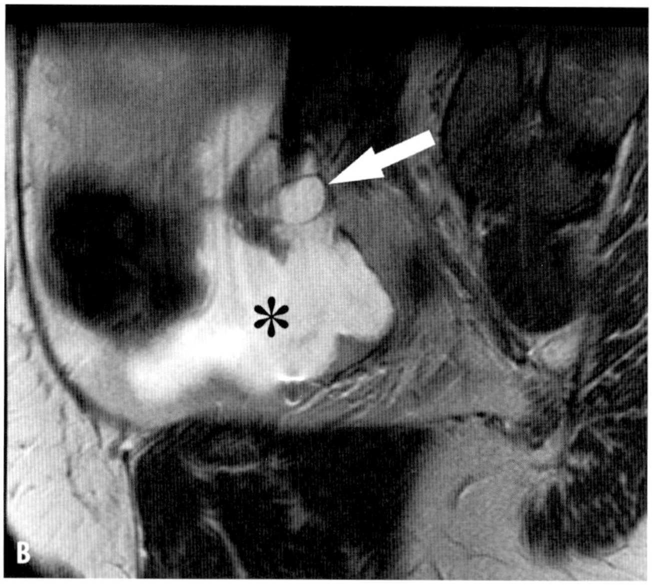

Figure 76.3 **A.** Sagittal ultrasound image of the right lower quadrant in a 41 year old woman 3 years after ovarian transposition and external beam radiotherapy for cervical cancer shows a large fluid-filled structure (asterisk) surrounding an ovoid and predominantly solid structure (arrow). The combination of findings might suggest a complex mixed cystic and solid lesion, and raise the consideration of recurrent malignancy. **B.** Sagittal T2-weighted MR image shows the fluid (asterisk) conforms to the shape of the peritoneal cavity in the pelvis, and that the ovoid structure (arrow) surrounded by the fluid contains several small cystic spaces, likely follicles. The findings are those of a peritoneal inclusion cyst developing around a transposed ovary.

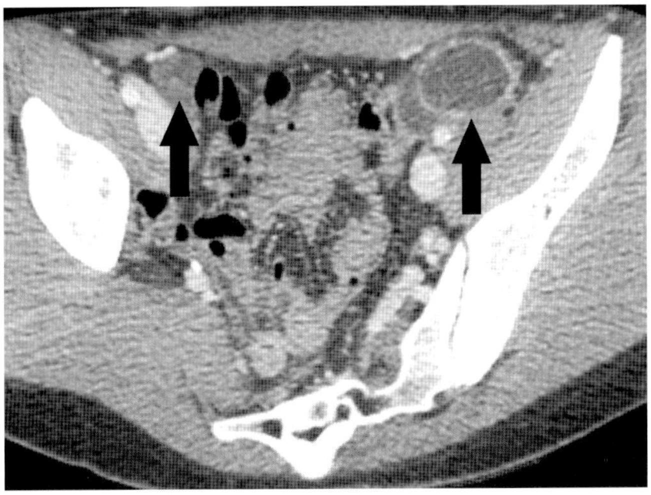

Figure 76.4 Axial curved planar reformatted contrast-enhanced CT image in a 26 year old woman 2 years after radical hysterectomy and bilateral lymphadenectomy for cervical cancer shows the post-hysterectomy ovaries (arrows) are located somewhat more laterally than might be expected, but ovarian transposition was not performed.

CASE 77

Massive ovarian edema

Imaging description

Massive ovarian edema is a rare and poorly understood condition characterized by marked unilateral (rarely bilateral) ovarian enlargement due to gross diffuse stromal edema [1]. It likely represents chronic or subacute vascular or lymphatic congestion related to incomplete torsion or other obstructive pathophysiology [2–4]. The condition is usually detected at pelvic ultrasound, where it manifests as asymmetric ovarian enlargement with echogenic stroma and peripherally displaced follicles (Figure 77.1). At MRI (Figures 77.1–77.3), the ovarian stroma characteristically demonstrates marked T2 signal hyperintensity and may demonstrate increased T1 signal intensity (possibly reflecting hemorrhage). The ovary may have a teardrop configuration [2]. The teardrop configuration supports the concept that massive ovarian edema reflects chronic vascular congestion of the ovary, with the ovarian pedicle being either torsed or compressed.

Importance

The optimal management of massive ovarian edema is unknown, because most cases have undergone surgery based on a preoperative assumption that the ovarian enlargement was due to tumor. Less aggressive surgical management for massive ovarian edema includes deep wedge resection for definitive diagnosis and detorsion or fixation [1, 5–7].

Typical clinical scenario

Massive ovarian edema may occur at any age, with a mean age at diagnosis of 20 years [5]. Presenting features include recurrent intermittent abdominal pain or distension, palpable pelvic mass, menstrual irregularity, or hormonal effects such as early puberty or virilization [6, 7]. Massive ovarian edema may occasionally complicate pregnancy [2].

Differential diagnosis

Asymmetry and stromal T2 hyperintensity help to distinguish massive ovarian edema from polycystic ovary syndrome, which can also result in stromal enlargement with peripherally displaced follicles, but which is usually bilateral with T2 hypointense stroma [8]. The non-specific solid appearance of ovarian enlargement due to massive ovarian edema on ultrasound may suggest any of several solid ovarian tumors, but the stromal T2 hyperintensity on MRI helps narrow the differential diagnosis. The only solid ovarian tumor that is markedly hyperintense on T2-weighted imaging is sclerosing stromal tumor of the ovary [9], but peripherally displaced follicles are not a feature of this tumor, allowing distinction from massive ovarian edema.

Teaching point

Unilateral solid ovarian enlargement with peripherally displaced follicles should suggest a diagnosis of massive ovarian edema, particularly if the ovarian stroma is of markedly increased T2 signal intensity at MRI. Imaging recognition of this entity may facilitate conservative management and ovarian preservation.

REFERENCES

1 Hall BP, Printz DA, Roth J. Massive ovarian edema: ultrasound and magnetic resonance characteristics. *J Comput Assist Tomogr* 1993; **17**: 477–479.

2 Anwar M, Poder L, Wang ZJ, et al. MRI of massive ovarian edema in pregnancy. *J Comput Assist Tomogr* (submitted).

3 Krasevic M, Haller H, Rupeie S, Behrem S. Massive edema of the ovary: a report of two cases due to lymphatic permeation by metastatic carcinoma from the uterine cervix. *Gynecol Oncol* 2004; **93**: 564–567.

4 Slotky B, Shrivastav, R, Lee BM. Massive edema of the ovary. *Obstet Gynecol* 1982; **59**: 92–94.

5 Geist RR, Rabinowitz R, Zuckerman B, et al. Massive edema of the ovary: a case report and review of the pertinent literature. *J Pediatr Adolesc Gynecol* 2005; **18**: 281–284.

6 Roberts CI, Weston MJ. Bilateral massive ovarian edema: a case report. *Ultrasound Obstet Gynecol* 1998; **11**: 65–67.

7 Gonzalez-Cejudo C, Martinez-Maestro MA, Peregrin-Alvares I, Daza-Manzano C. Unilateral massive ovarian edema: a diagnostic dilemma. *Eur J Obstet Gynecol Reprod Biol* 2006; **136**: 264–265.

8 Mitchell DG, Gefter WB, Spritzer CE, et al. Polycystic ovaries: MR imaging. *Radiology* 1986; **160**: 425–429.

9 Ihara N, Togashi K, Todo G, et al. Sclerosing stromal tumor of the ovary: MRI. *J Comput Assist Tomogr* 1999; **23**: 555–557.

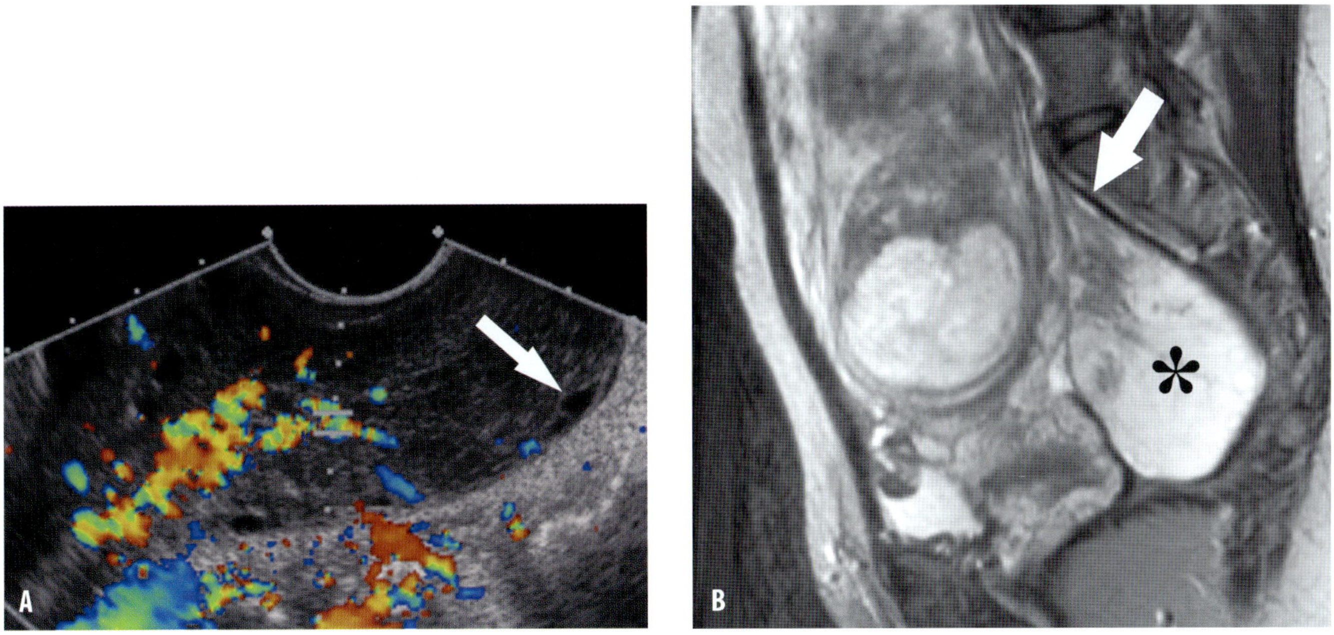

Figure 77.1 A. Endovaginal ultrasound in a 32 year old woman with mild right-sided discomfort at 26 weeks of gestation shows solid enlargement of the right ovary with peripherally displaced follicles (arrow) and preserved vascular flow on Doppler insonation. The left ovary (not shown) was unremarkable. **B.** Sagittal T2-weighted fast spin-echo MR image shows the right ovary (asterisk) is of markedly increased T2 signal intensity. The ovary has a teardrop configuration, with a narrow "beak" or pedicle (arrow) superiorly.

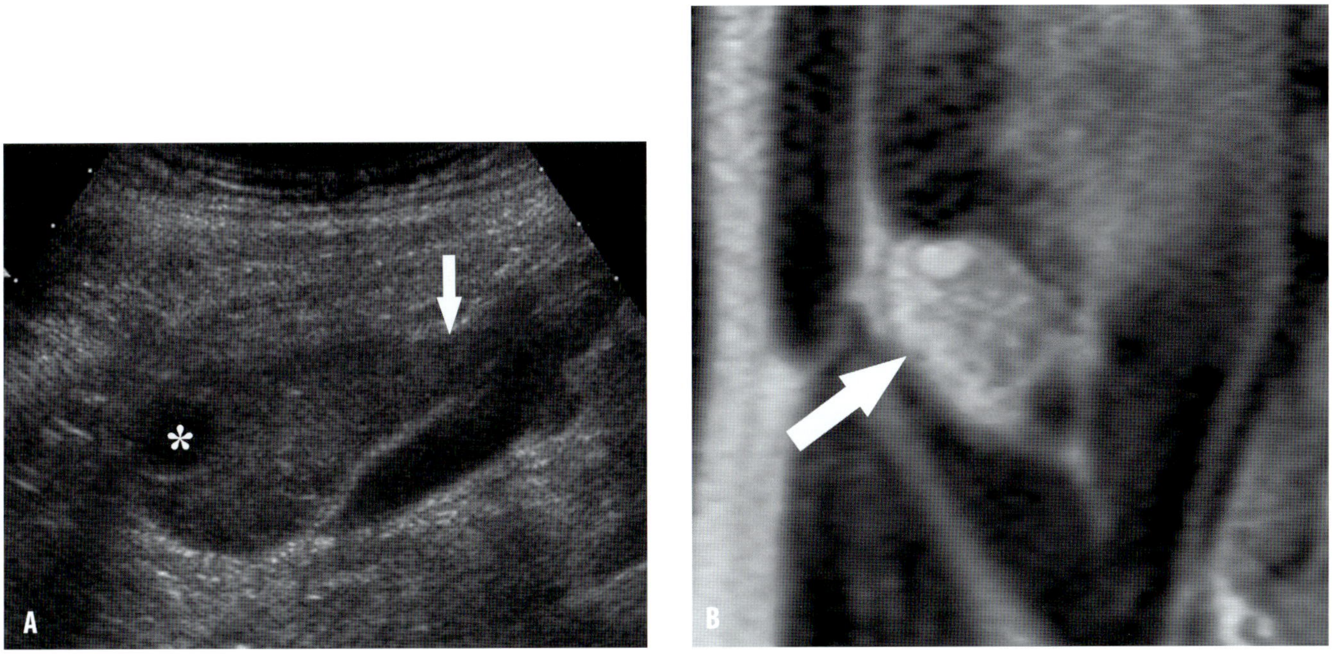

Figure 77.2 A. Transabdominal ultrasound in a 31 year old woman with right lower quadrant pain at 28 weeks of gestation shows solid enlargement of the right ovary with a peripherally displaced follicle (asterisk). The ovary has a teardrop configuration, with a narrow "beak" or pedicle (arrow). **B.** Coronal single-shot fast spin-echo MR image shows the right ovary (arrow) is of markedly increased T2 signal intensity.

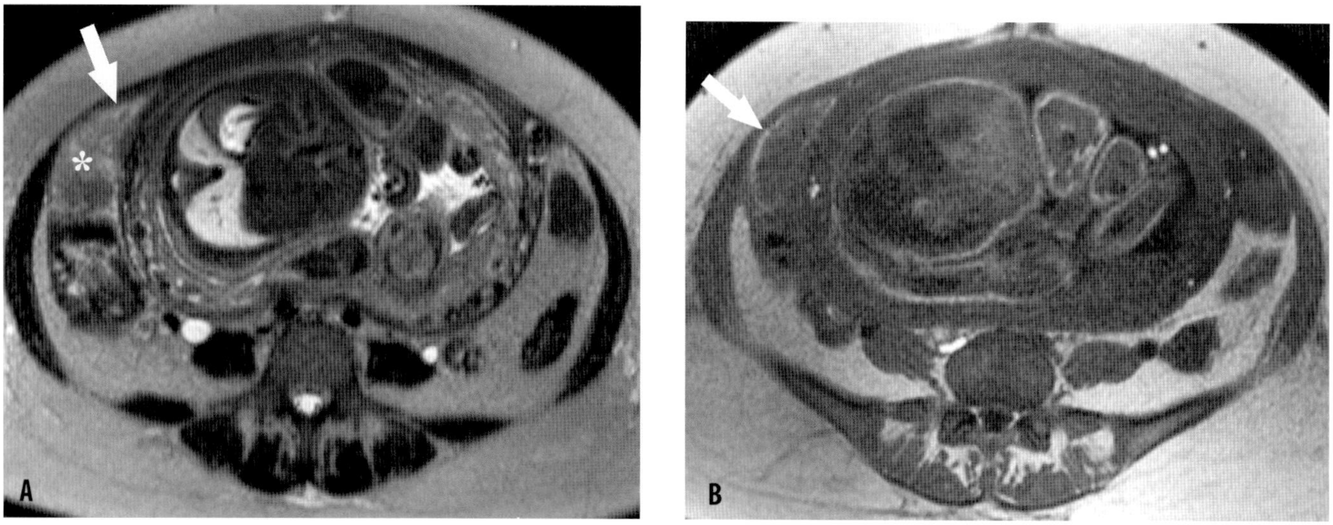

Figure 77.3 A. Axial single-shot fast spin-echo MR image in a 36 year old woman with intermittent abdominal pain at 31 weeks of gestation shows T2 hyperintense (asterisk) enlargement of the right ovary with a teardrop configuration. A narrow "beak" or pedicle (arrow) is visible between the gravid uterus and anterior abdominal wall. **B.** Axial T1-weighted spoiled gradient-echo MR image shows a peripheral rim of T1 hyperintensity (arrow) in the right ovary. The MRI findings are consistent with massive ovarian edema.

CASE 78 Decidualized endometrioma

Imaging description

Decidualization is the physiological transformation of endometrial tissue into the dense highly vascularized cellular matrix known as decidua that occurs in the endometrium of the uterus during pregnancy [1]. The same process can occur in the ectopic endometrium within endometriomas, and may result in the development of solid nodules in endometriomas during pregnancy (Figure 78.1).

Importance

Decidualization of endometriomas may be misdiagnosed as ovarian cancer or malignant transformation, resulting in unnecessary surgery during pregnancy.

Typical clinical scenario

Adnexal masses (excluding physiological corpus luteal cysts of early pregnancy) are seen in 0.5 to 1.2% of pregnancies, and 11% of these are endometriomas [2]. Despite the relative frequency of endometriomas in pregnancy, decidualization resulting in an appearance that mimics malignancy seems rare, with only a handful of reported cases [3–7]. Most of these cases have resulted in surgery during pregnancy because of the suspicious imaging findings [3–7].

Differential diagnosis

The development of solid nodules in an endometrioma could also indicate superimposed malignancy. While malignant transformation of an endometrioma during pregnancy has not been reported, this remains a potential concern when solid nodules develop within an endometrioma irrespective of whether the patient is pregnant or not. Suggested criteria for the prospective diagnosis of decidualization rather than malignant transformation of endometrioma are the presence of solid nodules in an endometrioma from early in pregnancy that do not progress substantially on serial studies, have a smoothly lobulated appearance, are markedly vascular at Doppler ultrasound, and are isointense to the decidualized endometrium in the uterus on all sequences at MRI [8].

Teaching point

Decidualized endometrioma can mimic ovarian malignancy during pregnancy, but a prospective diagnosis may be possible when solid smoothly lobulated nodules with prominent internal vascularity are seen from early in pregnancy within an endometrioma, and the nodules demonstrate marked similarity in signal intensity and texture to the decidualized endometrium in the uterus at MRI.

REFERENCES

1 Zaytsev P, Taxy JB. Pregnancy-associated ectopic decidua. *Am J Surg Pathol* 1987; **11**: 526–530.

2 Bromley B, Benacerraf B. Adnexal masses during pregnancy: accuracy of sonographic diagnosis and outcome. *J Ultrasound Med* 1997; **16**: 447–452.

3 Miyakoshi K, Tanaka M, Gabionza D, et al. Decidualized ovarian endometriosis mimicking malignancy. *Am J Roentgenol* 1998; **171**: 1625–1626.

4 Tanaka YO, Shigemitsu S, Nagata M, et al. A decidualized endometrial cyst in a pregnant woman: a case observed with a steady-state free precession imaging sequence. *Magn Reson Imaging* 2002; **20**: 301–304.

5 Fruscella E, Testa AC, Ferrandina G, et al. Sonographic features of decidualized ovarian endometriosis suspicious for malignancy. *Ultrasound Obstet Gynecol* 2004; **24**: 578–580.

6 Sammour RN, Leibovitz Z, Shapiro I, et al. Decidualization of ovarian endometriosis during pregnancy mimicking malignancy. *J Ultrasound Med* 2005; **24**: 1289–1294.

7 Iwamoto H, Suzuki M, Watanabe N, et al. Case study of a pregnant woman with decidualized ovarian endometriosis whose preoperative findings suggested malignant transformation. *Eur J Gynaecol Oncol* 2006; **27**: 301–303.

8 Poder L, Coakley FV, Rabban JT, et al. Decidualized endometrioma during pregnancy: recognizing an imaging mimic of ovarian malignancy. *J Comput Assist Tomogr* 2008; **32**: 555–558.

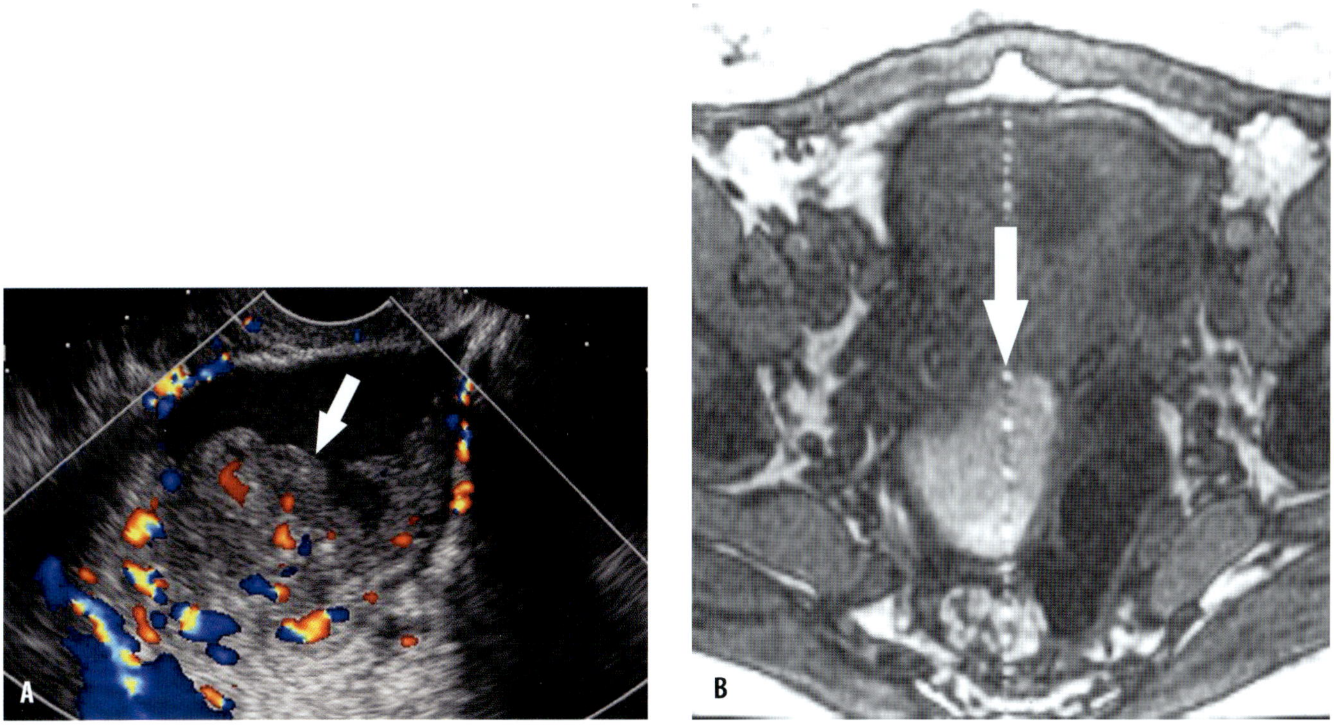

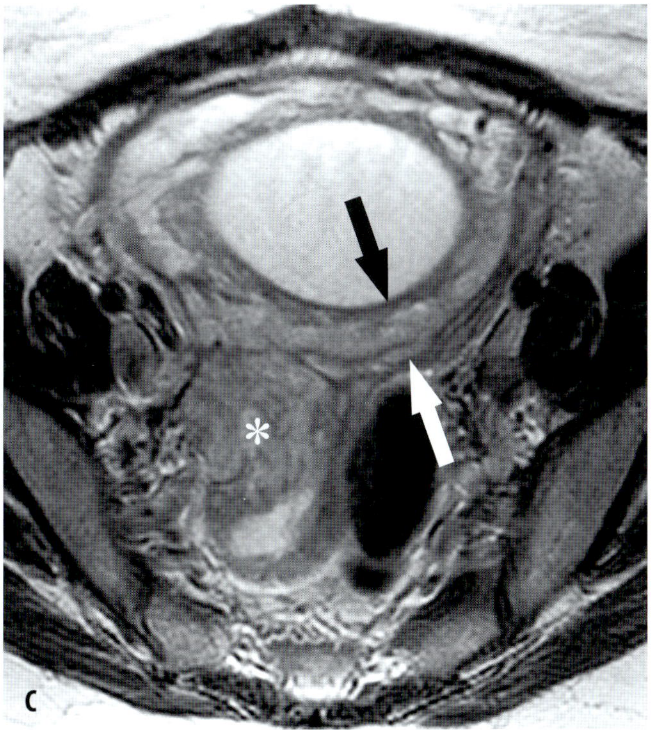

Figure 78.1 **A.** Ultrasound image in a 34 year old woman at 21 weeks' gestation shows a 6.2 cm cystic right ovarian mass containing smoothly lobulated mural nodules with prominent internal vascularity on color Doppler insonation. **B.** Axial T1-weighted MR image shows the fluid within the cystic part of the mass is of high T1 signal intensity (arrow), consistent with blood. **C.** Axial T2-weighted MR image shows the solid component (asterisk) of the mass is strikingly similar to the decidualized endometrium (between arrows) in the uterus, both with respect to signal intensity and texture.

CASE 79 Pseudotumor due to differential enhancement of the cervix

Imaging description

The cervix is more fibrous than the vascular myometrium of the uterine body, and so the cervix enhances more slowly than the rest of the uterus after the administration of intravenous contrast material at CT or MRI (Figures 79.1 and 79.2) [1]. The resultant differential enhancement of the cervix may result in the less enhancing cervix being mistaken for a hypovascular mass (Figure 79.3).

Importance

Misidentification of the normal cervix as a mass may result in unnecessary additional testing and patient anxiety.

Typical clinical scenario

In my experience, this pseudotumor has become more frequent as modern multidetector CT scanners have become progressively faster, because the pelvis is now imaged sooner after the administration of contrast material (i.e., in the phases when the differential hypoenhancement of the cervix is more marked).

Differential diagnosis

Awareness of this pseudotumor is usually sufficient for correct identification. In problematic cases, identification of the central endocervical canal or Nabothian cysts at the external os may be a useful diagnostic clue. Inspection of sagittal reformatted images can also be helpful.

Teaching point

On contrast-enhanced CT or MRI images obtained early after contrast material administration, a circular hypovascular pelvic mass inferior to the uterine body and in or adjacent to the midline is likely to represent normal differential hypoenhancement of the enhanced cervix.

REFERENCE

1 Kaur H, Loyer EM, Minami M, Charnsangavej C. Patterns of uterine enhancement with helical CT. *Eur J Radiol* 1998; **28**: 250–255.

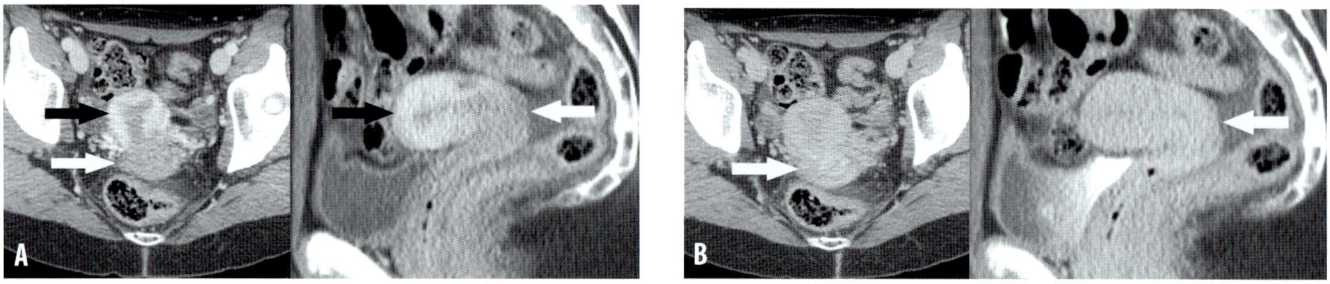

Figure 79.1 **A.** Photomontage of axial (left image) and sagittal reformatted (right image) contrast-enhanced CT images obtained during the portal venous phase of enhancement in a 28 year old woman with hematuria. The myometrium of the uterine body (black arrows) is more enhanced than the cervix (white arrows), so that the cervix could be mistaken for a hypovascular mass, particularly on the axial image. **B.** Photomontage of axial (left image) and sagittal reformatted (right image) contrast-enhanced CT images obtained during the delayed phase of enhancement. Note that the cervix (white arrows) now demonstrates isodense enhancement with the uterine body.

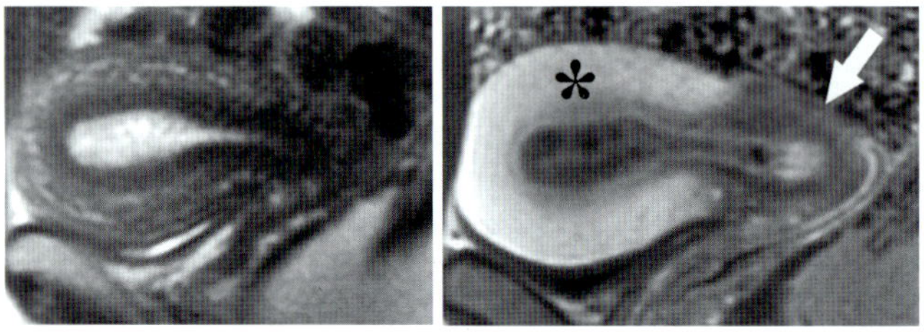

Figure 79.2 Photomontage of sagittal T2-weighted (left image) and early phase sagittal T1-weighted post-gadolinium (right image) MR images in a 35 year old woman with chronic pelvic pain shows that the cervix is of lower T2 signal intensity (presumably reflecting more fibrous stroma) and demonstrates reduced enhancement (white arrow) relative to the myometrium of the uterine body (asterisk) after contrast material administration.

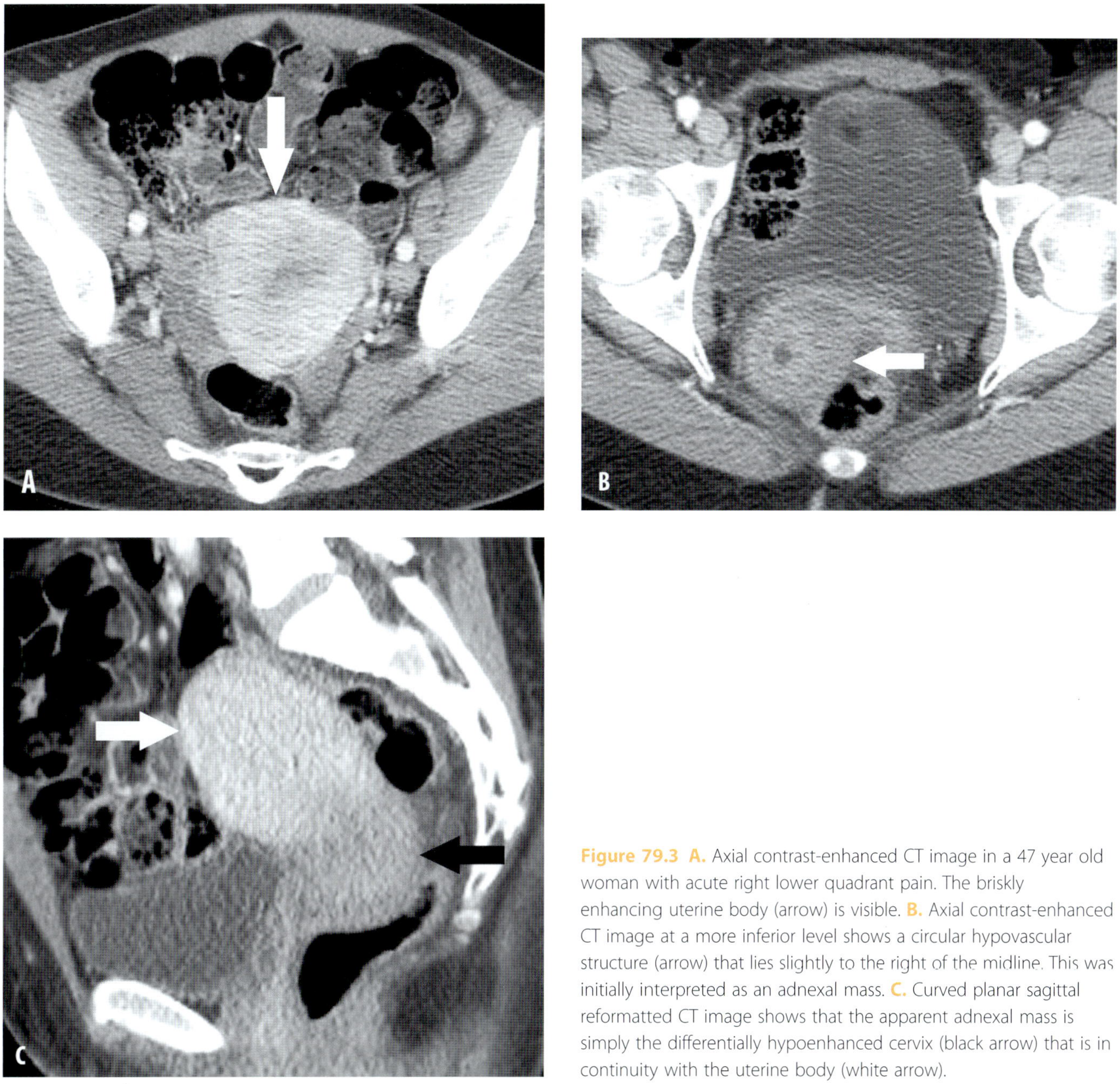

Figure 79.3 **A.** Axial contrast-enhanced CT image in a 47 year old woman with acute right lower quadrant pain. The briskly enhancing uterine body (arrow) is visible. **B.** Axial contrast-enhanced CT image at a more inferior level shows a circular hypovascular structure (arrow) that lies slightly to the right of the midline. This was initially interpreted as an adnexal mass. **C.** Curved planar sagittal reformatted CT image shows that the apparent adnexal mass is simply the differentially hypoenhanced cervix (black arrow) that is in continuity with the uterine body (white arrow).

CASE 80 Early intrauterine pregnancy on CT and MRI

Imaging description

The appearance of early pregnancy on ultrasound is well known, but the CT and MRI findings are less well described. In the first trimester, pregnancy results in an endometrial cyst-like structure (the gestational sac) of variable size at CT or MRI (Figures 80.1–80.6) [1, 2]. Fetal parts are usually not visible until the late first trimester. The developing placenta may be seen as a peripheral curvilinear enhancing structure. Later in gestation, the diagnosis of pregnancy is straightforward as the fetus becomes obviously visible. The gestational sac of early pregnancy may be small and inconspicuous at CT and MRI, but the co-existence of an ovarian corpus luteum cyst can be an important clue to the diagnosis of pregnancy (Figures 80.1 and 80.3–80.5). Remember the corpus luteum (which normally regresses in the second half of the menstrual cycle) persists as a critical source of progesterone in early pregnancy. A corpus luteum can nearly always be identified between five and eight weeks' gestation, but after that the corpus luteum shrinks (as the placenta becomes the source of progesterone in later pregnancy) and may not be visible at imaging [3–5].

Importance

It is important to know the appearance of early pregnancy on CT or MRI because occasionally the patient is unaware of being pregnant and the radiologist is the first to make the diagnosis. Failure to consider the possibility of pregnancy for an endometrial cyst-like structure might result in unnecessary workup, or even repeat irradiation. With respect to the latter, it should be remembered that the radiation dose to the fetus from a diagnostic abdominopelvic CT scan rarely merits consideration of pregnancy termination [6, 7].

Typical clinical scenario

Early pregnancy may be seen on CT or MRI when a patient with a known pregnancy requires imaging for acute abdominal pain, trauma, or another compelling indication or when the pregnancy is unsuspected and imaging is performed unintentionally. In the latter case, consultation with a medical physicist for calculation of fetal dose may be appropriate. In general, termination of pregnancy is not considered appropriate until the fetal dose exceeds 10 rads [6], while the typical fetal dose from an abdominopelvic CT scan is about 3 rads.

Differential diagnosis

An endometrial cyst-like structure in a woman of reproductive age carries a wide differential. With a positive pregnancy test, possibilities include gestational trophoblastic disease (Figure 80.7) or missed abortion. With a negative pregnancy test, possibilities include retained products of conception or intracavitary pathology such as a submucosal fibroid (Figures 80.8 and 80.9).

Teaching point

The possibility of early pregnancy should be considered for an endometrial cyst-like structure seen at CT or MRI in the uterus of a woman of reproductive age, especially if there is a co-existent corpus luteum cyst; serum β-human chorionic gonadotropin testing or ultrasound correlation should clarify.

REFERENCES

1 Magarik DE, Dunne MG, Weksberg AP. CT appearance of intrauterine pregnancy. *J Comput Assist Tomogr* 1984; **8**: 469–472.

2 Lowdermilk C, Gavant ML, Qaisi W, West OC, Goldman SM. Screening helical CT for evaluation of blunt traumatic injury in the pregnant patient. *Radiographics* 1999; **19**: S243–S255.

3 Durfee SM, Frates MC. Sonographic spectrum of the corpus luteum in early pregnancy: gray-scale, color, and pulsed Doppler appearance. *J Clin Ultrasound* 1999; **27**: 55–59.

4 Parsons AK. Imaging the human corpus luteum. *J Ultrasound Med* 2001; **20**: 811–819.

5 Rowan K, Meagher S, Teoh M, et al. Corpus luteum across the first trimester: size and laterality as observed by ultrasound. *Fertil Steril* 2008; **90**: 1844–1847.

6 Hammer-Jacobsen, E. Therapeutic abortion on account of x-ray examination during pregnancy. *Dan Med Bull* 1959; **6**: 113–122.

7 Patel SJ, Reede DL, Katz DS, Subramaniam R, Amorosa JK. Imaging the pregnant patient for nonobstetric conditions: algorithms and radiation dose considerations. *Radiographics* 2007; **27**: 1705–1722.

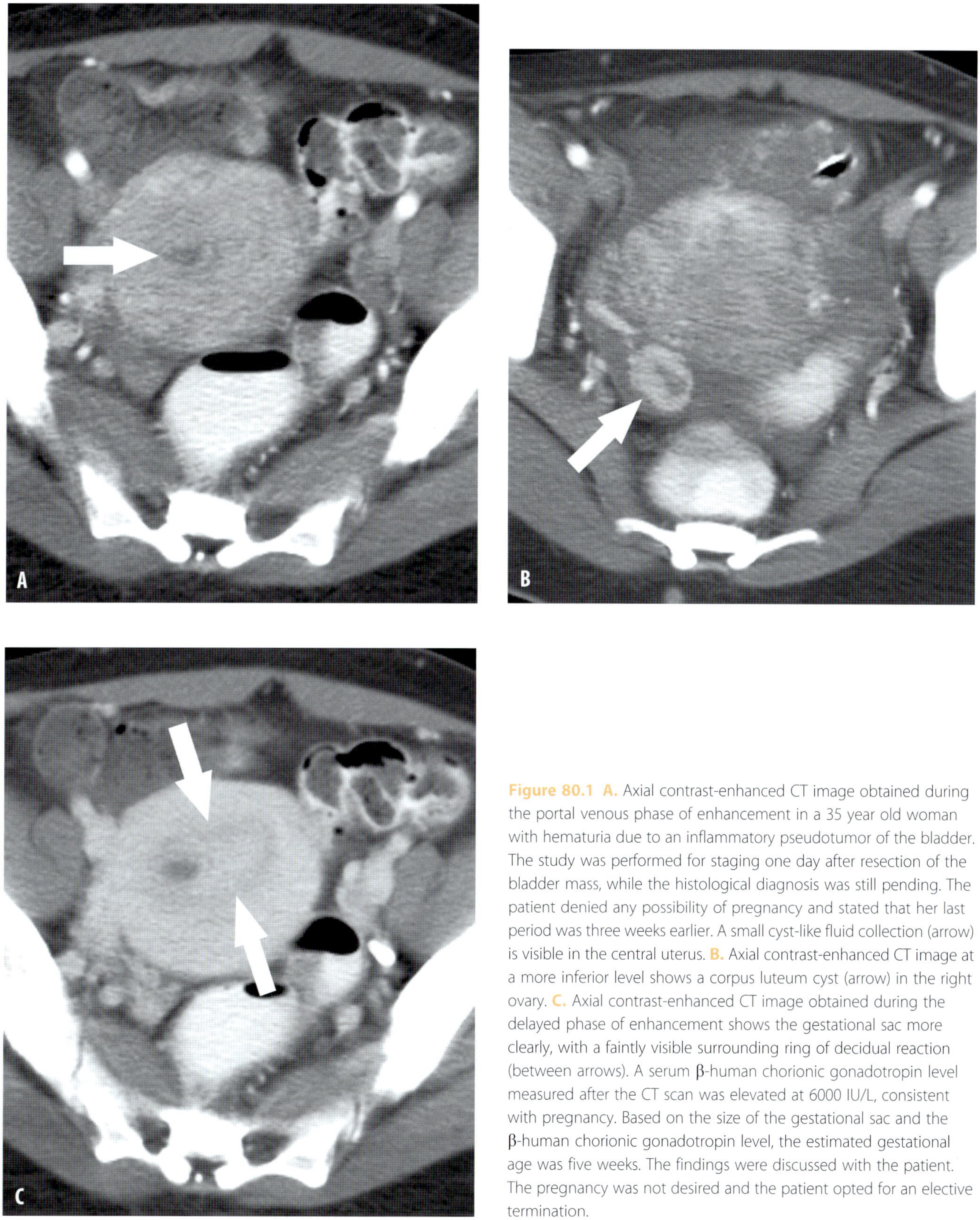

Figure 80.1 A. Axial contrast-enhanced CT image obtained during the portal venous phase of enhancement in a 35 year old woman with hematuria due to an inflammatory pseudotumor of the bladder. The study was performed for staging one day after resection of the bladder mass, while the histological diagnosis was still pending. The patient denied any possibility of pregnancy and stated that her last period was three weeks earlier. A small cyst-like fluid collection (arrow) is visible in the central uterus. B. Axial contrast-enhanced CT image at a more inferior level shows a corpus luteum cyst (arrow) in the right ovary. C. Axial contrast-enhanced CT image obtained during the delayed phase of enhancement shows the gestational sac more clearly, with a faintly visible surrounding ring of decidual reaction (between arrows). A serum β-human chorionic gonadotropin level measured after the CT scan was elevated at 6000 IU/L, consistent with pregnancy. Based on the size of the gestational sac and the β-human chorionic gonadotropin level, the estimated gestational age was five weeks. The findings were discussed with the patient. The pregnancy was not desired and the patient opted for an elective termination.

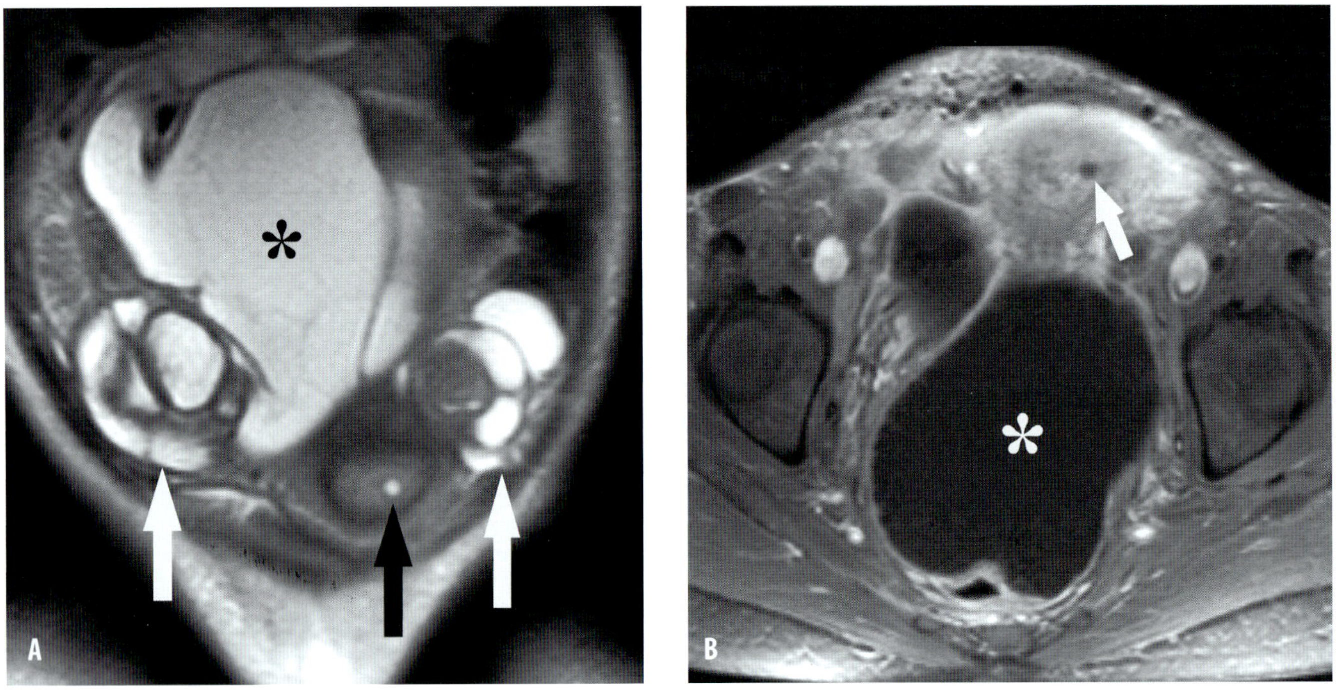

Figure 80.2 A. Coronal T2-weighted MR image in a pregnant 29 year old woman at 5 weeks' gestation with acute renal impairment due to obstruction by theca lutein cysts (white arrows) and a peritoneal inclusion cyst (asterisk). The theca lutein and peritoneal inclusion cysts were related to a combination of assisted ovarian hyperstimulation for in vitro fertilization assisted fertility and prior surgeries for ectopic pregnancy. The gestational sac is visible as a small cyst-like structure (black arrow) in the endometrial cavity. B. Axial post-gadolinium T1-weighted image shows the non-enhancing gestational sac (arrow) and the rim-enhancing peritoneal inclusion cyst (asterisk). The patient was managed by percutaneous nephrostomy with resolution of renal impairment and delivered a healthy baby girl nine months later.

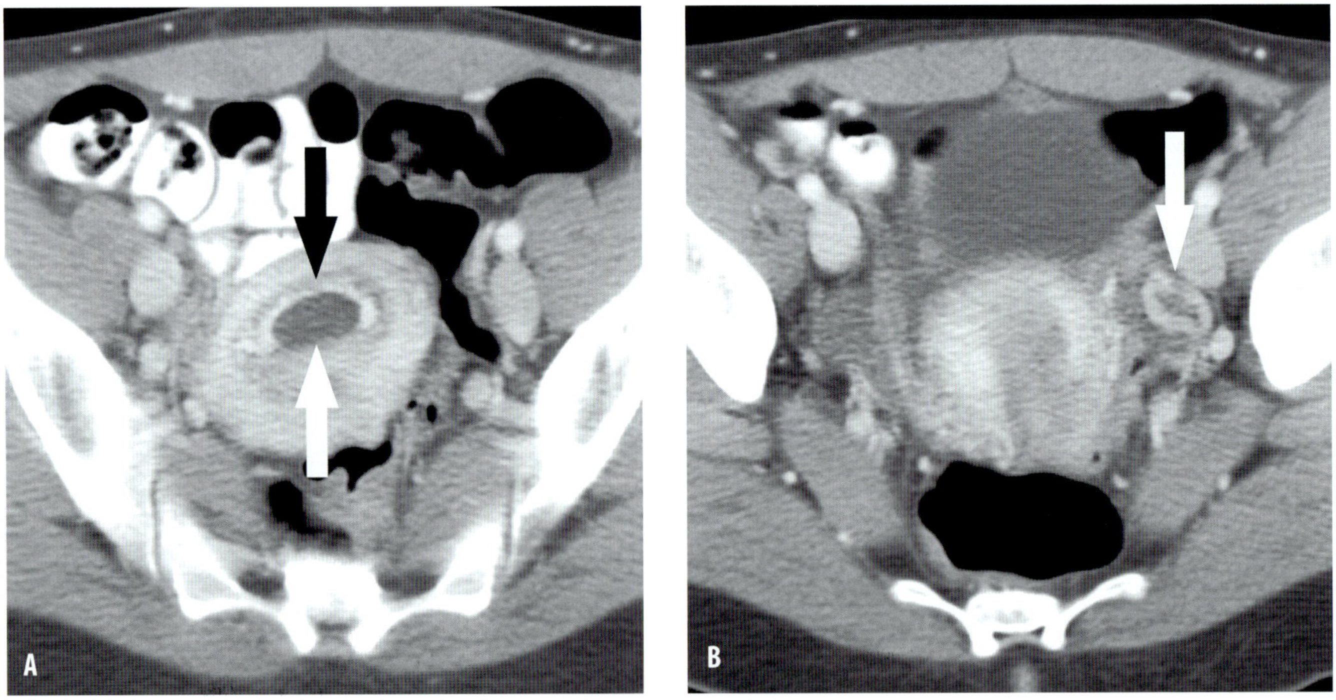

Figure 80.3 A. Axial contrast-enhanced CT image in a pregnant 20 year old woman at 7 weeks' gestation with right lower quadrant pain. The pregnancy was not desired and the patient underwent an elective termination after the study. The gestational sac is visible as a fluid-filled cyst-like structure (white arrow) in the uterus, and the placenta is visible as an eccentric briskly enhancing curvilinear structure at the periphery of the gestational sac (black arrow). B. Axial contrast-enhanced CT image at a more inferior level shows a corpus luteum cyst (arrow) in the left ovary.

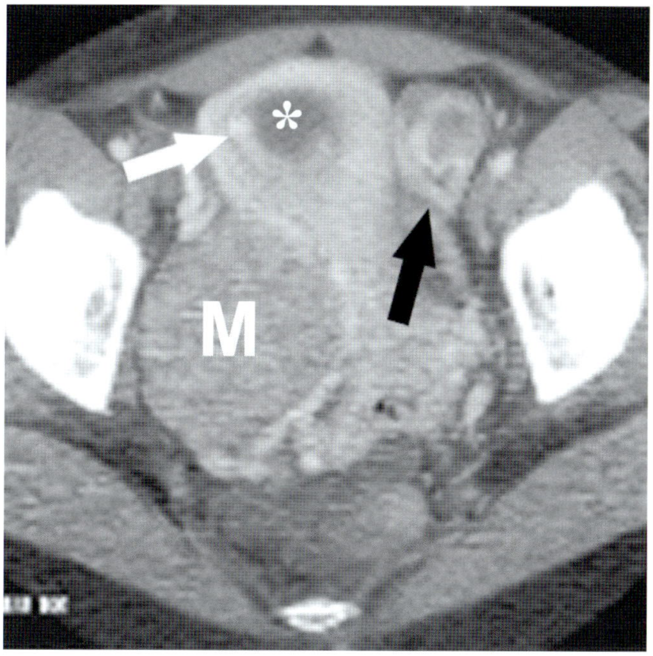

Figure 80.4 Axial contrast-enhanced CT image in a pregnant 36 year old woman at 7.5 weeks' gestation referred after an attempt at elective termination was unsuccessful secondary to cervical compression by a previously unrecognized pelvic mass. A gestational sac (asterisk) is partially surrounded by enhancing placenta (white arrow). A corpus luteum cyst (black arrow) is visible in the left ovary. A solid mass (M) adjacent to the cervix was found to be an exophytic fibroid at surgery.

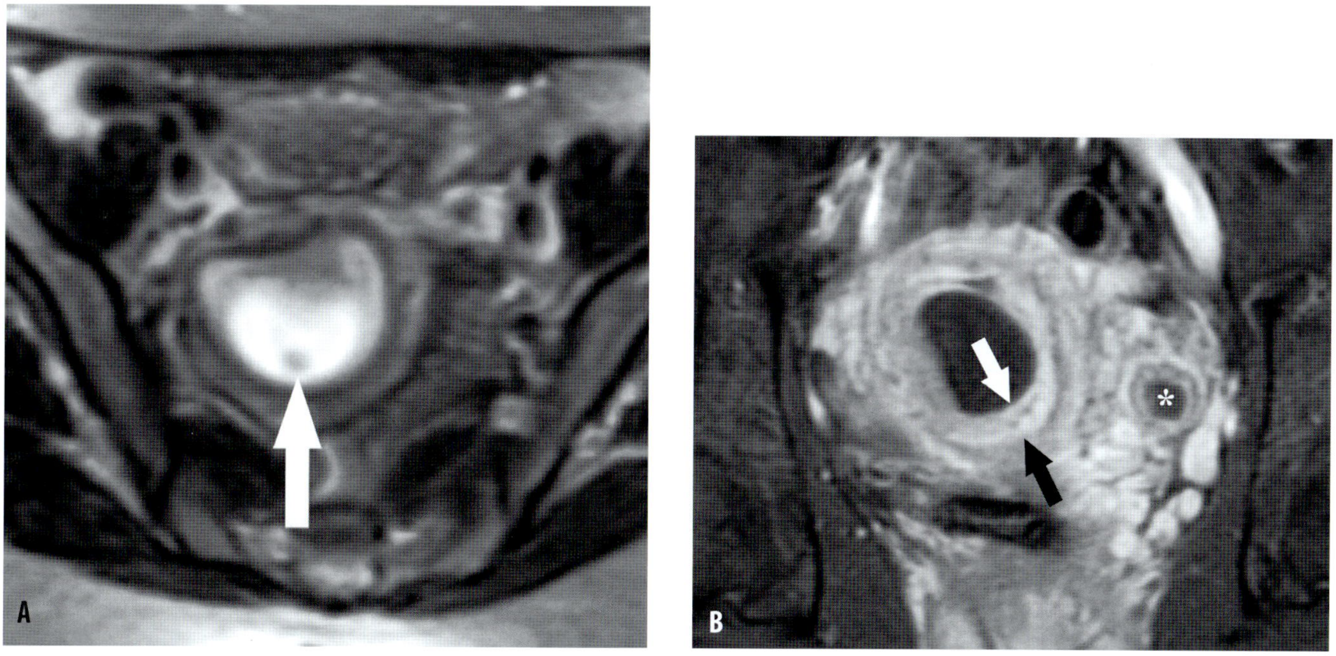

Figure 80.5 **A.** Axial T2-weighted MR image obtained in a pregnant 37 year old woman at 10 weeks' gestation because of newly diagnosed squamous cell carcinoma of the cervix. A gestational sac containing fetal parts (arrow) is visible. **B.** Coronal post-gadolinium T1-weighted image shows the gestational sac encased by enhancing placenta (between arrows) and a corpus luteum cyst (asterisk) in the left ovary.

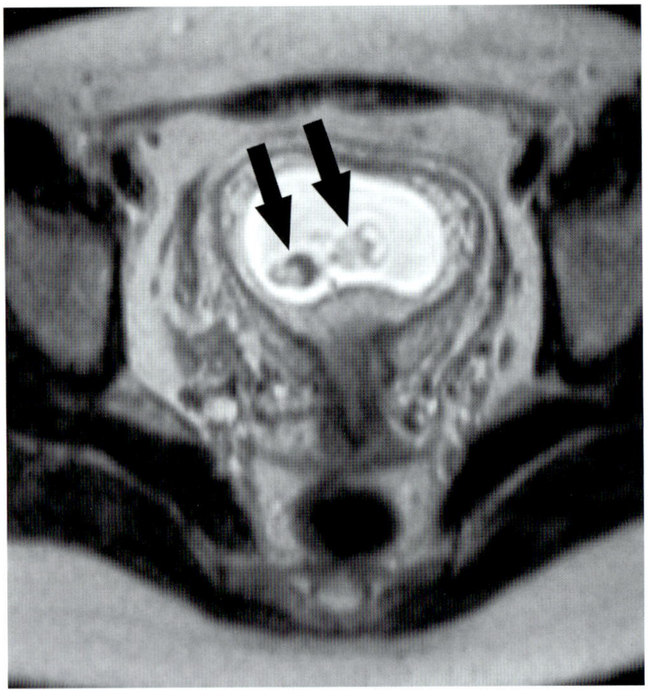

Figure 80.6 Axial T2-weighted MR image obtained in a pregnant 24 year old woman at 11 weeks' gestation because of acute right lower quadrant pain concerning for appendicitis. A gestational sac containing unequivocal fetal parts (arrows) is visible. A normal appendix was seen on other images. A corpus luteum was not seen.

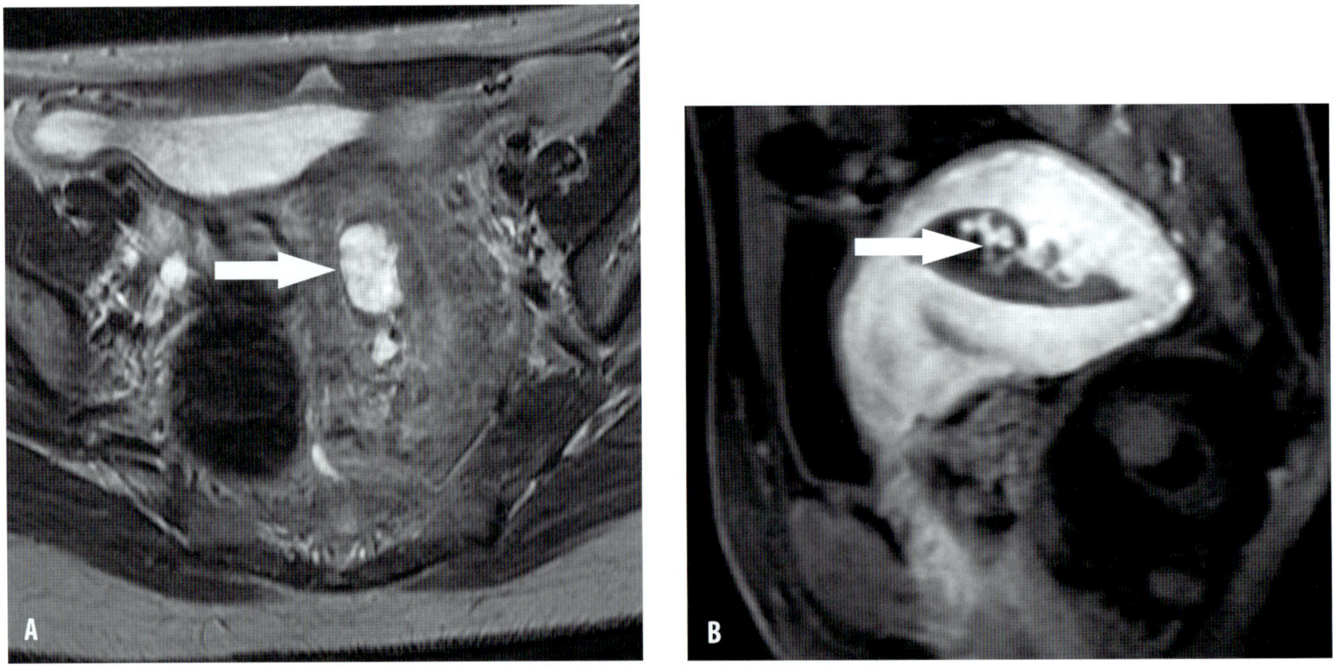

Figure 80.7 A. Axial T2-weighted MR image obtained in a 29 year old woman with persistent vaginal bleeding due to recurrent gestational trophoblastic disease two weeks after a dilatation and curettage for molar pregnancy shows a fluid-filled cyst-like structure in the endometrial cavity. **B.** Sagittal post-gadolinium T1-weighted MR image shows enhancing material within the cyst-like structure in the endometrial cavity. A contemporaneous ultrasound showed heterogeneous material in the endometrial cavity, without an identifiable fetus. A serum β-human chorionic gonadotropin level was grossly elevated at 25,488 IU/L.

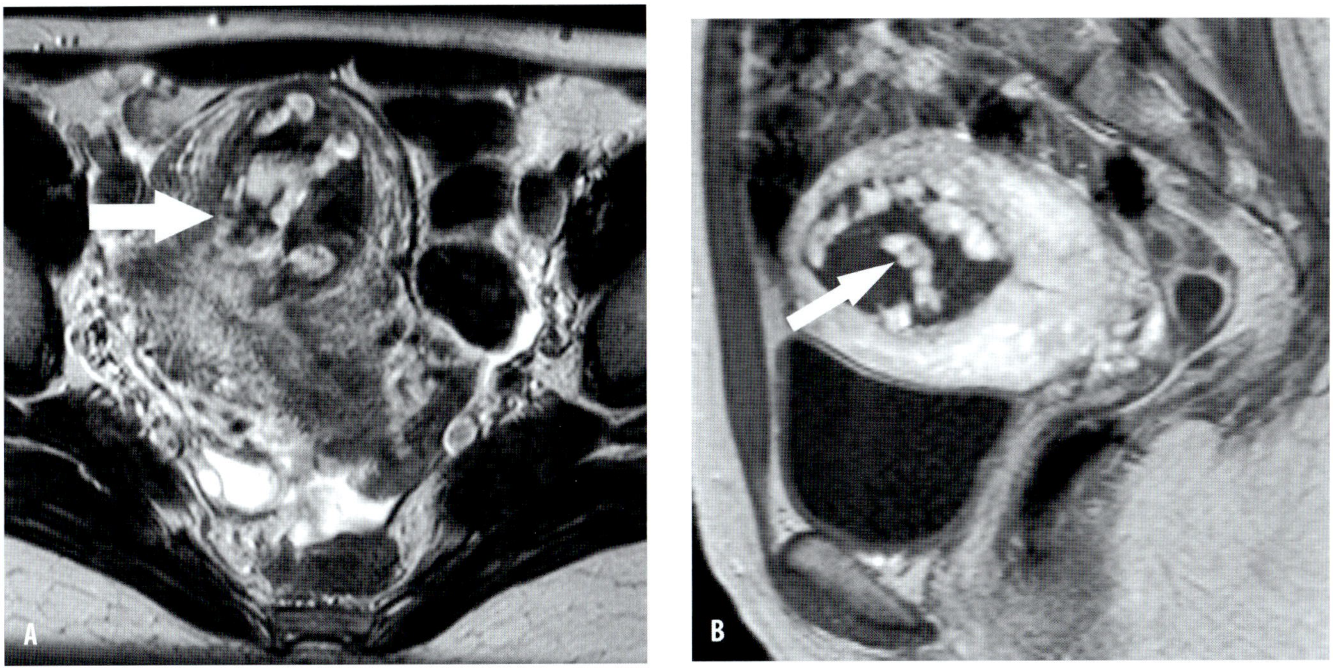

Figure 80.8 A. Axial T2-weighted MR image obtained in a 32 year old woman 2 weeks after spontaneous abortion with vaginal bleeding shows a heterogeneous uterine mass (arrow). B. Sagittal post-gadolinium T1-weighted MR image shows enhancing material (arrow) within the cyst-like structure in the endometrial cavity. A contemporaneous serum β-human chorionic gonadotropin was 33 mIU/mL, and dilatation and curettage revealed retained products of conception only.

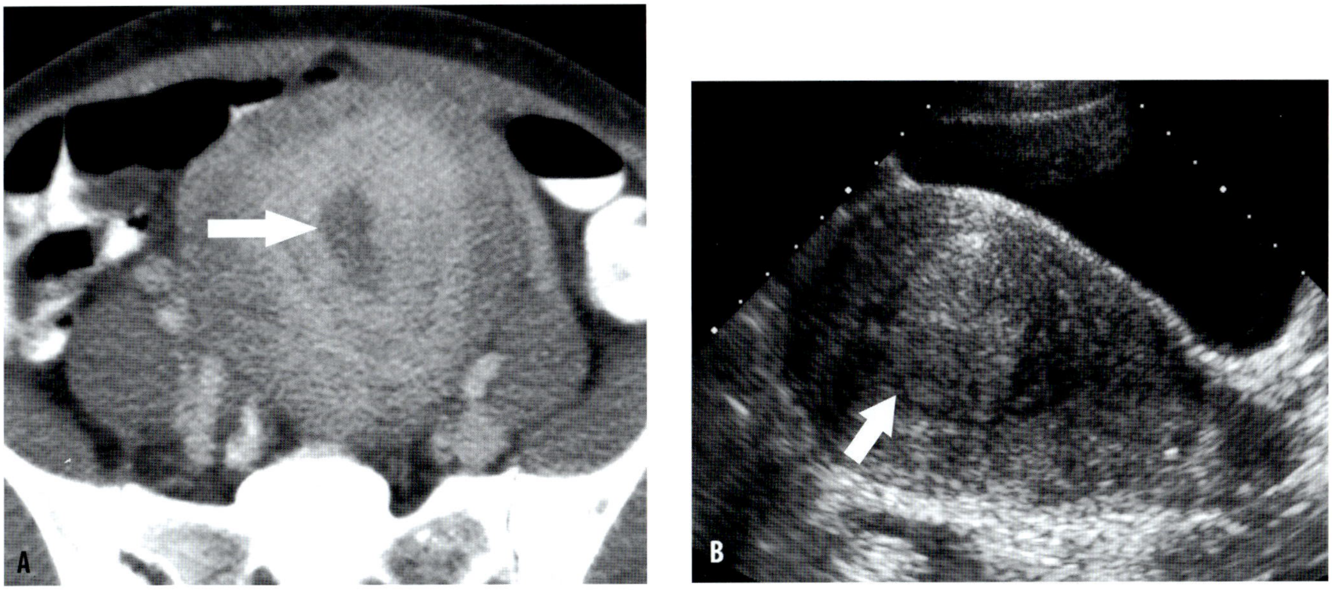

Figure 80.9 A. Axial contrast-enhanced CT image in a 45 year old woman with irregular vaginal bleeding and a negative pregnancy test shows a low-density structure (arrow) in the central uterus that resembles a gestational sac (no corpus luteum cyst was visible). B. Contemporaneous endovaginal ultrasound image shows a solid mass (between arrows) in the uterus. Hysterectomy revealed a large submucosal fibroid in the endometrial cavity.

CASE **81**

Prolapsed uterine tumor mimicking cervical cancer

Imaging description

Prolapsed uterine tumors can extend inferiorly through the cervical canal, and then the cervical component of these tumors can falsely suggest a primary cervical malignancy at imaging (Figures 81.1–81.6) [1–3]. Generally, close inspection of the images will demonstrate stalk-like continuity with intracavitary tumor within the uterine body – this morphological clue has been called the "broccoli sign" [4].

Importance

Prolapsed uterine tumors may be benign or malignant, and misdiagnosis as a primary cervical cancer might result in inappropriate management.

Typical clinical scenario

Vaginal bleeding is probably the commonest symptom associated with prolapsed uterine tumors. The diagnosis is not always clear-cut at vaginal examination, and imaging may be critical in establishing the true origin of an apparent cervical mass. In a recent study of 1785 women undergoing abdominal surgery for uterine leiomyomas, the prevalence of prolapsing submucosal leiomyomas was 2.5% [5].

Differential diagnosis

Several uterine tumors may be intracavitary and prolapse into the cervix, including leiomyoma, adenomyoma, adenosarcoma, and endometrial carcinoma [1–4]. The distinction of these different pathologies will generally require tissue sampling, but the low T2 signal intensity and well-circumscribed margin of prolapsing leiomyoma should allow differentiation of this entity from other possibilities [3].

Teaching point

Not every mass in the cervix arises from the cervix; prolapsed uterine tumors can be predominantly cervical in location. Close examination of sagittal images for the presence of a stalk (broccoli sign) should allow for accurate diagnosis of these prolapsed tumors, which may be benign or malignant.

REFERENCES

1 Horn LC, Fischer U, Höckel M. Polypoid endometrial adenomyoma in a postmenopausal woman, mimicking cervical cancer, associated with ovarian thecoma. *Arch Gynecol Obstet* 2002; **267**: 101–103.

2 McCluggage WG, Alderdice JM, Walsh MY. Polypoid uterine lesions mimicking endometrial stromal sarcoma. *J Clin Pathol* 1999; **52**: 543–546.

3 Panageas E, Kier R, McCauley TR, McCarthy S. Submucosal uterine leiomyomas: diagnosis of prolapse into the cervix and vagina based on MR imaging. *Am J Roentgenol* 1992; **159**: 555–558.

4 Kim JW, Lee CH, Kim KA, Park CM. Spontaneous prolapse of pedunculated uterine submucosal leiomyoma: usefulness of broccoli sign on CT and MR imaging. *Clin Imaging* 2008; **32**: 233–235.

5 Ben-Baruch G, Schiff E, Menashe Y, Menczer J. Immediate and late outcome of vaginal myomectomy for prolapsed pedunculated submucous myoma. *Obstet Gynecol* 1988; **72**: 858–861.

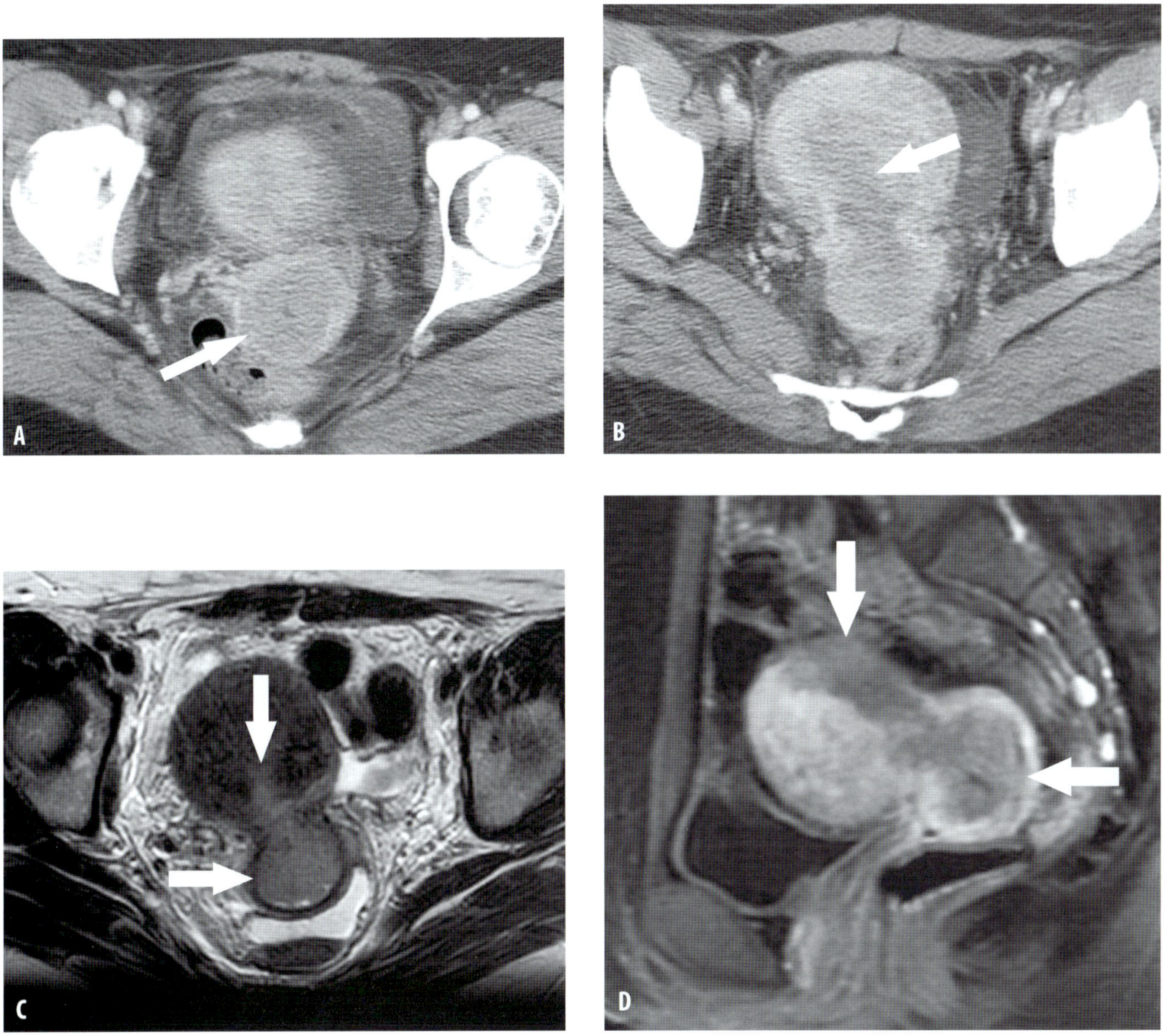

Figure 81.1 **A.** Axial contrast-enhanced CT image in a 37 year old woman with recent onset of abdominal pain and irregular vaginal bleeding shows a hypodense mass (arrow) in the cervix, that resembles a primary cervical cancer. **B.** Axial contrast-enhanced CT image at a more superior level shows the mass is continuous with infiltrative low-density tissue (arrow) in the endometrial cavity. **C.** Axial T2-weighted MR image also shows the mass (horizontal arrow) expanding the cervix and connecting to the endometrial cavity (arrow). **D.** Sagittal T1-weighted gadolinium enhanced MR image also shows the mass (horizontal arrow) expanding the cervix and connecting to infiltrative abnormal tissue (arrow) in the uterus. Surgical pathology demonstrated an endometrioid endometrial adenocarcinoma with extensive involvement of the cervix by prolapsed tumor.

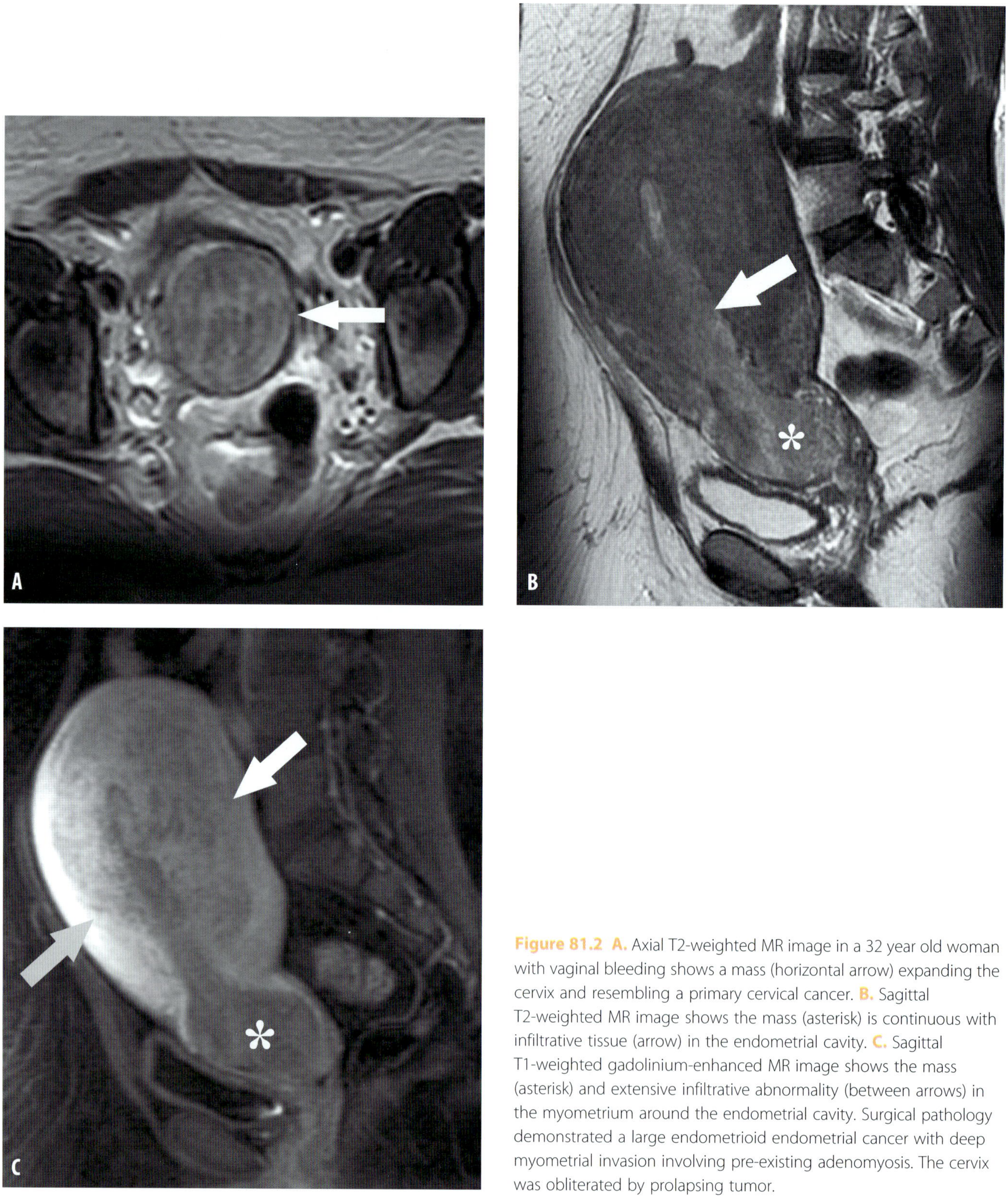

Figure 81.2 **A.** Axial T2-weighted MR image in a 32 year old woman with vaginal bleeding shows a mass (horizontal arrow) expanding the cervix and resembling a primary cervical cancer. **B.** Sagittal T2-weighted MR image shows the mass (asterisk) is continuous with infiltrative tissue (arrow) in the endometrial cavity. **C.** Sagittal T1-weighted gadolinium-enhanced MR image shows the mass (asterisk) and extensive infiltrative abnormality (between arrows) in the myometrium around the endometrial cavity. Surgical pathology demonstrated a large endometrioid endometrial cancer with deep myometrial invasion involving pre-existing adenomyosis. The cervix was obliterated by prolapsing tumor.

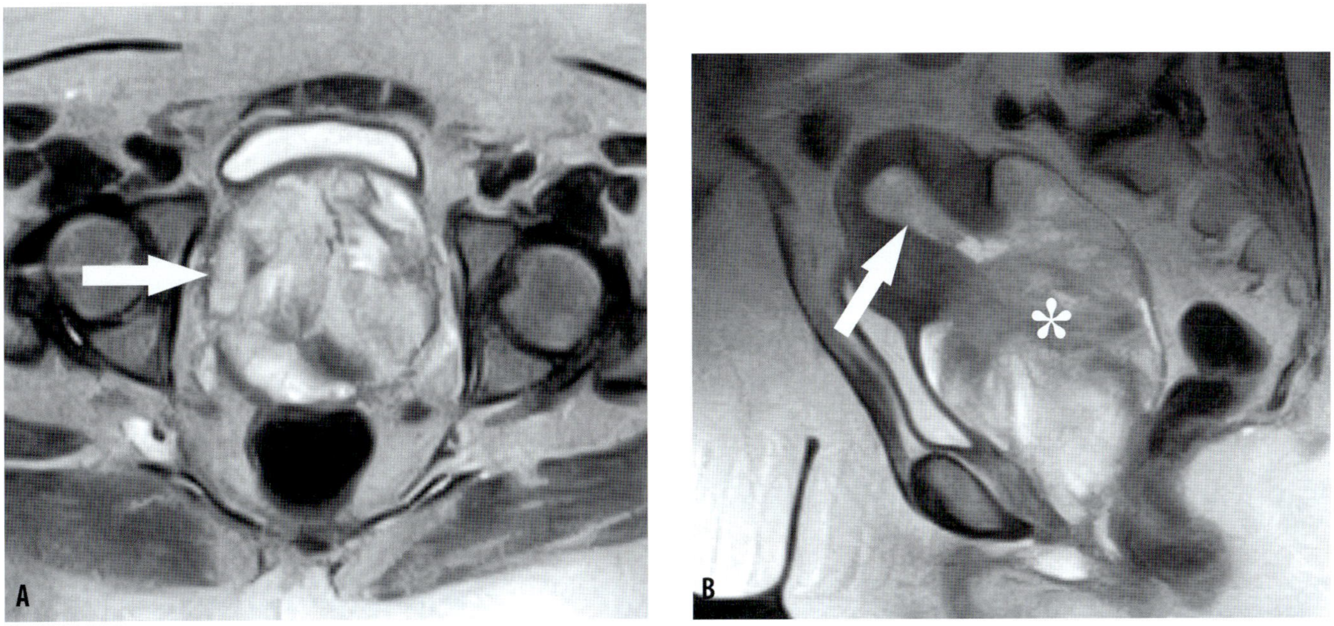

Figure 81.3 A. Axial T2-weighted MR image in a 60 year old woman with postmenopausal vaginal bleeding shows a large mass (arrow) in the expected location of the cervix. **B.** Sagittal T2-weighted MR image shows the mass (asterisk) prolapsing into the cervix. Note the stalk (arrow) attaching the mass is continuous with infiltrative tissue (arrow) in the endometrial cavity. Surgical pathology established a diagnosis of prolapsing adenosarcoma of the uterus.

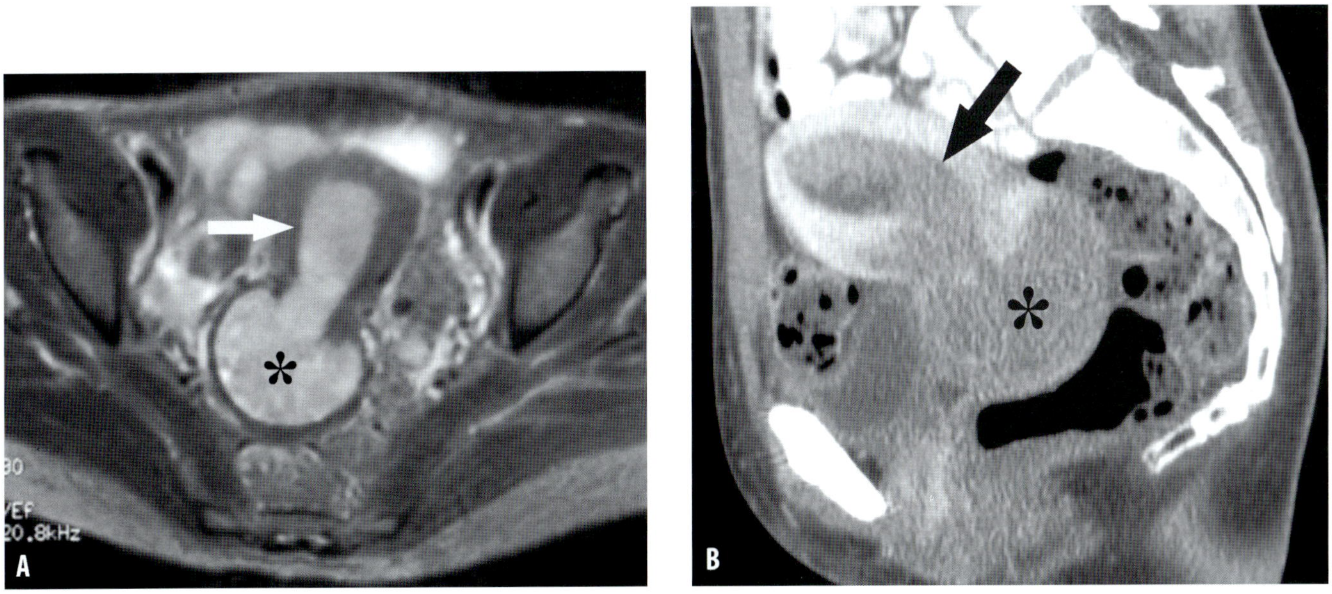

Figure 81.4 A. Axial T2-weighted MR image in a 41 year old woman with irregular vaginal bleeding shows a large mass (asterisk) in the cervix that is continuous with infiltrative tissue (arrow) in the endometrial cavity. **B.** Sagittal reformatted contrast-enhanced CT image shows the mass (asterisk) is continuous with infiltrative low-density tissue (arrow) in the endometrial cavity. Surgical pathology demonstrated a large endometrioid endometrial cancer growing inferiorly to invade the cervix.

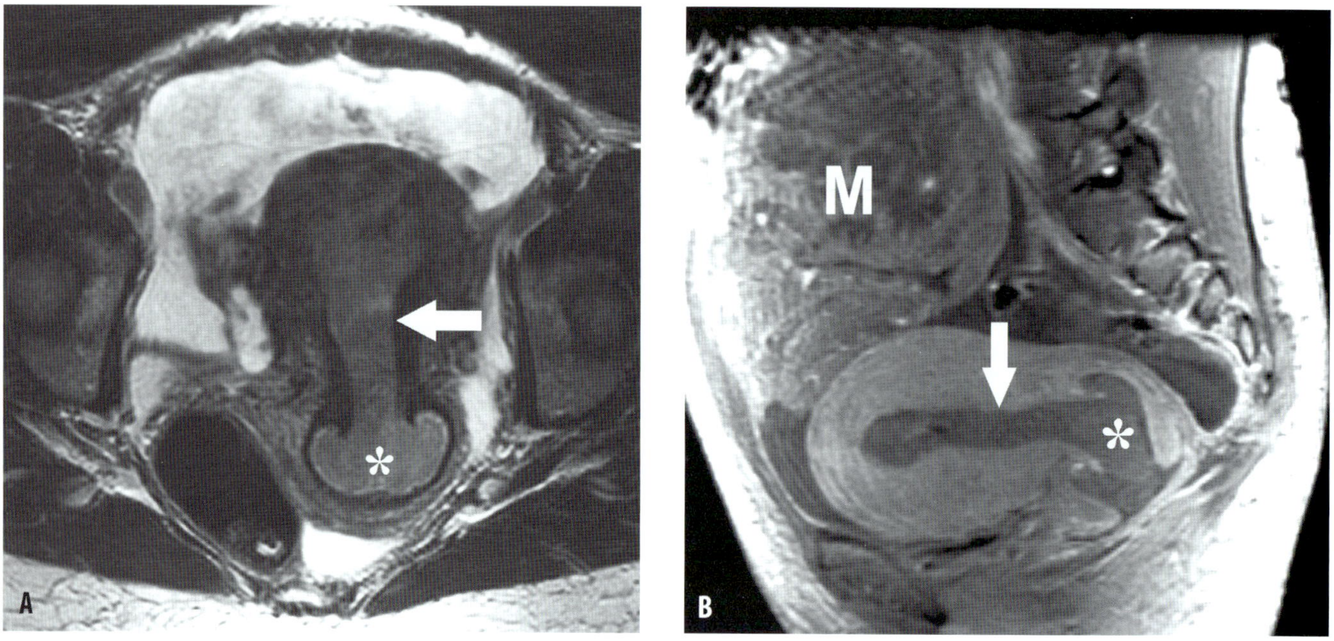

Figure 81.5 **A.** Axial T2-weighted MR image in a 48 year old woman with abdominal distension and intermenstrual vaginal bleeding shows a mass (asterisk) in the cervix that is continuous with infiltrative tissue (arrow) in the endometrial cavity. **B.** Sagittal T1-weighted gadolinium-enhanced MR image shows the mass (asterisk) is continuous with infiltrative tissue (arrow) in the endometrial cavity. A large mass (M) with ascites is present above the uterus. Surgical pathology demonstrated an endometrioid endometrial cancer with superficial myometrial invasion and cervical involvement by prolapsing tumor. The mass above the uterus was found to be a synchronous endometrioid ovarian cancer with peritoneal spread.

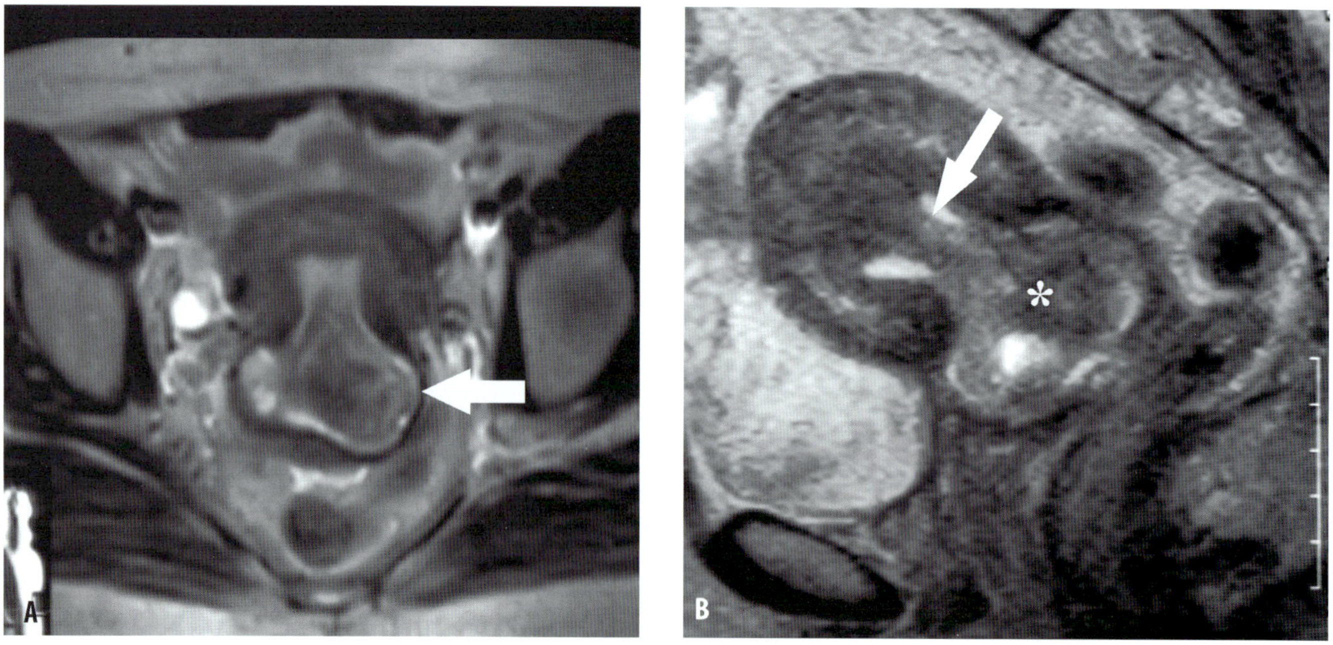

Figure 81.6 **A.** Axial T2-weighted MR image in a 46 year old woman with vaginal bleeding shows a mass (arrow) in the cervix. **B.** Sagittal T2-weighted MR image shows a stalk (arrow) attaching the mass to the uterine fundus. Hysteroscopic resection revealed the mass to be a prolapsed submucosal fibroid.

CASE **82**

Nabothian cysts

Imaging description

Nabothian cysts, named after the German anatomist Martin Naboth who first described them in 1707 [1], are retention or inclusion cysts of the uterine cervix that probably develop when chronic cervicitis results in blockage to the ducts of the mucus-secreting glands of the endocervix [2, 3]. At imaging, Nabothian cysts are seen as one or more well-circumscribed unilocular cysts abutting the endocervical canal that are usually a few mm in diameter but occasionally reach 4 cm or more (Figure 82.1) [4, 5]. The term "tunnel cluster" has been applied when multiple small Nabothian cysts are grouped together around the endocervical canal (Figure 82.2) [6], but the value of this term is questionable.

Importance

Incidentally detected Nabothian cysts, particularly at CT (Figure 82.3), may give rise to unjustified concerns of cervical carcinoma, resulting in unnecessary workup and patient anxiety.

Typical clinical scenario

Nabothian cysts are usually easily diagnosed clinically by visual inspection, or are incidental findings at imaging performed for unrelated reasons. Occasionally, extensive cyst formation or large cysts located deeper in the cervix may produce unexplained cervical enlargement and then imaging with ultrasound or MRI may be helpful in further assessment [7].

Differential diagnosis

Solitary Nabothian cysts are distinctive and generally do not carry an imaging differential. When multiple (i.e., tunnel cluster), the primary differential consideration for Nabothian cysts is adenoma malignum, a rare and aggressive mucinous form of cervical cancer that appears as a multicystic mass within the cervix, in contrast to the solid infiltrative appearance of standard non-mucinous cervical cancer. A solid component surrounding or separating multiple cysts favors a diagnosis of adenoma malignum, but exact differentiation is difficult as the MRI findings of both diseases overlap [8–11]. Clinically, adenoma malignum is frequently associated with a profuse watery or mucinous vaginal discharge, and this may be an important clue to the correct diagnosis [8]. It should be remembered that Nabothian cysts are far commoner than adenoma malignum, and so, even in the setting of proven malignancy, multiple cysts in the cervix may still just be coincidental Nabothian cysts (Figure 82.4).

Teaching point

Simple cysts seen at ultrasound, CT, or MRI in the uterine cervix are nearly always incidental benign Nabothian cysts. When multiple, the possibility of adenoma malignum, a rare multicystic form of cervical cancer usually associated with watery or mucinous vaginal discharge, should be considered.

REFERENCES

1 Naboth M. *De Sterilitate Mulierum*. Leipzig: A. Zeidler, 1707 [cited in: Morton LT. *A medical bibliography, Garrison and Morton*, 3rd edition. London: A. Deutsch, 1970; entry 1216].

2 Danforth DN, Scott JR. *Obstetrics and gynecology*. Philadelphia, PA: Lippincott, 1986.

3 Novak ER, Woodruff JD. *Novak's gynecologic and obstetric pathology with clinical and endocrine relations*, 8th edition. Philadelphia, PA: Saunders, 1979.

4 Okamoto Y, Tanaka YO, Nishida M, et al. MR imaging of the uterine cervix: imaging-pathologic correlation. *Radiographics* 2003; **23**: 425–445.

5 Fogel SR, Slasky BS. Sonography of Nabothian cysts. *Am J Roentgenol* 1982; **138**: 927–930.

6 Kurman RJ, Norris HJ, Wilkinson E. Tumors of the cervix. In: *Atlas of tumor pathology: tumors of the cervix, vagina, and vulva*. Washington, DC: Armed Forces Institute of Pathology, 1992.

7 Sosnovski V, Barenboim R, Cohen HI, Bornstein J. Complex Nabothian cysts: a diagnostic dilemma. *Arch Gynecol Obstet* 2009; **279**: 759–761.

8 Yamashita Y, Takahashi M, Katabuchi H, et al. Adenoma malignum: MR appearances mimicking Nabothian cysts. *Am J Roentgenol* 1994; **162**: 649–650.

9 Doi T, Yamashita Y, Yasunaga T, et al. Adenoma malignum: MR imaging and pathologic study. *Radiology* 1997; **204**: 39–42.

10 Yoden E, Mikami Y, Fujiwara K, Kohno I, Imajo Y. Florid endocervical glandular hyperplasia with pyloric gland metaplasia: a radiologic pitfall. *J Comput Assist Tomogr* 2001; **25**: 94–97.

11 Sugiyama K, Takehara Y. MR findings of pseudoneoplastic lesions in the uterine cervix mimicking adenoma malignum. *Br J Radiol* 2007; **80**: 878–883.

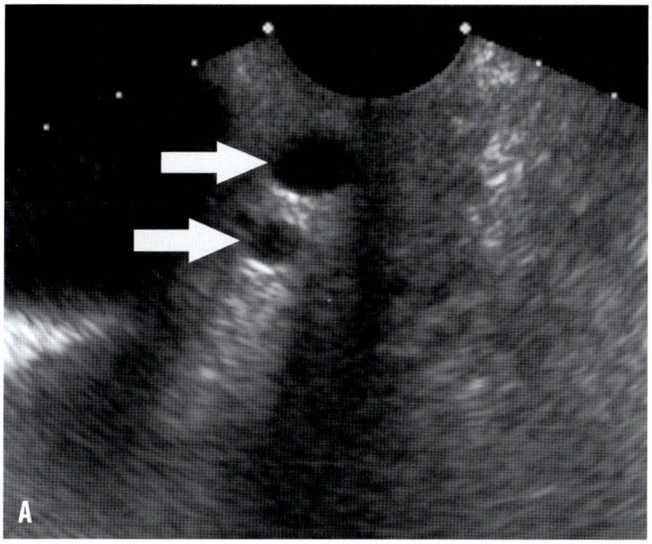

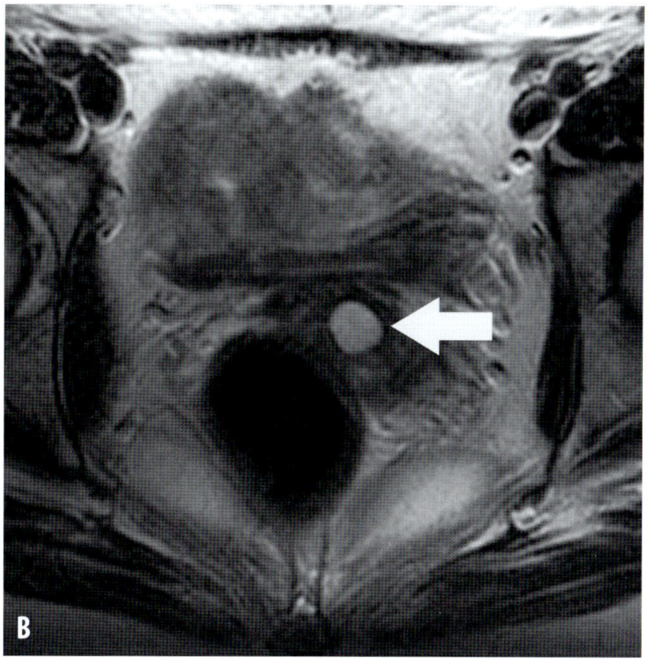

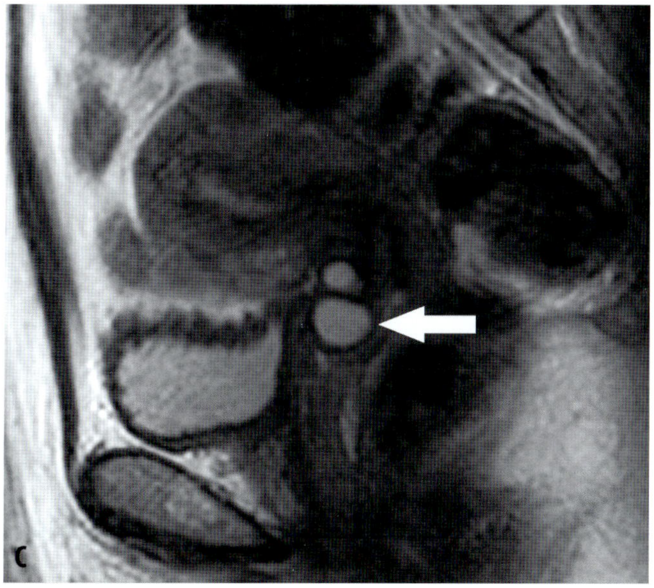

Figure 82.1 **A.** Axial endovaginal ultrasound image of the cervix in a 50 year old woman with irregular periods shows two adjacent Nabothian cysts (arrows) as an incidental finding. **B.** Axial T2-weighted MR image shows a well-circumscribed lesion (arrow) of fluid signal intensity in the cervix, the typical appearance of a Nabothian cyst. **C.** Sagittal T2-weighted MR image shows a Nabothian cyst (arrow) in the cervix, with a second smaller Nabothian cyst more superiorly.

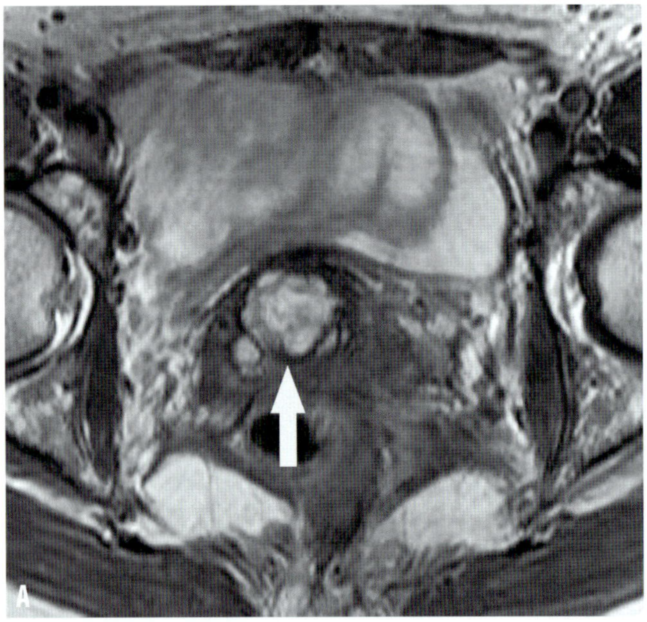

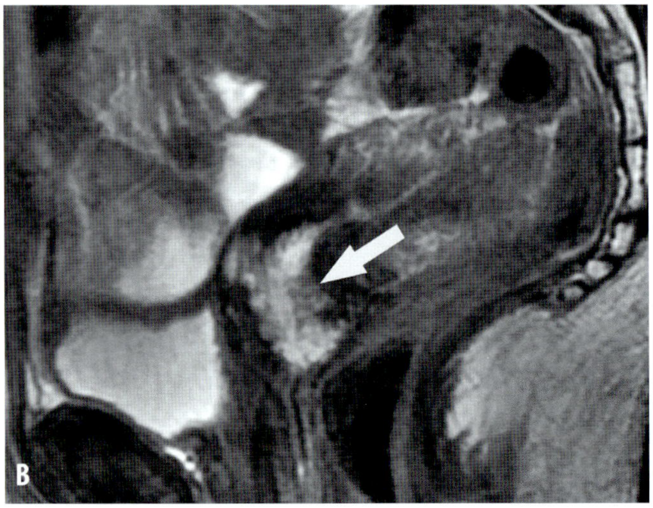

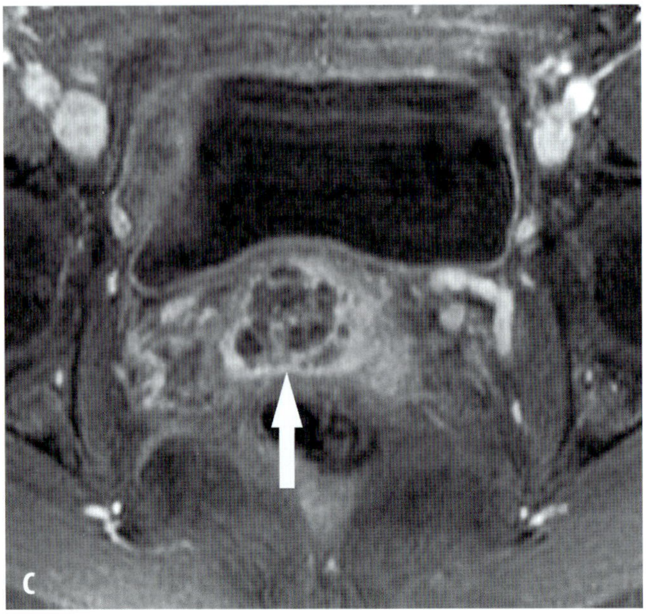

Figure 82.2 **A.** Axial T2-weighted MR image obtained in a 60 year old woman with pelvic floor weakness shows a group of Nabothian cysts (arrow) around the endocervix, the so-called tunnel cluster. **B.** Sagittal T2-weighted MR image shows the arrangement of the Nabothian cysts (arrow) around the endocervix. **C.** Axial gadolinium-enhanced T1-weighted MR image shows the cluster of cysts (arrow) as multiple non-enhancing foci.

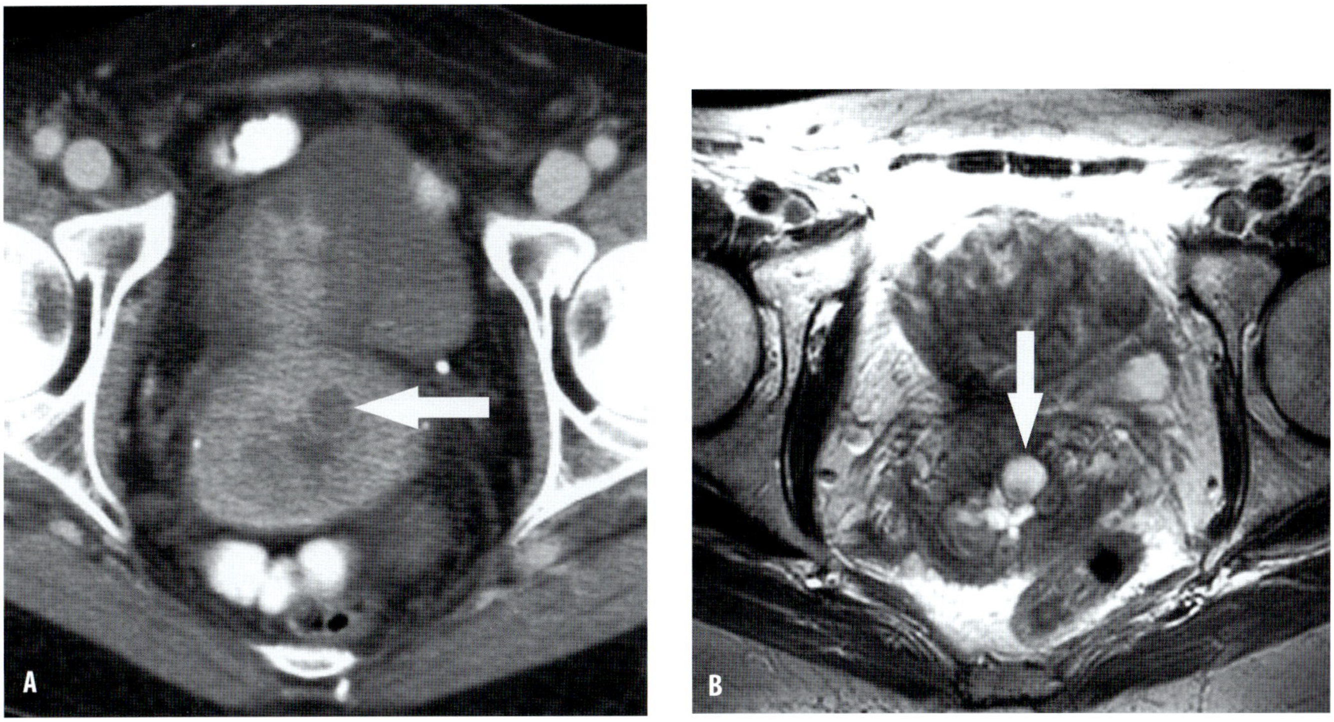

Figure 82.3 **A.** Axial contrast-enhanced CT image in a 76 year old woman with abdominal pain shows a well-circumscribed hypodense lesion in the cervix that was initially interpreted as a possible cervical cancer. **B.** Axial T2-weighted MR image shows the lesion is of fluid signal intensity, consistent with a Nabothian cyst.

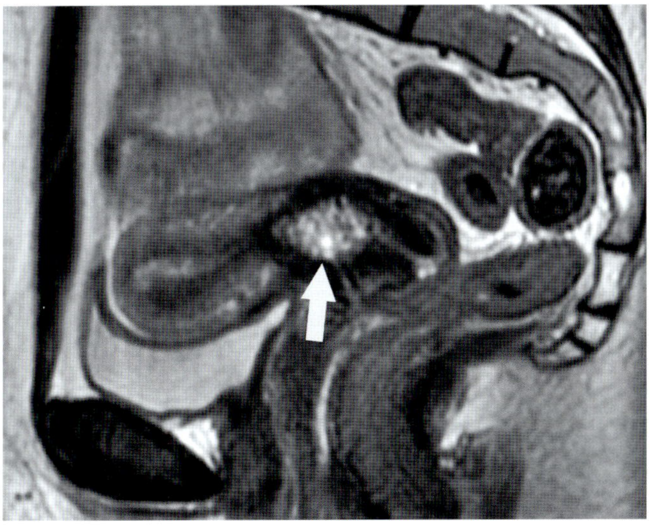

Figure 82.4 Sagittal T2-weighted MR image in a 44 year old woman with atypical glandular cells suggestive of endocervical adenocarcinoma detected at an endocervical curettage and biopsy performed after an abnormal routine pap smear shows a cluster of small cysts (arrow) centered on the endocervix. The conjunction of biopsy and imaging findings was considered suggestive of adenoma malignum. Pathological examination after radical hysterectomy showed prominent Nabothian cysts clustered in the endocervix adjacent to a separate smaller focus of invasive adenocarcinoma with a conventional appearance. The final interpretation was that of an adenocarcinoma arising adjacent to a coincidental group of Nabothian cysts and not adenoma malignum.

CASE 83 Vaginal pessary

Imaging description

A vaginal pessary is an object, usually made of rubber or plastic, inserted into the vagina to support the uterus and pelvic floor (the term can also be used more widely to include contraceptive vaginal devices or medicated vaginal suppositories) [1]. The commonest appearance is of a radiopaque ring-like structure in the vagina seen at CT (Figure 83.1), although a wide variety of shapes is available. Pessaries are usually only slightly opaque on plain radiographs. In 2001, a hormone-releasing vaginal contraceptive device (NuvaRing®) that appears as a radiolucent 5 cm diameter ring, was approved for use by the United States Food and Drug Administration (Figures 83.2 and 83.3) [2].

Importance

It is important to recognize vaginal pessaries so that they are not mistaken for unintentionally retained foreign bodies in the vagina. At MRI, the signal void created by a pessary may be misdiagnosed as a bowel loop or even suggest a sigmoidocele (Figure 83.4). Pessaries should be properly fitted and removed every few months for cleaning [1]. Occasionally a vaginal pessary may impact, fistulize or migrate, particularly if neglected, and imaging may contribute to the evaluation of such complications [3–6].

Typical clinical scenario

Pessaries are widely used by gynecologists as first-line therapy for pelvic organ prolapse, which typically affects older multiparous women. Pessaries may be divided into ring-shaped support pessaries and variously shaped space-filling pessaries, such as the distinctive T-shaped Gellhorn pessary (Figure 83.5). While pessaries of different shapes and sizes are designed and marketed for different forms of pelvic organ prolapse, there is no evidence or consensus on whether such tailored strategies are appropriate [7]. In contrast to pessaries used for pelvic floor weakness, the vaginal contraceptive ring is used in women of reproductive age.

Differential diagnosis

The imaging features of pessaries used for prolapse are distinctive and should not cause confusion provided the interpreting radiologist is aware of their appearance. The radiolucent ring seen at CT in patients with a vaginal NuvaRing® is a little more problematic, and could suggest a diagnosis of emphysematous vaginitis, a benign, self-limiting bacterial infection that manifests as discrete or confluent gas-filled cysts in the vaginal wall (Figure 83.6) [8]. Misdiagnosis can be avoided by examining the shape and location of the radiolucencies. In emphysematous vaginitis, the gas is typically bubbly and in the vaginal wall while the radiolucency of the contraceptive vaginal ring is curvilinear and in the vaginal lumen.

Teaching point

Ring-shaped or other foreign bodies seen at imaging in the vaginas of older women are usually pessaries placed to treat pelvic floor prolapse. A radiolucent ring-like structure seen in the vagina of a younger woman is likely to be a vaginal contraceptive device (NuvaRing®).

REFERENCES

1 Margolis M, McLennan MK. Radiology rounds. Intrauterine contraceptive device. *Can Fam Physician* 1993; **39**: 2283–2284.

2 Blackmon GB, Paulson EK. Multidetector CT appearance of a novel contraceptive device: the vaginal ring. *Am J Roentgenol* 2003; **180**: 1659–1660.

3 Poma PA. Management of incarcerated vaginal pessaries. *J Am Geriatr Soc* 1981; **29**: 325–327.

4 Powers K, Grigorescu B, Lazarou G, Greston WM, Weber T. Neglected pessary causing a rectovaginal fistula: a case report. *J Reprod Med* 2008; **53**: 235–237.

5 Hanavadi S, Durham-Hall A, Oke T, Aston N. Forgotten vaginal pessary eroding into rectum. *Ann R Coll Surg Engl* 2004; **86**: W18–W19.

6 Lim BK, Collaris RRJ. Migration of a Hodge pessary into the abdominal cavity; a rare complications. *J Obstet Gynaecol Res* 2008; **34**: 436–438.

7 Cundiff GW, Weidner AC, Visco AG, Bump RC, Addison WA. A survey of pessary use by members of the American Urogynecologic Society. *Obstet Gynecol* 2000; **95**: 931–935.

8 Leder RA, Paulson EK. Vaginitis emphysematosa: CT and review of the literature. *Am J Roentgenol* 2001; **176**: 623–625.

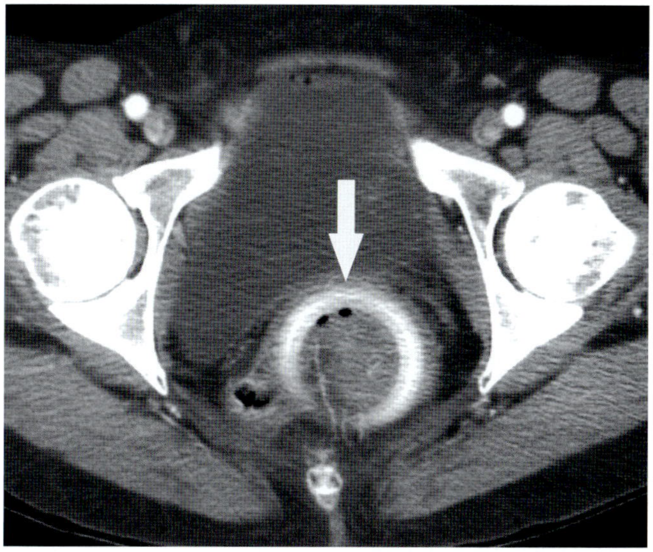

Figure 83.1 Axial contrast-enhanced CT image obtained in a 61 year old woman with primary peritoneal carcinoma shows the typical radiopaque ring-like appearance (arrow) of a support pessary in the vagina.

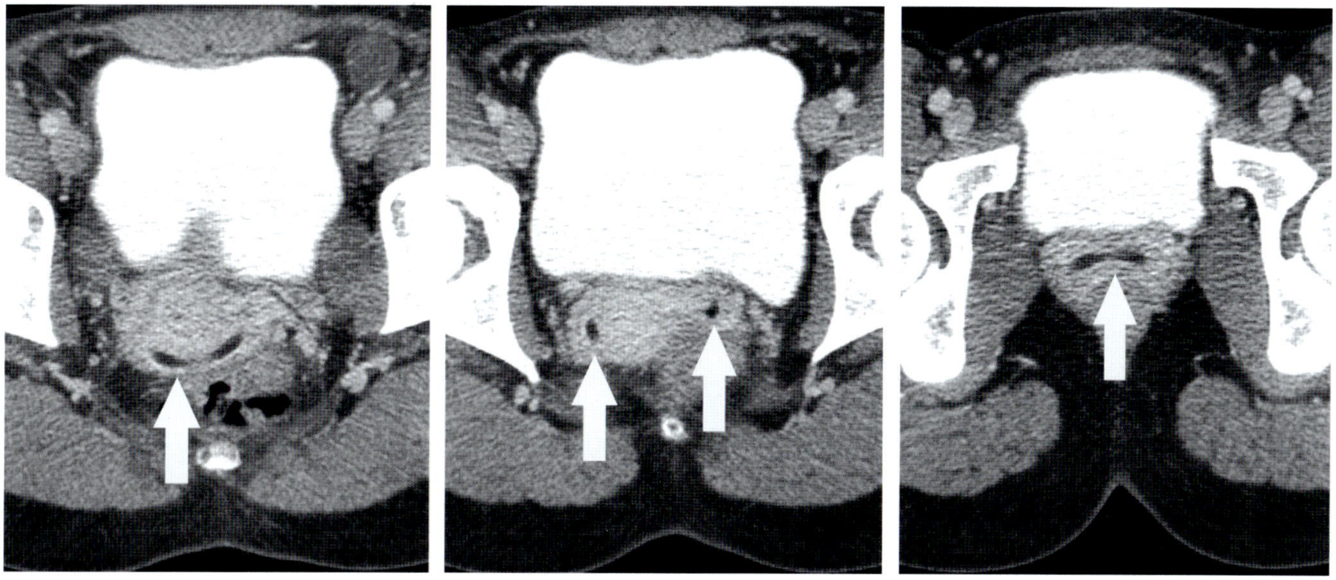

Figure 83.2 Photomontage of three axial delayed phase contrast-enhanced CT images arranged from superior to inferior in a 26 year old woman with congenital ureteropelvic junction obstruction. A vaginal contraceptive ring (NuvaRing®) forms a circular structure in the vaginal lumen (arrows).

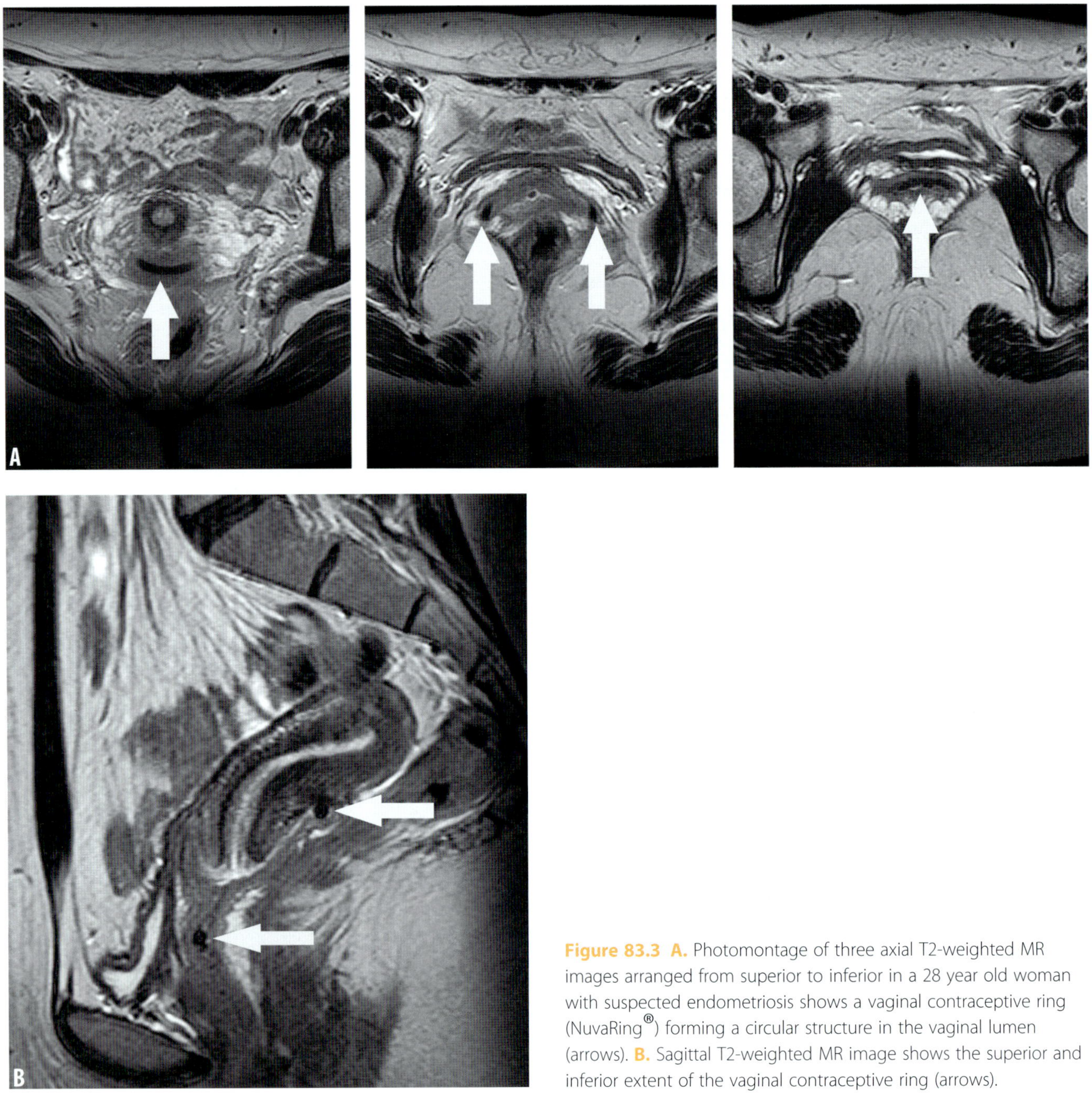

Figure 83.3 **A.** Photomontage of three axial T2-weighted MR images arranged from superior to inferior in a 28 year old woman with suspected endometriosis shows a vaginal contraceptive ring (NuvaRing®) forming a circular structure in the vaginal lumen (arrows). **B.** Sagittal T2-weighted MR image shows the superior and inferior extent of the vaginal contraceptive ring (arrows).

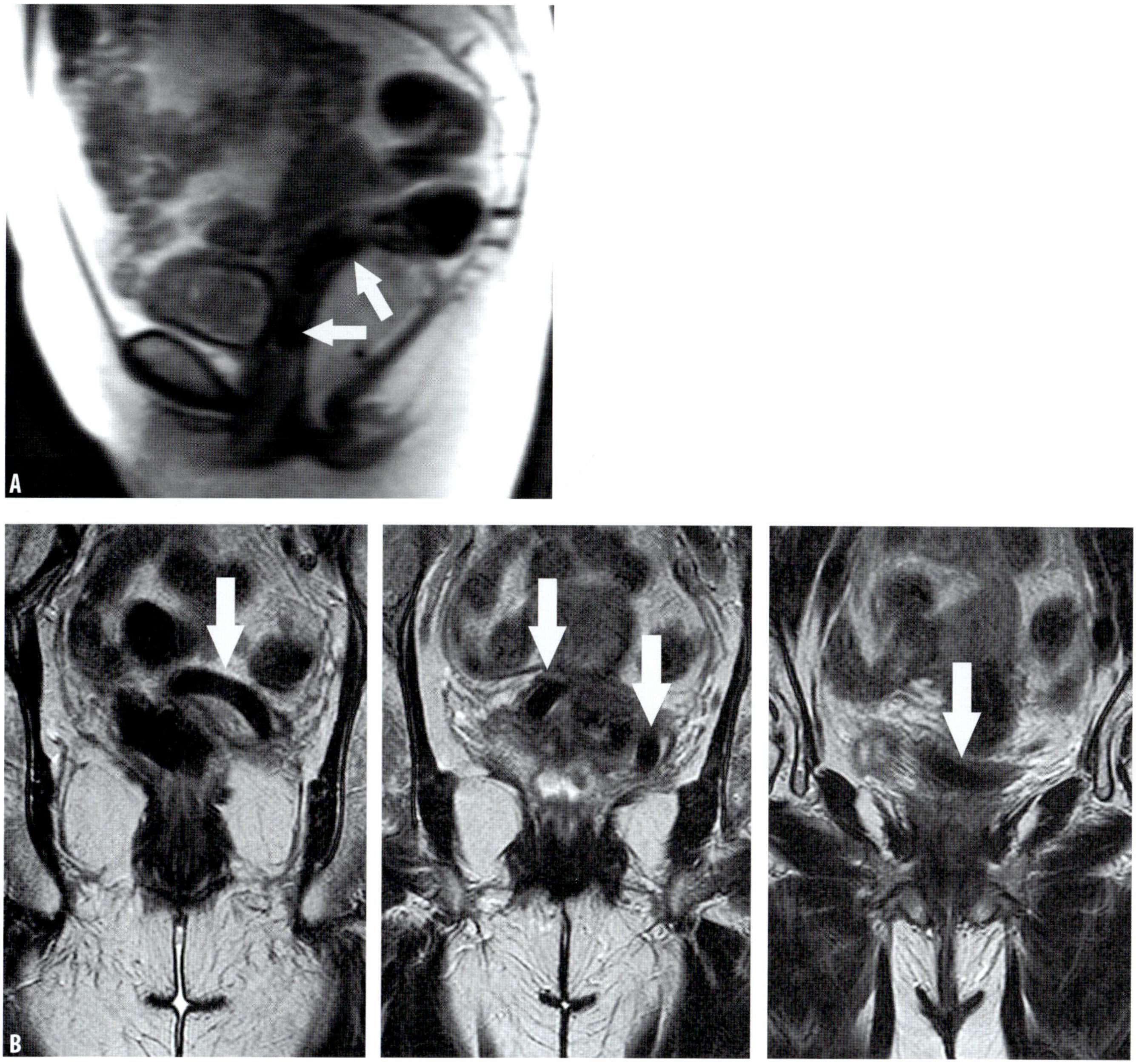

Figure 83.4 **A.** Sagittal T2-weighted MR image in a 70 year old woman with lower abdominal pain and difficulty with defecation shows a signal void posterior to the bladder and uterus that was initially interpreted as a sigmoidocele. **B.** Photomontage of three coronal T2-weighted MR images arranged from posterior to anterior shows the signal void has the typical circular morphology of a ring support pessary.

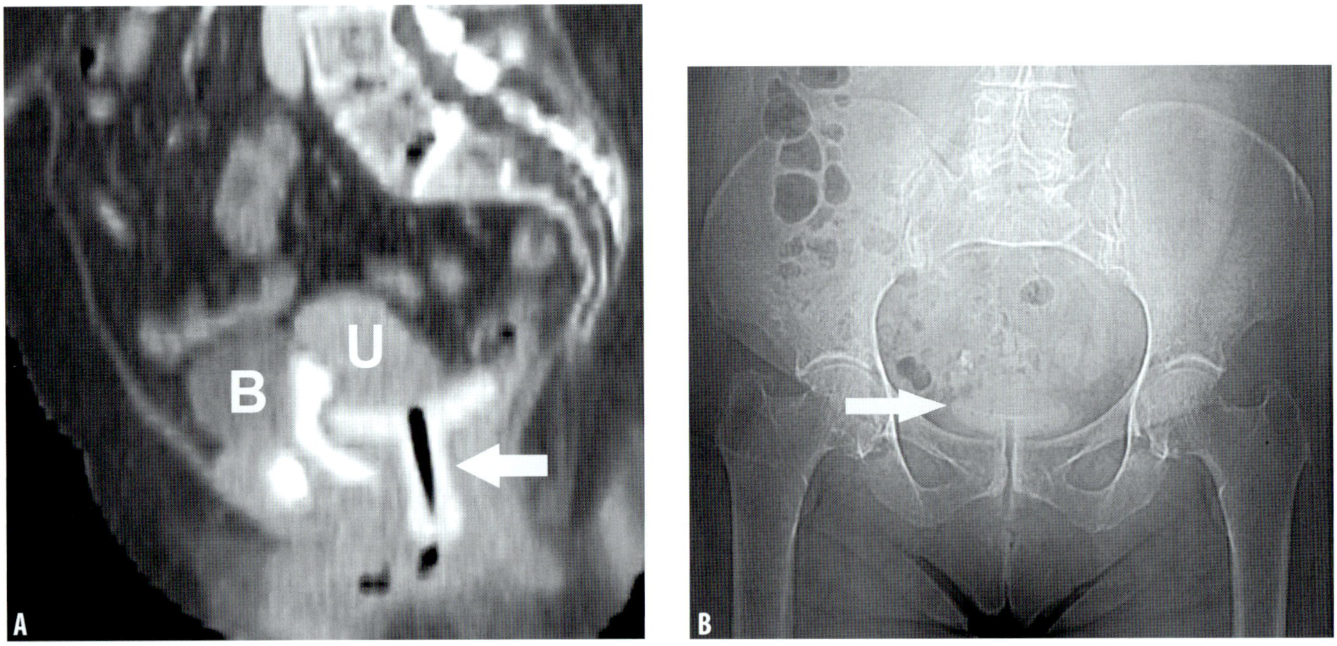

Figure 83.5 **A.** Sagittal reformatted contrast-enhanced CT image in a 67 year old woman with abdominal pain shows the distinctive T-shaped morphology of a Gellhorn pessary. Excreted contrast is layering in the posterior part of the bladder. **B.** Scout view from the same CT scan shows that the pessary (arrow) is only faintly visible on a plain radiograph.

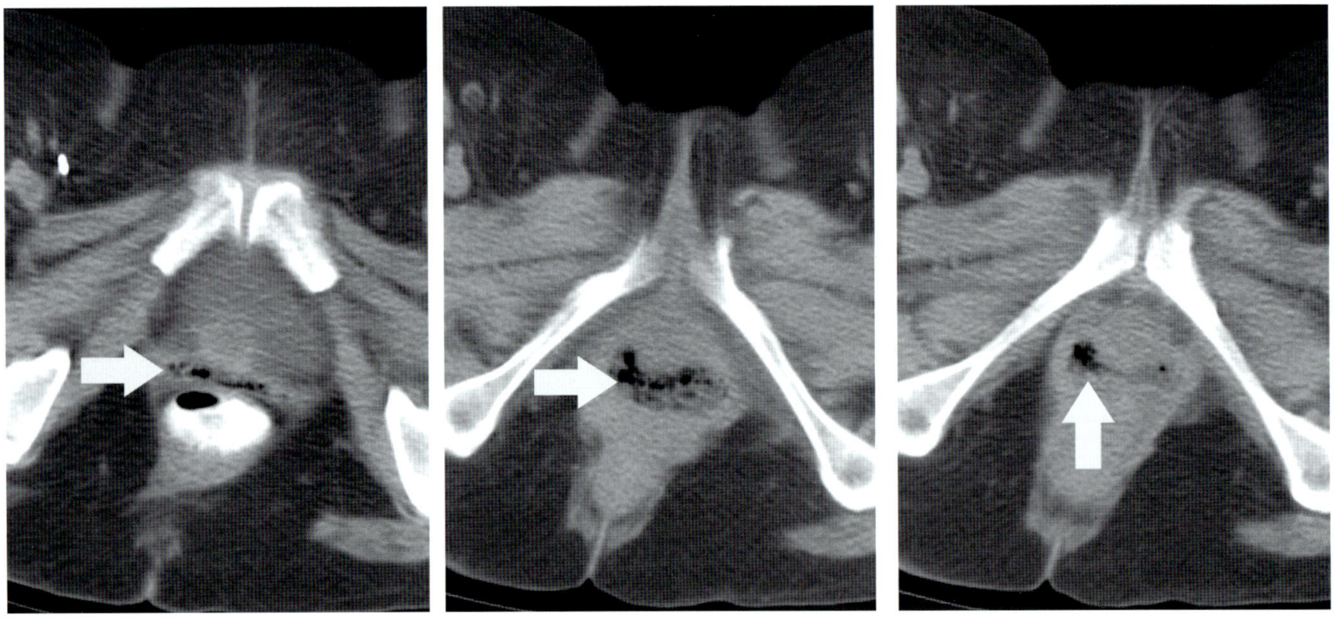

Figure 83.6 Photomontage of three axial contrast-enhanced CT images arranged from superior to inferior in a 63 year old woman being treated with steroids for autoimmune hemolytic anemia shows bubbly gaseous lucencies in the vagina consistent with emphysematous vaginitis. A subsequent culture demonstrated infection with Trichomonas and the patient was treated with oral metronidazole.

CASE 84 Pseudobladder

Imaging description

Midline cystic tumors or fluid collections in the pelvis may be mistaken for the urinary bladder, and the term pseudobladder has been applied to the resulting pitfall which can occur at ultrasound, CT, or MRI (Figures 84.1–84.4) [1–6].

Importance

Misinterpretation of a pelvic cystic mass or fluid collection as the bladder may result in a missed opportunity for timely surgery or drainage, and potentially increased patient morbidity or mortality.

Typical clinical scenario

The pitfall occurs most frequently when a large benign cystic adnexal mass that is unilocular or has one large dominant locule occupies a midline position and therefore closely simulates the urinary bladder. As such, the misdiagnosis occurs primarily in women, although cystic pelvic masses or collections can also simulate the bladder in men.

Differential diagnosis

This diagnostic pitfall can generally be avoided provided the possibility is considered, because close examination usually allows identification of the collapsed or compressed bladder as a separate structure. In problematic cases, delayed phase contrast-enhanced CT images can be helpful in demonstrating the opacified bladder as being distinct from the pseudobladder.

Teaching point

The possibility of a pseudobladder due to a pelvic cystic tumor or fluid collection should be considered for any midline fluid-filled structure that appears to be the urinary bladder.

REFERENCES

1 Gaeta M, Blandino A, Scribano E, et al. Diagnostic pitfalls of breath-hold MR urography in obstructive uropathy. *J Comput Assist Tomogr* 1999; **23**: 891–897.

2 Vick CW, Viscomi GN, Mannes E, Taylor KJ. Pitfalls related to the urinary bladder in pelvic sonography: a review. *Urol Radiol* 1983; **5**: 253–259.

3 Fiske CE, Callen PW. Fluid collections ultrasonically simulating urinary bladder. *J Can Assoc Radiol* 1980; **31**: 254–255.

4 Elyaderani MK, Gabriele OF. Large ovarian cyst simulating urinary bladder on ultrasonography. *South Med J* 1980; **73**: 829.

5 Haney AF, Trought WS. Parovarian cysts resembling a filled urinary bladder. *J Clin Ultrasound* 1978; **6**: 53–54.

6 Cooperberg MR, Chambers SK, Rutherford TJ, Foster HE Jr. Cystic pelvic pathology presenting as falsely elevated post-void residual urine measured by portable ultrasound bladder scanning: report of 3 cases and review of the literature. *Urology* 2000; **55**: 590.

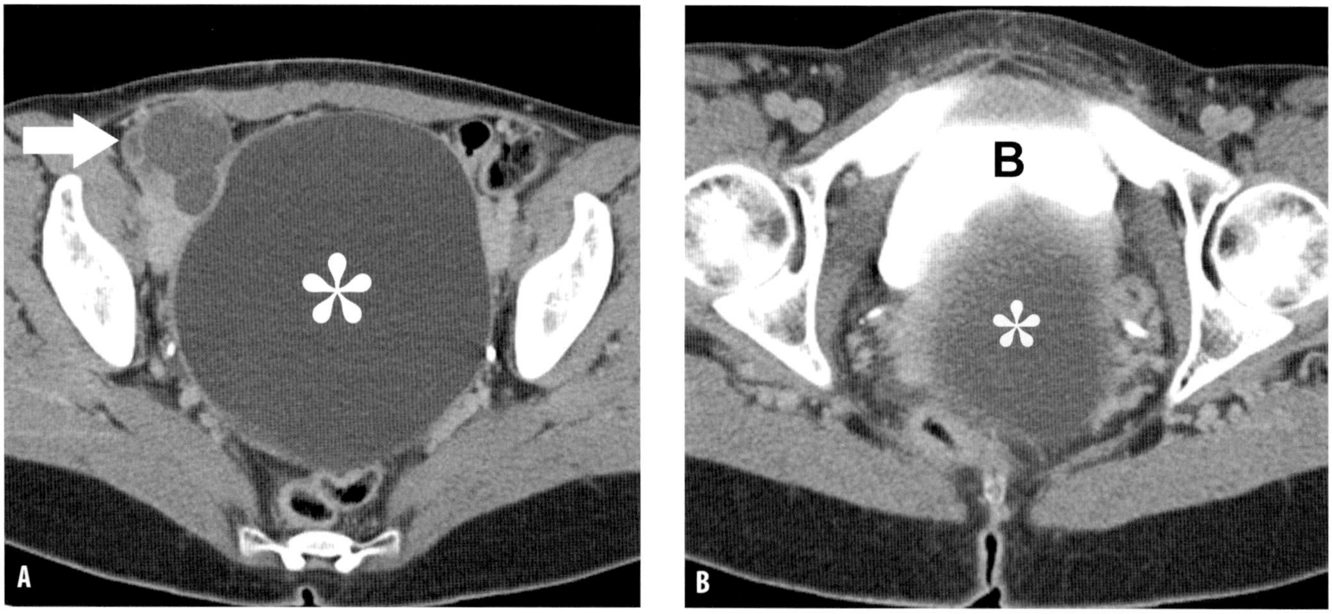

Figure 84.1 A. Axial contrast-enhanced CT image in a 52 year old woman with a 6-month history of abdominal discomfort, constipation, and urinary frequency shows a multiloculated cystic mass (arrow) on the right side of what appears to be a distended urinary bladder (asterisk). **B.** Axial delayed phase contrast-enhanced CT image shows that the true bladder (B) is visible as a contrast-filled structure distinct from the pseudobladder (asterisk). Surgical pathology demonstrated a 15 cm left cystadenofibroma consisting of a single unilocular cyst and a multiloculated 5 cm right cystadenofibroma.

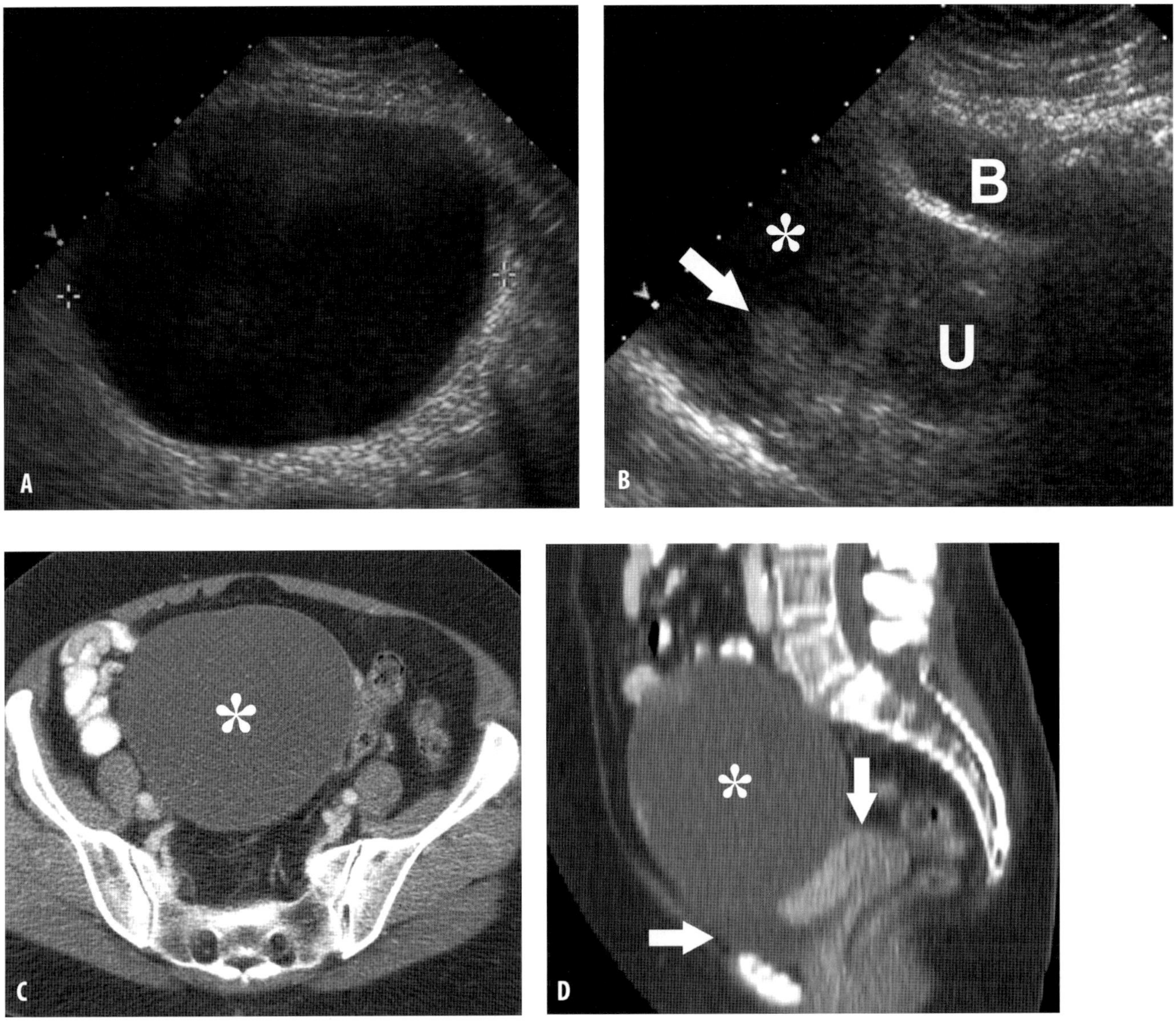

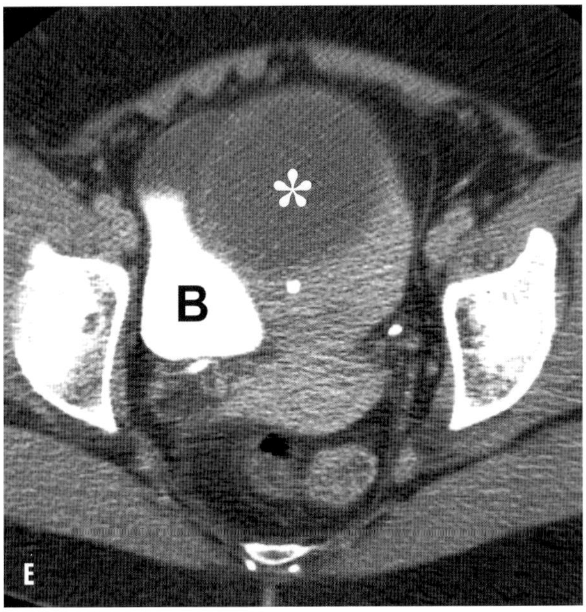

Figure 84.2 **A.** Transvaginal ultrasound image in a 57 year old woman with a one-month history of bloating, abdominal distention, pressure and urinary frequency. A midline fluid-filled structure (between calipers) could be mistaken for the urinary bladder. **B.** Transvaginal ultrasound image shows the pseudobladder (asterisk) is distinct from both the true bladder (B) and the uterus (U). A mural nodule (arrow) is visible in the pseudobladder. **C.** Axial contrast-enhanced CT image shows a midline fluid-filled structure (asterisk) in the pelvis that resembles the urinary bladder. **D.** Sagittal reformatted contrast-enhanced CT image shows the pseudobladder is superior to the uterus (vertical arrow) and distinct from the true bladder (horizontal arrow). **E.** Axial delayed phase contrast enhanced CT image confirms the true bladder (B) is distinct from the pseudobladder (asterisk). Surgical resection demonstrated a 12.5 cm cystadenocarcinoma of the left ovary.

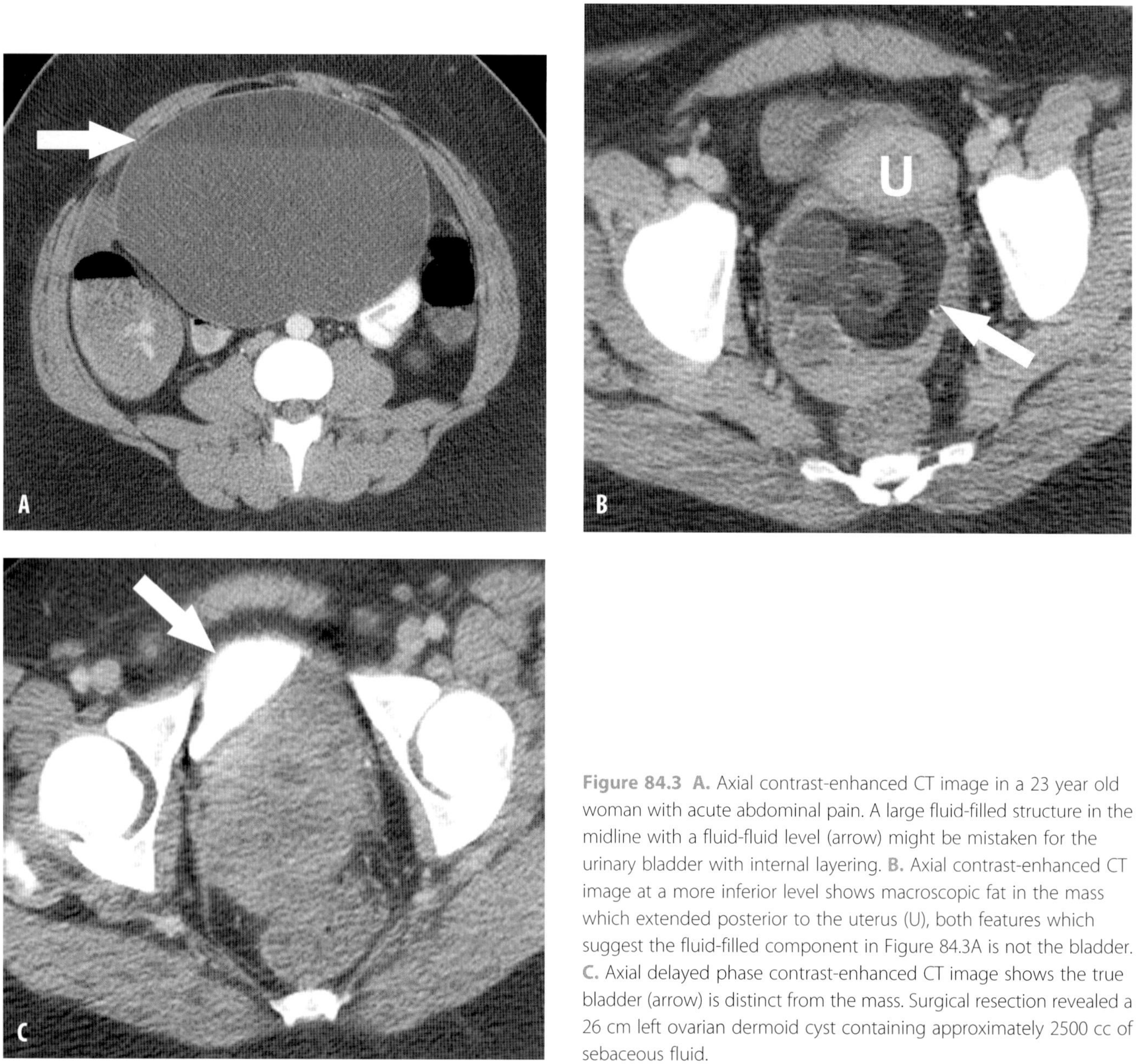

Figure 84.3 A. Axial contrast-enhanced CT image in a 23 year old woman with acute abdominal pain. A large fluid-filled structure in the midline with a fluid-fluid level (arrow) might be mistaken for the urinary bladder with internal layering. **B.** Axial contrast-enhanced CT image at a more inferior level shows macroscopic fat in the mass which extended posterior to the uterus (U), both features which suggest the fluid-filled component in Figure 84.3A is not the bladder. **C.** Axial delayed phase contrast-enhanced CT image shows the true bladder (arrow) is distinct from the mass. Surgical resection revealed a 26 cm left ovarian dermoid cyst containing approximately 2500 cc of sebaceous fluid.

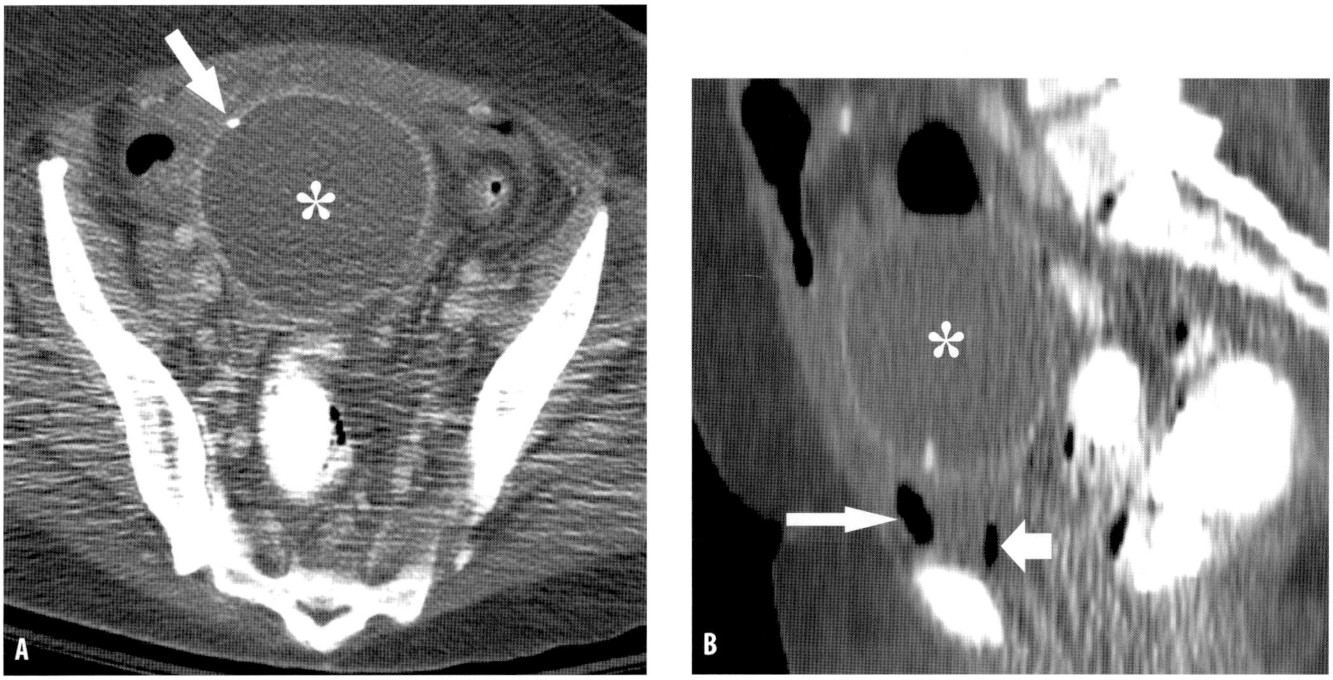

Figure 84.4 A. Axial contrast-enhanced CT image in a 65 year old woman with altered mental status and a history of a ventriculoperitoneal shunt placement for normal pressure hydrocephalus. A large fluid-filled structure in the midline resembles the urinary bladder. The ventriculoperitoneal shunt (arrow) abuts the wall of this structure. B. Sagittal reformatted contrast-enhanced CT image shows the pseudobladder (asterisk) is superior to the true bladder which contains air (long thin arrow) due to the presence of a Foley catheter. Air (short thick arrow) is visible in the balloon of the catheter. Surgical exploration demonstrated a midline pelvic abscess related to the catheter tip.

CASE **85** Urachal remnant disorders

Imaging description

In early fetal life, the urachus connects the cloaca (which gives rise to the bladder) to the yolk sac (which is outside the fetal abdomen) [1]. As the yolk sac regresses, the urachus closes but persists into postnatal life as a fibrous band connecting the dome of the bladder to the umbilicus. This vestigial structure is known as the median umbilical ligament, and is sometimes seen incidentally at CT when outlined by fluid or air in patients with ascites or pneumoperitoneum (Figure 85.1). A normal urachal remnant may also be seen as a hypoechoic area at the dome of the bladder in up to 60% of children [2]. Occasionally, the urachus fails to regress and remains patent either partially or entirely. Depending on which part remains patent, such urachal remnant disorders are classified as patent urachus (fistulous track connecting the bladder and umbilicus), urachal umbilical sinus (blind ending track connecting to the umbilicus), vesicourachal diverticulum (upward tubular extension from the bladder), or urachal cyst (non-communicating cyst along the course of the urachus). At imaging, urachal remnants are seen as fluid-filled structures along the course of the urachus, i.e., in the midline anteriorly between the bladder and umbilicus (Figures 85.2–85.4). Eggshell mural calcification is sometimes seen in urachal cysts [3]. Urachal remnants may become infected or develop malignancy (Figures 85.5 and 85.6), and such complications should be suspected if a urachal remnant is thick-walled, echogenic, or contains solid elements. Urachal carcinoma should also be considered for a midline adenocarcinoma arising anteriorly in the bladder dome (Figure 85.7); this location is suggestive and the histology is atypical for bladder cancer (the majority of which are transitional cell carcinomas). Most urachal cancers are adenocarcinomas, and many are mucin-producing [4–6]. Calcification may occur, and calcification in a midline supravesical mass is nearly diagnostic of a urachal carcinoma [1].

Importance

Recognition that a midline supravesical abnormality is a urachal remnant disorder helps guide management. While infected urachal cysts can be initially treated conservatively, total surgical resection is ultimately recommended because there is a 30% reinfection rate [7] and carcinoma may develop in an incompletely resected urachal remnant [8, 9]. Urachal adenocarcinoma of the bladder may be adequately treated by partial cystectomy, whereas surgery for standard bladder transitional cell carcinoma usually requires a radical cystectomy.

Typical clinical scenario

Microscopic urachal remnants are found at autopsy in approximately 3% of the population [10, 11]. Congenital urachal anomalies are twice as common in men as in women [8]. The relative frequencies of the four types of urachal anomalies are patent urachus – 50%, urachal umbilical sinus – 15%, vesicourachal diverticulum–3 to 5%, and urachal cyst – 30% [9]. The majority of patients with urachal abnormalities (except those with a patent urachus) are asymptomatic and symptoms only develop if there is superimposed infection or malignancy.

Differential diagnosis

The location of urachal remnants is relatively distinctive, with only limited differential considerations. The primary differential diagnosis for an apparent patent urachus is a vesicostomy, a surgical anastomosis of the bladder to an opening in the skin below the umbilicus, creating a stoma for bladder drainage. Sometimes the appendix (Mitrofanoff procedure) or another tubular structure is used in an effort to create a continent stoma (Figures 85.7 and 85.8). The procedure is primarily used to protect the kidneys in children with neurogenic bladder or other disorders of bladder emptying [12, 13]. Differential considerations for complex or solid masses between the bladder and umbilicus include metastases (including the Sister Mary Joseph's nodule – a metastatic lesion of the umbilicus originating from intra-abdominal or pelvic malignancy) and periumbilical varices in cirrhosis (Figure 85.9) [14, 15].

Teaching point

A fluid-filled or complex structure seen at imaging in or near the midline anteriorly between the bladder and umbilicus should suggest the diagnosis of a urachal remnant disorder.

REFERENCES

1 Yu JS, Kim KW, Lee HJ, et al. Urachal remnant diseases: spectrum of CT and US findings. *Radiographics* 2001; **21**: 451–461.

2 Cacciarelli A, Kass EJ, Yang SS. Urachal remnants: sonographic demonstration in children. *Radiology* 1990; **174**: 473–475.

3 Leyson JFJ. Calcified urachal cyst. *Br J Urol* 1984; **56**: 438.

4 Sheldon CA, Clayman RV, Gonzalez R, Williams RD, Fraley EE. Malignant urachal lesions. *J Urol* 1984; **131**: 1–8.

5 Ravi R, Shrivastava BR, Chandrasekhar GM, et al. Adenocarcinoma of the urachus. *J Surg Oncol* 1992; **50**: 201–203.

6 Beck AD, Gaudin JH, Bonham DG. Carcinoma of the urachus. *Br J Urol* 1970; **42**: 555–562.

7 Goldman IL, Caldamone AA, Gauderer M, et al. Infected urachal cysts: a review of 10 cases. *J Urol* 1988; **140**: 375–378.

8 Friedland GW, Devries PA, Matilde NM, Cohen R, Rifkin MD. Congenital anomalies of the urinary tract. In: Pollack HM, ed. *Clinical Urography.* Philadelphia, PA: Saunders, 1990; 559–787.

9 Blichert-Toft M, Nielsen OV. A congenital patent urachus and acquired variants. *Acta Chir Scand* 1971; **137**: 807–814.

10 Jarzylo SV, Challis TW, Bruce AW. Urachal abnormalities. *J Can Assoc Radiol* 1965; **16**: 175–180.

11 Schubert GE, Pavkovic MB, Bethke-Bedurftig BA. Tubular urachal remnants in adult bladders. *J Urol* 1982; **127**: 40–42.

12 Morrisroe SN, O'Connor RC, Nanigian DK, Kurzrock EA, Stone AR. Vesicostomy revisited: the best treatment for the hostile bladder in myelodysplastic children? *BJU Int* 2005; **96**: 397–400.
13 Mollard P, Gauriau L, Bonnet JP, Mure PY. Continent cystostomy (Mitrofanoff's procedure) for neurogenic bladder in children and adolescent (56 cases: long-term results). *Eur J Pediatr Surg* 1997; **7**: 34–37.
14 Abu-Hilal M, Newman JS. Sister Mary Joseph and her nodule: historical and clinical perspective. *Am J Med Sci* 2009; **337**: 271–273.
15 Ishikawa T, Tsukune Y, Ohyama Y, et al. Venous abnormalities in portal hypertension demonstrated by CT. *Am J Roentgenol* 1980; **134**: 271–276.

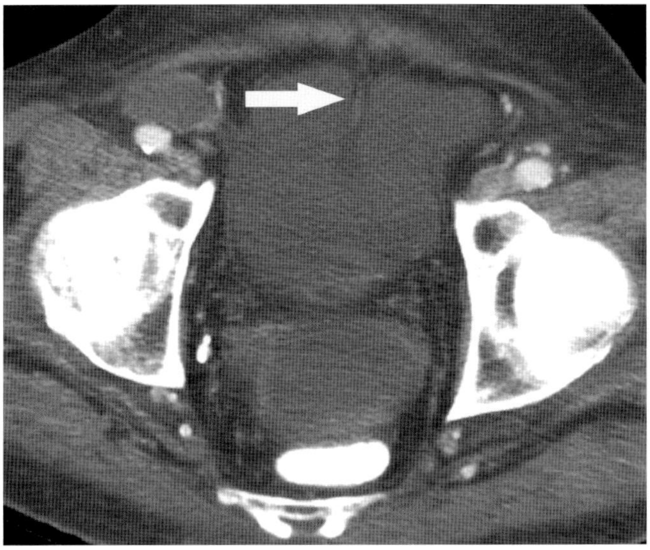

Figure 85.1 Axial contrast-enhanced CT image in an 87 year old man with malignant ascites shows the median umbilical ligament (arrow) outlined by fluid. The median umbilical ligament is the normal postnatal vestigial remnant of the obliterated urachus.

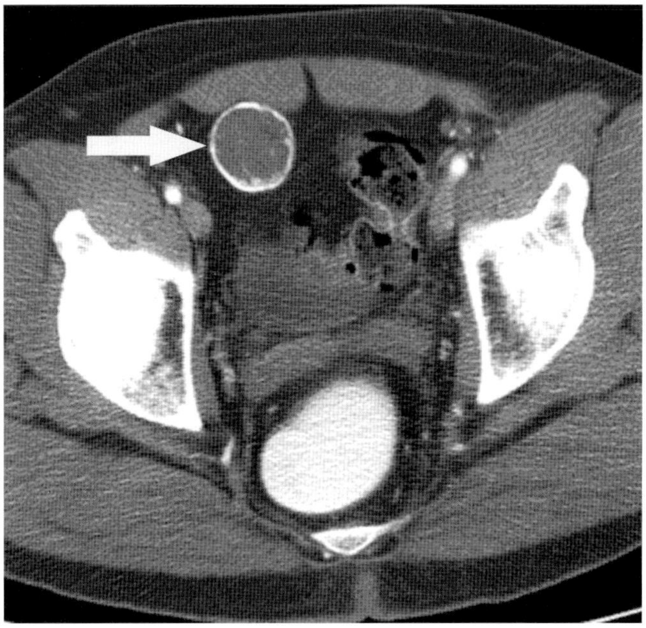

Figure 85.2 Axial contrast-enhanced CT image in a 64 year old man with colorectal cancer shows a rim-calcified cystic structure (arrow) anteriorly near the midline above the bladder. Surgical resection demonstrated a benign urachal cyst.

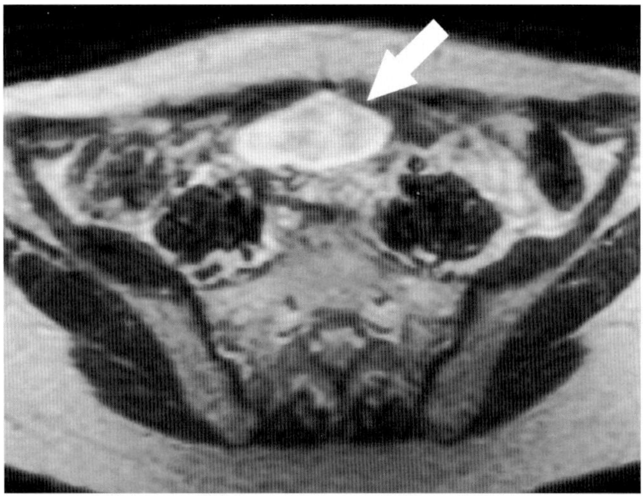

Figure 85.3 Axial T2-weighted MR image in a 56 year old woman with pelvic floor weakness shows an incidental urachal cyst (arrow).

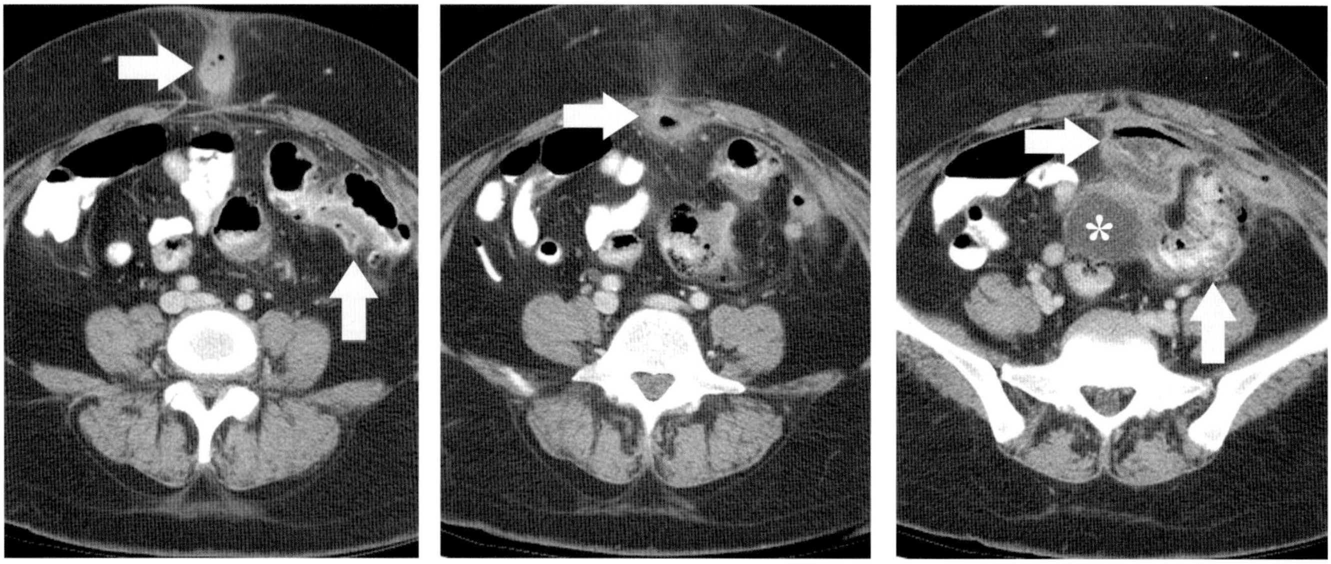

Figure 85.4 Photomontage of three axial contrast-enhanced CT images arranged from superior to inferior in a 39 year old otherwise healthy woman with a one-week history of abdominal pain and a one-day history of purulent discharge from the umbilicus. An infected urachal remnant is seen as a thick-walled air and fluid-containing structure (horizontal arrows) passing from the umbilicus to the bladder (asterisk). Infection of the urachal remnant probably occurred secondary to co-existent acute diverticulitis (vertical arrows).

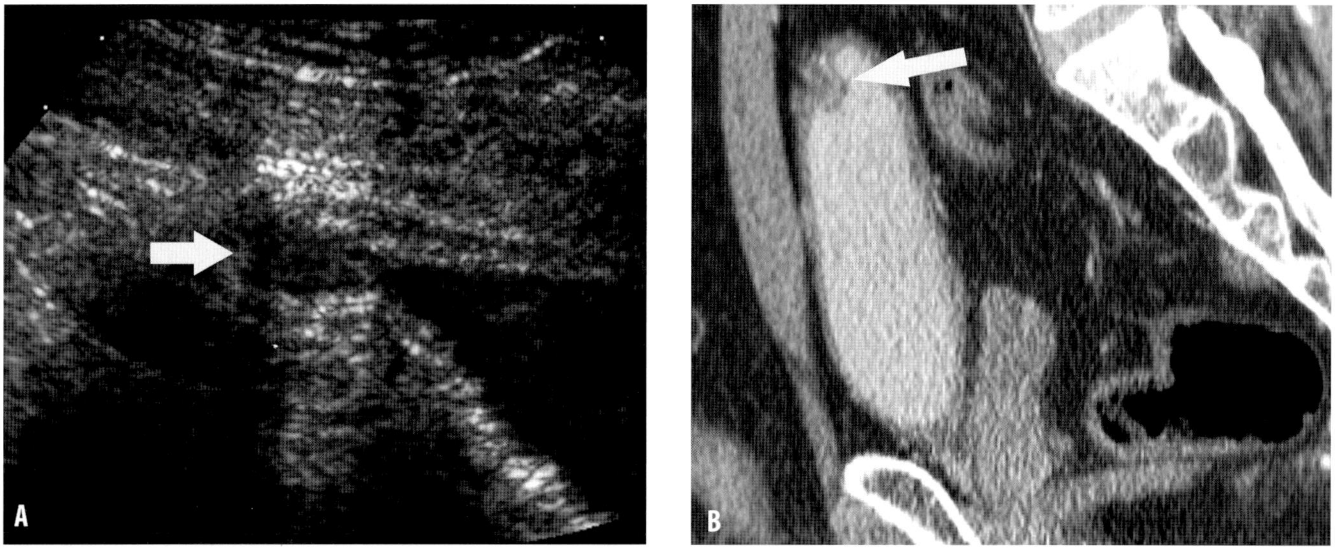

Figure 85.5 A. Sagittal transabdominal ultrasound image in a 50 year old man with painless gross hematuria shows a solid nodule (arrow) in the dome of the bladder. B. Sagittal reformatted CT image confirms the presence of a soft-tissue nodule (arrow) in the midline anteriorly. Partial cystectomy confirmed a urachal adenocarcinoma.

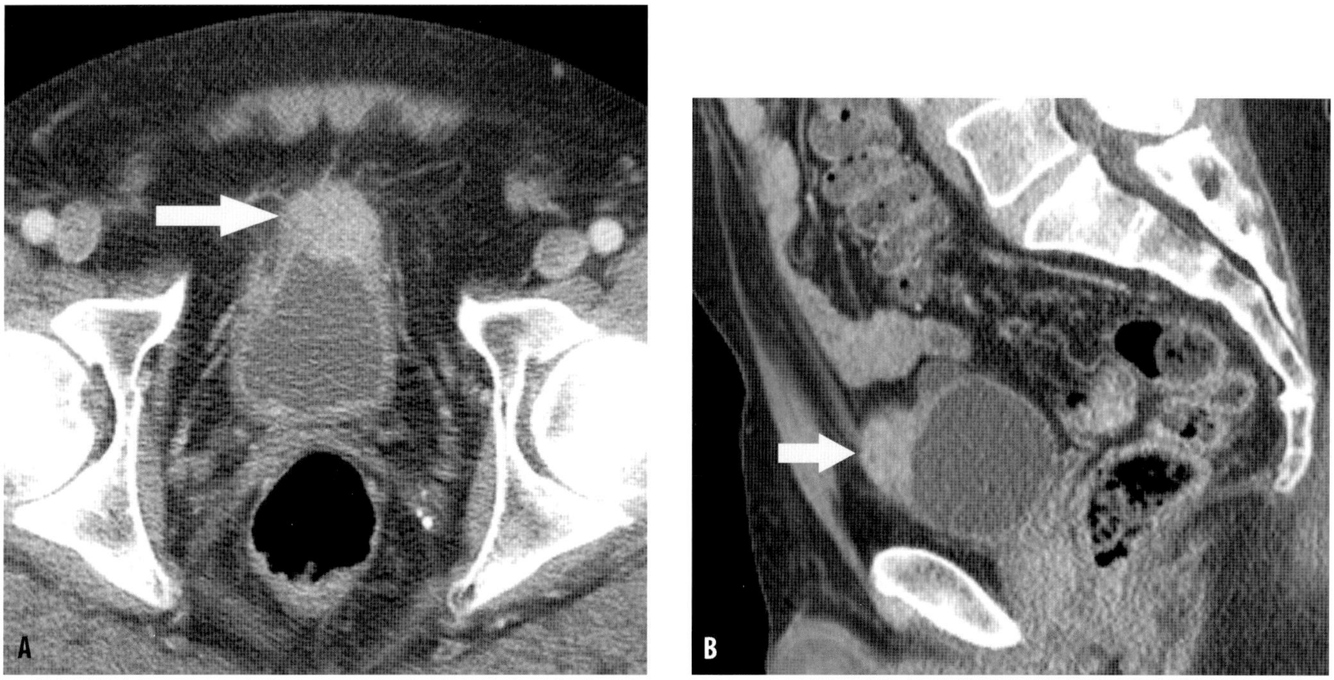

Figure 85.6 A. Axial contrast-enhanced CT image in a 68 year old man with a urachal transitional cell carcinoma shows a midline enhancing mass (arrow) in the anterior bladder. B. Sagittal reformatted contrast-enhanced CT image shows the typical midline anterior location of the urachal carcinoma (arrow) on the bladder dome.

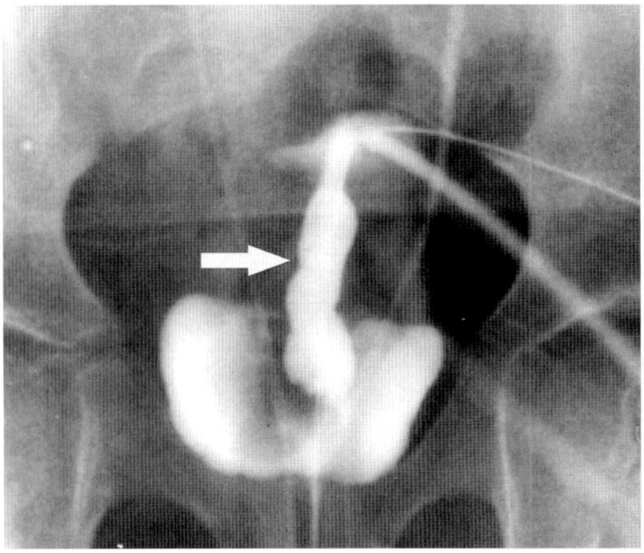

Figure 85.7 Antegrade cystogram in a child with a Mitrofanoff appendicovesicostomy shows a tubular communication (arrow) between the umbilicus and the bladder that resembles a patent urachus.

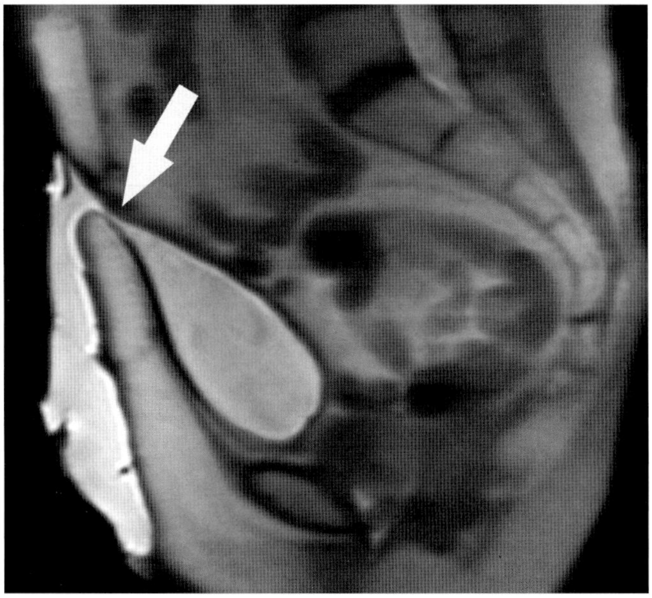

Figure 85.8 Sagittal T2-weighted MR image in a 77 year old woman with a vesicostomy created after resection of a urethral carcinoma shows a surgically created communication (arrow) between the bladder and the umbilicus that could be mistaken for a patent urachus.

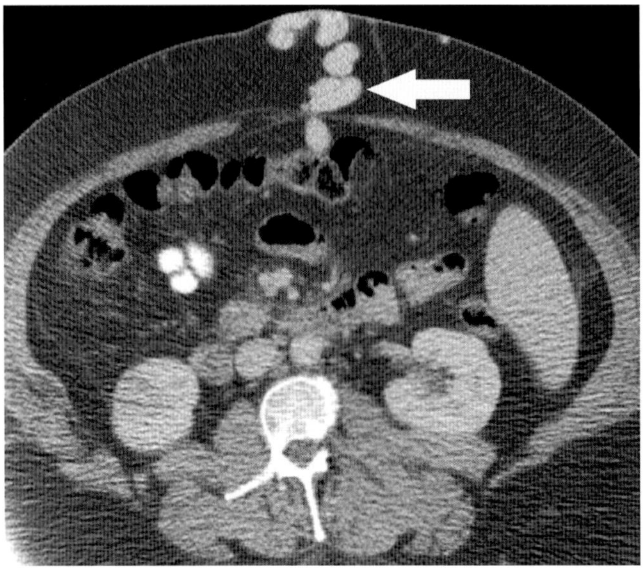

Figure 85.9 Axial contrast-enhanced CT image in a 50 year old man with cirrhosis and portosystemic collaterals, the so-called caput medusa. Such periumbilical varices should not be mistaken for a urachal remnant.

CASE 86 Pseudotumor due to ureteral jet

Imaging description

The ureter is a peristalsing structure, and on contrast-enhanced studies, this results in intermittent jetting of dense excreted contrast from the ureter into the less dense urine in the bladder. The ureteral jet phenomenon was first recognized on intravenous urography, but is also seen on CT. The typical jet or plume of contrast emanating from a ureteral orifice is usually easy to recognize on CT (Figure 86.1) but occasionally, the mixing effect may be atypical and simulate a bladder mass (Figures 86.2 and 86.3) [1].

Importance

Misinterpretation of a ureteral jet as a bladder mass may result in unnecessary workup and patient anxiety.

Typical clinical scenario

In my experience, this pseudotumor is usually seen when the typical appearance of a ureteral jet is distorted by ureteral surgery such as transplantation or by the presence of a nephroureteral stent.

Differential diagnosis

This pseudotumor can usually be recognized once the possibility is considered, but delayed images can be helpful if the appearance is problematic.

Teaching point

The possibility of a pseudotumor due to a ureteral jet should be considered for an apparent mass in the bladder, particularly in the setting of renal transplantation or nephroureteral stenting.

REFERENCE

1 Akbar SA, Shirkhoda A, Jafri SZ. Normal variants and pitfalls in CT of the gastrointestinal and genitourinary tracts. *Abdom Imaging* 2003; **28**: 115–128.

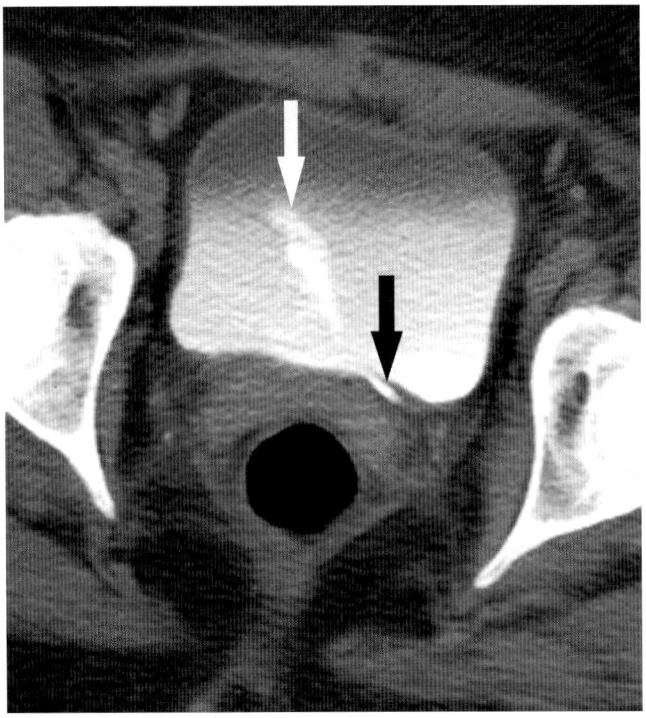

Figure 86.1 Axial contrast-enhanced CT image in a 70 year old woman with abdominal pain shows the typical appearance of a ureteral jet (white arrow) emanating from the left ureter (black arrow).

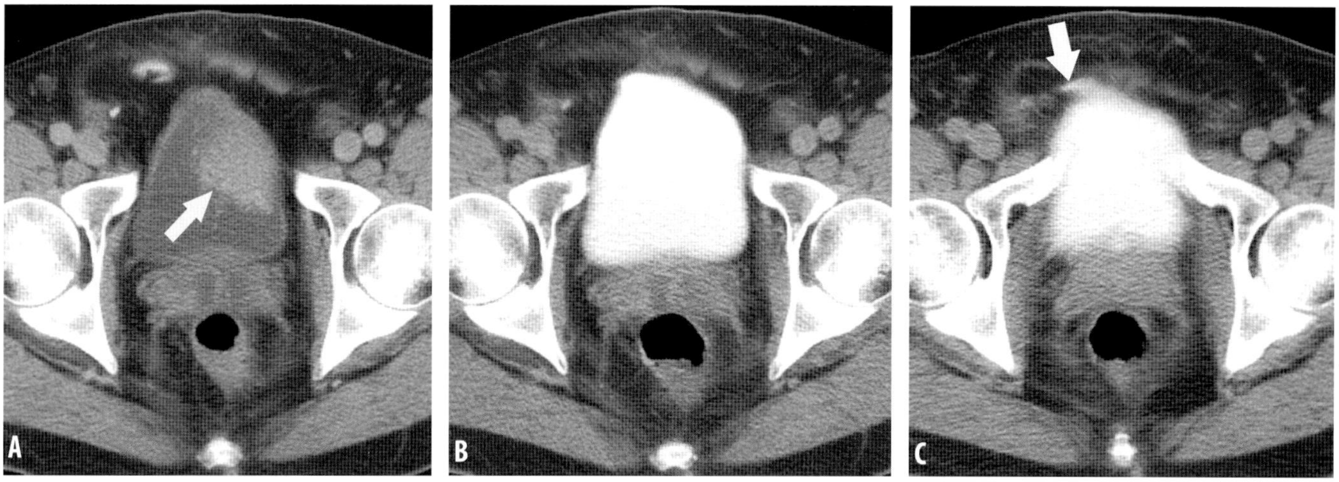

Figure 86.2 A. Axial contrast-enhanced CT image in a 39 year old man with polycystic kidneys and a renal transplant in the right lower quadrant (not shown) shows an apparent soft-tissue mass (arrow) in the left anterolateral aspect of the bladder. B. No mass is present on the corresponding axial delayed phase contrast-enhanced CT image. C. Axial delayed phase contrast-enhanced CT image at a more inferior level shows the transplant ureter (arrow) entering the bladder anteriorly. The apparent mass in Figure 86.1A was a pseudotumor due to an unusual pattern of ureteral jetting and mixing artifact from this ureter.

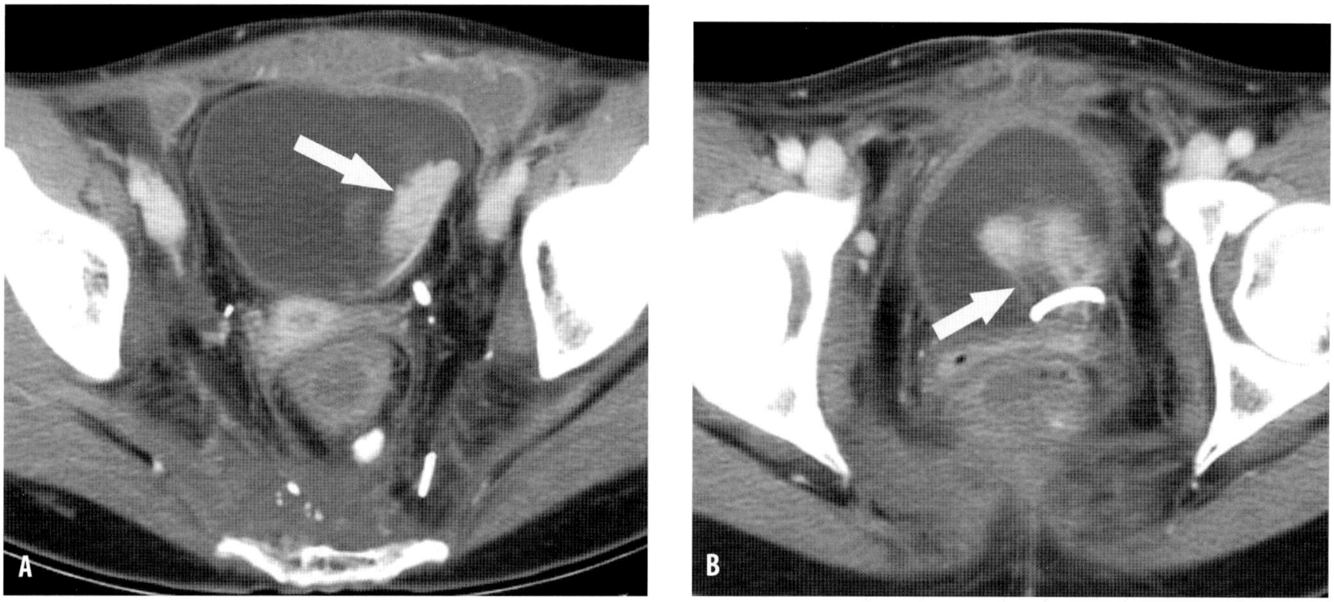

Figure 86.3 A. Axial contrast-enhanced CT image in a 69 year old woman with recurrent rectal cancer requiring placement of a left nephroureteral stent. High density along the left lateral aspect of the bladder could be mistaken for a mass (arrow). B. Axial contrast-enhanced CT image at a more inferior level shows the apparent mass is artifactual and due to jetting of contrast from the nephroureteral stent. Note the linear jets (arrow) of contrast emanating from the lower end of the stent.

CASE 87 Pelvic pseudotumor due to bladder outpouchings

Imaging description

Outpouchings of the bladder may mimic cystic or fluid-filled pelvic masses at cross-sectional imaging (Figures 87.1–87.3) [1–4]. Accumulation of excreted FDG in such outpouchings may suggest the diagnosis of a malignant mass at PET [5–7] (Figure 87.4). Such outpouchings may be true diverticula (either congenital or acquired) or bladder ears. The term bladder ear strictly refers to lateral outpouchings of the bladder in infants, sometimes extending into the inguinal canal or femoral ring, which are usually detected incidentally during cystography or intravenous urography and are thought to be a normal developmental variant [8]. In practice, the term bladder ear is used less precisely to refer to any lateral protrusion or extension of the bladder that simulates a cystic mass adjacent to the bladder.

Importance

Misinterpretation of a bladder diverticulum as a cystic mass at cross-sectional imaging or as a metastatic focus at PET may result in needless patient anxiety and unnecessary treatment or workup such as cyst aspiration, biopsy, or surgery [1].

Typical clinical scenario

Bladder diverticula are usually acquired and due to chronic bladder outlet obstruction, most commonly in older men with benign prostatic hyperplasia. Congenital bladder diverticula are rare and most occur in boys immediately adjacent to the ureteral orifice (so-called Hutch diverticula). Bladder herniation into the inguinal canal is usually seen in men over 50 years of age [5].

Differential diagnosis

This pseudotumor can often be recognized by close inspection of cross-sectional images to establish continuity between a cystic perivesical mass and the bladder (Figure 87.4). Delayed phase CT images can also be very helpful by demonstrating accumulation of excreted contrast within such outpouchings (Figures 87.1–87.3). Occasionally, cystography may be required to make the diagnosis [1]. It should also be remembered that recognition of a bladder outpouching is not absolute proof of benignity, because occasionally malignancy can arise in a bladder diverticulum (Figure 87.5) [9].

Teaching point

The possibility of a bladder outpouching should be considered for any fluid-filled or FDG-positive mass adjacent to the bladder or in the inguinal canal.

REFERENCES

1 Wang CW, Chang YL, Horng SG, et al. Pitfalls in the differential diagnosis of a pelvic cyst: lessons from a post-menopausal woman with bladder diverticulum. *Int J Clin Pract* 2004; **58**: 894–896.

2 Farhi J, Dicker D, Goldman JA. Giant diverticulum of the bladder simulating ovarian cyst. *Int J Gynaecol Obstet* 1991; **36**: 55–57.

3 Vick CW, Viscomi GN, Mannes E, Taylor KJ. Pitfalls related to the urinary bladder in pelvic sonography: a review. *Urol Radiol* 1983; **5**: 253–259.

4 Maynor CH, Kliewer MA, Hertzberg BS, et al. Urinary bladder diverticula: sonographic diagnosis and interpretive pitfalls. *J Ultrasound Med* 1996; **15**: 189–194.

5 Pirson AS, Krug B, Lacrosse M, et al. Bladder hernia simulating metastatic lesion on FDG PET study. *Clin Nucl Med* 2004; **29**: 767.

6 Park SH, Kim MJ, Kim JH, et al. FDG uptake in PET by bladder hernia simulating inguinal metastasis. *Yonsei Med J* 2007; **48**: 886–890.

7 Aygen M, Akduman IE, Osman MM. Bladder ear: a potential source of false interpretation on F-18 FDG PET. *Clin Nucl Med* 2008; **33**: 721–722.

8 Curry NS. Hernias of the urinary tract. In: Pollack HM, McClennan BL. *Clinical urography,* 3rd edition. Philadelphia, PA: Saunders, 2000; 2981–2991.

9 Dondalski M, White EM, Ghahremani GG, Patel SK. Carcinoma arising in urinary bladder diverticula: imaging findings in six patients. *Am J Roentgenol* 1993, **161**: 817–820.

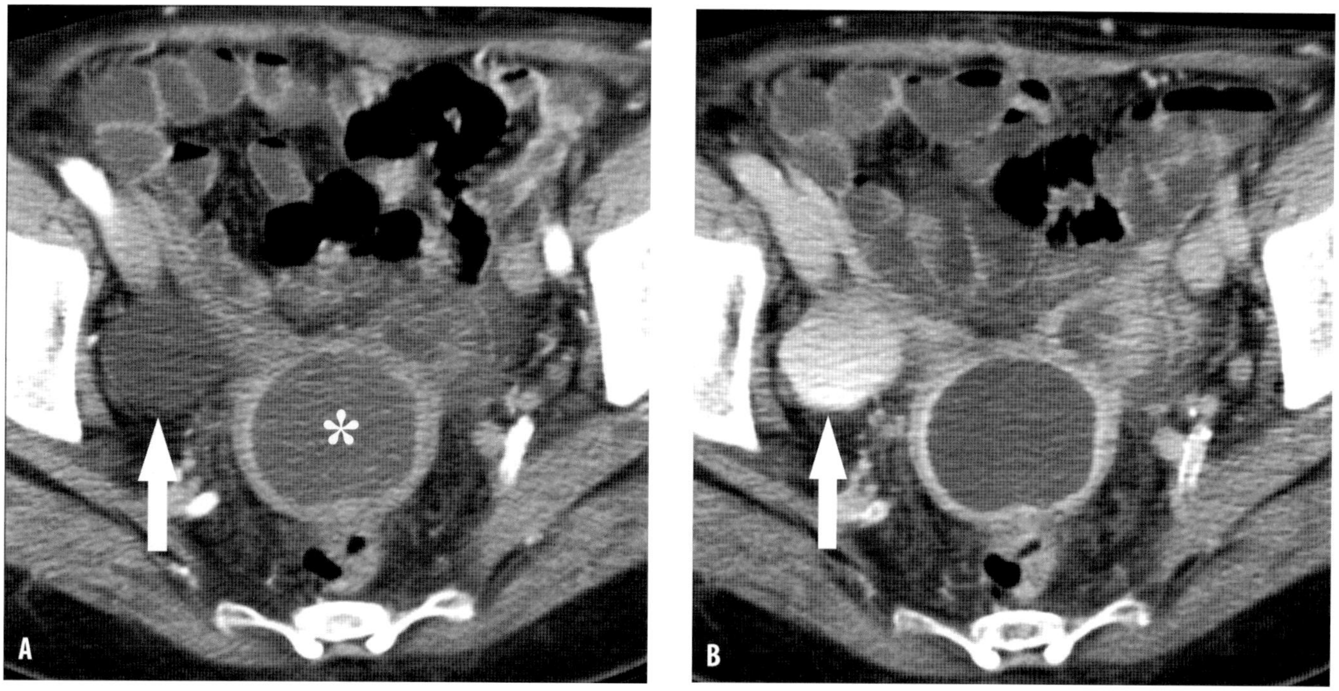

Figure 87.1 **A.** Axial contrast-enhanced CT image in a 72 year-old woman with a postoperative pelvic collection (asterisk) after reversal of a descending colostomy shows an apparent cystic adnexal mass (arrow) on the right side of the pelvis. **B.** Axial delayed contrast-enhanced CT image shows excreted contrast filling the apparent adnexal mass. More inferior images (not shown) confirmed continuity with the bladder, confirming the mass as a pseudotumor due to a bladder ear.

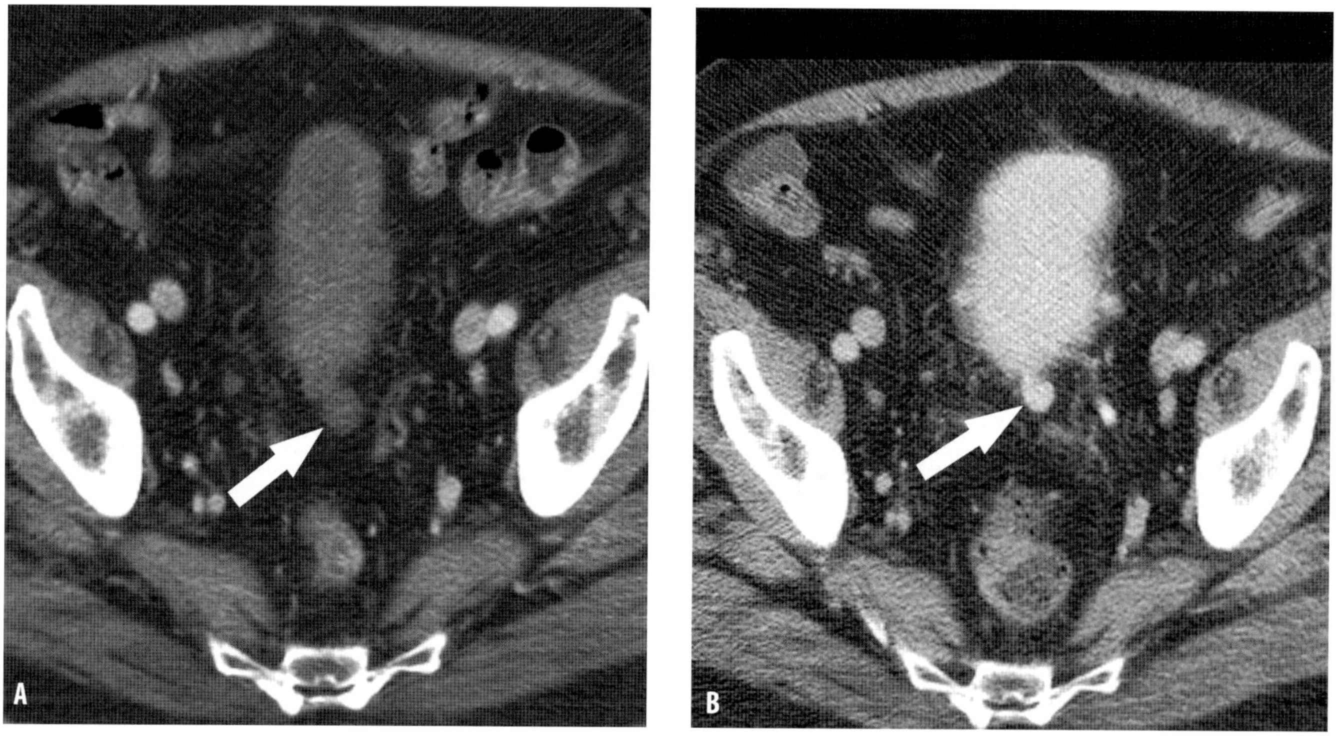

Figure 87.2 **A.** Axial contrast-enhanced CT image in a 76 year old man receiving chemotherapy for colorectal cancer metastatic to the peritoneal cavity shows a small nodule (arrow) posterior to the bladder that could be interpreted as a peritoneal implant. **B.** Axial delayed contrast-enhanced CT image shows excreted contrast filling the apparent peritoneal implant (arrow) through a small neck communicating with the bladder, confirming the mass is a pseudotumor due to a bladder diverticulum.

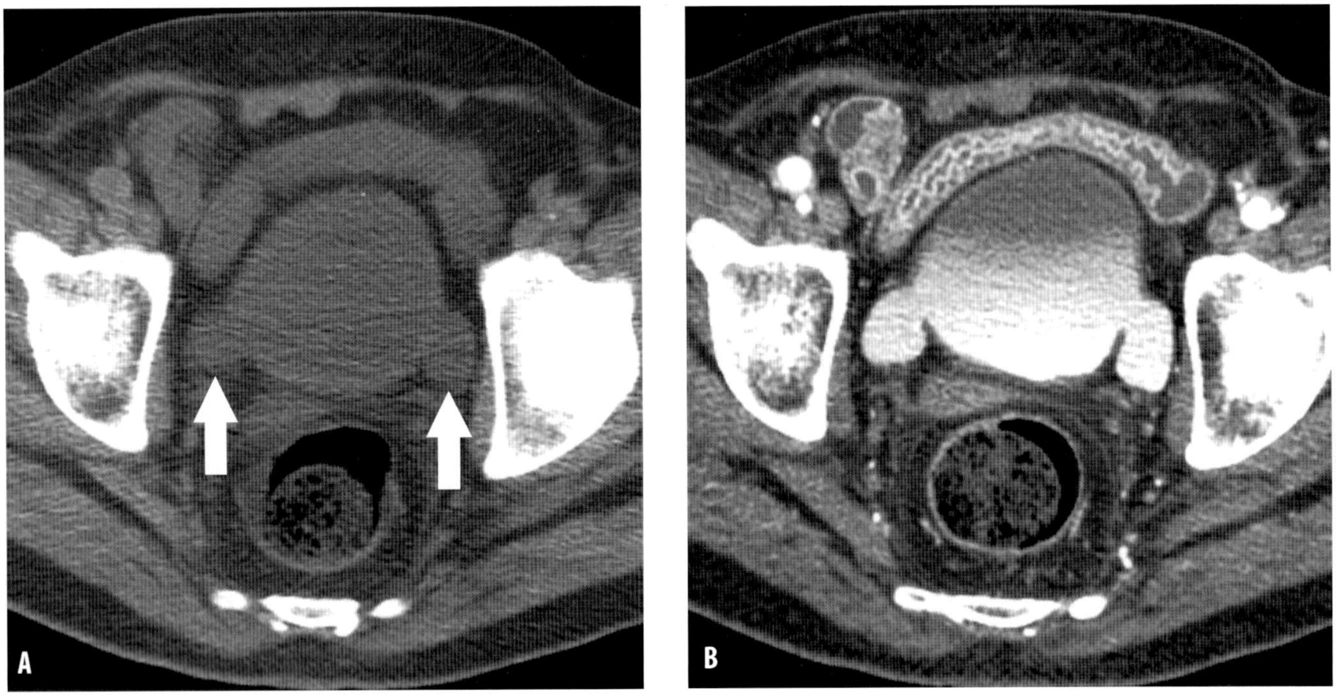

Figure 87.3 A. Axial non-enhanced CT image in an 80 year old man with nausea and vomiting and a history of both peripheral vascular disease and prostate cancer treated by radiotherapy shows bilateral low-density masses adjacent to the bladder that could be interpreted as pelvic sidewall adenopathy. B. Axial delayed contrast-enhanced CT image shows excreted contrast filling the apparent masses (arrow) which clearly communicate with the bladder, confirming the masses are pseudotumors due to bilateral bladder diverticula.

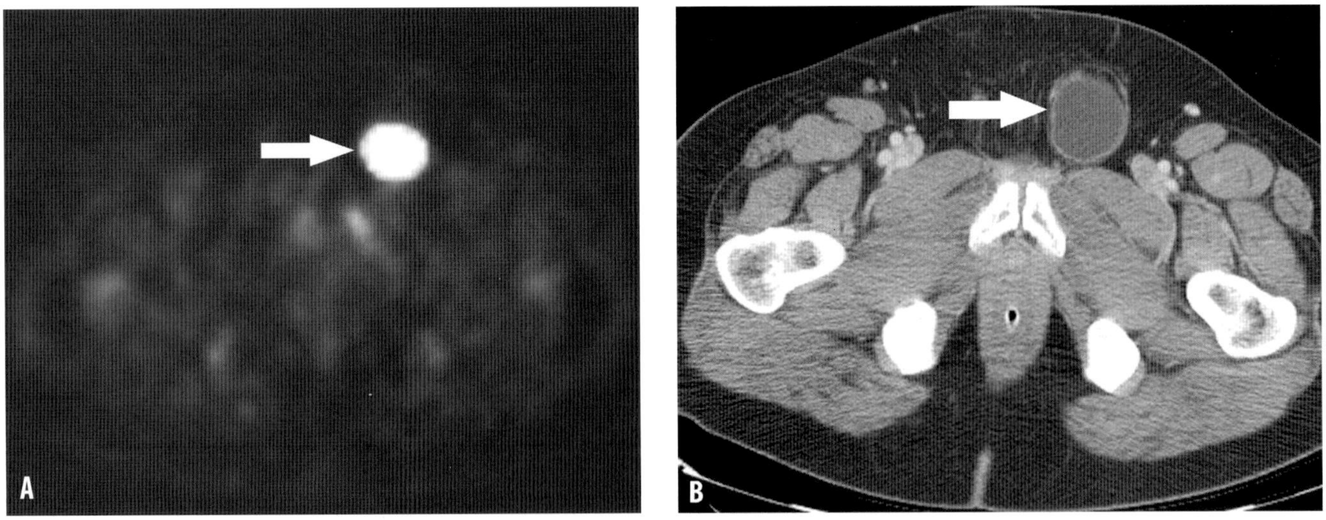

Figure 87.4 A. Axial PET image in a 40 year old man undergoing PET/CT for surveillance of melanoma shows a large focus (arrow) of increased FDG uptake in the left groin, which could be interpreted as a probable metastatic deposit. B. Corresponding axial contrast-enhanced CT image shows a fluid-filled structure (arrow) in the location of the focus of increased FDG uptake. C. Axial contrast-enhanced CT image at a more superior level shows continuity (arrow) between the fluid-filled structure and the bladder, confirming it is inguinal herniation of the bladder and that increased FDG uptake is due to urinary excretion of FDG accumulating in the herniated portion of the bladder.

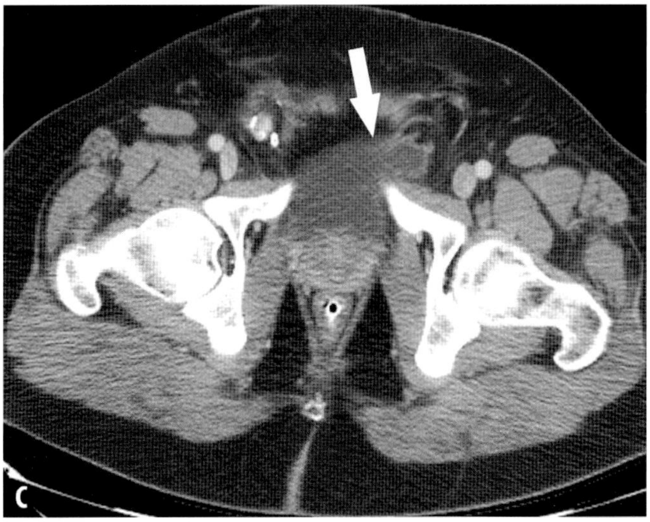

Figure 87.4 (cont.)

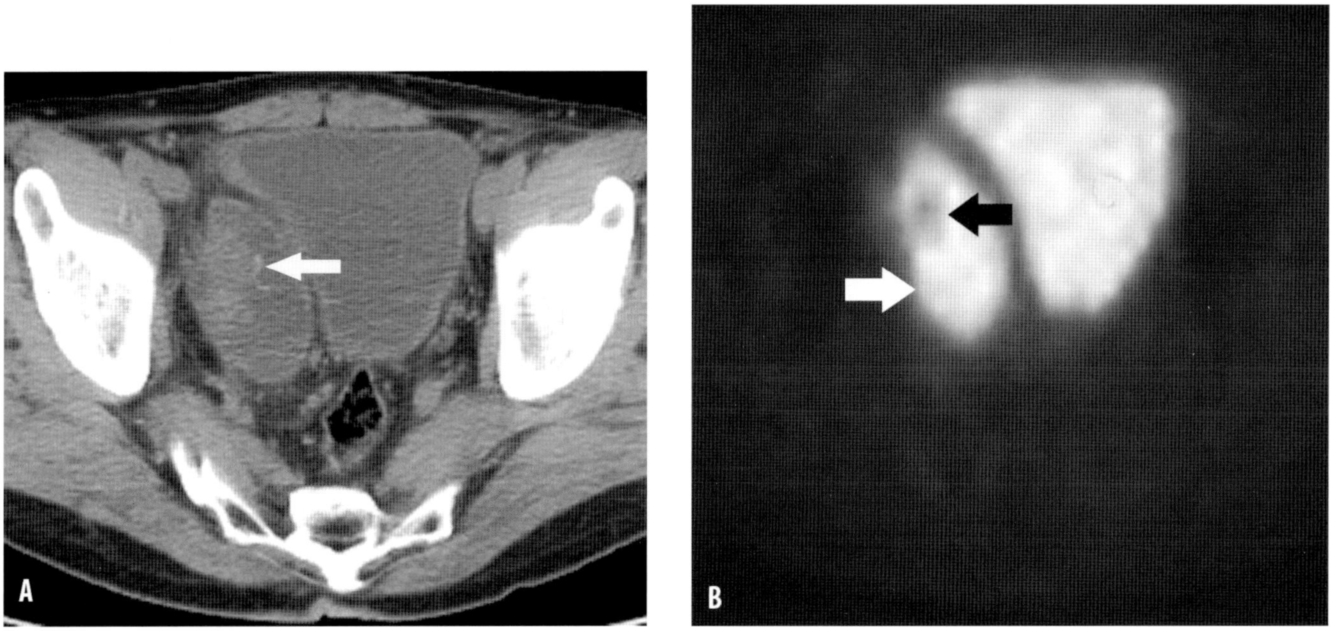

Figure 87.5 A. Axial non-enhanced CT image from a staging PET/CT study performed in a 65 year old man with newly diagnosed transitional cell carcinoma arising within a bladder diverticulum shows a subtle soft-tissue mass in the right-sided diverticulum associated with focal surface calcification (arrow), suggestive of transitional cell carcinoma. B. Co-registered axial PET image shows diffuse increased FDG activity (white arrow) due to excreted FDG in the bladder diverticulum. The tumor (black arrow) is seen as a focus of relatively reduced activity within the diverticulum.

CASE

88 Inflammatory pseudotumor of the bladder

Imaging description

Inflammatory pseudotumors are rare benign masses of uncertain etiology composed of spindle cells mixed with variable amounts of extracellular collagen, lymphocytes, and plasma cells [1]. Inflammatory pseudotumor most commonly affects the lungs and orbits, but can occur virtually anywhere. The bladder is perhaps the commonest site of involvement in the abdomen and pelvis. Inflammatory pseudotumor forms a polypoid or infiltrative bladder mass, which may be ulcerated or extend into the perivesical fat (Figures 88.1–88.3) [1–4]. A wide variety of terms have been used to describe this condition, including inflammatory myofibroblastic tumor, plasma cell granuloma or pseudotumor, and pseudosarcomatous myofibroblastic proliferation. The term inflammatory myofibroblastic tumor has recently become increasingly common in pathological usage, based on electron microscopic and immunohistochemical findings [5, 6].

Importance

Inflammatory pseudotumor of the bladder can mimic transitional cell carcinoma, both clinically and radiologically. Awareness of this entity and its inclusion in the differential diagnosis may prevent unnecessary radical surgery, since inflammatory pseudotumor of the bladder can usually be managed by conservative surgery such as transurethral resection or partial cystectomy [7].

Typical clinical scenario

Most cases of inflammatory pseudotumor of the bladder occur in childhood or early adulthood. Women are affected about twice as often as men [8]. Patients may be asymptomatic or have constitutional symptoms such as fever or weight loss. Local symptoms may include pain, hematuria, urinary frequency, or dysuria. While the condition can recur, the prognosis is usually good if the mass is completely resected [1].

Differential diagnosis

Transitional cell carcinoma or other bladder malignancies such as squamous cell carcinoma, adenocarcinoma, or sarcoma are the primary considerations for a bladder mass seen at imaging. Rarely, other tumors such as pheochromocytoma or leiomyoma occur in the bladder. However, a wide variety of non-neoplastic or inflammatory masses can affect the bladder, including endometriosis (Figures 88.4 and 88.5), nephrogenic adenoma, malacoplakia, cystitis cystica/glandularis, polypoid cystitis (Figure 88.6), eosinophilic cystitis, tuberculosis, schistosomiasis, Crohn's disease, diverticulitis, radiation cystitis, and chemotherapy-related cystitis [9–11].

Teaching point

Transitional cell carcinoma is the most common cause of a polypoid or infiltrative bladder mass at imaging, but a variety of other neoplastic and non-neoplastic processes can mimic this appearance, including inflammatory pseudotumor of the bladder. All bladder masses should undergo histological examination.

REFERENCES

1 Park SB, Cho KS, Kim JK, et al. Inflammatory pseudotumor (myoblastic tumor) of the genitourinary tract. *Am J Roentgenol* 2008; **191**: 1255–1262.

2 Fujiwara T, Sugimura K, Imaoka I, Igawa M. Inflammatory pseudotumor of the bladder: MR findings. *J Comput Assist Tomogr* 1999; **23**: 558–561.

3 Sugita R, Saito M, Miura M, Yuda F. Inflammatory pseudotumour of the bladder: CT and MRI findings. *Br J Radiol* 1999; **72**: 809–811.

4 Park SJ, Lee HK, Yi BH, et al. Pseudotumour in the bladder as a complication of total hip replacement: ultrasonography, CT and MR findings. *Br J Radiol* 2007; **80**: e119–e121.

5 Coffin CM, Watterson J, Priest JR, Dehner LP. Extrapulmonary inflammatory myofibroblastic tumor (inflammatory pseudotumor): a clinicopathologic and immunohistochemical study of 84 cases. *Am J Surg Pathol* 1995; **19**: 859–872.

6 Coffin CM, Dehner LP, Meis-Kindblom JM. Inflammatory myofibroblastic tumor, inflammatory fibrosarcoma, and related lesions: an historical review with differential diagnostic considerations. *Semin Diagn Pathol* 1998; **15**: 102–110.

7 Kumar A, Bhatti SS, Sharma S, Gupta SD, Kumar R. Inflammatory pseudotumor of urinary bladder – a diagnostic and management dilemma. *Int Urol Nephrol* 2007; **39**: 799–802.

8 Young RH. Pseudoneoplastic lesions of the urinary bladder and urethra: a selective review with emphasis on recent information. *Semin Diagn Pathol* 1997; **14**: 133–146.

9 Wong-You-Cheong JJ, Woodward PJ, Manning MA, Davis CJ. From the archives of the AFIP: inflammatory and nonneoplastic bladder masses: radiologic-pathologic correlation. *Radiographics* 2006; **26**: 1847–1868.

10 Norlén LJ, Ekelund P, Hedelin H, Johansson SL. Effects of indwelling catheters on the urethral mucosa (polypoid urethritis). *Scand J Urol Nephrol* 1988; **22**: 81–86.

11 Kim SH, Yang DM, Kim NR. Polypoid and papillary cystitis mimicking a large transitional carcinoma in a patient without a history of catheterization: computed tomography and magnetic resonance findings. *J Comput Assist Tomogr* 2004; **28**: 485–487.

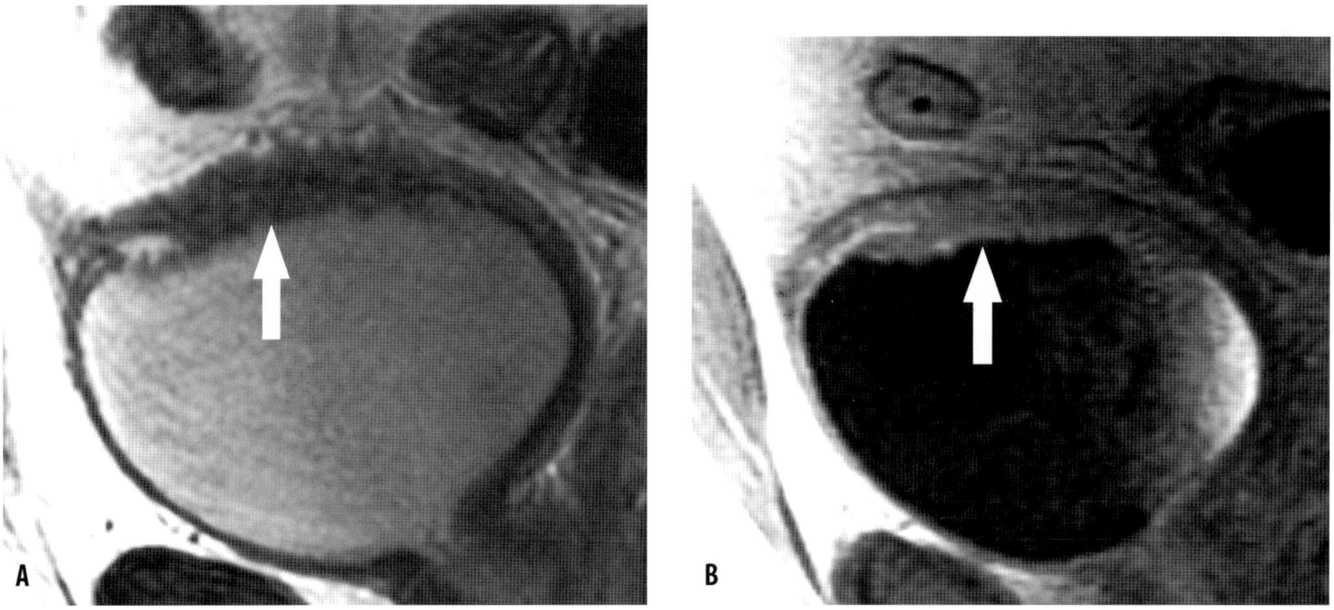

Figure 88.1 **A.** Sagittal T2-weighted MR image in a 42 year old man with abdominal pain and hematuria shows infiltrative thickening (arrow) of the bladder wall superiorly. **B.** Sagittal post-gadolinium T1-weighted MR image also demonstrates infiltrative thickening (arrow) of the bladder wall superiorly. Partial cystectomy was performed. No tumor was found in the specimen, which demonstrated benign spindle cell proliferation consistent with inflammatory pseudotumor of the bladder.

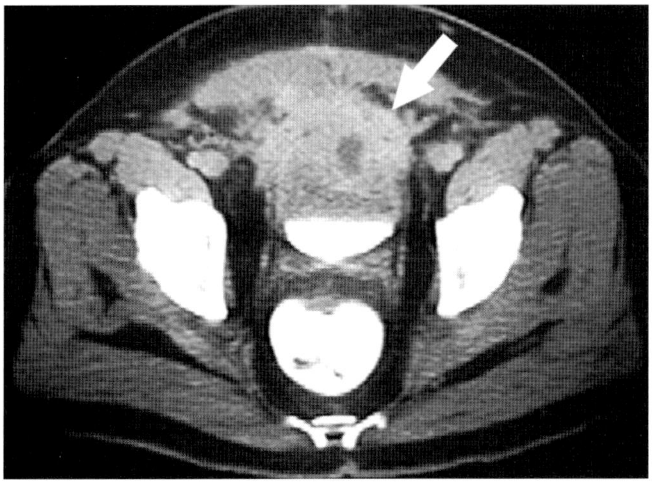

Figure 88.2 Axial contrast-enhanced CT image in a 55 year old man with hematuria and suprapubic discomfort shows a large centrally necrotic mass at the anterior aspect of the bladder. The mass was resected and pathological examination showed it was composed primarily of spindle cells with a chronic inflammatory infiltrate, consistent with inflammatory pseudotumor of the bladder.

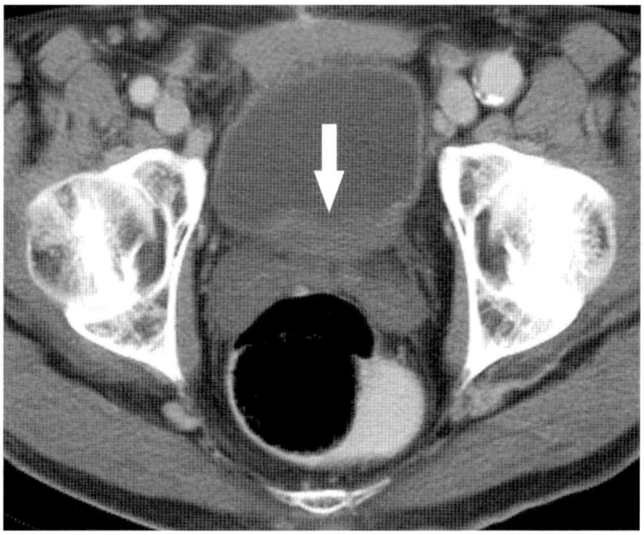

Figure 88.3 Axial contrast-enhanced CT image in a 61 year old man being staged for a recently diagnosed gastrointestinal stromal tumor of the jejunum. The bladder base demonstrates diffuse infiltrative thickening (arrow). Biopsy demonstrated a benign spindle cell proliferation consistent with inflammatory pseudotumor.

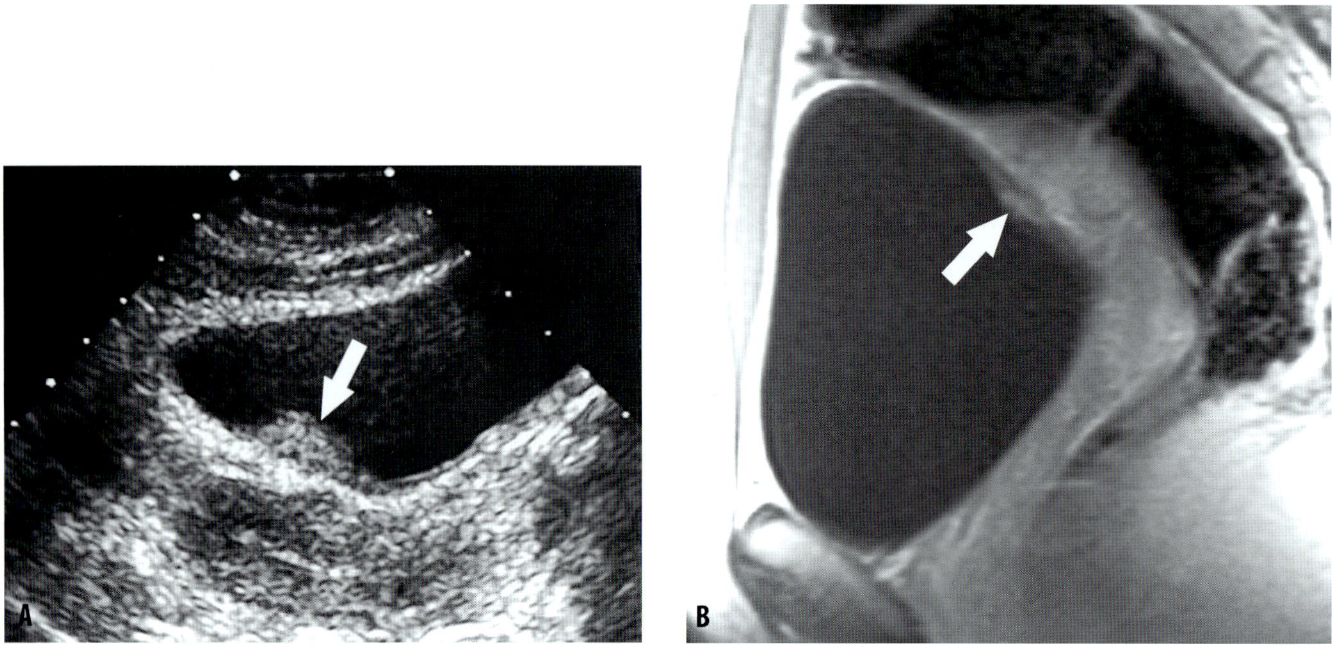

Figure 88.4 **A.** Ultrasound image in a 29 year old woman with dysuria and other irritative voiding symptoms shows a focal non-specific sessile mass (arrow). **B.** Sagittal post-gadolinium T1-weighted MR image shows a small enhancing sessile mass on the posterior wall of the bladder. The appearance is non-specific. Cystoscopic biopsy demonstrated endometriosis.

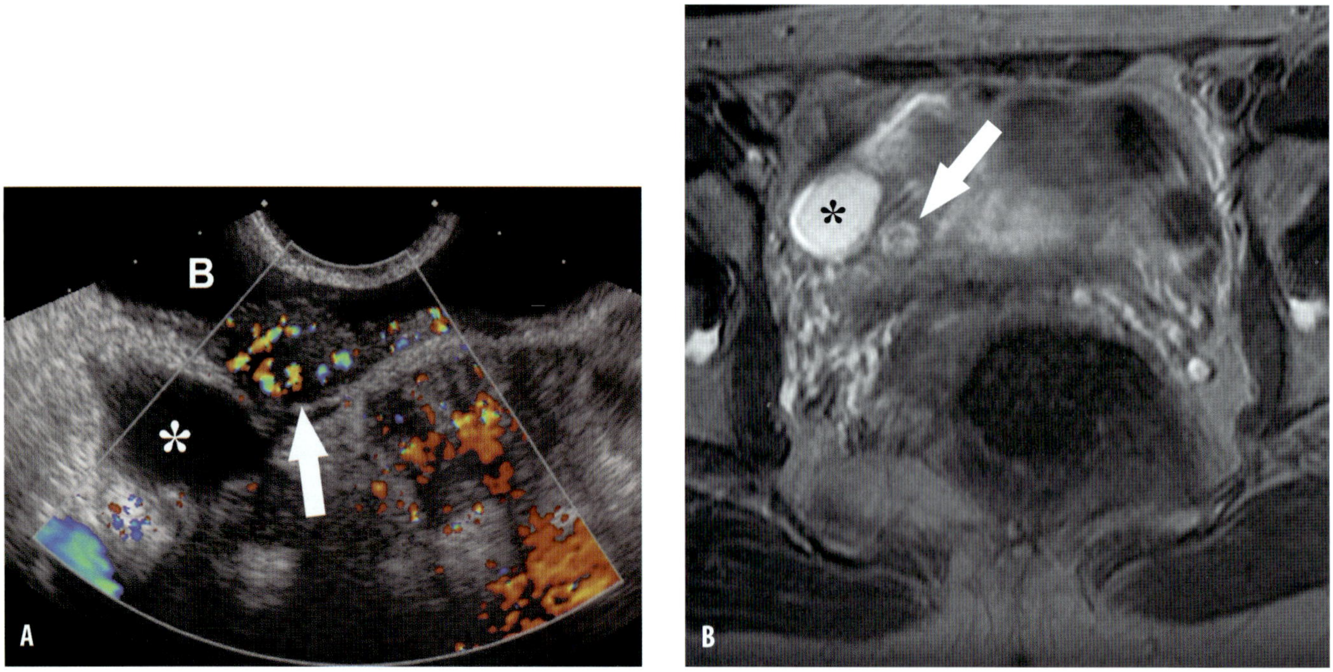

Figure 88.5 **A.** Ultrasound image in a 29 year old woman with a one-week history of pelvic pain and hematuria. **B.** Axial T2-weighted MR image shows a non-specific nodule (arrow) of high T2 signal intensity related to the right side of the bladder, and appearing to be separate to a cyst (asterisk) in the right ovary. **C.** Axial post-gadolinium T1-weighted MR image shows peripheral enhancement in the nodule (arrow) and non-enhancement in the right ovarian cyst (asterisk). Cystoscopic biopsy demonstrated endometriosis involving the bladder.

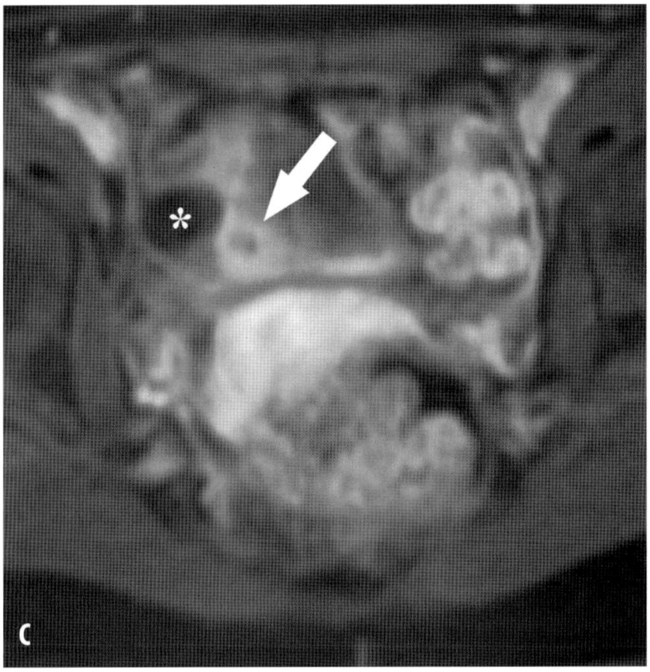

Figure 88.5 (cont.)

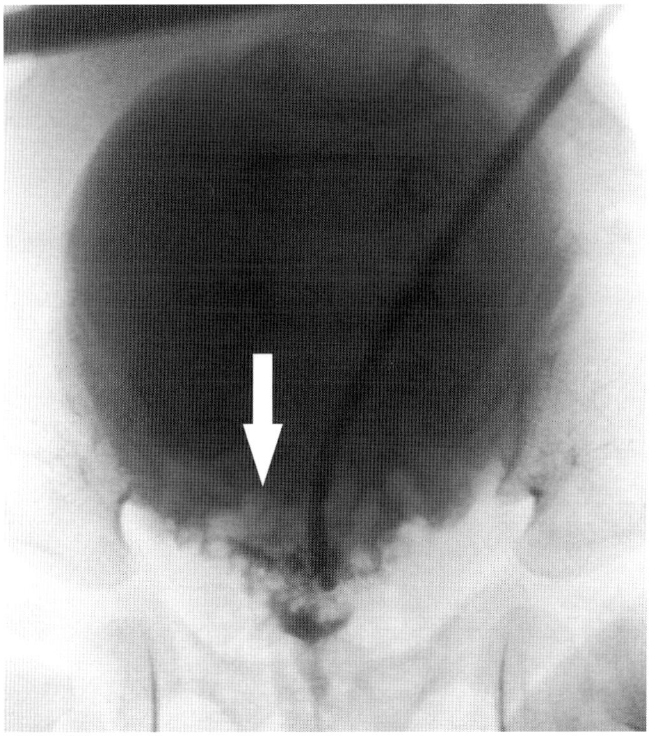

Figure 88.6 Image from a voiding cystourethrogram in an 11 month old boy with an indwelling suprapubic catheter due to voiding difficulties as part of VATER syndrome. Lobulated lucent filling defects in the bladder inferiorly were due to polypoid cystitis.

CASE

89 Urethral diverticulum

Imaging description

A urethral diverticulum is a variably sized outpouching of the urethra. The condition occurs primarily in middle-aged and elderly women, in whom it is thought to arise from rupture or dilatation secondary to infection or obstruction of normal paraurethral glands. Urethral diverticula appear at ultrasound, CT, or MRI as fluid-filled cyst-like structures wrapped around the urethra and protruding into the anterior vaginal wall (Figures 89.1 and 89.2). Urethral diverticula may be seen during voiding cystourethrogram (Figure 89.3), but only if they fill with contrast. If strongly suspected clinically, special techniques to generate high pressure in the urethra may help drive contrast into the diverticulum. This can be achieved by occluding the meatal opening with a finger during voiding, or by using a special double-balloon catheter technique that effectively seals the urethra at both ends and forces contrast into the diverticulum from a hole in the catheter between the balloons (Figure 89.4). In a study of 32 women with symptoms of a urethral diverticulum and using surgery as the gold standard, the sensitivity of double-balloon urethrography was 100% compared to 44% for standard voiding cystourethrography [1]. Transvaginal ultrasound and endovaginal MRI are more advanced techniques that can also be used when standard imaging has failed to demonstrate a suspected diverticulum [2–4].

Importance

While the appearances are distinctive, limited awareness of this entity may result in urethral diverticula going unrecognized, being confused with the bladder neck, or being mistaken for a Bartholin cyst. At PET scanning, excreted urinary FDG may accumulate in the diverticulum and result in increased activity that could be mistaken for malignancy, unless correlated with CT or other cross-sectional imaging (Figure 89.5). Failure to make the diagnosis may result in a missed opportunity for treatment of symptoms that can be distressing and may be the cause of substantial morbidity.

Typical clinical scenario

Urethral diverticula reportedly occur in 0.6 to 6.0% of women [4–6]. Some of the wide variation in frequency may reflect selection bias in the study populations or the assiduousness of the investigators in diagnosing very small diverticula. Urethral diverticula may be asymptomatic or cause chronic genitourinary symptoms such as frequency, urgency, dysuria, recurrent urinary tract infections, postmicturition dribbling, dyspareunia, haematuria, tender vaginal mass, urinary incontinence, discharge of pus per urethra, or retention. Physical examination may be unremarkable or reveal a mass or tenderness on the anterior vaginal wall. Urethral diverticula may be complicated by infection, stone formation, malignancy, or endometriosis. Therefore, a urethral diverticulum seen at imaging should be inspected for calcifications that might suggest stones (Figure 89.6) or soft-tissue masses that might suggest malignancy (Figure 89.7) or endometriosis. The frequency of these complications is not well known, but in one series, malignant change was identified in 5 of 90 (6%) urethral diverticulectomy specimens [7]. In another series of 46 patients with urethral diverticula, complications were detected in 6, and consisted of cancer (n = 2), stones (n = 2), and endometriosis (n = 2) [8].

Differential diagnosis

Urethral diverticula can be distinguished from a Bartholin cyst because a urethral diverticulum is usually midline and at least partially surrounds the urethra, while a Bartholin cyst lies just left or right of midline and does not conform to the urethra. Another distinguishing feature seen at MRI is that Bartholin cysts are often of increased T1 signal intensity, presumably due to proteinaceous or hemorrhagic content (Figure 89.8). Some authors have described benign paraurethral cysts as an additional differential consideration, allegedly distinguished from urethral diverticula by lack of communication with the urethral lumen [9]. Whether these are truly distinct lesions or whether the distinction has any real therapeutic implication is unclear.

Teaching point

A cyst-like structure surrounding the urethra just inferior to the bladder neck in a middle-aged or elderly woman is likely to be a urethral diverticulum. Correct recognition may potentially allow for surgical alleviation of distressing genitourinary symptoms. The diverticulum should also be inspected for calcified or soft-tissue filling defects that might suggest superimposed stones or malignancy, respectively.

REFERENCES

1 Jacoby K, Rowbotham RK. Double balloon positive pressure urethrography is a more sensitive test than voiding cystourethrography for diagnosing urethral diverticulum in women. *J Urol* 1999; **162**: 2066–2069.

2 Wu YY, Li YZ, Wei Q, et al. Transvaginal sonographic diagnosis of female urethral diverticula. *J Clin Ultrasound* 2009; **37**: 40–42.

3 Blander DS, Rovner ES, Schnall MD, et al. Endoluminal magnetic resonance imaging in the evaluation of urethral diverticula in women. *Urology* 2001; **57**: 660–665.

4 Keefe B, Warshauer DM, Tucker MS, Mittelstaedt CA. Diverticula of the female urethra: diagnosis by endovaginal and transperineal sonography. *Am J Roentgenol* 1991; **156**: 1195–1197.

5 Andersen MJ. The incidence of diverticula in the female urethra. *J Urol* 1967; **98**: 96–98.

6 Davis HJ, Telinde RW. Urethral diverticula: an assay of 121 cases. *J Urol* 1958; **80**: 34–39.

7 Thomas AA, Rackley RR, Lee U, et al. Urethral diverticula in 90 female patients: a study with emphasis on neoplastic alterations. *J Urol* 2008; **180**: 2463–2467.

8 Romanzi LJ, Groutz A, Blaivas JG. Urethral diverticulum in women: diverse presentations resulting in diagnostic delay and mismanagement. *J Urol* 2000; **164**: 428–433.

9 Lucioni A, Rapp DE, Gong EM, Fedunok P, Bales GT. Diagnosis and management of periurethral cysts. *Urol Int* 2007; **78**: 121–125.

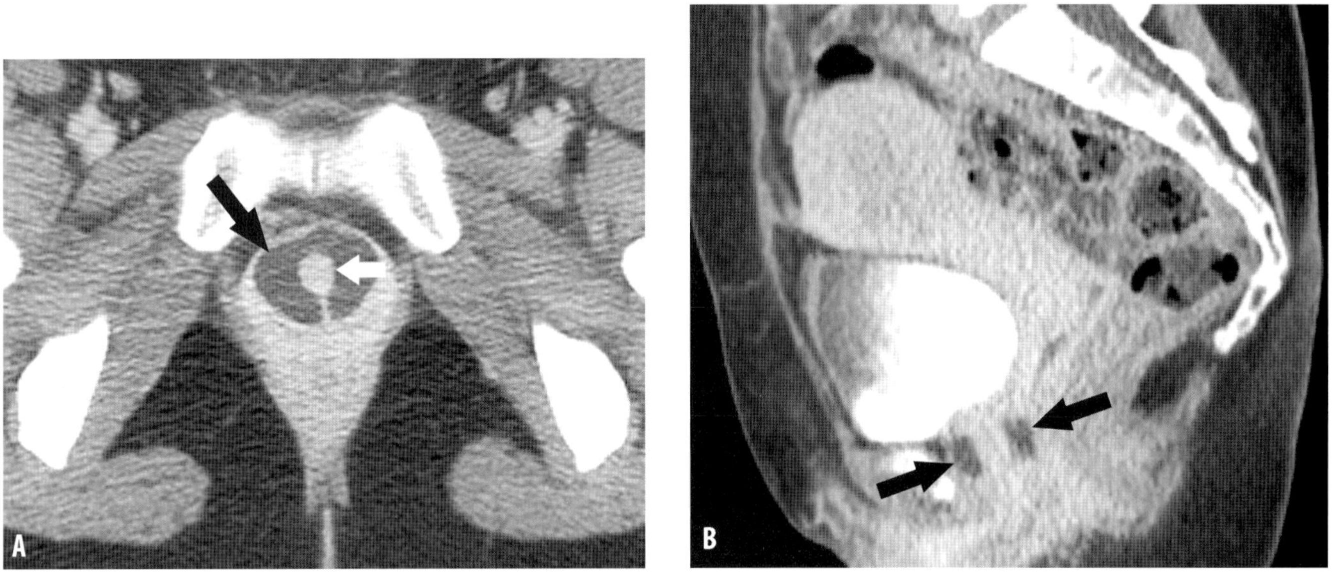

Figure 89.1 A. Axial contrast-enhanced CT image in a 31 year old woman with recurrent urinary tract infections and dyspareunia shows a urethral diverticulum (black arrow) surrounding the urethra (white arrow). **B.** Sagittal reformatted CT image confirms the diverticulum (black arrows) is separate to the contrast-filled bladder, and therefore should not be mistaken for the bladder neck.

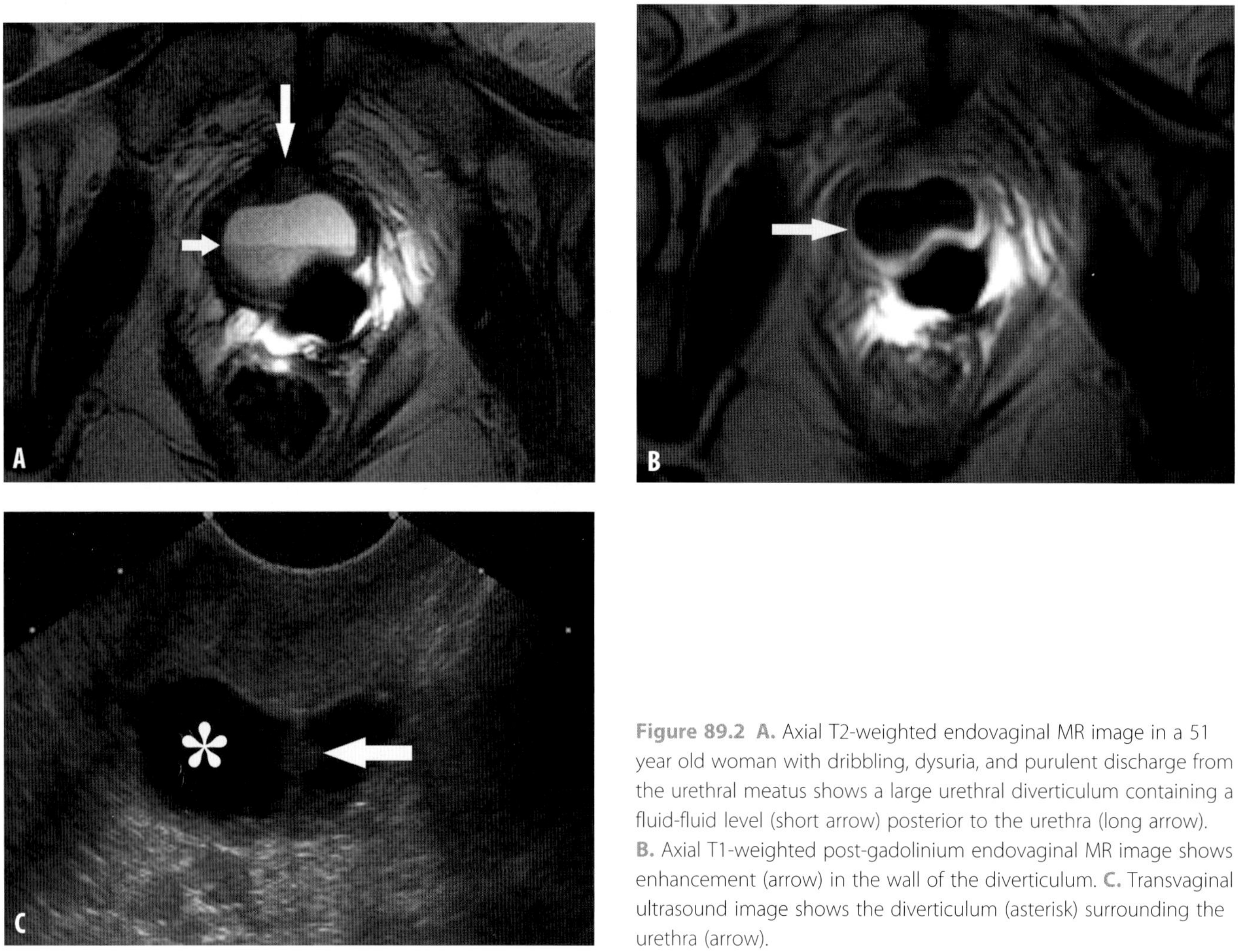

Figure 89.2 A. Axial T2-weighted endovaginal MR image in a 51 year old woman with dribbling, dysuria, and purulent discharge from the urethral meatus shows a large urethral diverticulum containing a fluid-fluid level (short arrow) posterior to the urethra (long arrow). B. Axial T1-weighted post-gadolinium endovaginal MR image shows enhancement (arrow) in the wall of the diverticulum. C. Transvaginal ultrasound image shows the diverticulum (asterisk) surrounding the urethra (arrow).

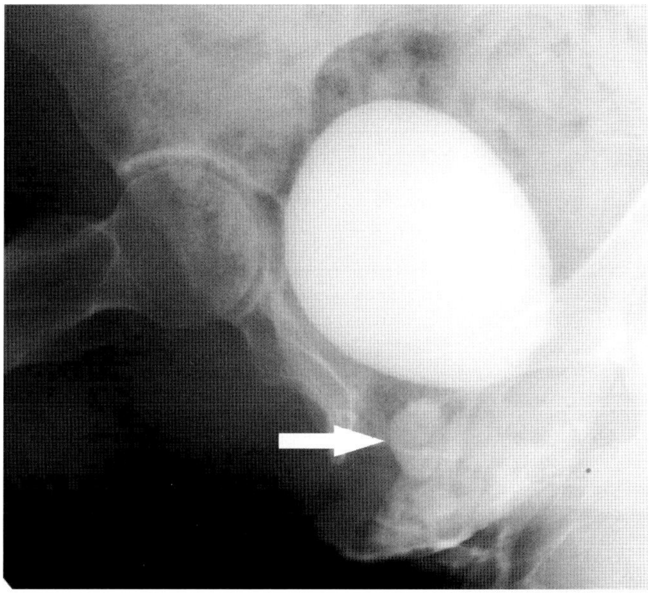

Figure 89.3 Image from a voiding cystourethrogram in a 46 year old woman with recurrent urinary tract infections shows a large urethral diverticulum (arrow).

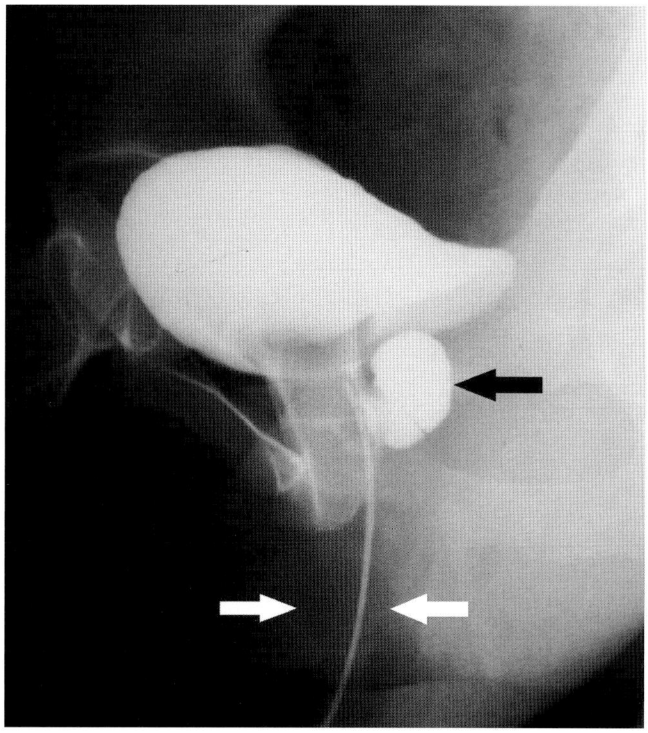

Figure 89.4 Image from a double-balloon positive pressure urethrogram in a 53 year old woman with irritative voiding symptoms shows contrast filling a large urethral diverticulum (black arrow). Note there is a balloon in the bladder and a second balloon abutting the introitus (white arrows). This special double-balloon catheter technique allows contrast to be injected into the urethra with greater pressure, and therefore improves filling and depiction of urethral diverticula.

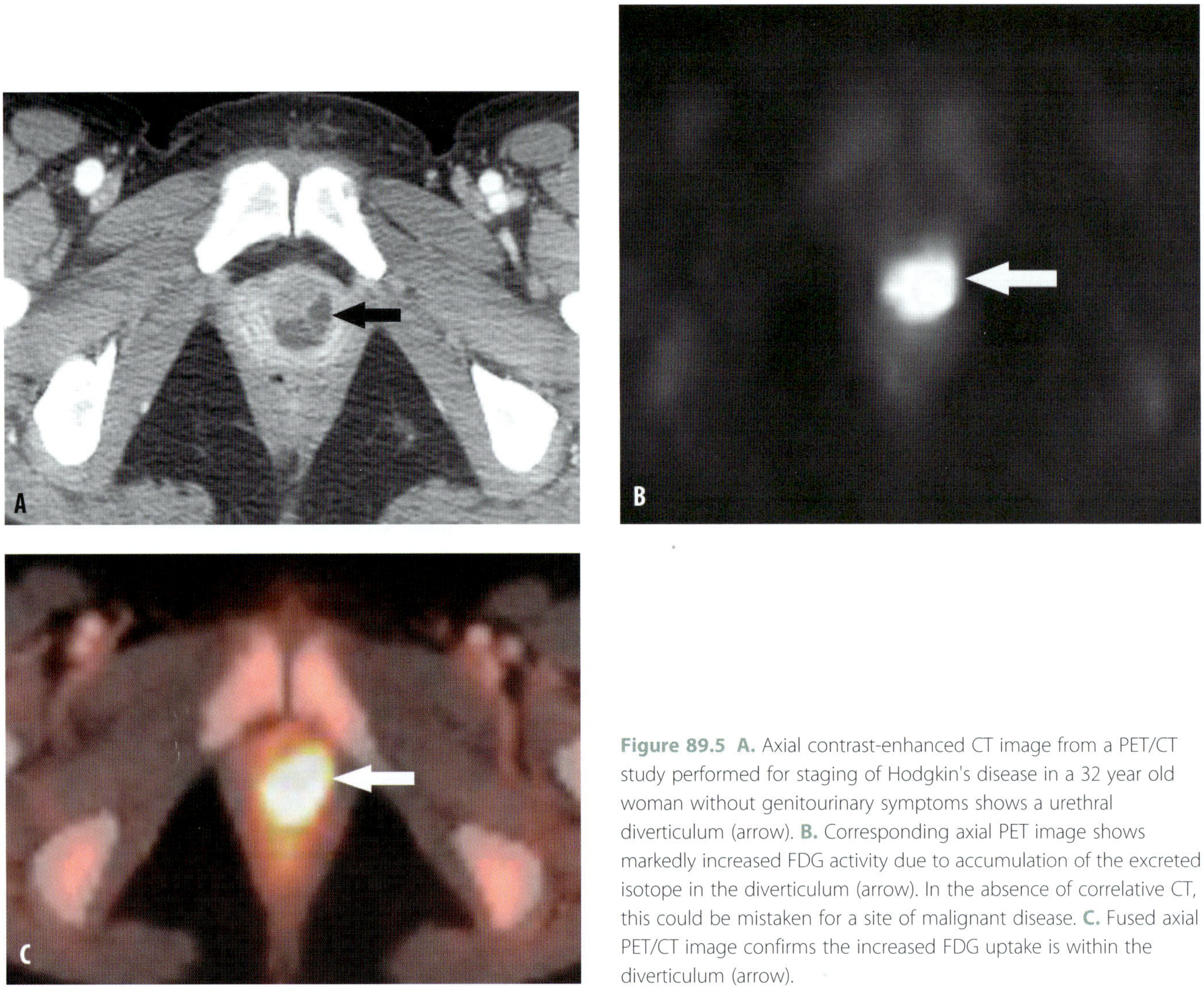

Figure 89.5 **A.** Axial contrast-enhanced CT image from a PET/CT study performed for staging of Hodgkin's disease in a 32 year old woman without genitourinary symptoms shows a urethral diverticulum (arrow). **B.** Corresponding axial PET image shows markedly increased FDG activity due to accumulation of the excreted isotope in the diverticulum (arrow). In the absence of correlative CT, this could be mistaken for a site of malignant disease. **C.** Fused axial PET/CT image confirms the increased FDG uptake is within the diverticulum (arrow).

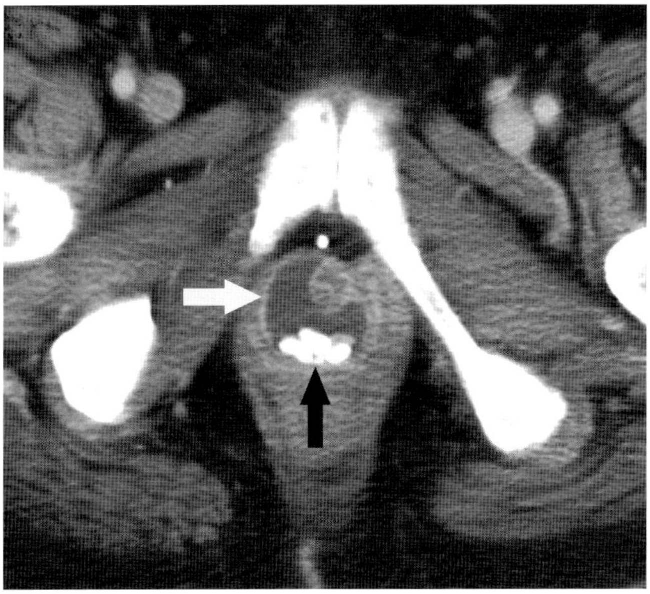

Figure 89.6 Axial contrast-enhanced CT image in a 44 year old woman with frequency and dysuria shows a urethral diverticulum (white arrow) with dependently layering stones (black arrow) posteriorly.

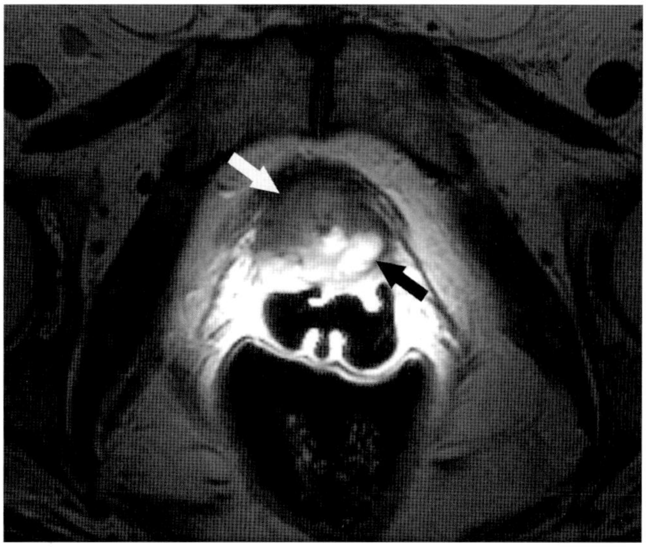

Figure 89.7 Axial T2-weighted endovaginal MR image in a 70 year old woman complaining of recent inability to pass a Foley catheter despite a long history of clean intermittent self-catheterization for atonic bladder secondary to multiple sclerosis. A fluid-filled urethral diverticulum (black arrow) is visible on the left side but infiltrative soft tissue (white arrow) is seen on the right side. Biopsy demonstrated squamous cell carcinoma complicating the urethral diverticulum.

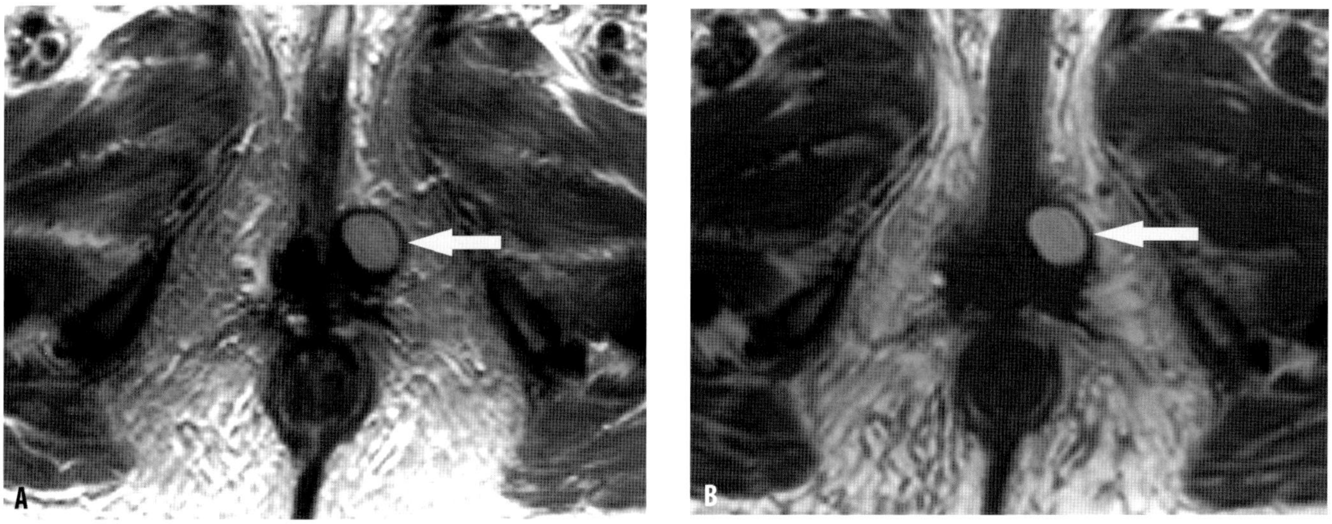

Figure 89.8 **A.** Axial T2-weighted MR image in a 61 year old woman with ovarian cancer showing an incidental Bartholin cyst (arrow) in the left side of the introitus. Note the cyst does not surround the urethra and lies entirely lateral to the midline. **B.** Corresponding axial T1-weighted MR image shows high signal intensity in the cyst (arrow). This is a common finding, presumably due to proteinaceous or hemorrhagic content, and is one feature that helps distinguish a urethral diverticulum from a Bartholin cyst.

CASE 90 Post-proctectomy presacral pseudotumor

Imaging description

After surgical resection of the rectum with creation of a colostomy, the prostate and seminal vesicles in men and the uterus in women are displaced posteriorly into the presacral space, where they may be mistaken for a mass (Figures 90.1 and 90.2) [1].

Importance

This pseudotumor is most problematic in patients who have had abdominoperineal resection of the rectum for cancer, since the pseudomass may be mistaken for recurrent malignancy.

Typical clinical scenario

Post-proctectomy presacral pseudotumor is commonly seen in patients after abdominoperineal resection of the rectum for cancer (which necessarily includes creation of colostomy since the anal sphincter is resected), or proctectomy with creation of colostomy (usually after total proctocolectomy for refractory ulcerative colitis or proctectomy for perianal fistulizing Crohn's disease).

Differential diagnosis

Awareness is the key to diagnosing this pseudotumor. In men, recognition of the symmetric seminal vesicles, which are in continuity with the vas deferens, is a useful clue (Figure 90.1), while in women recognition of the central endometrial or endocervical canal can be helpful (Figure 90.2). The uterus is also easily identifiable on MRI (Figure 90.2). In the absence of such unequivocal findings, PET or biopsy may be required [2].

Teaching point

The diagnosis of a presacral mass after proctectomy with creation of a colostomy should be made only after the possibility of a pseudotumor due to posterior displacement of the prostate and seminal vesicles in men or the uterus in women has been considered.

REFERENCES

1 Lee JK, Stanley RJ, Sagel SS, Levitt RG, McClennan BL. CT appearance of the pelvis after abdomino-perineal resection for rectal carcinoma. *Radiology* 1981; **141**: 737–741.

2 Butch RJ, Wittenberg J, Mueller PR, et al. Presacral masses after abdominoperineal resection for colorectal carcinoma: the need for needle biopsy. *Am J Roentgenol* 1985; **144**: 309–312.

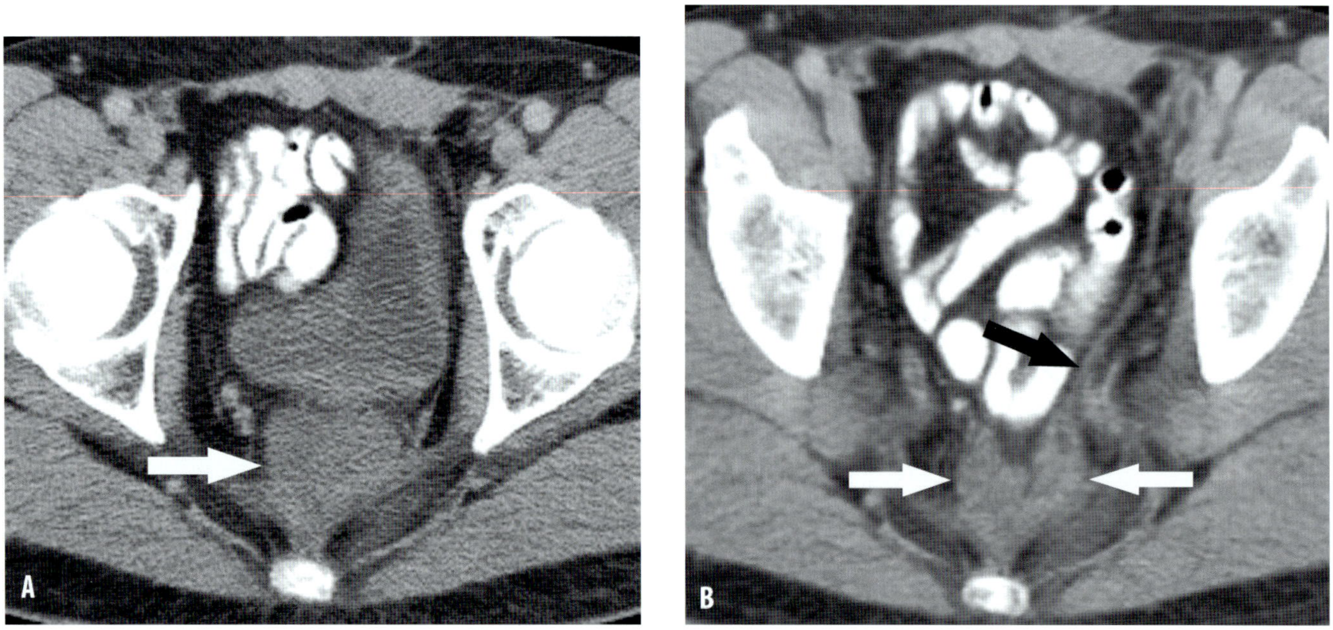

Figure 90.1 A. Axial contrast-enhanced CT image in a 64 year old man 6 months after abdominoperineal resection of the rectum shows a posteriorly displaced prostate (arrow) that could be mistaken for tumor. B. Curved planar axial contrast-enhanced CT image at a more superior level shows the symmetric appearance of the seminal vesicles (white arrows), as confirmed by continuity with the vas deferens (black arrow).

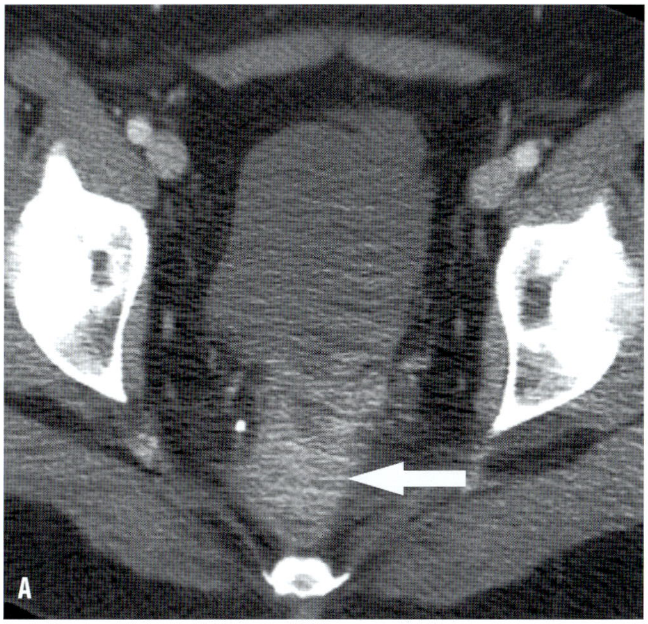

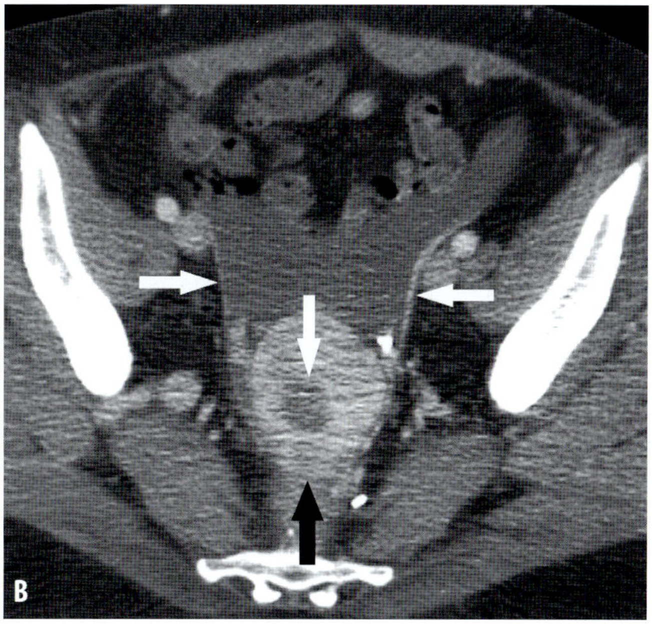

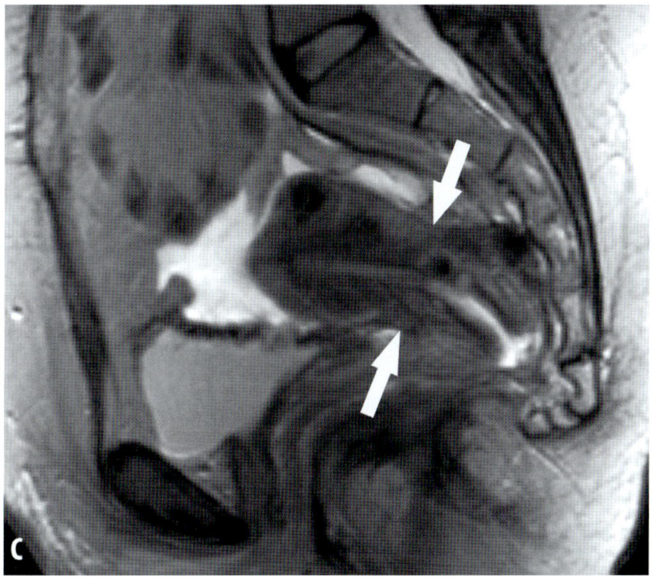

Figure 90.2 A. Axial contrast-enhanced CT image in a 43 year old woman 5 years after abdominoperineal resection of the rectum for Crohn's disease shows an enhancing presacral mass that could be mistaken for a tumor. **B.** Axial contrast-enhanced CT image at a more superior level shows the mass (black arrow) is due to posterior displacement of the uterus, as confirmed by visualization of the endometrial cavity (vertical white arrow) and continuity with the round ligaments (horizontal white arrows). Physiologic free fluid is present as an incidental finding anterior to the uterus. **C.** Sagittal T2-weighted MR image also confirms the presacral mass is due to posterior displacement of the uterus (between arrows).

CASE 91 Pelvic pseudotumor due to perineal muscle flap

Imaging description

Perineal wound complications are a substantial source of morbidity after major perineal surgery such as abdominoperineal resection or pelvic exenteration [1]. A variety of muscular interposition flaps, including rectus abdominis, gracilis, and gluteus maximus, can be used to help in the reconstruction of the perineum after such surgery in order to reduce these complications [1–3]. Other uses of perineal myocutaneous flaps include formation of a neovagina or replacement of the anal sphincter [4–6]. Imaging findings vary with the type of flap. In general, the transposed muscle appears as a bandlike soft-tissue density mass, abutting the pelvic side wall or sacrum (Figure 91.1). Transposed subcutaneous adipose tissue may be seen as a fat density mass within the perineum (Figure 91.2). Transposed skin, if harvested to form a neovagina, may be seen as linear soft-tissue density surrounded by fat posterior to the symphysis pubis. When the rectus abdominis is used, the donor site can be identified by unilateral thinning of the rectus sheath. Findings of complications may also be seen, including flap necrosis, infection, fistula, muscle atrophy, or tumor recurrence [4, 5].

Importance

This pseudotumor is most problematic in patients who have had perineal surgery for malignancy, since in this setting the pseudomass created by the transposed muscle may be mistaken for recurrent cancer.

Typical clinical scenario

The two main indications for perineal muscle flap transposition are reconstruction after pelvic exenteration for recurrent gynecologic malignancy or reconstruction after abdominoperineal resection for anorectal cancer.

Differential diagnosis

Pelvic pseudotumor due to a perineal muscle flap can usually be identified by awareness of the surgical procedure and recognizing the clinical scenario. The curvilinear morphology of the transposed muscle may also be a helpful clue, as may recognition of the vascular pedicle formed during harvest of such flaps. The primary differential consideration is a true pelvic tumor or mass, particularly recurrent malignancy in those with a history of cancer. In cases where this distinction is unclear, imaging follow-up may be helpful since the pelvic pseudotumor of muscle transposition will not progress and may even regress, due to muscle atrophy, while a recurrent tumor will progress.

Teaching point

The diagnosis of recurrent tumor in a patient who has pelvic exenteration or abdominoperineal resection should be made only after the possibility of a perineal muscle flap has been considered and correlated with the surgical history.

REFERENCES

1 Nisar PJ, Scott HJ. Myocutaneous flap reconstruction of the pelvis after abdominoperineal excision. *Colorectal Dis* 2009; **11**: 806–816.

2 Sunesen KG, Buntzen S, Tei T, et al. Perineal healing and survival after anal cancer salvage surgery: 10-year experience with primary perineal reconstruction using the vertical rectus abdominis myocutaneous (VRAM) flap. *Ann Surg Oncol* 2009; **16**: 68–77.

3 Weiwei L, Zhifei L, Ang Z, et al. Vaginal reconstruction with the muscle-sparing vertical rectus abdominis myocutaneous flap. *J Plast Reconstr Aesthet Surg* 2009; **62**: 335–340.

4 Willing SJ, Pursell SH, Koch SR, Tobin GR. Vaginal reconstruction with rectus abdominis myocutaneous flap: CT findings. *Am J Roentgenol* 1991; **156**: 1001–1004.

5 Epstein DM, Arger PH, LaRossa D, Mintz MC, Coleman BG. CT evaluation of gracilis myocutaneous vaginal reconstruction after pelvic exenteration. *Am J Roentgenol* 1987; **148**: 1143–1146.

6 Ho KS, Seow-Choen F. Dynamic graciloplasty for total anorectal reconstruction after abdominoperineal resection for rectal tumour. *Int J Colorectal Dis* 2005; **20**: 38–41.

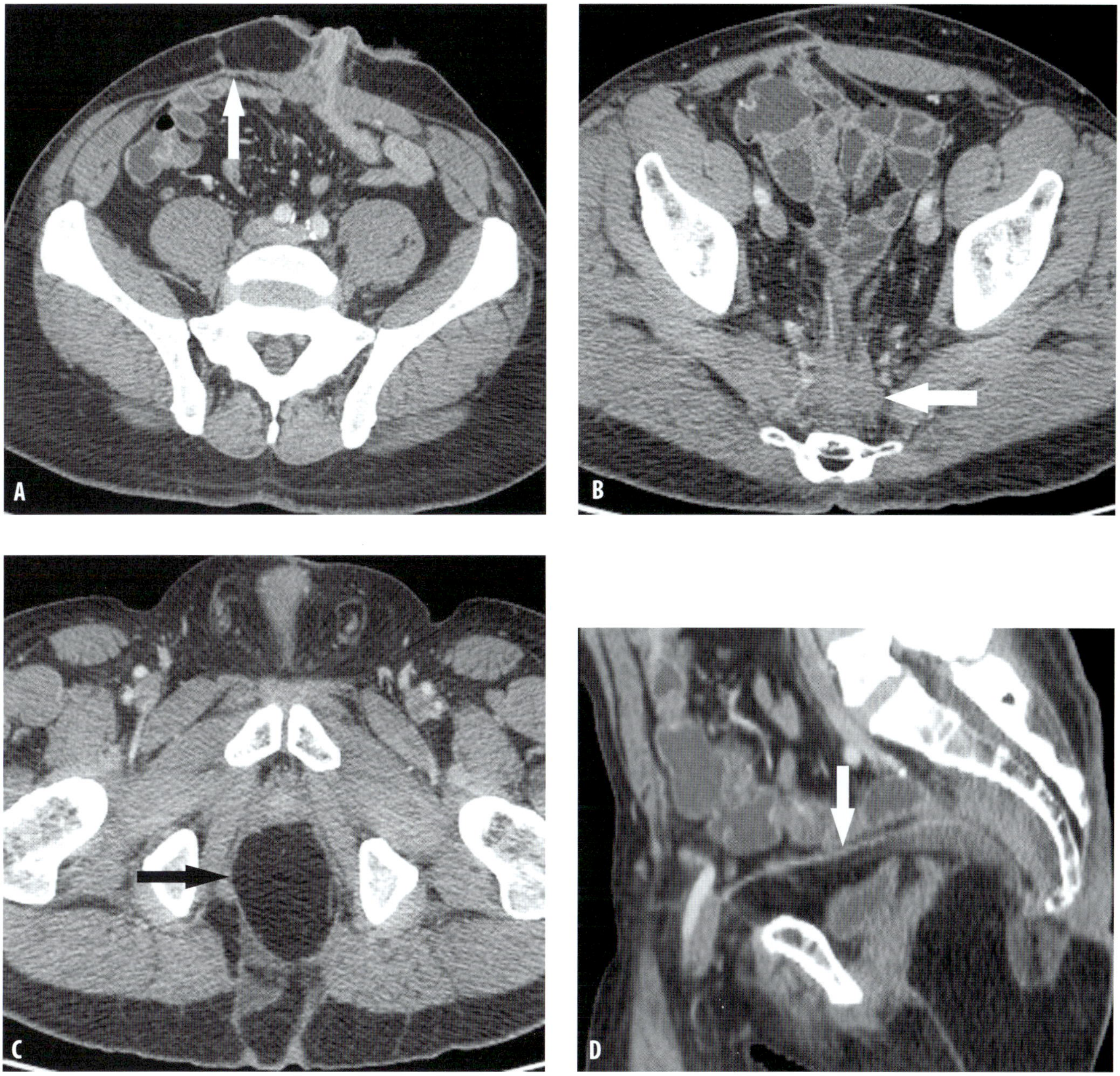

Figure 91.1 **A.** Axial contrast-enhanced CT image in a 64 year old man 6 months after abdominoperineal resection for anal cancer shows absence of the right rectus abdominis (arrow). **B.** Axial contrast-enhanced CT image at a more inferior level shows a masslike soft-tissue density (arrow) in the presacral space, due to the transposed rectus abdominis. **C.** Axial contrast-enhanced CT image at a more inferior level shows transposed fat (arrow) filling the perineal defect created by resection of the anal sphincter. **D.** Curved planar sagittal reformatted contrast-enhanced CT image shows the vascular pedicle of the inferior epigastric artery (arrow) coursing through the pelvis and entering the presacral soft tissue created by harvest of the rectus abdominis muscle.

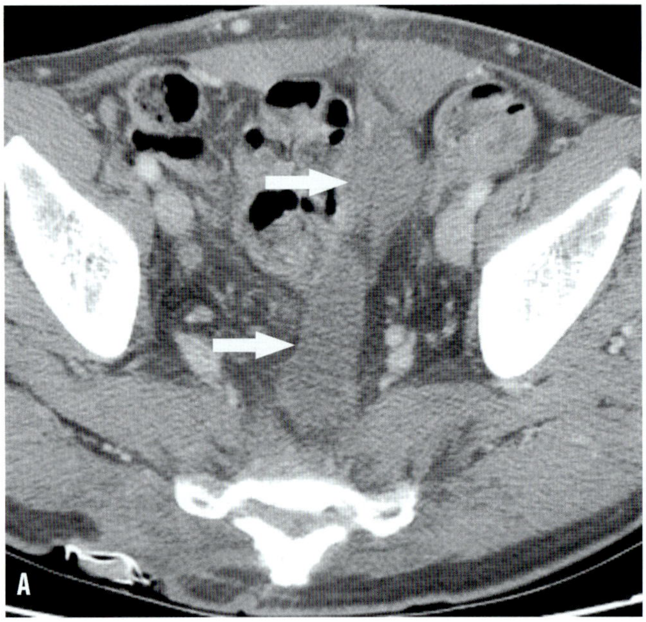

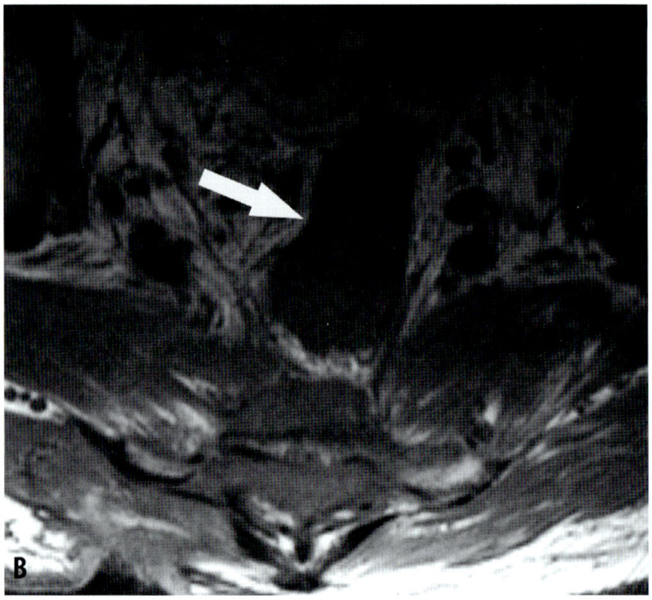

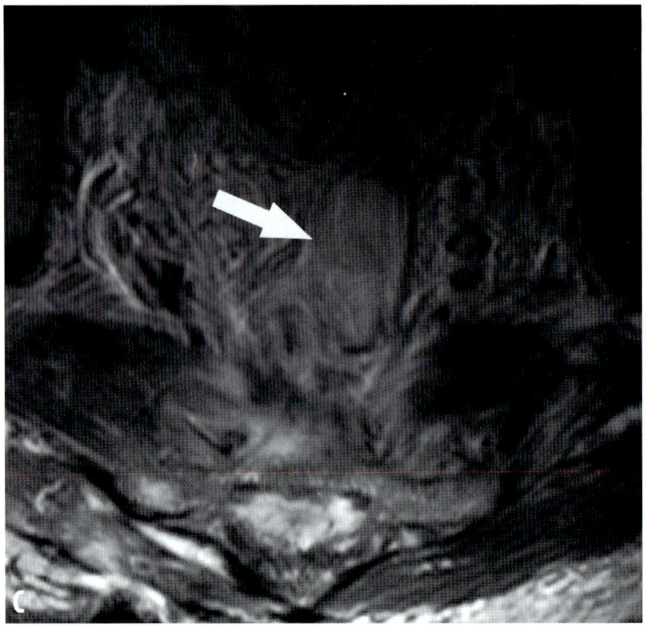

Figure 91.2 **A.** Axial contrast-enhanced CT image in a 50 year old man 2 months after abdominoperineal resection for complex perianal fistulae secondary to Crohn's disease shows a bandlike soft-tissue density (arrows) coursing through the pelvis due to transposition of the left rectus abdominis muscle. **B.** Axial spin-echo T1-weighted MR image shows the transposed muscle (arrow) is isointense to other pelvic skeletal muscles. **C.** Axial fast spin-echo T2-weighted MR image shows the transposed muscle (arrow) is moderately hyperintense relative to other pelvic skeletal muscles, likely secondary to denervation or edema. **D.** Axial spoiled gradient-echo fat-saturated post-gadolinium T1-weighted MR image shows the transposed muscle (arrow) is relatively non-enhancing, again resembling nearby pelvic skeletal muscles. **E.** Axial contrast-enhanced CT image obtained 11 months later shows the transposed muscle (arrow) has reduced in size (compare to Figure 91.1A), presumably due to a combination of denervation and disuse atrophy.

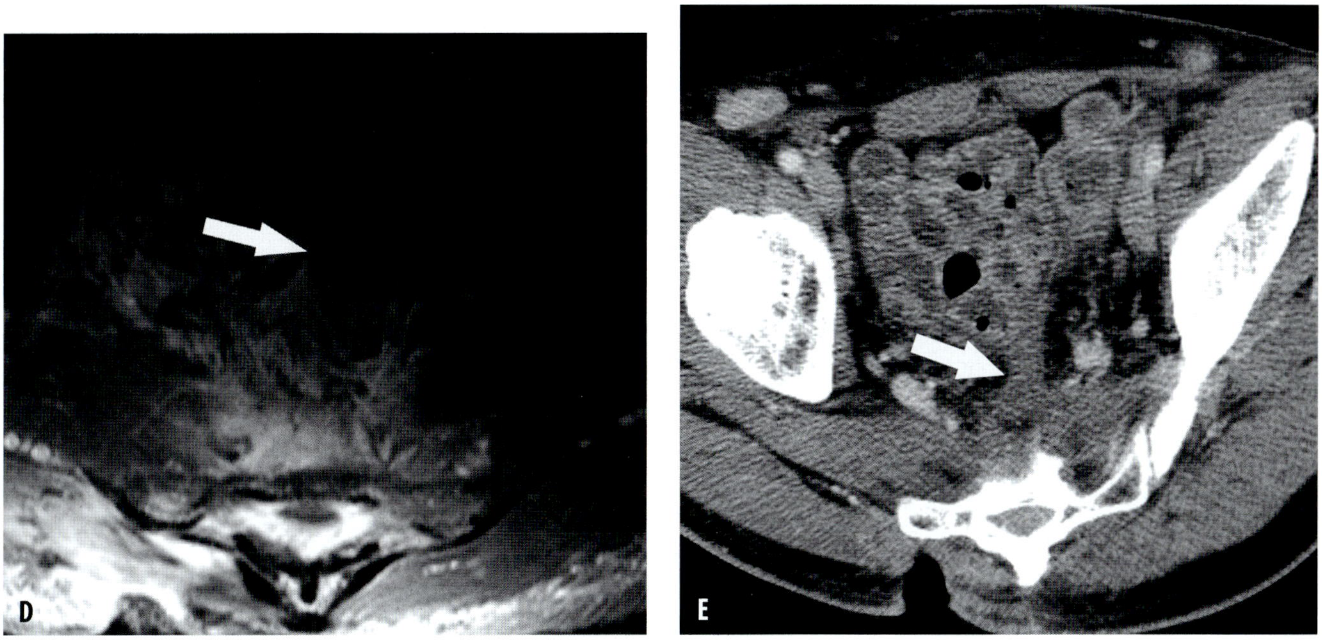

Figure 91.2 (cont.)

CASE 92

Pseudotumor due to failed renal transplant

Imaging description

Chronically failed renal allografts typically appear shrunken and heavily calcified, and the finding of a small rounded soft-tissue mass in the iliac fossa may simulate a tumor at cross-sectional imaging or be mistaken for bowel loops containing oral contrast at CT (Figures 92.1–92.4) [1, 2]. Other changes that can occur in failed allografts include cyst formation or renal enlargement due to infarction or hemorrhage. These latter changes may also simulate a tumor mass.

Importance

Misdiagnosis of a pelvic or adnexal mass may result in unnecessary additional testing, procedures, and patient anxiety.

Typical clinical scenario

This pseudotumor may develop secondary to transplant failure in any patient with end-stage renal disease who has received a prior transplant.

Differential diagnosis

The most important clue that an iliac fossa mass is a failed renal transplant is recognizing that the native kidneys are abnormal. In end-stage renal disease, the native kidneys are usually markedly atrophic and sometimes contain multiple small cysts due to acquired cystic disease. In the specific setting of polycystic disease, the native kidneys are enlarged and replaced by innumerable cysts of varying sizes. The presence of a functioning renal allograft in the opposite iliac fossa can be another important clue in determining the true nature of an iliac fossa mass [2].

Teaching point

An iliac fossa mass in a patient with abnormal native kidneys or a contralateral functioning renal transplant is likely to be a failed transplant kidney.

REFERENCES

1 Gayer G, Apter S, Katz R, et al. CT findings in ten patients with failed renal allografts: comparison with findings in functional grafts. *Eur J Radiol* 2000; **36**: 133–138.

2 Daly BD, Goldberg PA, Krebs TL, Wong-You-Cheong JJ, Drachenberg CI. End stage renal transplant failure: allograft appearances on CT. *Clin Radiol* 1997; **52**: 849–853.

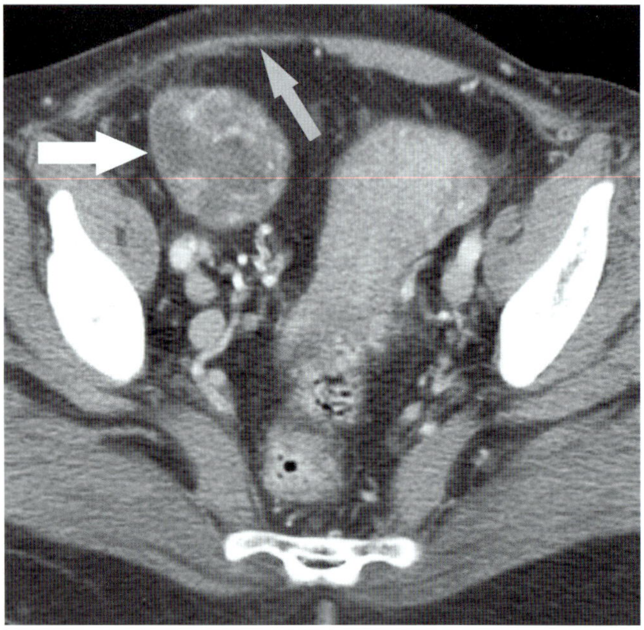

Figure 92.1 Axial contrast-enhanced CT image in a 53 year old woman with idiopathic end-stage renal disease and a failed renal transplant 9 years earlier shows the failed allograft as a non-specific soft-tissue mass (white arrow) in the right iliac fossa. Note asymmetric thinning of the right rectus abdominis muscle (grey arrow) due to prior transplant surgery. Axial images of the native kidneys (not shown) showed marked atrophy.

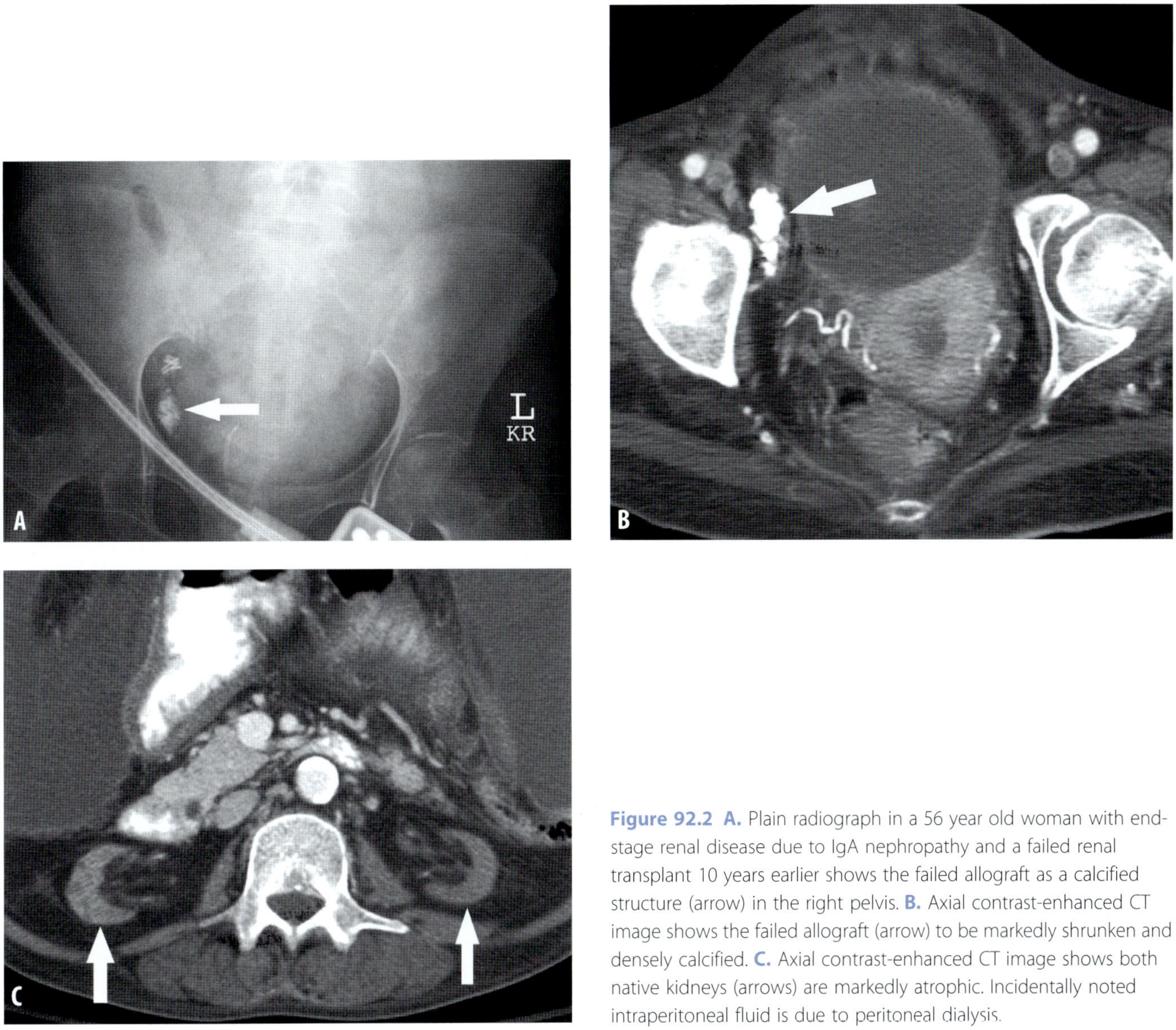

Figure 92.2 A. Plain radiograph in a 56 year old woman with end-stage renal disease due to IgA nephropathy and a failed renal transplant 10 years earlier shows the failed allograft as a calcified structure (arrow) in the right pelvis. **B.** Axial contrast-enhanced CT image shows the failed allograft (arrow) to be markedly shrunken and densely calcified. **C.** Axial contrast-enhanced CT image shows both native kidneys (arrows) are markedly atrophic. Incidentally noted intraperitoneal fluid is due to peritoneal dialysis.

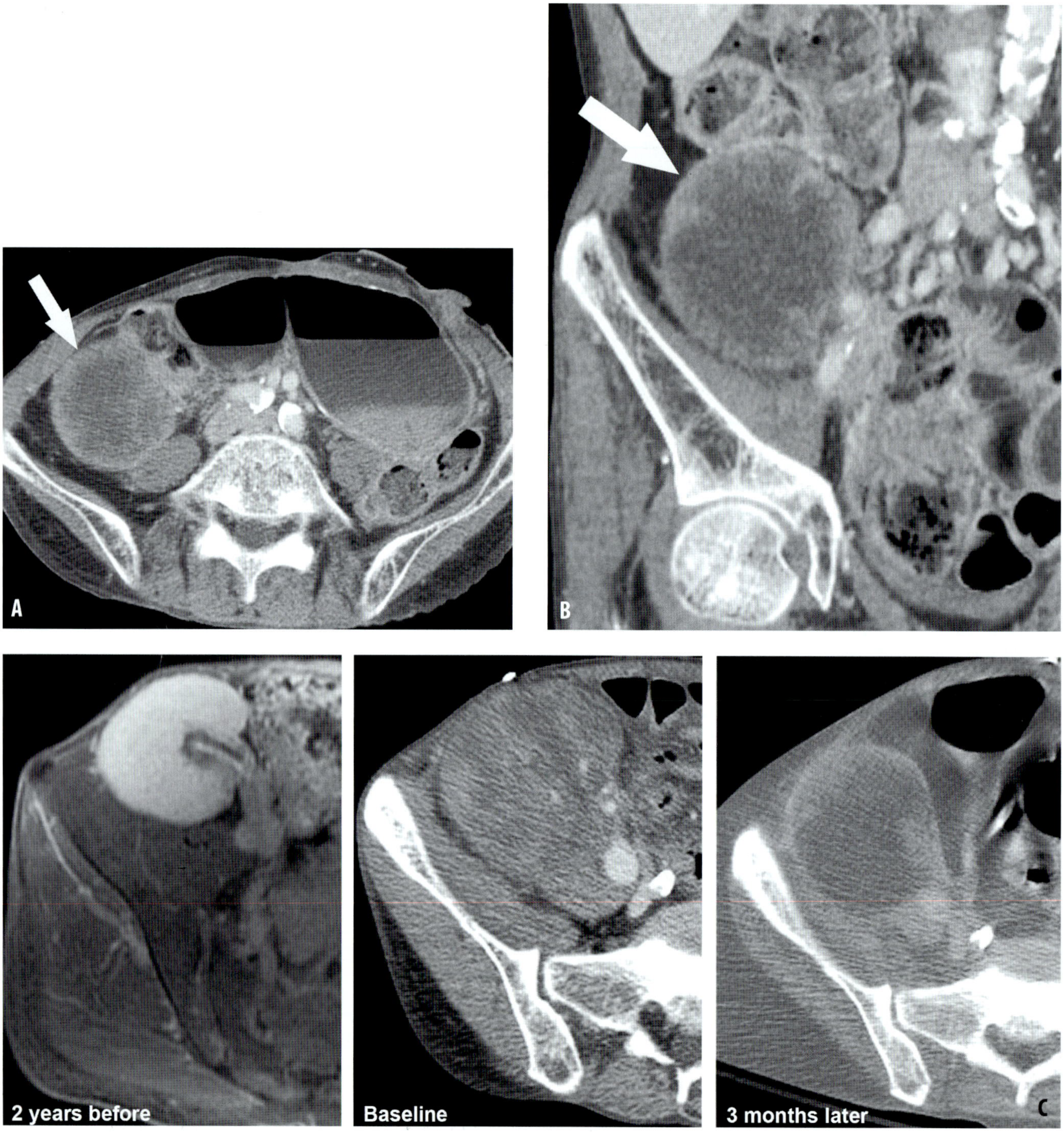

Figure 92.3 A. Axial contrast-enhanced CT image in a 54 year old man with end-stage renal disease due to IgA nephropathy and a failed renal transplant 6 years earlier shows the allograft as a non-specific mass (arrow) in the right iliac fossa (incidentally noted dilated small bowel was due to recurrent adhesive obstruction secondary to prior perforated diverticulitis). **B.** Coronal contrast-enhanced CT image shows the failed allograft as a rounded soft-tissue mass (arrow) in the right iliac fossa. **C.** Photomontage of axial gadolinium-enhanced T1-weighted MR image (left panel), axial contrast-enhanced CT images (middle and right panels) obtained 2 years before, at baseline, and 3 months after Figures 92.3A and 92.3B, respectively. Renal allograft biopsy at the time of the MRI scan showed a mixture of IgA and transplant glomerulopathy. Subsequent serial imaging is consistent with superimposed transplant infarction (note progressive reduced enhancement over time). Tissue confirmation was not clinically indicated.

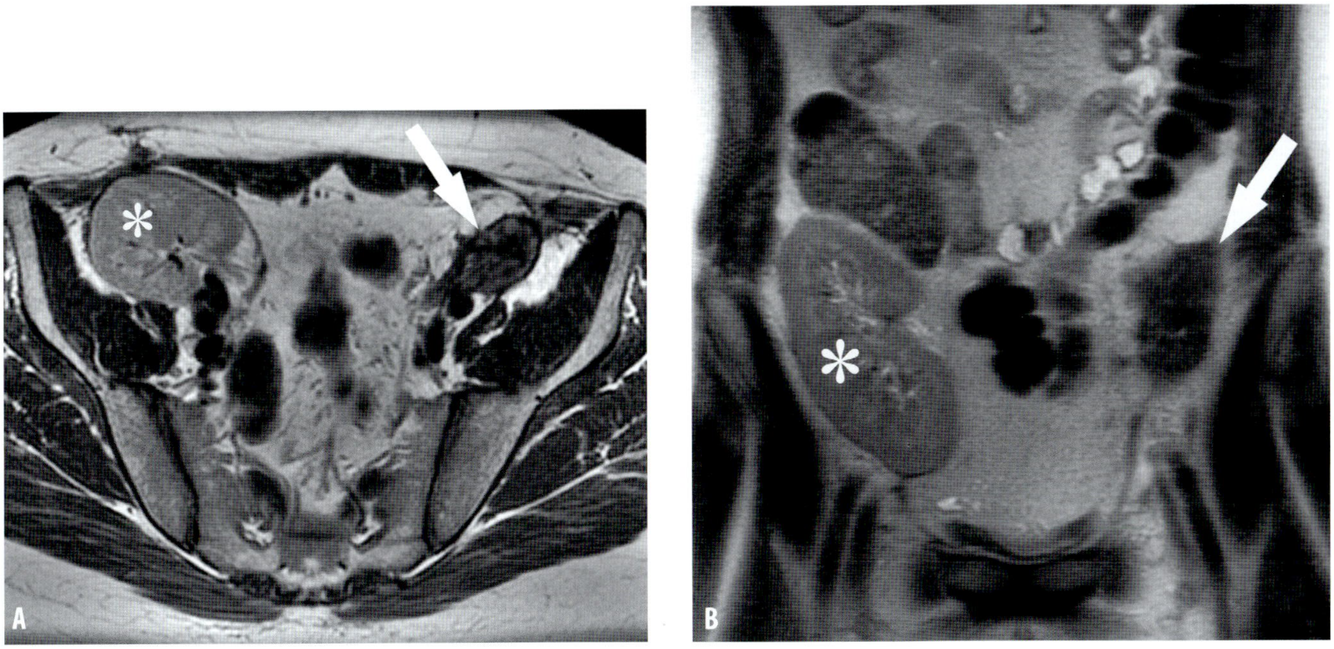

Figure 92.4 A. Axial T2-weighted MR image in a 28 year old woman with a history of two prior renal transplants for end-stage renal disease due to glomerulosclerosis associated with Frasier syndrome. A normal-appearing renal transplant (asterisk) in the right iliac fossa can be easily identified. A chronically failed transplant (arrow) in the left iliac fossa has a more non-specific appearance. **B.** Coronal T2-weighted MR image shows the normal-appearing renal transplant (asterisk) and the chronically failed transplant (arrow) both have similar kidney-shaped outlines, although the failed allograft is smaller.

CASE 93 Pseudotumor due to hernia repair device

Imaging description

Flat meshes or mesh plugs, or a combination of both, used to seal inguinal hernial orifices may produce a variety of groin pseudotumors on postoperative CT, consisting of soft-tissue to fluid density nodules (resembling lymphadenopathy), focal stranding or streakiness, or a soft tissue rim around fat density (resembling epiploic appendagitis) (Figures 93.1–93.3) [1, 2].

Importance

Misidentification of hernia mesh repair as lymphadenopathy or epiploic appendagitis may result in unnecessary workup and treatment.

Typical clinical scenario

Prosthetic meshes are increasingly used for inguinal hernia repair, because this practice reduces the hernia recurrence rate [3–5]. Most surgeons in the United States use a flat overlying polypropylene mesh with a mesh plug placed in the hernia orifice. This pitfall is typically seen when a surgical patient is scanned for evaluation of postoperative fever or other complications, but may also be seen when a patient with a prior hernia repair is scanned for other indications such as pain or malignancy.

Differential diagnosis

Groin pseudotumors are commoner after use of a mesh plug than after a flat mesh. In one study [1], a pseudotumor was seen at postoperative CT in 63 of 71 mesh plug repairs (with a mean size of 2.5 cm) compared to 6 of 25 flat mesh repairs (with a mean size of 1.9 cm, significantly smaller than the pseudotumors after mesh plug repairs). The findings that help differentiate a mesh pseudotumor from other pathology include clinical history, location, low density (less than 20 HU in 49 of 69 cases [1]), and margin. The typical serrated or lobulated margin of the pseudotumor conforms to the ex vivo appearance of the mesh plug (Figure 93.4), suggesting the pseudotumor reflects fluid or debris trapped within the mesh plug. Correlation with the operative record or direct discussion with the surgeon may also be helpful.

Teaching point

The possibility of mesh pseudotumor should be considered when an apparent mass or fluid collection is seen deep to the internal inguinal ring on a post-hernioplasty CT scan, especially when the lesion is low-density or has a serrated or lobulated margin.

REFERENCES

1 Chernyak V, Rozenblit AM, Patlas M, et al. Pelvic pseudolesions after inguinal hernioplasty using prosthetic mesh: CT findings. *J Comput Assist Tomogr* 2007; **31**: 724–727.

2 Yeung VH, Pearl JM, Coakley FV, et al. Computed tomographic appearance of Prolene Hernia System and polypropylene mesh plug inguinal hernia repair. *J Comput Assist Tomogr* 2008; **32**: 529–532.

3 Seleem MI. Open mesh-plug technique in inguinal hernia repair–short-term results. *S Afr J Surg* 2003; **41**: 44–47.

4 Rutkow IM. The PerFix plug repair for groin hernias. *Surg Clin North Am* 2003; **83**: 1079–1098.

5 Dieudonne G. Plug repair of groin hernias: a 10-year experience. *Hernia* 2001; **5**: 189–191.

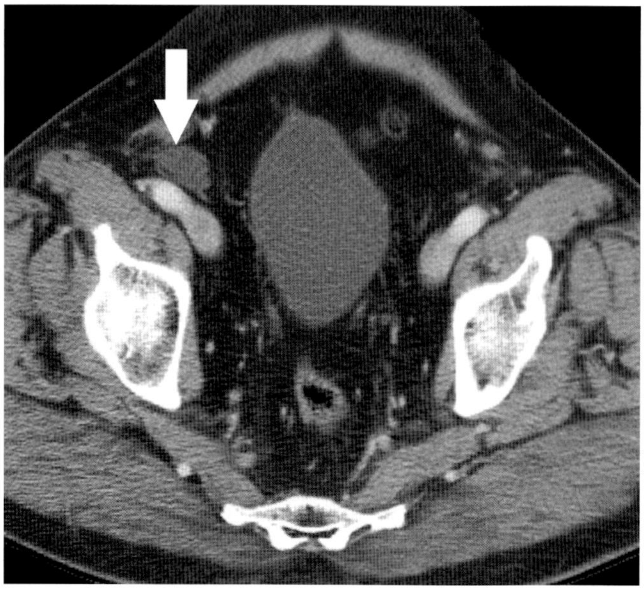

Figure 93.1 Axial contrast-enhanced CT image in a 54 year old man with fever 4 days after right inguinal hernioplasty with a mesh plug shows a low-density mass (arrow) in the right groin due to the mesh plug that resembles lymphadenopathy.

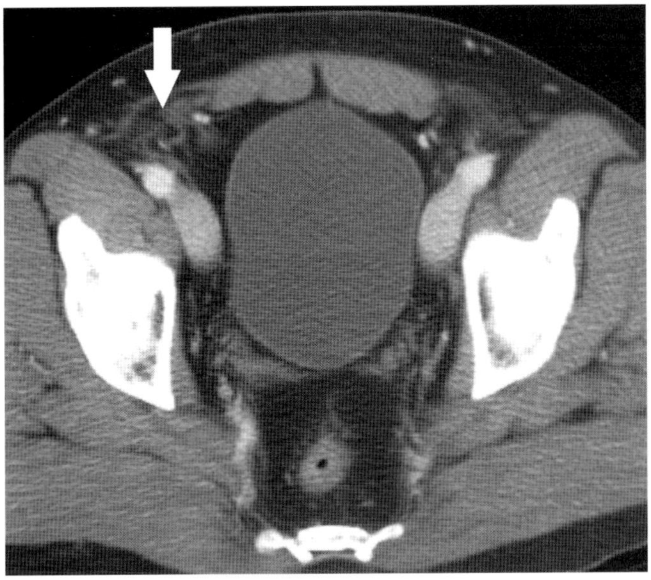

Figure 93.2 Axial contrast-enhanced CT image in a 42 year old man with a history of right inguinal hernioplasty with a combination of flat and plug meshes shows focal fat stranding (arrow) in the internal inguinal orifice.

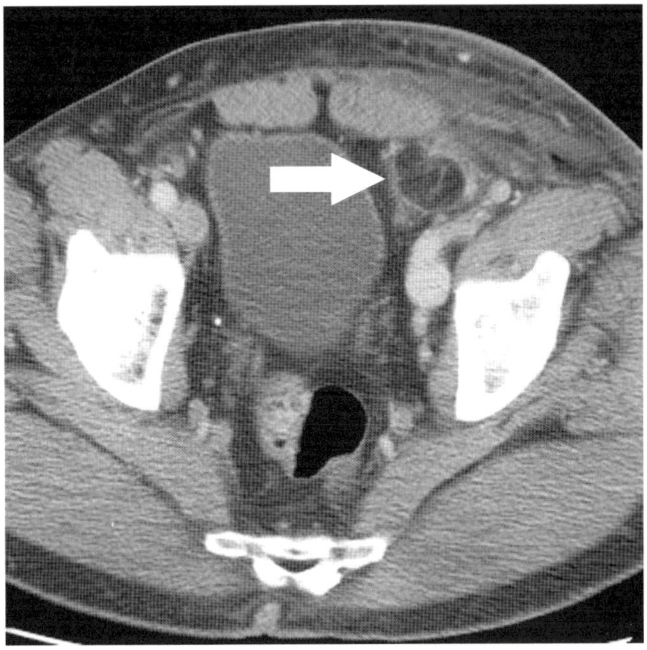

Figure 93.3 Axial contrast-enhanced CT image in a 48 year old man with fever one day after left inguinal hernioplasty with a plug mesh repair shows a mass (arrow) due to the mesh plug. The mass consists of a soft-tissue rim around fat density and mimics epiploic appendagitis.

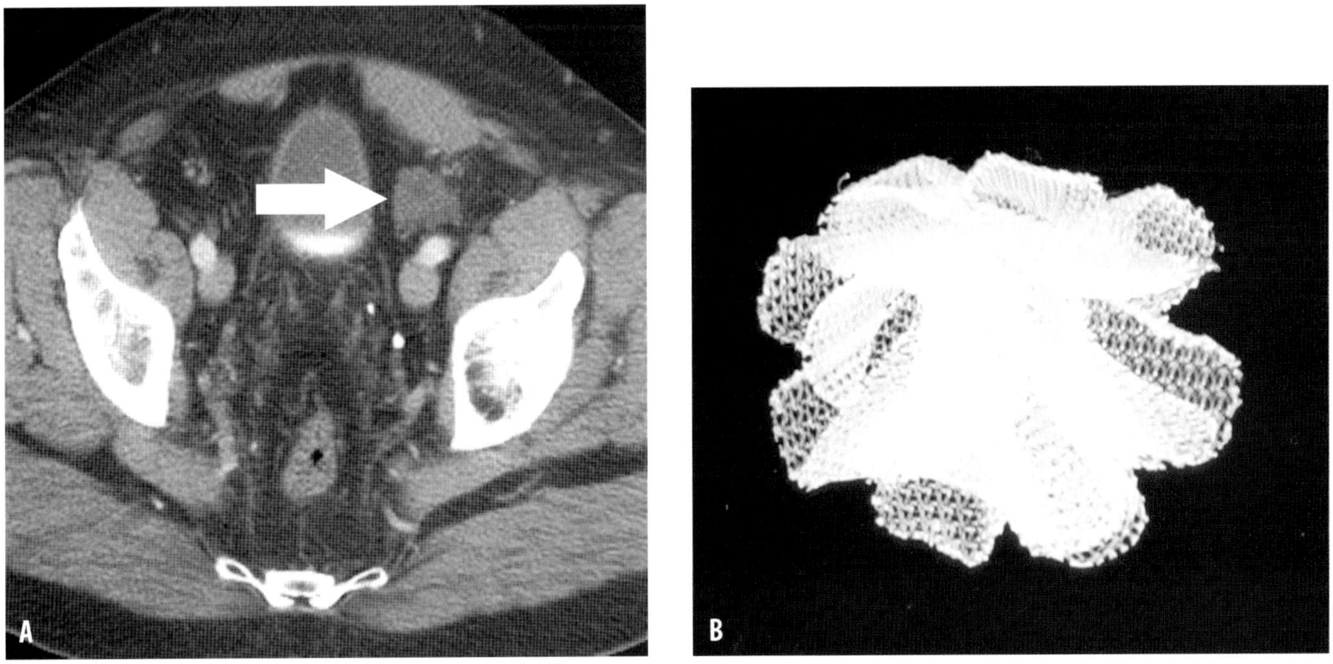

Figure 93.4 A. Axial contrast-enhanced CT image in a 63 year old man with a history of left inguinal hernioplasty and newly diagnosed lung cancer shows a low-density mass (arrow) in the left groin consistent with a pseudotumor due to hernia repair with a mesh plug. Note the mass is of low density and has a finely serrated or lobulated margin. **B.** Ex vivo photograph of a mesh plug (measures 2.5 cm in diameter). Note the serrated or lobulated periphery of the mesh conforms to the appearance of the pseudotumor margin at CT.

CASE 94 Pseudotumor due to muscle transposition

Imaging description

Muscle flaps may be used to protect surgically debrided femoral vessels in the groin. The transposed muscle, commonly the sartorius or rectus abdominis, results in a soft-tissue mass anterolateral or anterior to the femoral vessels on postoperative imaging that can be potentially confused for a postoperative collection or tumor recurrence (Figures 94.1 and 94.2) [1].

Importance

Misidentification of a transposed muscle as a postoperative collection or recurrence may result in unnecessary workup and treatment.

Typical clinical scenario

Muscle transposition may be performed after radical inguinal lymphadenectomy, debridement of open groin wounds, or repair of infected femoral artery grafts [1–3]. This pitfall is typically seen when a surgical patient is scanned for evaluation of postoperative fever or other complications, or when a patient who has had a radical lymphadenectomy for malignancy such as melanoma undergoes surveillance imaging.

Differential diagnosis

Absence of the normal muscle is a critical observation in establishing the imaging diagnosis of muscle transposition. The contralateral side can be used as an internal control for this purpose. In the early postoperative period, the lack of a rim-enhancing capsule helps distinguish a muscle flap from an abscess. Later, continuity and isodensity with the sartorius muscle inferiorly or the rectus abdominis superiorly and lack of typical masslike features help distinguish muscle transposition from tumor recurrence. While the MRI findings in muscle transposition to the groin have not been specifically reported, experience with other flaps and muscular denervation suggests that the transposed muscle is likely to show increased T2 signal intensity [2, 3], so that T2 isointensity with adjacent muscle may not be a reliable discriminating feature.

Teaching point

The possibility of muscle transposition should be considered when an apparent mass or fluid collection is seen adjacent to the femoral vessels in the groin in a patient who has had extensive soft-tissue or vascular surgery in this region.

REFERENCES

1 Wong JJ, Daly B, Krebs TL, Elias EG, Jacobs SC. Surgical transfer of the sartorius muscle to the groin after lymphadenectomy or debridement: CT findings. *Am J Roentgenol* 1996; **166**: 109–112.

2 Fox MG, Bancroft LW, Peterson JJ, et al. MRI appearance of myocutaneous flaps commonly used in orthopedic reconstructive surgery. *Am J Roentgenol* 2006; **187**: 800–806.

3 Coakley FV, Finlay DB, Harper WM, Allen MJ. Direct and indirect MRI findings in ganglion cysts of the common peroneal nerve. *Clin Radiol* 1995; **50**: 168–169.

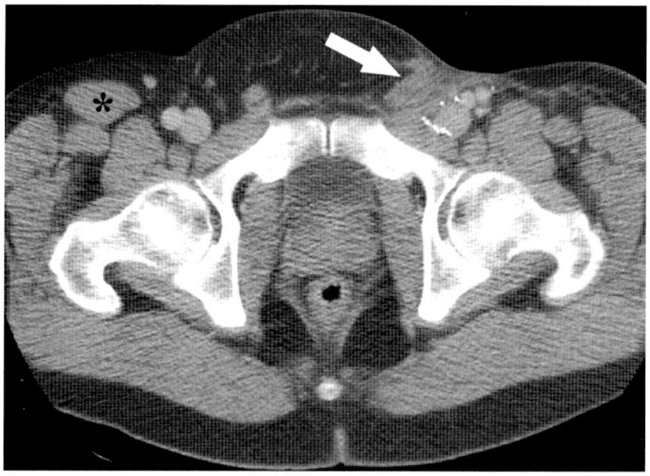

Figure 94.1 Axial contrast-enhanced CT image in a 45 year old man 5 years after radical inguinal node dissection with sartorial transposition for melanoma shows the transposed sartorius muscle as an infiltrative mass (arrow) anterior to the femoral vessels in the groin. Note asymmetric absence of the left sartorius muscle in the expected location when compared to the right sartorius muscle (asterisk).

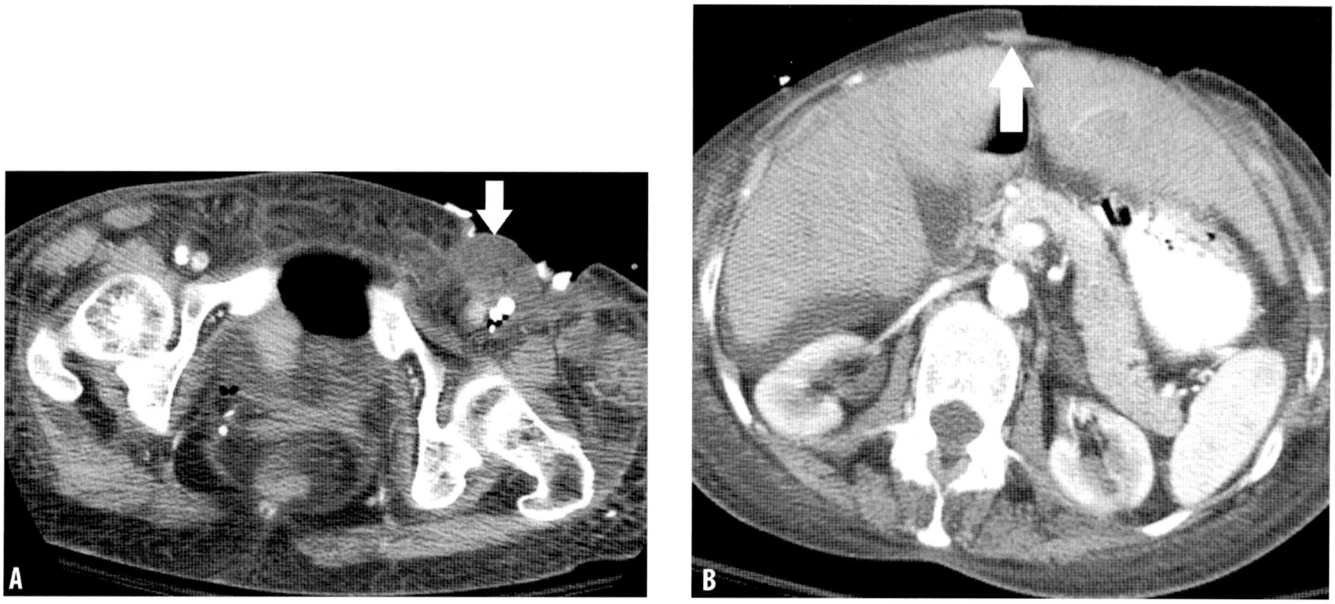

Figure 94.2 **A.** Axial contrast-enhanced CT image in a 68 year old woman with a complex vascular history including ilioprofunda bypass graft infection requiring extensive debridement of the left groin repaired with a rectus abdominis myocutaneous flap shows the transposed muscle as a low-density lesion (arrow) anterior to the left femoral vessels. **B.** Axial contrast-enhanced CT image at a more superior level confirms asymmetric absence of the left rectus abdominis (and overlying tissue) when compared to the right rectus abdominis (arrow).

CASE **95**

Distended iliopsoas bursa

Imaging description

The iliopsoas (also termed the iliopectineal, iliofemoral, iliac, or subpsoas) bursa is the largest normal bursa in the body, and is present in 98% of adults [1]. The iliopsoas bursa is located between the iliopsoas tendon and the anterior aspect of the hip joint, and serves to reduce tendon friction over the hip joint [2]. The bursa is lined by synovium and is normally collapsed. When distended with fluid, due either to effusive hip disease or primary inflammation of the bursal lining, the bursa becomes enlarged and appears as a fluid-filled structure in the groin at ultrasound, CT, or MRI (Figures 95.1–95.4) [1–3]. When the bursa is distended, communication with the hip joint is visible at US and CT in about 50% of cases, but is seen in virtually all cases at MRI [3]. Typically, the mass lies adjacent to the psoas muscle at the level of the hip joint, with variable proximal (into the pelvic retroperitoneum) and distal (into the upper portion of the thigh) extension [1, 4, 5].

Importance

A distended iliopsoas bursa may be mistaken for a cystic neoplasm in the groin or pelvis.

Typical clinical scenario

A distended iliopsoas bursa may be detected incidentally as an asymptomatic mass or cause symptoms due to compression of adjacent structures such as lower extremity edema or femoral neuropathy [6–11].

Differential diagnosis

The location and fluid-filled appearance of a distended iliopsoas bursa are relatively distinctive. In problematic cases, MRI may help make the diagnosis by demonstrating communication with the hip joint.

Teaching point

A fluid-filled cyst-like structure in the groin anterior to the hip is likely to be a distended iliopsoas bursa, and suggests underlying hip or synovial disease. MRI may be helpful to confirm the diagnosis by showing communication between the bursa and the hip joint.

REFERENCES

1 Sartoris DJ, Danzig L, Gilula L, Greenway G, Resnick D. Synovial cysts of the hip joint and iliopsoas bursitis: a spectrum of imaging abnormalities. *Skeletal Radiol* 1985; **14**: 85–94.

2 Bianchi S, Martinoli C, Keller A, Bianchi-Zamorani MP. Giant iliopsoas bursitis: sonographic findings with magnetic resonance correlations. *J Clin Ultrasound* 2002; **30**: 437–441.

3 Wunderbaldinger P, Bremer C, Schellenberger E, et al. Imaging features of iliopsoas bursitis. *Eur Radiol* 2002; **12**: 409–415.

4 Penkava RR. Iliopsoas bursitis demonstrated by computed tomography. *Am J Roentgenol* 1980; **135**: 175–176.

5 Peters JC, Coleman BG, Turner ML, et al. CT evaluation of enlarged iliopsoas bursa. *Am J Roentgenol* 1980; **135**: 392–394.

6 Pellman E, Kumari S, Greenwald R. Rheumatoid iliopsoas bursitis presenting as unilateral leg edema. *J Rheumatol* 1986; **13**: 197–200.

7 Yang SS, Bronson MJ. Cystic enlargement of the iliopsoas bursa causing venous obstruction as a complication of total hip arthroplasty. A case report. *J Arthroplasty* 1993; **8**: 657–661.

8 Yoon TR, Song EK, Chung JY, Park CH. Femoral neuropathy caused by enlarged iliopsoas bursa associated with osteonecrosis of femoral head–a case report. *Acta Orthop Scand* 2000; **71**: 322–324.

9 Binek R, Levinsohn EM. Enlarged iliopsoas bursa. An unusual cause of thigh mass and hip pain. *Clin Orthop Relat Res* 1987; **224**: 158–163.

10 Janus C, Hermann G. Enlargement of the iliopsoas bursa: unusual cause of cystic mass on pelvic sonogram. *J Clin Ultrasound* 1982; **10**: 133–135.

11 Leekam RN, Matzinger MA, Mustard RA, Grosman H. Enlarged iliopsoas bursa simulating neoplasm on sonographic examination. *J Ultrasound Med* 1985; **4**: 493–494.

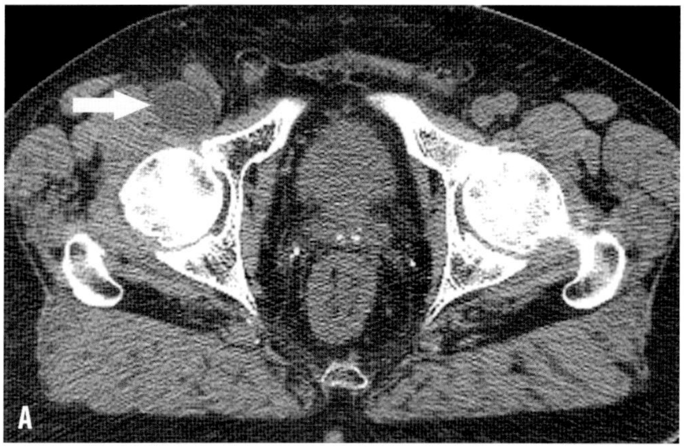

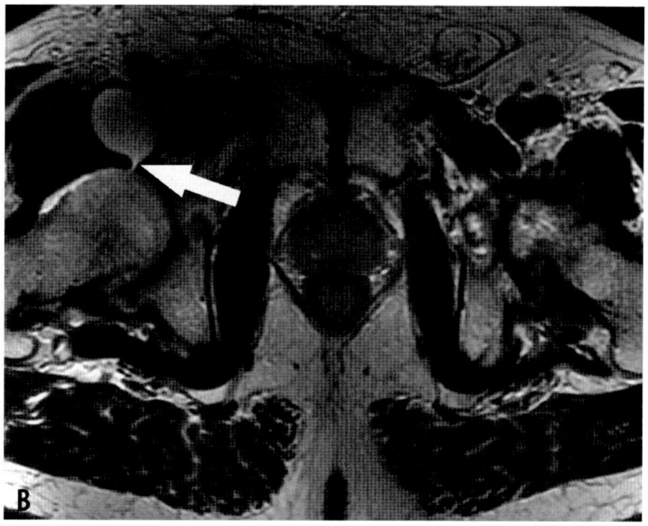

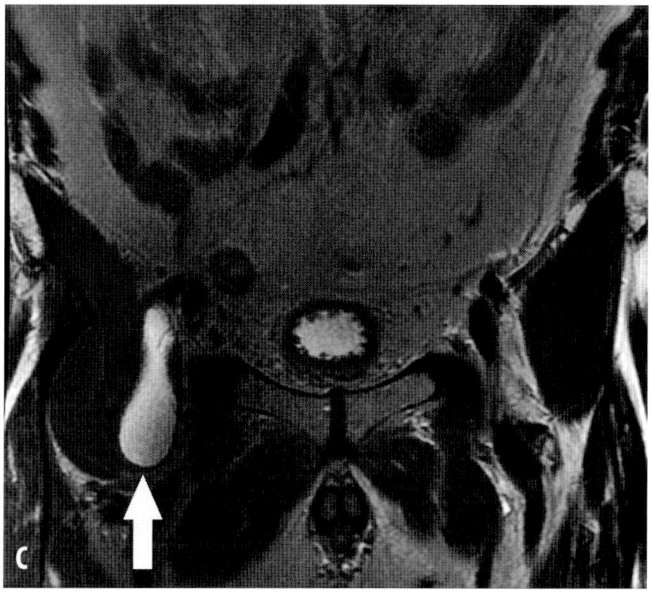

Figure 95.1 **A.** Axial non-enhanced CT image performed for radiation treatment planning in a 78 year old man with prostate cancer shows a cyst-like structure (arrow) anterior to the right hip. **B.** Axial T2-weighted MR image shows the structure is of fluid signal intensity and has a narrow neck (arrow) connecting to the hip joint. The appearances are those of a distended iliopsoas bursa. **C.** Coronal T2-weighted MR image shows the craniocaudad extent of the bursa. Note the inferior extension (arrow) into the proximal thigh.

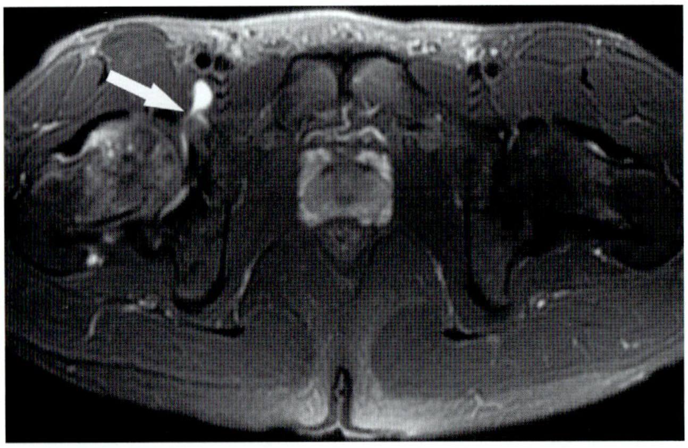

Figure 95.2 Axial T2-weighted MR image in a 21 year old man with idiopathic bilateral avascular necrosis of the hips shows a small right iliopsoas bursa communicating (arrow) with the hip joint.

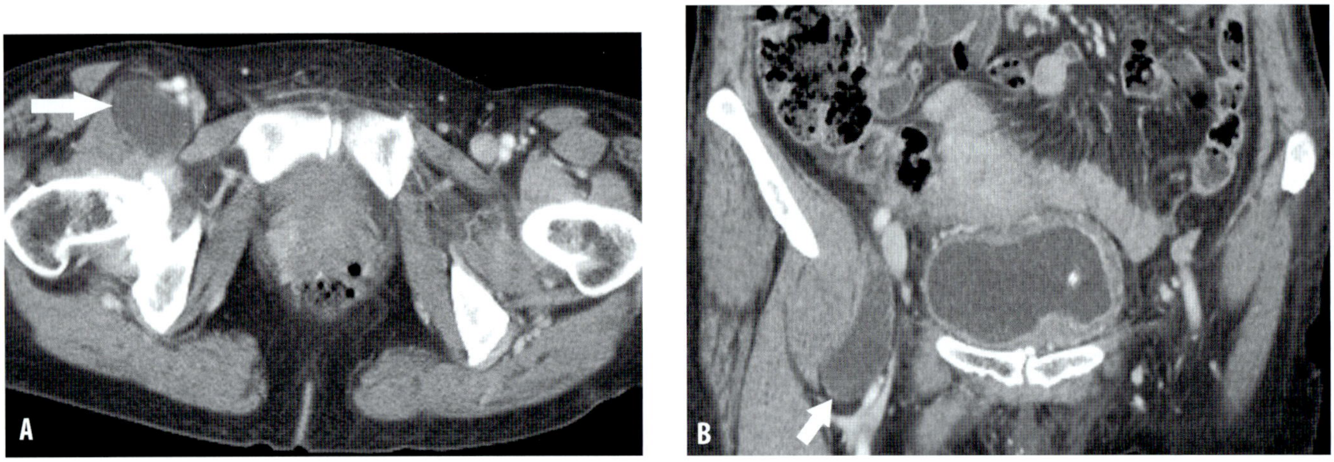

Figure 95.3 A. Axial contrast-enhanced CT image performed for surveillance of rectal cancer in an 82 year old woman shows a distended iliopsoas bursa in the right groin anterior to the hip. **B.** Coronal reformatted contrast-enhanced CT image shows inferior extension (arrow) of the bursa into the proximal thigh.

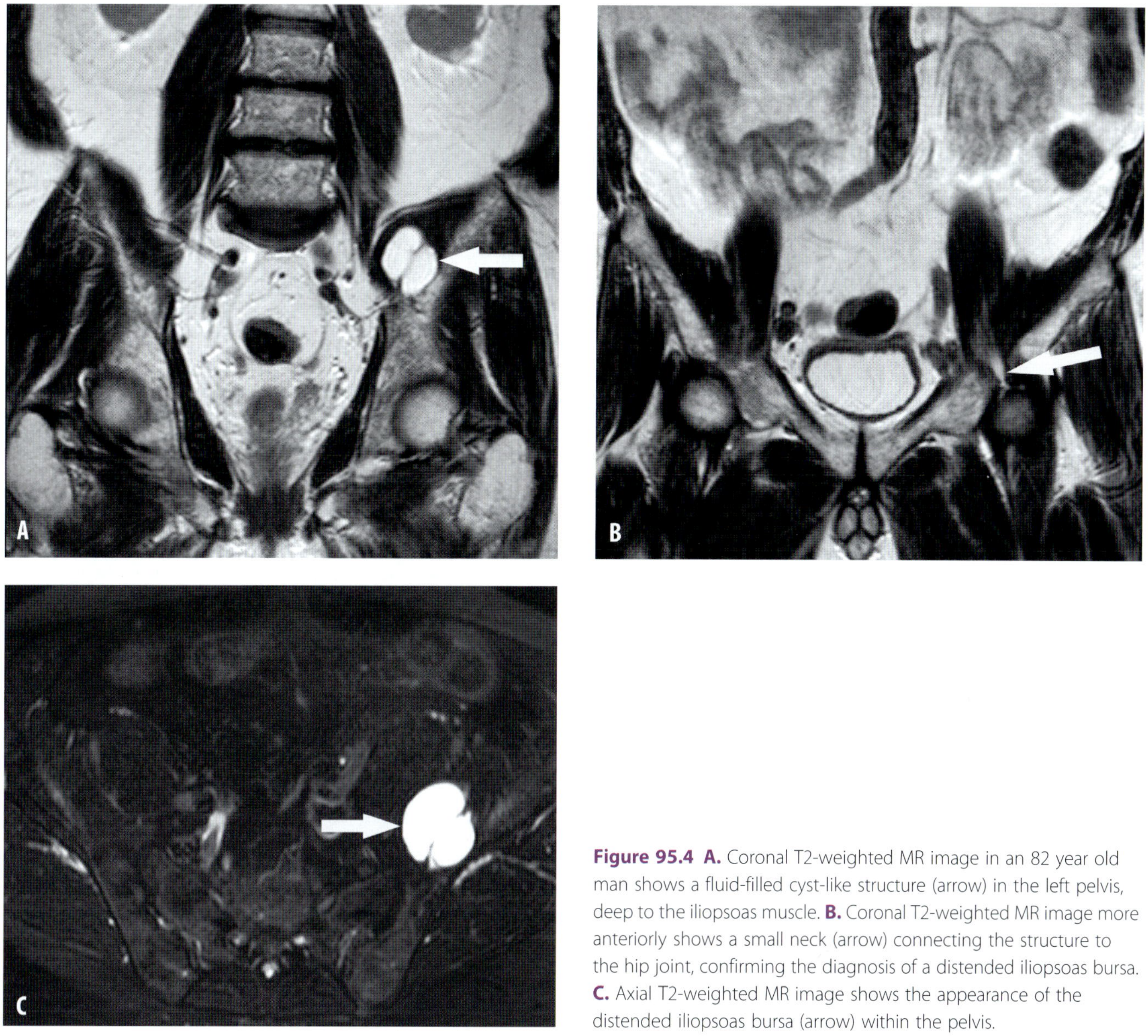

Figure 95.4 A. Coronal T2-weighted MR image in an 82 year old man shows a fluid-filled cyst-like structure (arrow) in the left pelvis, deep to the iliopsoas muscle. **B.** Coronal T2-weighted MR image more anteriorly shows a small neck (arrow) connecting the structure to the hip joint, confirming the diagnosis of a distended iliopsoas bursa. **C.** Axial T2-weighted MR image shows the appearance of the distended iliopsoas bursa (arrow) within the pelvis.

CASE **96**

Pseudothrombosis of the iliofemoral vein

Imaging description

On early post-contrast CT or MRI studies of the abdomen and pelvis, mixing artifact of unenhanced blood from the lower extremity and enhanced blood from a renal transplant (Figure 96.1) or asymmetric enhancement due to flow through collateral venous channels (Figures 96.2 and 96.3) may create the appearance of a filling defect that resembles a thrombus in an iliofemoral vein [1, 2].

Importance

Pseudothrombosis of an iliofemoral vein may be mistaken for a true thrombus, potentially resulting in inappropriate treatment or unnecessary follow-up investigation.

Typical clinical scenario

Conditions that may result in mixing of enhanced and unenhanced blood or asymmetric enhancement in an iliofemoral vein include renal transplantation or downhill venous collateral formation in subclavian vein occlusion or portosystemic collateral formation in portal hypertension.

Differential diagnosis

The primary differential is bland thrombus. Close examination of axial and reformatted CT images or delayed images to show resolution of the filling defect is usually sufficient to confirm the artifactual nature of this pseudolesion, but ultrasound or flow-sensitive MRI sequences can be used if the distinction cannot be made with certainty on CT.

Teaching point

Mixing artifact or asymmetric enhancement should be considered for an apparent thrombus in the iliofemoral vein, especially in the setting of renal transplantation or venous collateral formation.

REFERENCES

1 Gupta R, Rottenberg G, Taylor J. Pseudothrombosis of the iliac vein in patients following combined kidney and pancreas transplantation. *Br J Radiol* 2002; **75**: 692–694.

2 Desai G, Poder L, Wang ZJ, et al. CT findings in pseudothrombosis of the iliofemoral vein. *J Comput Assist Tomogr* 2010; **34**: 146–148.

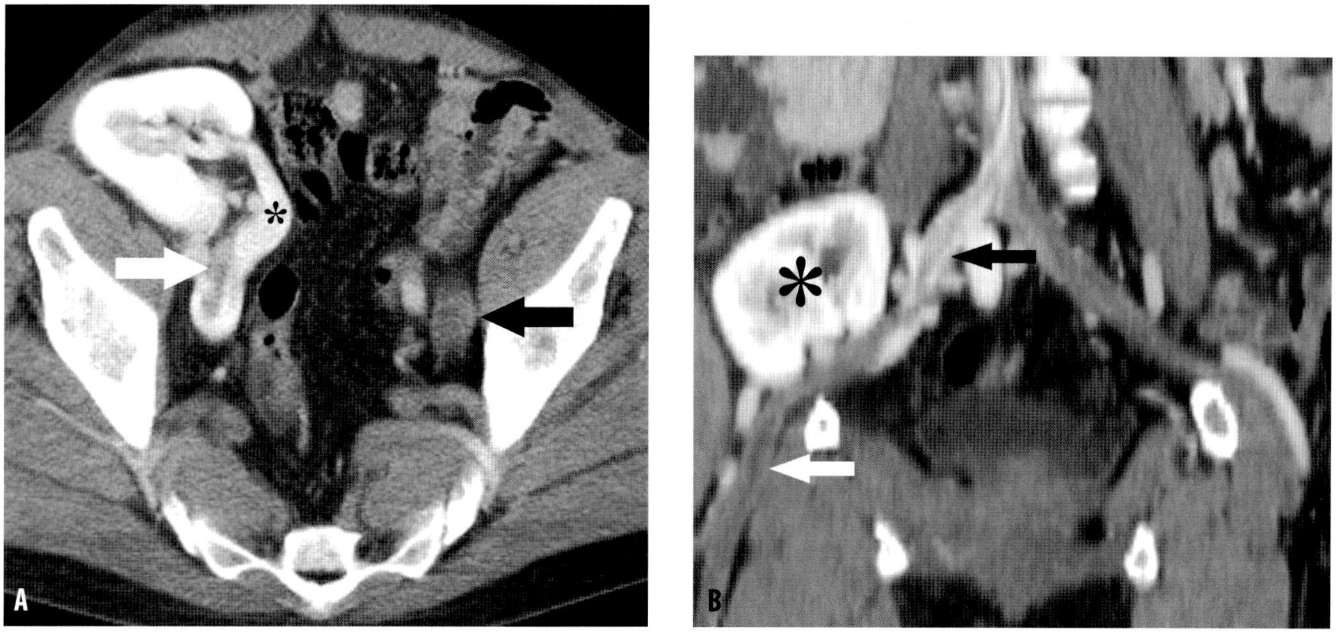

Figure 96.1 **A.** Axial contrast-enhanced CT image obtained in a 52 year old man after renal transplantation shows an apparent hypodense filling defect (white arrow) in the right external iliac vein. Note that the transplanted renal vein (asterisk) is enhanced and the left common iliac vein (black arrow) is unenhanced. **B.** Coronal curved planar reformatted image demonstrates the mechanism of this pseudothrombosis; the artifact (black arrow) is due to the laminar mixing of enhanced blood from the renal transplant (asterisk) with unopacified blood returning from the lower extremity in the right femoral vein (white arrow).

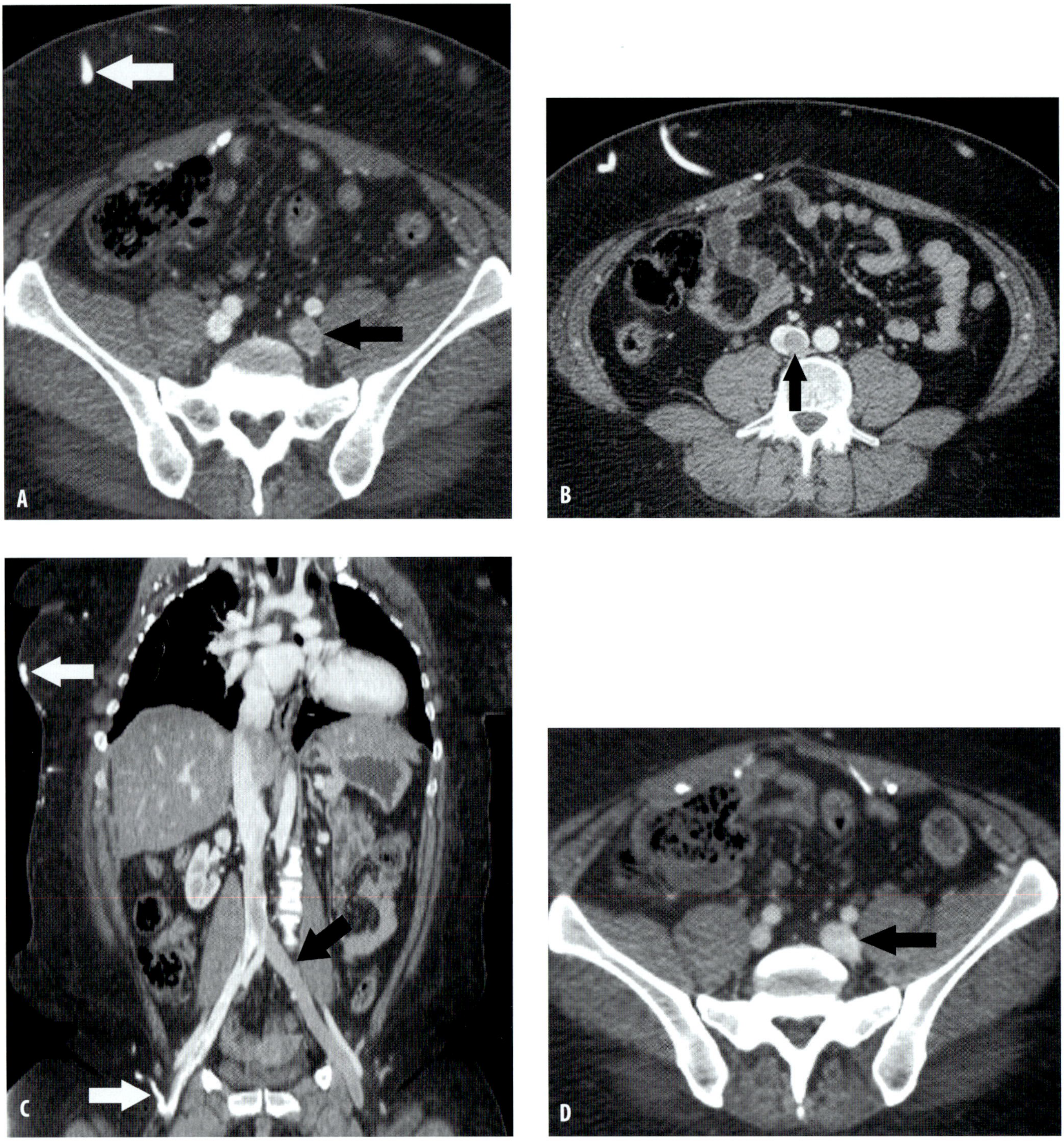

Figure 96.2 **A.** Axial contrast-enhanced CT image in a 22 year old woman with metastatic germ cell tumor and prior right subclavian vein thrombosis related to central line placement. Contrast was administrated through a peripheral line in the right arm. The left common iliac vein (black arrow) shows asymmetrically reduced enhancement, suggesting thrombus. Note the presence of enhancing venous collaterals (white arrow) in the right anterior abdominal wall. **B.** Axial contrast-enhanced CT image obtained at a more superior level shows continuation of an apparent filling defect (arrow) into the inferior vena cava, again suggesting thrombus. **C.** Coronal curved planar reformatted image demonstrates the apparent thrombus is artifactual and due to asymmetric early enhancement of the right iliofemoral vein relative to the left iliofemoral vein (black arrow) from the venous collaterals (white arrows) that have developed as a response to the prior right subclavian vein occlusion. **D.** Axial contrast-enhanced CT image obtained four months later shows normal symmetric opacification of the left common iliac vein (arrow). On this occasion, contrast was administrated through a peripheral line in the left arm.

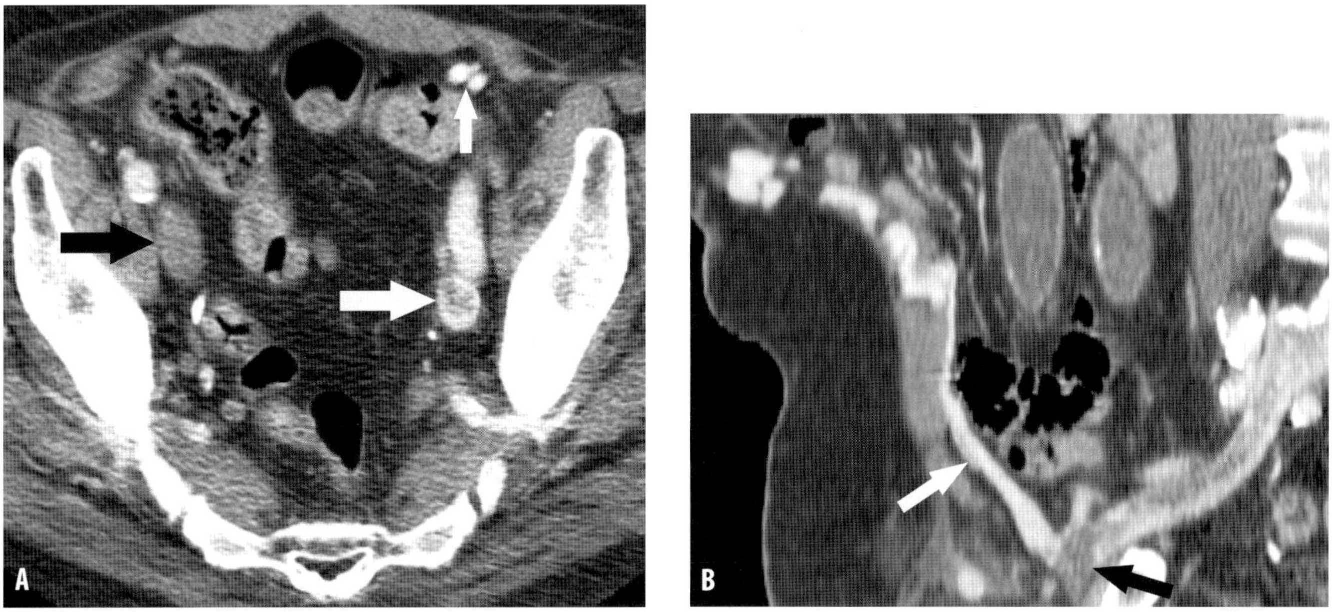

Figure 96.3 A. Axial contrast-enhanced CT image obtained in a 67 year old man with cirrhosis and recent endovascular stenting of an abdominal aortic aneurysm. Apparent thrombus is seen in the left external iliac vein (horizontal white arrow). Note that there are enhancing left-sided venous collaterals (vertical white arrow) just deep to the anterior abdominal wall and that the right external iliac vein (black arrow) is relatively unenhanced. **B.** Sagittal curved planar reformatted image demonstrates the apparent thrombus is artifactual and due to laminar mixing of enhanced blood passing inferiorly through portosystemic venous collaterals (white arrow) with unenhanced blood in the left femoral vein (black arrow).

CASE **97**

Postradiation pelvic insufficiency fracture

Imaging description

Pelvic insufficiency fractures are a form of stress fracture in which the physiologic load of weight bearing is sufficient to cause pelvic fractures in bone that is weakened by demineralization and decreased elastic resistance. Pelvic radiation is one cause of bone weakening that may lead to insufficiency fracture. At CT, pelvic insufficiency fractures appear as linear sclerotic lesions with or without cortical discontinuity in the sacral body parallel to the sacro-iliac joints. This results in the characteristic H or Honda sign in the sacrum at bone scintigraphy. MRI shows reduced T1 and increased T2 signal intensity. The pubic rami adjacent to the symphysis pubis and the acetabulum can also be affected. PET findings are variable and likely related to timing; that is, in the acute phase FDG uptake may be increased but it may decrease as the fracture heals (Figures 97.1–97.4) [1–4].

Importance

Postradiation pelvic insufficiency fractures are often misdiagnosed as bony metastases because of the combination of a known prior malignancy and the detection of a new bone lesion. Negative consequences of such an erroneous diagnosis include unnecessary biopsy and unwarranted further irradiation or chemotherapy.

Typical clinical scenario

The reported frequency of insufficiency fracture after pelvic radiation varies between publications, likely reflecting selection bias and different methodologies. Despite this variation, it is clearly a common occurrence. In a study of 18 patients with advanced cervical cancer studied by serial MRI before and after radiation therapy, 16 (89%) developed pelvic insufficiency fractures [5]. Fractures typically developed in the first year after treatment and many showed evidence of healing by 30 months. In a larger study of 510 patients undergoing irradiation for cervical cancer, the 5-year cumulative prevalence of pelvic insufficiency fracture was 45%, with a median interval between radiation and diagnosis of 17 months [6]. Patients with pelvic insufficiency fractures may have pain or may be asymptomatic [6, 7].

Differential diagnosis

It is possible for a new bone lesion developing in a patient who has been radiated for pelvic malignancy to be a metastasis, but this seems to be a rare event. In a study of 514 such patients, only 3 developed sacral metastases while 100 developed pelvic insufficiency fractures [6]. In the absence of an extra-osseous soft-tissue component or bony destruction, the diagnosis of postradiation insufficiency fracture can be made with high accuracy and biopsy is not recommended [6]. Biopsy may be misleading because histological findings of hemorrhage, fibrosis, necrosis, and trabecular bone and cartilage growth may be confused for osteomyelitis, enchondroma, osteosarcoma, or chondrosarcoma.

Teaching point

A new bone lesion developing in the sacrum, pubic ramus, or acetabulum of a patient who has been radiated for pelvic malignancy is likely to be a postradiation pelvic insufficiency fracture.

REFERENCES

1 Schneider R, Yacovone J, Ghelman B. Unsuspected sacral fractures: detection by radionuclide bone scanning. *Am J Roentgenol* 1985; **144**: 337–341.

2 Cooper KL, Bebout JW, Swee RG. Insufficiency fractures of the sacrum. *Radiology* 1985; **156**: 15–20.

3 DeSmet AA, Neff JR. Pubic and sacral insufficiency fractures: clinical course and radiographic findings. *Am J Roentgenol* 1985; **145**: 601–606.

4 Fayad LM, Cohade C, Wahl RL, Fishman EK. Sacral fractures: a potential pitfall of FDG positron emission tomography. *Am J Roentgenol* 2003; **181**: 1239–1243.

5 Blomlie V, Rreofstad EK, Talle K, et al. Incidence of radiation-induced insufficiency fractures of the female pelvis: evaluation with MR imaging. *Am J Roentgenol* 1996; **167**: 1205–1210.

6 Kwon JW, Huh SJ, Yoon CY, et al. Pelvic bone complications after radiation therapy of uterine cervical cancer: evaluation with MRI. *Am J Roentgenol* 2008; **191**: 987–994.

7 Ogino I, Okamoto N, Ono Y, Kitamura T, Nakayama H. Pelvic insufficiency fractures in postmenopausal woman with advanced cervical cancer treated radiation therapy. *Radiother Oncol* 2003; **68**: 61–67.

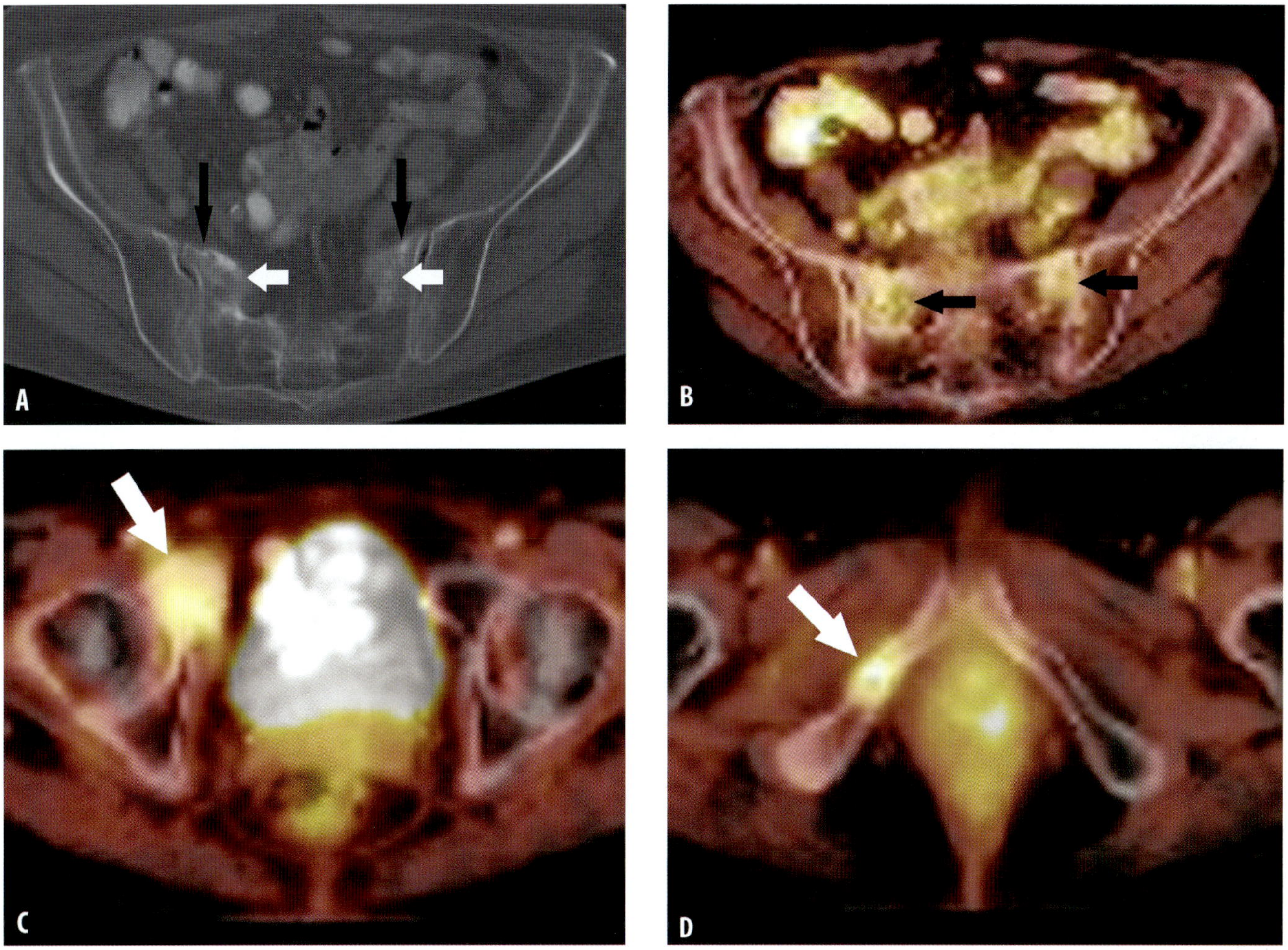

Figure 97.1 A. Axial contrast-enhanced CT image in a 77 year old woman 6 months after pelvic radiotherapy for a urethral melanoma shows typical pelvic insufficiency fractures in the sacrum bilaterally, with linear sclerotic changes (white arrows) running parallel to the sacro-iliac joints and cortical discontinuity anteriorly (black arrows). B. Fused PET/CT image at the same level shows symmetric increased FDG uptake (arrows) related to the fractures. C. Fused PET/CT image at a lower level shows increased FDG uptake related to a co-existent insufficiency fracture in the right acetabulum (arrow). D. Fused PET/CT image at a lower level shows increased FDG uptake related to a co-existent insufficiency fracture in the right inferior pubic ramus (arrow). The sacrum, acetabulum, and pubic rami are the typical sites of pelvic insufficiency fracture.

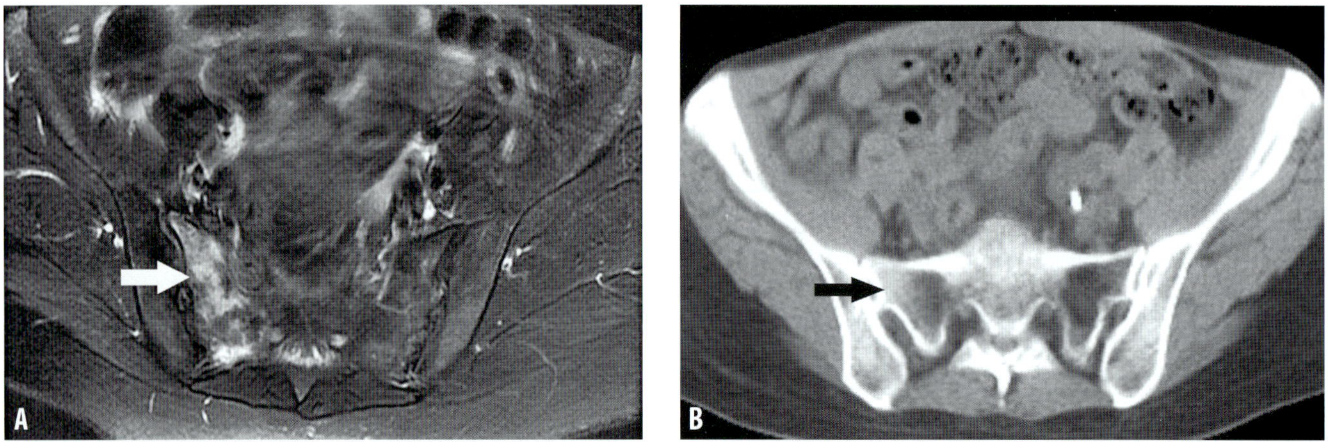

Figure 97.2 **A.** Axial T2-weighted MR image in a 41 year old woman 10 months after external beam pelvic radiotherapy for cervical cancer shows increased signal intensity (arrow) in the right sacral body parallel to the sacro-iliac joint, typical of a postradiation pelvic insufficiency fracture. **B.** Axial non-enhanced CT image at the same level shows only very subtle increased sclerosis (arrow) in the right side of the sacrum. Pelvic insufficiency fractures are frequently more obvious on MRI than CT.

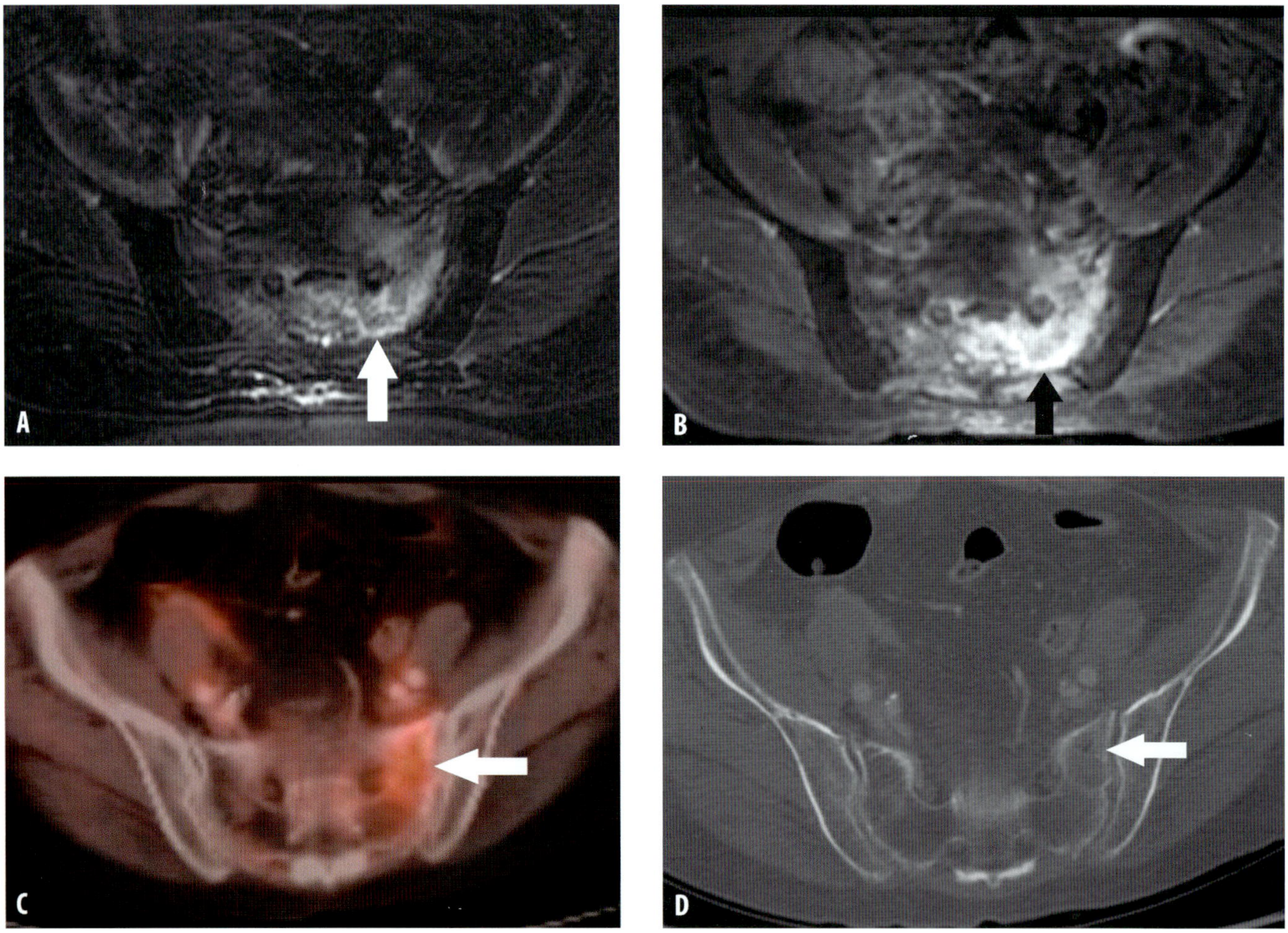

Figure 97.3 **A.** Axial T2-weighted MR image in a 73 year old woman one year after external beam pelvic radiotherapy for endometrial carcinoma shows increased signal intensity (arrow) primarily in the left sacrum, typical of a postradiation pelvic insufficiency fracture. **B.** Axial fat-suppressed post-gadolinium T1-weighted MR image at the same level shows prominent increased enhancement (arrow) related to the fracture. **C.** Fused PET/CT image at the same level shows increased FDG uptake in the left side of the sacrum. **D.** Corresponding axial contrast-enhanced CT image shows only very subtle increased sclerosis (arrow) in the left side of the sacrum.

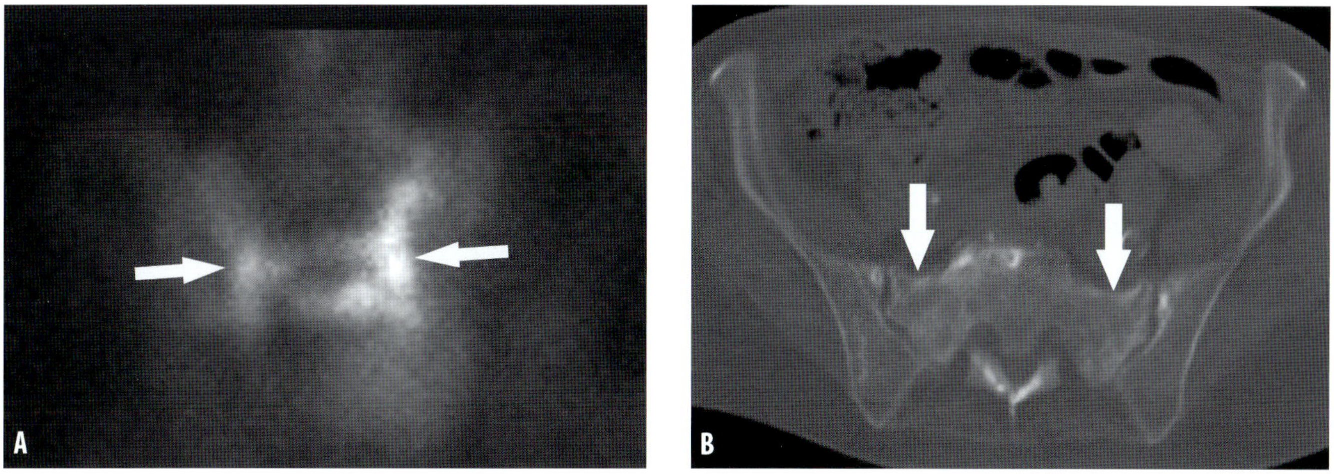

Figure 97.4 **A.** Planar bone scan image in an 89 year old woman with new bony pain in the pelvis and a remote history of pelvic radiation for endometrial cancer shows the typical H-shaped pattern (arrows) of increased uptake in the sacrum due to pelvic insufficiency fracture. **B.** Corresponding axial non-enhanced CT image shows subtle increased sclerosis (arrows) in both sides of the sacrum.

CASE 98 Iliac pseudotumor due to bone harvesting

Imaging description

The posterior iliac crest is an excellent site to obtain cancellous bone for bone grafting at other locations or bone marrow for autologous bone marrow transplantation. These procedures can result in a focus of lucency or heterogeneity in the posterior iliac crest sometimes associated with a cortical break at CT or MRI or increased uptake on bone scintigraphy that can simulate a tumor (Figures 98.1–98.3) [1, 2].

Importance

Misinterpretation of a focal iliac bone lesion due to bone harvesting as a potential malignancy may result in unnecessary anxiety or biopsy [1]. However, this pseudotumor has other consequences. Chronic donor site pain occurs in up to 39% of patients and may be related to degenerative changes induced in the sacro-iliac joint due to destabilization [3, 4]. Iliac bone growth disturbance has been reported in children [5]. Because of these donor site complications, some surgeons backfill the donor site [6], which may result in a sclerotic appearance that mimics osteoblastic malignancy (Figure 98.2).

Typical clinical scenario

Iliac pseudotumor due to bone harvesting is typically seen in patients who have undergone spinal fusion surgery or who have had autologous bone marrow transplantation for treatment of high-risk or recurrent pediatric solid tumors such as neuroblastoma.

Differential diagnosis

The primary differential consideration is a true tumor of bone, either benign or malignant. Awareness of this pseudotumor combined with correlation with the surgical history is usually sufficient to confirm the presumptive diagnosis. If followed over time, the pseudotumor of bone harvesting will typically show healing (Figures 98.1 and 98.3), although progression has been described, possibly due to accumulation of synovial fluid from the sacro-iliac joint [1].

Teaching point

The possibility of a bone harvest site should be considered whenever a focal lucent, sclerotic, or heterogeneous lesion is seen in the posterior iliac crest.

REFERENCES

1 Kavanagh E, Roth C, O'Connell M, Eustace S. Sacral pseudotumor complicating iliac bone harvest: radiographic, CT and MRI appearances. *Eur Radiol* 2003; **13**(Suppl 4): L178–L180.

2 Ortiz SS, Miller JH, Villablanca JG, Seeger RC. Bone abnormalities detected with skeletal scintigraphy after bone marrow harvest in patients with childhood neuroblastoma. *Radiology* 1994; **192**: 755–758.

3 Kurz LT, Garfin SR, Booth RE. Harvesting autogenous iliac bone grafts. A review of complications and techniques. *Spine* 1989; **14**: 1324–1331.

4 Ebraheim NA, Elgafy H, Semaan HB. Computed tomographic findings in patients with persistent sacroiliac pain after posterior iliac graft harvesting. *Spine* 2000; **25**: 2047–2051.

5 Rossillon R, Desmette D, Rombouts JJ. Growth disturbance of the ilium after splitting the iliac apophysis and iliac crest bone harvesting in children: a retrospective study at the end of growth following unilateral Salter innominate osteotomy in 21 children. *Acta Orthop Belg* 1999; **65**: 295–301.

6 Bojescul JA, Polly DW Jr, Kuklo TR, Allen TW, Wieand KE. Backfill for iliac-crest donor sites: a prospective, randomized study of coralline hydroxyapatite. *Am J Orthop* 2005; **34**: 377–382.

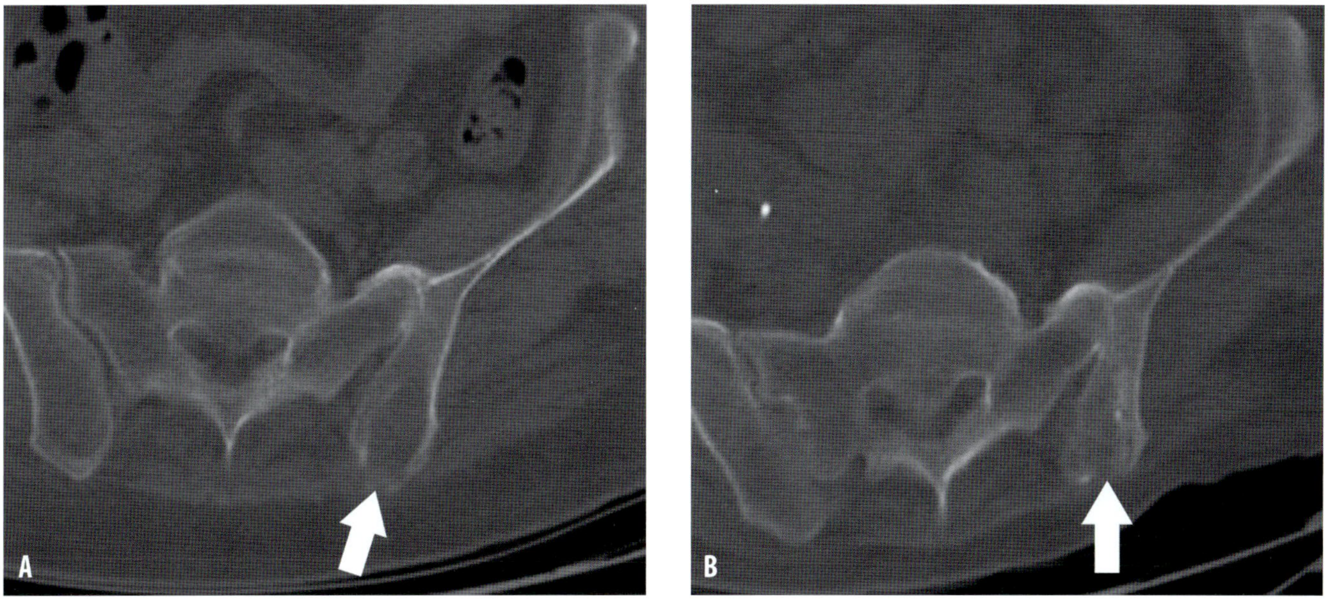

Figure 98.1 **A.** Axial CT image in a 56 year old woman one month after lumbar arthrodesis with autologous bone grafting using bone harvested from the left posterior iliac crest shows the harvest site as a lucent lesion in the left posterior iliac bone associated with a cortical break (arrow). The CT findings mimic those of malignancy. **B.** Axial CT image obtained two years later shows the defect has largely healed with only a small focal cortical interruption (arrow) still visible.

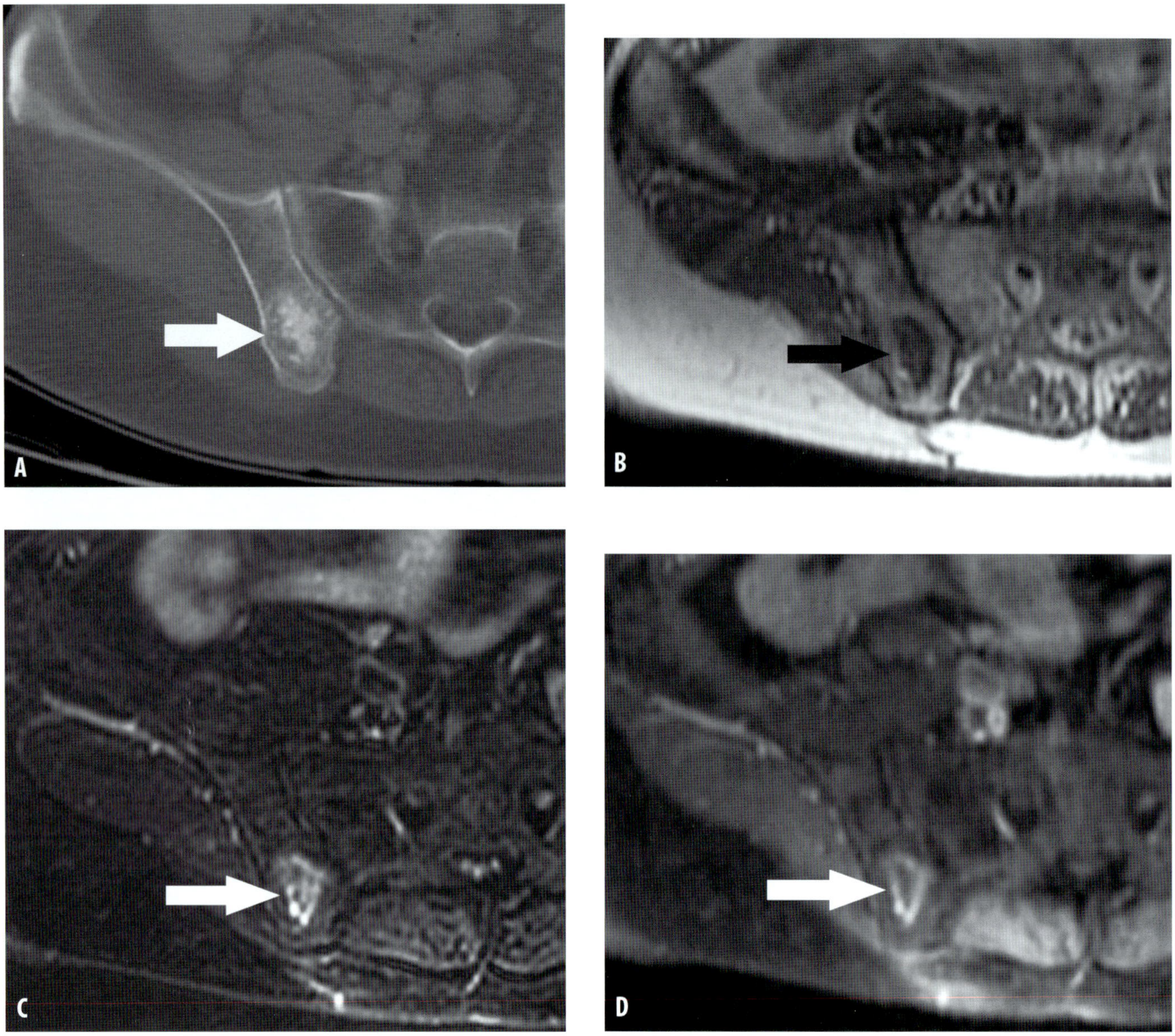

Figure 98.2 A. Axial CT image in a 62 year old man after T10 vertebrectomy for a solitary spinal plasmacytoma with autologous bone grafting using bone harvested from the right posterior iliac crest. The harvest site was packed with allograft bone. The resulting ill-defined sclerotic focus (arrow) resembles an osteoblastic metastasis. B. Axial T1-weighted MR image shows a focus (arrow) of reduced signal intensity in the bone harvest site. C. Axial T2-weighted MR image shows a focus (arrow) of increased signal intensity in the bone harvest site. D. Axial post-gadolinium fat-saturated T1-weighted MR image shows peripheral enhancement (arrow) in the bone harvest site. The MR findings simulate malignancy.

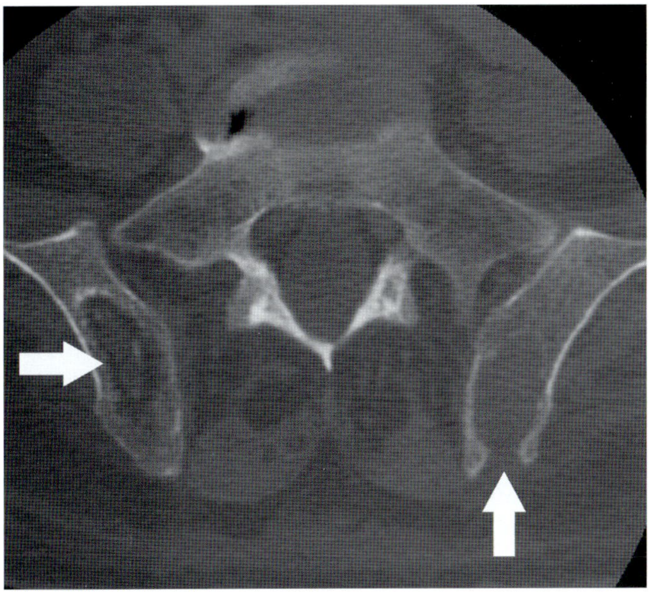

Figure 98.3 Axial CT image in a 67 year old man after multiple spinal surgeries showing a recent (6 weeks earlier) bone harvest site in the left iliac crest (vertical arrow) and an older (3 years earlier) bone harvest site in the right iliac crest (horizontal arrow). Note that the older lesion shows substantial healing by sclerosis.

CASE 99

Pseudoprogression due to healing of bone metastases by sclerosis

Imaging description

Pseudoprogression of bone metastases occurs when a previously undetectable bone metastasis on plain radiography or CT heals by sclerosis as a response to successful treatment and becomes newly evident as an osteoblastic metastasis (Figures 99.1 and 99.2) [1–3]. Such a new lesion may be incorrectly interpreted as indicating progressive disease.

Importance

Pseudoprogression due to healing of bone metastases by sclerosis may be misdiagnosed as treatment failure instead of treatment response, and this misdiagnosis may result in an unwarranted change of treatment.

Typical clinical scenario

In my experience, this pitfall occurs primarily in patients with metastatic breast cancer, probably reflecting the frequency of osteolytic metastases from this disease and the increasing efficacy of chemotherapy for breast cancer. In a study of 24 patients receiving chemotherapy for breast cancer and undergoing serial radiographs and bone scintigraphy, 52 apparently new sclerotic lesions were seen on plain radiography during therapy. On previous bone scintigraphy, 17 of the 52 apparently new sclerotic lesions (33%) had positive uptake, suggesting they had become radiographically visible due to pseudoprogression [1]. Note that occult bony metastases at CT are relatively common. In a series of 359 patients undergoing PET/CT, 49 of 133 metastases seen at PET were occult at CT [4].

Differential diagnosis

The primary consideration for an apparently new sclerotic bone lesion in a patient being treated for metastatic disease is true progression. Bone scintigraphy or PET may be helpful in determining whether such a new lesion reflects healing by sclerosis or true progression. Apparent discordant response between bone metastases and extra-osseous disease sites should also raise the consideration of pseudoprogression (Figure 99.1).

Teaching point

A new sclerotic bone lesion at radiography or CT in a patient being treated for metastatic disease may reflect healing by sclerosis (pseudoprogression) of a previously occult bone metastasis or true progression. Correlation with other disease sites or with bone scintigraphy or PET may be helpful in making the distinction between these two possibilities.

REFERENCES

1 Ciray I, Aström G, Andréasson I, et al. Evaluation of new sclerotic bone metastases in breast cancer patients during treatment. *Acta Radiol* 2000; **41**: 178–182.

2 Scher HI, Morris MJ, Kelly WK, Schwartz LH, Heller G. Prostate cancer clinical trial end points: "RECIST"ing a step backwards. *Clin Cancer Res* 2005; **11**: 5223–5232.

3 Smith PH, Bono A, Calais da Silva F, et al. Some limitations of the radioisotope bone scan in patients with metastatic prostatic cancer. *Cancer* 1990; **66**: 1009–1016.

4 Nakamoto Y, Cohade C, Tatsumi M, Hammoud D, Wahl RL. CT appearance of bone metastases detected with FDG PET as part of the same PET/CT examination. *Radiology* 2005; **237**: 627–634.

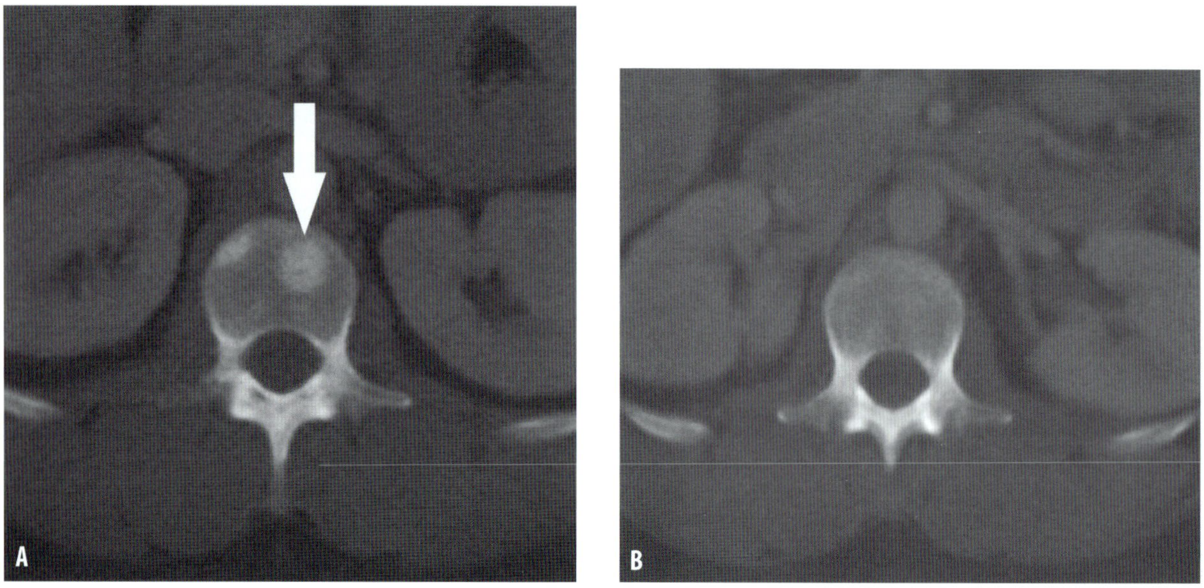

Figure 99.1 (cont.)

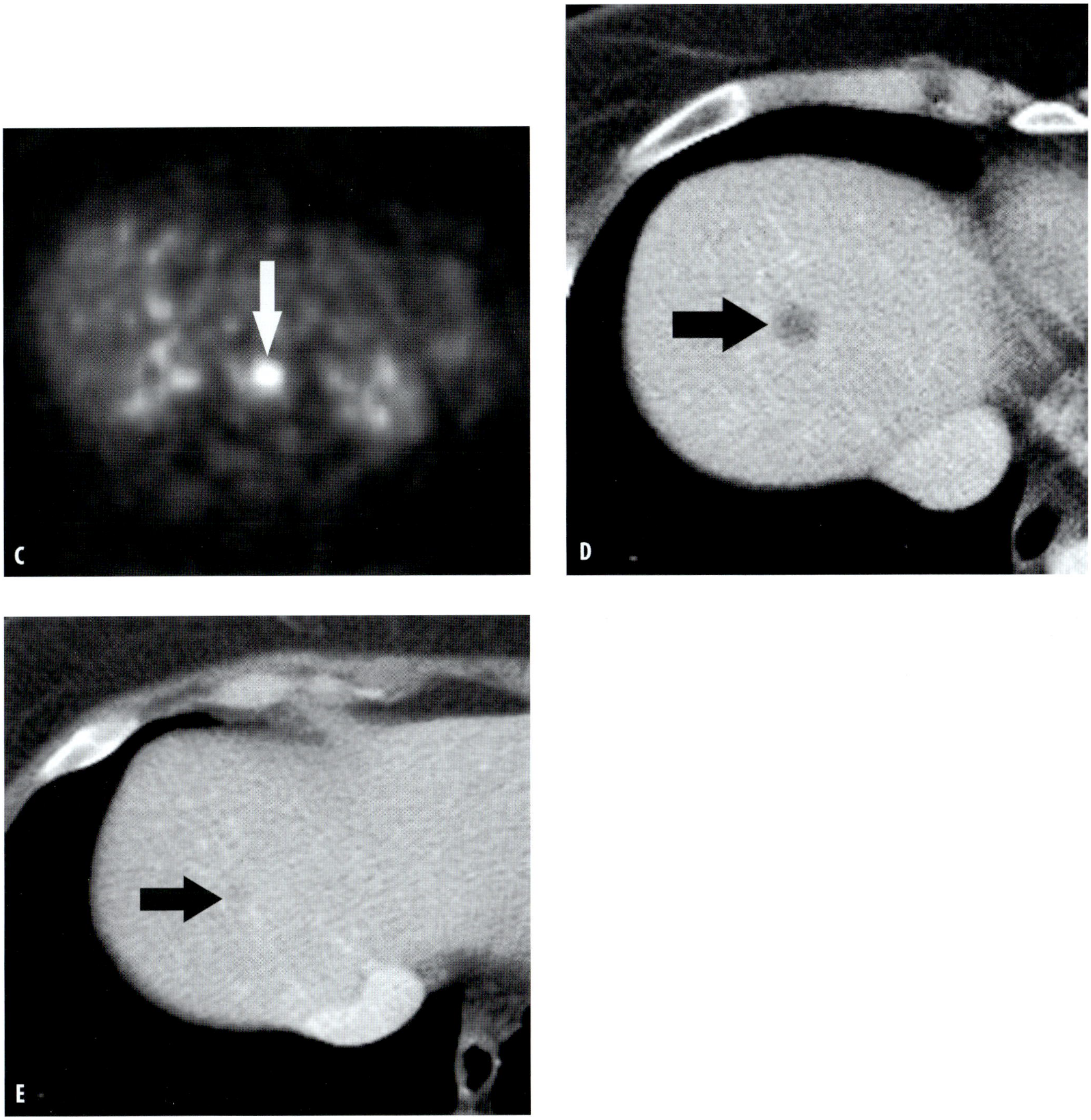

Figure 99.1 A. Axial CT image in a 44 year old woman with metastatic breast cancer shows a sclerotic focus in L1, consistent with a bone metastasis. B. Axial CT image obtained 2 months earlier, before commencing treatment with capecitabine, appears unremarkable. In isolation, the sequence of CT findings might be interpreted as treatment failure with disease progression manifested by the development of a new bone metastasis on therapy. C. Axial PET image obtained at the same time as the CT in Figure 99.1B shows a focus of increased uptake in L1, indicating the metastasis was present prior to therapy but occult at CT. The increased sclerosis in the lesion while on treatment represents an example of pseudoprogression due to healing by sclerosis. D. Axial contrast-enhanced CT image obtained at the same time as the image in Figure 99.1A shows a low-density lesion (arrow) in the liver that was new and associated with increased FDG uptake at PET, consistent with a metastasis. E. Axial contrast-enhanced CT image obtained at the same time as the image in Figure 99.1B shows shrinkage of the metastasis (arrow) consistent with treatment response. Discordant progression between different disease sites (i.e., apparent progression in bone metastasis with regression of liver metastasis) is rare and should raise the consideration of pseudoprogression.

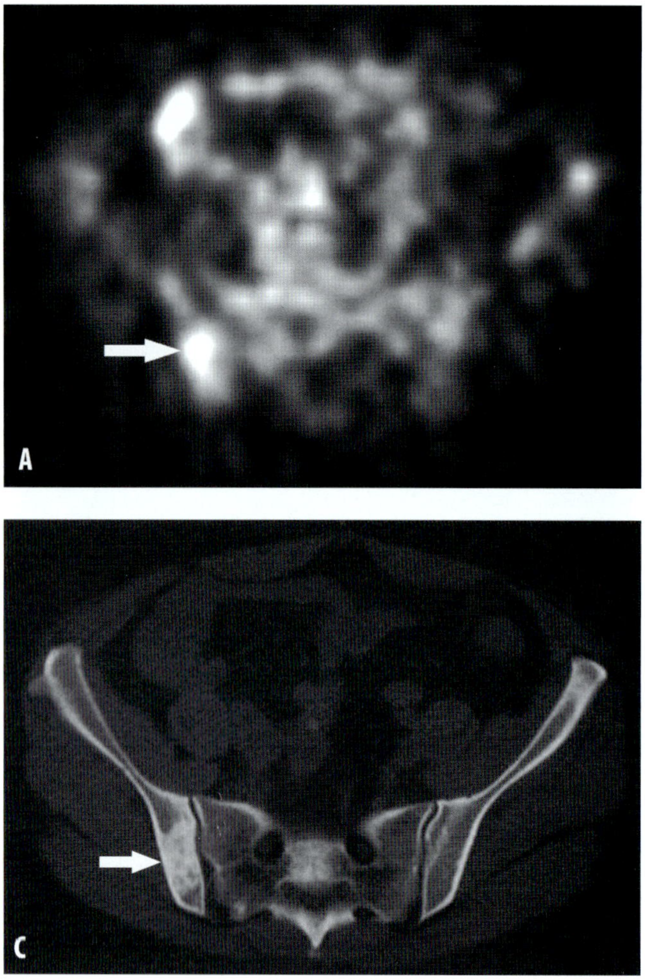

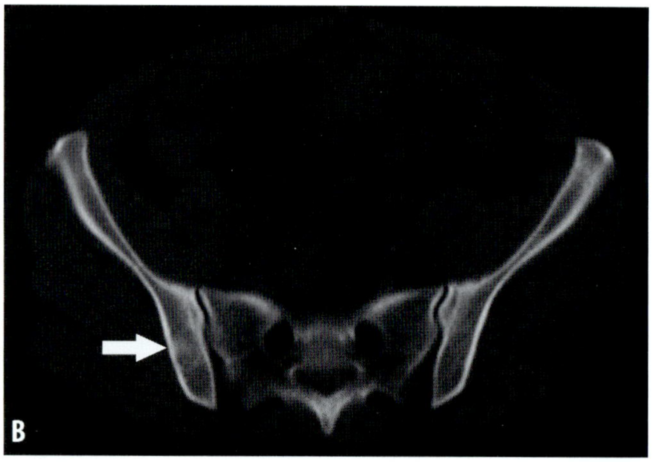

Figure 99.2 **A.** Axial PET image in a 54 year old woman undergoing surveillance for node positive breast cancer shows a focus of increased uptake in the right iliac bone. **B.** Axial CT image obtained at the same time as the PET in Figure 99.2A shows questionable minimal sclerosis (arrow) in the right iliac bone. CT-guided biopsy was performed and established a diagnosis of metastatic breast cancer. **C.** Axial CT image obtained 6 months later after treatment with tamoxifen and vorinostat (a chemotherapy agent) shows an obvious sclerotic lesion (arrow) in the right iliac bone. In the absence of the PET scan, the change between Figures 99.2B and 99.2C might be interpreted as showing progression. With the PET scan, a correct diagnosis of pseudoprogression due to healing by sclerosis can be established.

CASE 100 Pseudometastases due to red marrow conversion

Imaging description

Red bone marrow can become stimulated and metabolically active as a rebound phenomenon after chemotherapy, in response to severe or chronic hemorrhage, or by bone marrow stimulants used in oncology patients (e.g., granulocyte-colony stimulating factor, erythropoietin or interleukin-3) [1–5]. In such settings, bone marrow uptake of 18F-FDG at PET can be markedly increased and simulate diffuse metastatic disease (Figure 100.1).

Importance

An incorrect diagnosis of metastases due to increased FDG uptake at PET by red marrow conversion could lead to unnecessary additional treatment or inappropriate changes in management [4]. Conversely, it is also possible that this appearance could mask true bone metastases [5, 6].

Typical clinical scenario

Increased bone marrow activity at PET has been reported primarily in cancer patients treated with colony stimulating factors.

Differential diagnosis

The main differential consideration for widespread bone marrow uptake of FDG by converted red marrow is diffuse medullary metastases (Figure 100.2). In practice, medullary metastases are usually focal while red marrow conversion is usually diffuse, but this rule is not absolute – medullary metastases are occasionally diffuse [7] and red marrow conversion is occasionally focal [8, 9]. The evolution of these abnormalities over time may allow accurate differentiation. If critical to pending management decisions, biopsy may be required.

Teaching point

Widespread increased FDG uptake in the bone marrow at PET is more likely to reflect red marrow conversion than diffuse metastases. Biopsy may occasionally be required to make the distinction if correlation with clinical and imaging history does not help clarify.

REFERENCES

1 Daldrup-Link HE, Henning T, Link TM. MR imaging of therapy-induced changes of bone marrow. *Eur Radiol* 2007; **17**: 743–761.

2 Sonet A, Graux C, Nollevaux MC, et al. Unsuspected FDG-PET findings in the follow-up of patients with lymphoma. *Ann Hematol* 2007; **86**: 9–15.

3 Goerres GW, Von Schulthess GK, Hany TF. Positron emission tomography and PET CT of the head and neck: FDG uptake in normal anatomy, in benign lesions, and in changes resulting from treatment. *Am J Roentgenol* 2002; **179**: 1337–1343.

4 Hollinger EF, Alibazoglu H, Ali A, Green A, Lamonica G. Hematopoietic cytokine-mediated FDG uptake stimulates the appearance of diffuse metastatic disease on whole-body PET imaging. *Clin Nucl Med* 1998; **23**: 93–98.

5 Gorospe L, Raman S, Echeveste J, et al. Whole-body PET/CT: spectrum of physiological variants, artifacts and interpretative pitfalls in cancer patients. *Nucl Med Commun* 2005; **26**: 671–687.

6 Chhabra A, Batra K, Makler PT Jr. Obscured bone metastases after administration of hematopoietic factor on FDG-PET. *Clin Nucl Med* 2006; **31**: 328–330.

7 Hanna SL, Fletcher BD, Fairclough DL, Jenkins JH III, Le AH. Magnetic resonance imaging of disseminated bone marrow disease in patients treated for malignancy. *Skeletal Radiol* 1991; **20**: 79–84.

8 Bordalo-Rodrigues M, Galant C, Lonneux M, Clause D, Vande Berg BC. Focal nodular hyperplasia of the hematopoietic marrow simulating vertebral metastasis on FDG positron emission tomography. *Am J Roentgenol* 2003; **180**: 669–671.

9 Pui MH, Tan MH, Kuan JHY, Pho RWH. Haematopoietic marrow hyperplasia simulating transarticular skip metastasis in osteosarcoma. *Australas Radiol* 1995; **39**: 303–305.

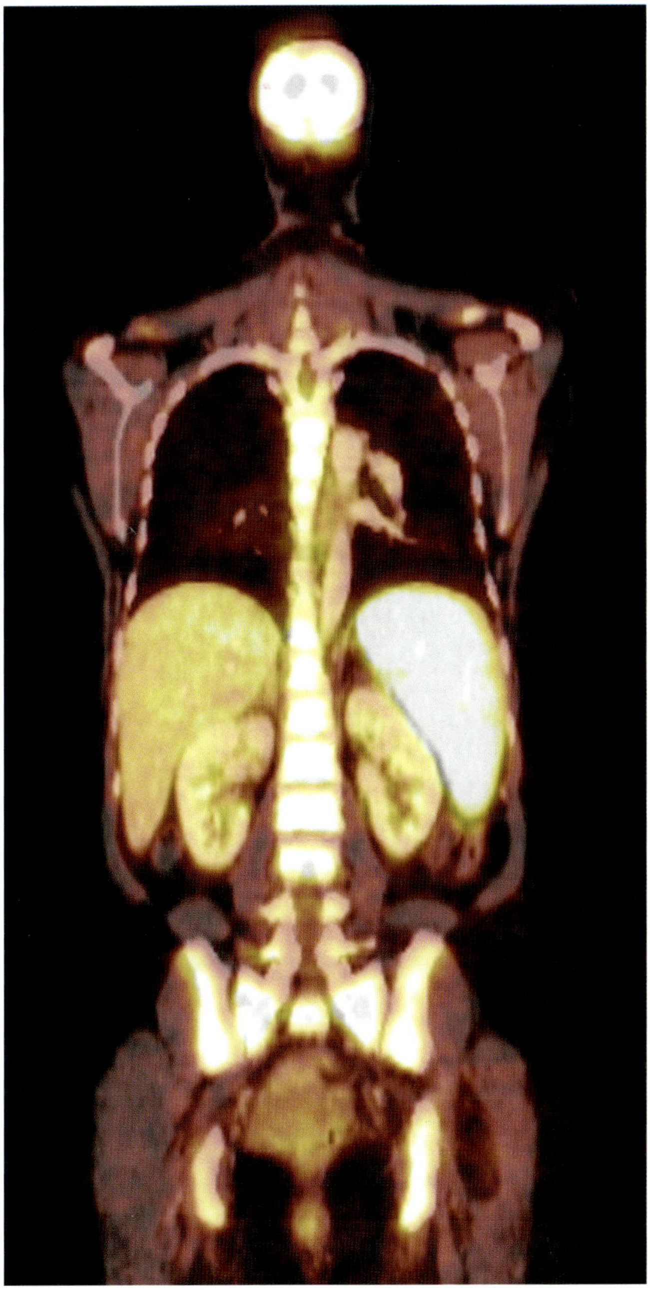

Figure 100.1 Coronal fused PET/CT image in a 44 year old woman with a history of widespread adenopathy due to non-Hodgkin's lymphoma. The study was performed after 5 cycles of CHOP chemotherapy with midcycle high-dose methotrexate. Diffusely increased bone marrow uptake of FDG is consistent with red marrow conversion as a response to chemotherapy. The bone marrow uptake resolved spontaneously on a follow-up PET scan performed 5 months later (not shown).

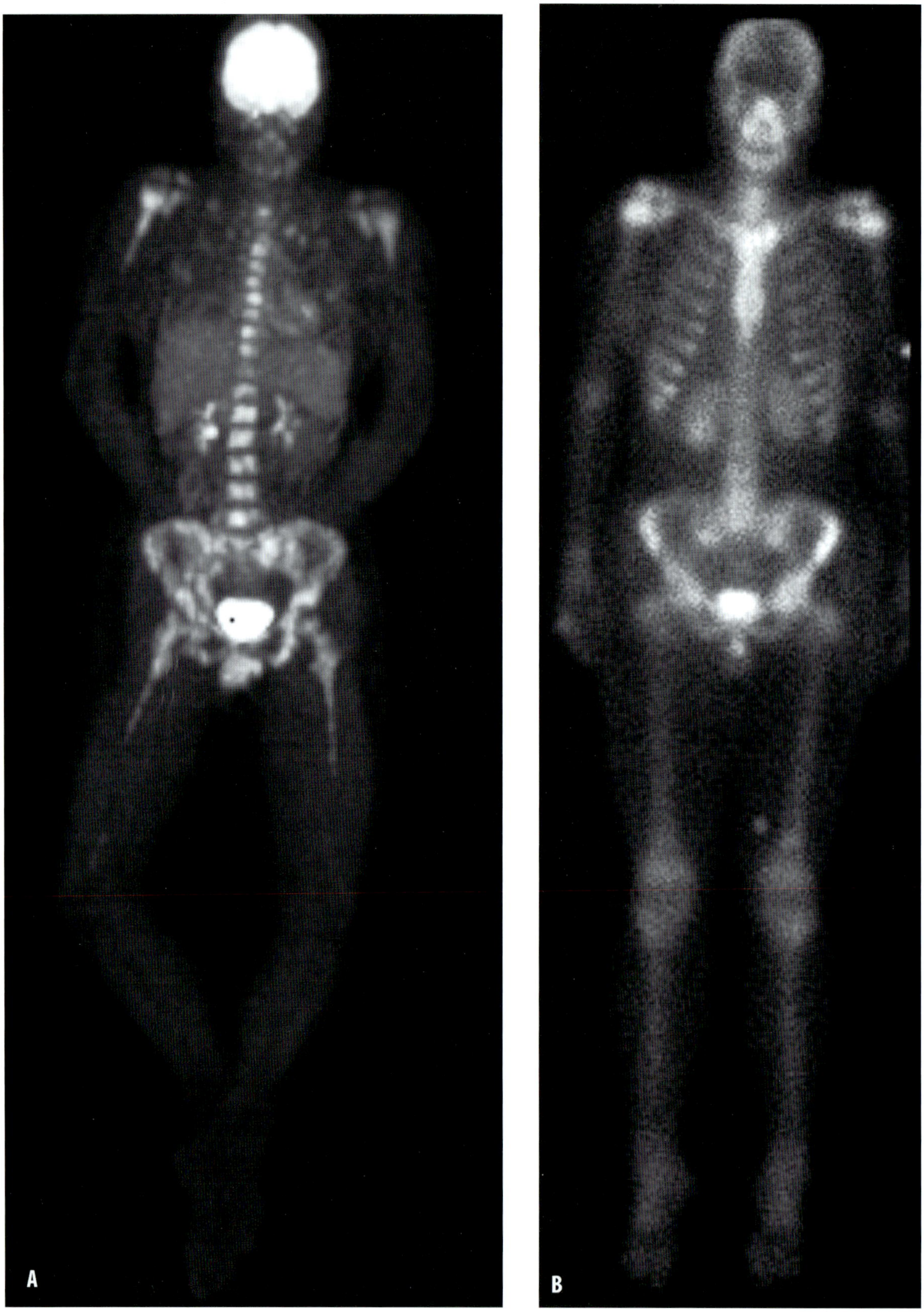

Figure 100.2

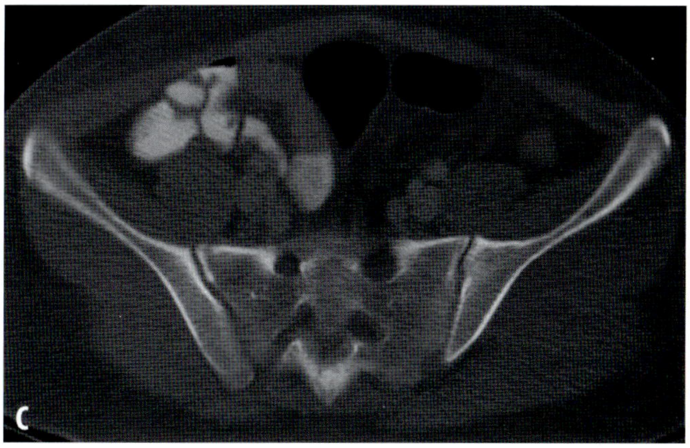

Figure 100.2 **A.** Coronal projection image of a PET scan in a 16 year old girl with a right perineal alveolar rhabdomyosarcoma shows widespread increased FDG uptake in the bone marrow. **B.** Whole body coronal projection bone scintigraphy image is unremarkable. **C.** Axial CT image of the pelvis shows no bony abnormalities. Bilateral bone marrow aspiration biopsies from the posterior iliac crests showed extensive infiltration by metastatic alveolar rhabdomyosarcoma, indicating the PET uptake in Figure 100.1A was due to widespread medullary metastases.

CASE 101

Iliac bone defect due to iliopsoas transfer

Imaging description

Subluxation of the hip in spina bifida, cerebral palsy, and other paralytic disorders is due to the pull of spastic hip adductors and flexors. Iliopsoas transfer is a surgical procedure designed to restore muscular balance and prevent hip dislocation by transposition of the iliopsoas muscle through a surgically created defect in the iliac bone with reattachment to the femur [1, 2]. The resulting window in the iliac bone is seen at imaging as a large lucent defect that may simulate tumor (Figure 101.1).

Importance

The iliac bone defect due to iliopsoas transfer might be mistaken for a lucent bone tumor at plain radiography, although the absence of a mass at CT or MRI should prevent this misdiagnosis.

Typical clinical scenario

This defect can be seen (either unilaterally or bilaterally) in patients affected by spastic paralytic disorders such as spina bifida or cerebral palsy.

Differential diagnosis

The appearance of an iliac bone defect due to iliopsoas transfer is characteristic, and awareness of the entity should be sufficient to ensure correct diagnosis. Correlation with the surgical history should help confirm in cases of uncertainty.

Teaching point

A large lucent window or defect in the iliac bone is characteristic of prior iliopsoas transfer.

REFERENCES

1 Lorente Moltó FJ, Martínez Garrido I. Retrospective review of L3 myelomeningocele in three age groups: should posterolateral iliopsoas transfer still be indicated to stabilize the hip? *J Pediatr Orthop B* 2005; **14**: 177–184.

2 Sharrard WJW, Burke J. Iliopsoas transfer in the management of established dislocation and refractory progressive subluxation of the hip in cerebral palsy. *Int Orthop* 1982; **6**: 149–154.

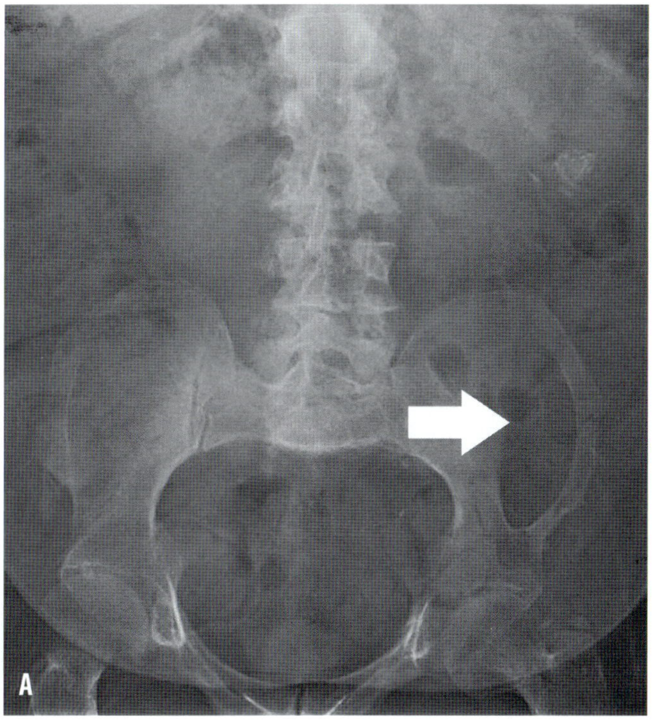

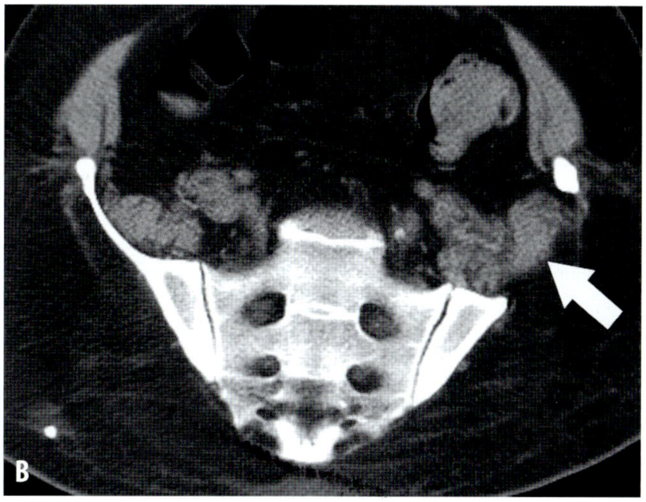

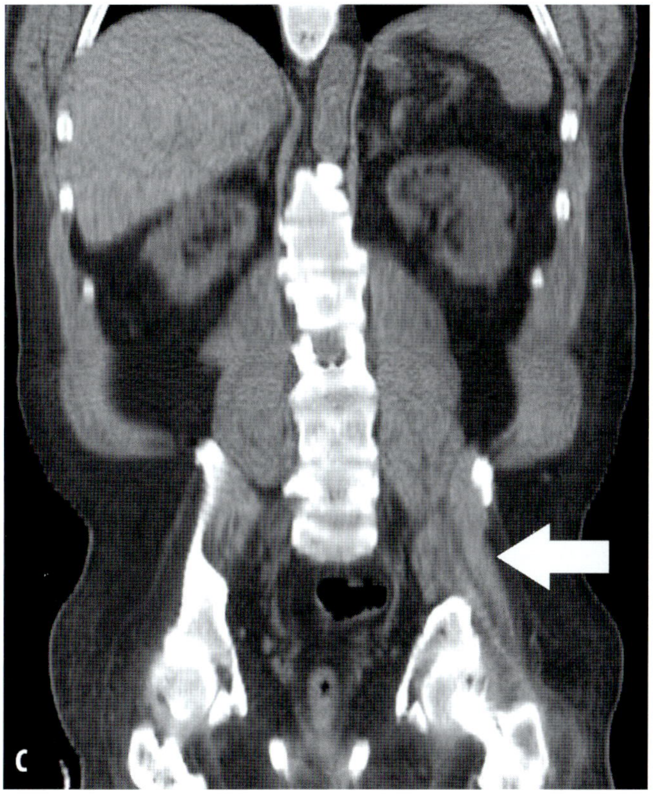

Figure 101.1 A. Plain abdominal radiograph in a 52 year old man with spastic paralysis due to a childhood gunshot wound. Left iliopsoas transfer was performed to prevent subluxation of the left hip. The resulting bone defect is visible as a well-circumscribed large lucent defect (arrow) in the left iliac bone. **B.** Axial CT image shows the lucency in the left iliac bone is due to absence of bone (arrow) rather than a destructive lesion. **C.** Curved planar coronal reformatted CT image shows the left iliopsoas muscle (arrow) passing through the surgically created defect in the left iliac bone.

Index